QC 351 .P76 v.21

Progress in optics

PROGRESS IN OPTICS

VOLUME XXI

EDITORIAL ADVISORY BOARD

L. ALLEN,	*Brighton, England*
M. FRANÇON,	*Paris, France*
E. INGELSTAM,	*Stockholm, Sweden*
A. KUJAWSKI,	*Warsaw, Poland*
A. LOHMANN,	*Erlangen, Fed. Rep. Germany*
M. MOVSESSIAN,	*Armenia, U.S.S.R.*
G. SCHULZ,	*Berlin, D.D.R.*
W. H. STEEL,	*Sydney, Australia*
J. TSUJIUCHI,	*Tokyo, Japan*
W. T. WELFORD,	*London, England*

PROGRESS IN OPTICS

VOLUME XXI

EDITED BY

E. WOLF
University of Rochester, N.Y., U.S.A.

Contributors

D. MAYSTRE, L. A. LUGIATO
H. H. BARRETT, N. M. CEGLIO, D. W. SWEENEY
J. C. ENGLUND, R. R. SNAPP, W. C. SCHIEVE

1984

NORTH-HOLLAND PHYSICS PUBLISHING
AMSTERDAM · OXFORD · NEW YORK · TOKYO

QC
351
.P76
v. 21

© ELSEVIER SCIENCE PUBLISHERS B.V., 1984

All Rights Reserved. No part of this publication may be reproduced, stored in a retrieval system, or transmitted, in any form or by any means, electronic, mechanical, photocopying, recording or otherwise, without the prior permission of the Copyright owner.

LIBRARY OF CONGRESS CATALOG CARD NUMBER: 61-19297
ISBN: 0444867619

PUBLISHERS:

NORTH-HOLLAND PHYSICS PUBLISHING

A DIVISION OF:

ELSEVIER SCIENCE PUBLISHERS B.V.
P.O. BOX 103
1000 AC AMSTERDAM
THE NETHERLANDS

SOLE DISTRIBUTORS FOR THE U.S.A. AND CANADA:

ELSEVIER SCIENCE PUBLISHING COMPANY, INC.
52 VANDERBILT AVENUE
NEW YORK, N.Y. 10017
U.S.A.

PRINTED IN THE NETHERLANDS

CONTENTS OF VOLUME I (1961)

CONTENTS OF VOLUME II (1963)

CONTENTS OF VOLUME III (1964)

CONTENTS OF VOLUME IV (1965)

CONTENTS OF VOLUME V (1966)

CONTENTS OF VOLUME VI (1967)

CONTENTS OF VOLUME VII (1969)

CONTENTS OF VOLUME VIII (1970)

CONTENTS OF VOLUME IX (1971)

CONTENTS OF VOLUME X (1972)

CONTENTS OF VOLUME XI (1973)

CONTENTS OF VOLUME XII (1974)

CONTENTS OF VOLUME XIII (1976)

CONTENTS OF VOLUME XIV (1977)

CONTENTS OF VOLUME XV (1977)

CONTENTS OF VOLUME XVI (1978)

CONTENTS OF VOLUME XVII (1980)

CONTENTS OF VOLUME XVIII (1980)

CONTENTS OF VOLUME XIX (1981)

CONTENTS OF VOLUME XX (1983)

PREFACE

Just as its twenty predecessors, this volume presents reviews of current developments in several areas of optics and related fields.

The first article gives an account of some of the theories that have been developed in recent years for the analysis of properties of diffraction gratings with very small groove spacing, of the order of the wavelength of light. Such gratings are now frequently encountered in the laboratory. Scalar diffraction theories are inadequate for the analysis of their performance, because polarization properties of light cannot be ignored under these circumstances. Vector theories of diffraction gratings, such as those described in this article, must then be used.

The second article is concerned with the theory of optically bistable systems, i.e., of systems whose transmittance has two stable states. This subject has attracted a good deal of attention in recent years, because bistable systems are likely to find useful applications, for example as optical transistors or as memory elements.

The third article deals with the basic theoretical aspects of a technique that has found important uses in connection with numerous inverse reconstruction schemes. The unifying mathematical concept that underlies these techniques is the so-called Radon transform. It is, for example, at the heart of computerized medical tomography, which makes it possible to determine the detailed distribution of the attenuation coefficient of selected portions of a patient's body from measurements of the intensity of transmitted X-ray beams.

The fourth article deals with the theory and application of a method of coded imaging. Coded imaging is a two-step process of image reconstruction. In the first stage information about a source is generated by geometrical shadowing through a coded aperture. In the second stage the image is reconstructed by means of numerical or optical techniques. This procedure is mainly used with sources that generate very short wavelength radiation (e.g., X-rays), because such radiation cannot easily be imaged or re-directed by reflection or refraction. In the last two decades such techniques have been frequently used in nuclear medicine, in nuclear engineering, in inertial confinement fusion research, and in X-ray astronomy. The article deals with one particular technique of this kind, which uses Fresnel zone plates as the aperture and in which the decoding is performed optically, with coherent light.

The concluding article deals with certain properties that in recent years have been found to occur in a variety of non-linear phenomena. They are discussed here with special reference to the laser-driven non-linear ring cavity. In particular, the onset of instability, deterministic switching in the bistable regime, fluctuation dynamics, and the generation of deterministic chaos from instabilities are discussed, and examples are given of the remarkable period-doubling sequences that accompany the transition to chaos.

Although the individual articles are self-contained, the reader will undoubtedly note that several of the topics covered in the different articles are interrelated.

EMIL WOLF

Department of Physics and Astronomy
University of Rochester
Rochester, NY 14627, USA

January 1984

CONTENTS

I. RIGOROUS VECTOR THEORIES OF DIFFRACTION GRATINGS

by D. Maystre (Marseille, France)

II. THEORY OF OPTICAL BISTABILITY

by L. A. LUGIATO (MILAN, ITALY)

III. THE RADON TRANSFORM AND ITS APPLICATIONS

by H. H. Barrett (Tucson, Arizona)

IV. ZONE PLATE CODED IMAGING: THEORY AND APPLICATIONS

by N. M. CEGLIO (LIVERMORE, CALIFORNIA)
and D. W. SWEENEY (WEST LAFAYETTE, INDIANA)

V. FLUCTUATIONS, INSTABILITIES AND CHAOS IN THE LASER-DRIVEN NONLINEAR RING CAVITY

by J. C. EGLUND, R. R. SNAPP and W. C. SCHIEVE (AUSTIN, TEXAS)

E. WOLF, PROGRESS IN OPTICS XXI
© ELSEVIER SCIENCE PUBLISHERS B.V., 1984

I

RIGOROUS VECTOR THEORIES OF DIFFRACTION GRATINGS

BY

D. MAYSTRE

Laboratoire d'Optique Electromagnétique, Equipe de Recherche Associée au C.N.R.S. n° 597,
Faculté des Sciences et Techniques,
Centre de Saint-Jérôme, 13397 MARSEILLE CEDEX 13, France.

CONTENTS

§ 1. Introduction

Diffraction grating was born almost two centuries ago. Its father, an American astronomer, saw a spectrum produced by hairs placed in the threads of two parallel screws (RITTENHOUSE [1786]). However, it was not until 1821 that the first metallic grating was ruled by FRAUNHOFER who found the famous grating formula giving the directions of diffraction (FRAUNHOFER [1821]). One century after the discovery of RITTENHOUSE, ROWLAND [1882] initiated the production of high quality ruled gratings. Recently there appeared a new type of diffraction grating, constructed by recording on a photoresist interference fringes produced by a laser source (RUDOLPH and SCHMAHL [1967], LABEYRIE and FLAMAND [1969]).

Indisputably, the diffraction grating is a valuable instrument for scientific research. Mainly, it is used for spectroscopy and filtering. Even though the holographic grating has not eliminated the classical, ruled grating which remains the most suitable for important applications, it has permitted a considerable extension of the use of gratings for industrial or scientific purposes: wavelength selectors for tunable lasers, selective surfaces for solar energy, masks for photolithography, beam sampling mirrors for high power lasers, spectrometers in extreme UV or X-ray regions for Space Optics (MAYSTRE, NEVIERE and PETIT [1980]).

Before the second world war, the pitch of the standard gratings was large with respect to the wavelengths of visible radiations, due to the difficulty of manufacturing high quality gratings having high space frequency. Under these circumstances, the properties of gratings do not depend, in practice, upon the polarization of the incident light. This explains why by far the major part of the theories dealing with these instruments are "scalar" and do not take the polarization into account (see for example BORN and WOLF [1965]). On the other hand, there exists experimental evidence that the distribution of energy among the various diffracted waves depends on polarization as soon as the wavelength to pitch ratio exceeds a number of the order of 0.2 (MADDEN and STRONG [1958]). Today, gratings having more than 1000 grooves per mm, i.e. with groove spacings lower than 1 μm, are commonplace. When these gratings are employed in the visible region, the wavelength of the light is of the order of the groove spacing. To our knowledge, there do not exist non-rigorous theories able to predict accurately the properties of gratings in this "resonance domain". As a consequence, it is necessary to achieve rigorous vector theories. Here, we call rigorous theory any theory able to reduce the physical problem to an equivalent classical problem of mathematics, using the elementary laws of electromagnetism and rigorous mathematical theorems. By classical problem of mathematics, we mean a problem whose solution can be directly obtained on a computer using numerical analysis. Approximations are made only in this numerical implementation of the classical mathematical problem, never before. To point out the difference between rigorous theories and the others, a comparison will be made between the numerical results obtained from a rigorous theory and from a famous non-rigorous theory (BECKMANN [1967], BECKMANN and SPIZZICHINO [1963]).

The first attempt at solving the electromagnetic problem of gratings was made at the beginning of the century (RAYLEIGH [1907]). The Rayleigh theory is based on an assumption, the so-called Rayleigh hypothesis, which remained unquestioned for almost 50 years, but provoked considerable controversies thereafter. Though the controversial aspect of the Rayleigh hypothesis has not yet died down, we have thought it useful to present a synthesis of the theoretical achievements on this topic in order to conclude whether or not the Rayleigh method is a rigorous vector theory.

It is impossible to describe in detail all the rigorous vector theories of gratings. Since a choice was necessary, we have selected the theories numerically implemented on a computer and giving accurate results for arbitrary gratings in a large domain of wavelengths. We have also taken into account the possibility of implementing these theories in other problems of optics or electromagnetism. Practically all the methods presented here can be generalized to other problems of scattering or propagation: diffraction by non-periodic objects, propagation of radio waves, etc. ... The reader will also find many references to studies in acoustics: from a mathematical point of view, the problems of periodic structures in acoustics and optics are very much akin. Mainly, three formalisms will be described in detail after the Rayleigh method: the integral and differential formalisms, the most widely used today, and the WATERMAN [1975] formalism, well known in electromagnetism.

It seems useful to point out the main advantages and shortcomings of each method for numerical application. Although the opinions expressed in this chapter are personal, they take into account the experience of many other researchers in the field of gratings, first of all my colleagues in the laboratory at Marseille.

§ 2. Presentation of the problem and preliminary results

The physical problem is presented and the fundamental laws of electromagnetism show that it reduces to a boundary value problem of mathematical physics. Some elementary properties of the diffracted field are deduced.

2.1. THE PHYSICAL PROBLEM (Fig. 2.1)

In the rectangular coordinate system Oxyz, the grating has a profile $\mathscr{P}$, of equation $y = f(x)$, invariant on the Oz axis and periodic in x, which separates a region R_+ ($y > f(x)$) from a region R_- ($y < f(x)$). The period $d = 2\pi/K$ of $f(x)$ will be called groove spacing or pitch or period of the grating. The maximum and minimum values of $f(x)$ will be denoted by y_{M} and y_{m}, respectively, $H = y_{\mathrm{M}} - y_{\mathrm{m}}$ being the groove depth. The unit normal to $\mathscr{P}$ is denoted by $\hat{\boldsymbol{n}}$. Region R_+ mostly contains air and thus its index will be assumed as being equal to 1. Region R_- is filled with a metal or a dielectric. An incident monochromatic homogeneous plane wave of wave-vector $\boldsymbol{k}$ (with $k = |\boldsymbol{k}| = 2\pi/\lambda$, λ being the wavelength), lying in the Oxy plane, illuminates the grating with the angle of incidence θ, algebraic number measured clockwise. Assuming a time dependence of the fields in $\exp(-\mathrm{i}\omega t)$, we denote by ν the complex index of region R_-. In fact, for historical reasons, and also for the sake

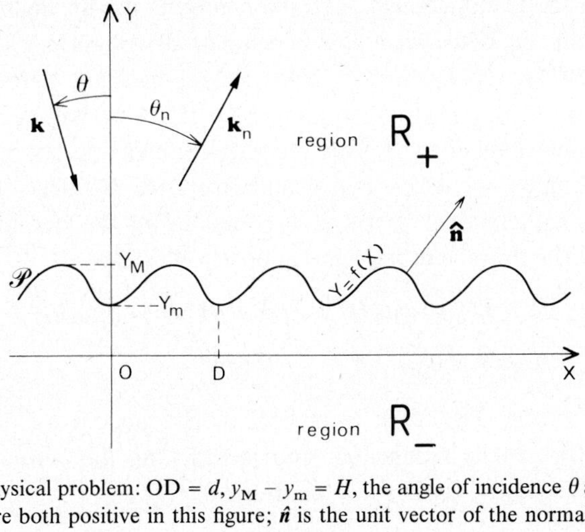

Fig. 2.1. The physical problem: OD = d, $y_{\mathrm{M}} - y_{\mathrm{m}} = H$, the angle of incidence θ and the angle of diffraction θ_n are both positive in this figure; $\hat{\boldsymbol{n}}$ is the unit vector of the normal to $\mathscr{P}$ oriented upward.

of simplicity, it is assumed, until § 7, that region R_- is perfectly conducting, i.e. that the electromagnetic field vanishes in this region.

The complex vector amplitude of the incident electric field, perpendicular to $\boldsymbol{k}$, is denoted by $\boldsymbol{a}$. So, this incident electric field can be written in the form:

$$\boldsymbol{E}^{\mathrm{i}} = \boldsymbol{a} \exp(\mathrm{i}\alpha x - \mathrm{i}\beta y), \tag{2.1}$$

where α and $-\beta$ are the two components of the incident wave-vector $\boldsymbol{k}$:

$$\alpha = k \sin \theta, \qquad \beta = k \cos \theta.$$

The incident magnetic field can be deduced easily from (2.1) and Maxwell's equations:

$$\boldsymbol{H}^{\mathrm{i}} = \boldsymbol{b} \exp(\mathrm{i}\alpha x - \mathrm{i}\beta y), \tag{2.2}$$

with $\boldsymbol{b} = \boldsymbol{k} \times \boldsymbol{a}/\omega\mu_0$.

The physical problem is to find the total fields $\boldsymbol{E}$ and $\boldsymbol{H}_{\mathrm{i}}$ at any point. The diffracted fields are defined in region R_+ by $\boldsymbol{E}^{\mathrm{d}} = \boldsymbol{E} - \boldsymbol{E}^{\mathrm{i}}$ and $\boldsymbol{H}^{\mathrm{d}} = \boldsymbol{H} - \boldsymbol{H}^{\mathrm{i}}$.

2.2. THE MATHEMATICAL BOUNDARY VALUE PROBLEM FOR THE PERFECTLY CONDUCTING GRATING

First, it is worth noting that the physical problem remains unchanged after translation of the grating along the Oz axis; in other words, neither the grating profile, nor the incident wave are dependent on z. As a consequence, if $\boldsymbol{E}(x, y, z)$ and $\boldsymbol{H}(x, y, z)$ are the total fields corresponding to a given grating and a given incident field, $\boldsymbol{E}(x, y, z + z_0)$ and $\boldsymbol{H}(x, y, z + z_0)$ will be solutions too, regardless of the value of z_0. If, from the physical meaning, we believe that the total field is unique, we derive that $\boldsymbol{E}$ and $\boldsymbol{H}$ are independent of z.

Now, the mathematical problem is stated using the harmonic Maxwell equations for the diffracted fields in region R_+:

$$\nabla \times \boldsymbol{E}^{\mathrm{d}} = \mathrm{i}\omega\mu_0 \boldsymbol{H}^{\mathrm{d}}, \qquad \nabla \times \boldsymbol{H}^{\mathrm{d}} = -\mathrm{i}\omega\varepsilon_0 \boldsymbol{E}^{\mathrm{d}},$$

$$\nabla \cdot \boldsymbol{E}^{\mathrm{d}} = 0, \qquad \nabla \cdot \boldsymbol{H}^{\mathrm{d}} = 0.$$

We note that the last two equations, consequences of the first two ($\nabla \cdot \nabla \times \boldsymbol{J} = 0$), can be eliminated, and introducing the components of the diffracted fields on the three axes, we then have:

$$\partial_y E_z^{\mathrm{d}} = \mathrm{i}\omega\mu_0 H_x^{\mathrm{d}}, \tag{2.3a}$$

$$\partial_x E_z^{\mathrm{d}} = -\mathrm{i}\omega\mu_0 H_y^{\mathrm{d}}, \tag{2.3b}$$

$$\partial_x E_y^{\mathrm{d}} - \partial_y E_x^{\mathrm{d}} = \mathrm{i}\omega\mu_0 H_z^{\mathrm{d}}, \tag{2.3c}$$

$$\partial_y H_z^{\mathrm{d}} = -\mathrm{i}\omega\varepsilon_0 E_x^{\mathrm{d}}, \tag{2.4a}$$

$$\partial_x H_z^{\mathrm{d}} = \mathrm{i}\omega\varepsilon_0 E_y^{\mathrm{d}}, \tag{2.4b}$$

$$\partial_x H_y^{\mathrm{d}} - \partial_y H_x^{\mathrm{d}} = -\mathrm{i}\omega\varepsilon_0 E_z^{\mathrm{d}}, \tag{2.4c}$$

where ε_0 and μ_0 are the electric permittivity and the magnetic permeability of vacuum, the symbol $\partial_y E_z^{\mathrm{d}}$ denoting the partial derivative $\partial E_z^{\mathrm{d}}/\partial y$. In the same way, the boundary conditions for the fields on the grating surface

$$\hat{\boldsymbol{n}} \times (\boldsymbol{E}^{\mathrm{d}} + \boldsymbol{E}^{\mathrm{i}}) = 0, \qquad \hat{\boldsymbol{n}} \cdot (\boldsymbol{H}^{\mathrm{d}} + \boldsymbol{H}^{\mathrm{i}}) = 0,$$

are linked to each other, and it suffices to write the first:

$$E_z^{\mathrm{d}} = -E_z^{\mathrm{i}}, \tag{2.5a}$$

$$n_x E_y^{\mathrm{d}} - n_y E_x^{\mathrm{d}} = -(n_x E_y^{\mathrm{i}} - n_y E_x^{\mathrm{i}}). \tag{2.5b}$$

It is worth noting that the linkage between these two boundary conditions is a typical example of an elementary property which is very difficult to establish, at least for those who are not acquainted with the theory of distributions.

The fundamental feature of equations (2.3a) to (2.5b) is that they can be separated into two independent sets. The first set, viz. (2.3a), (2.3b), (2.4c) and (2.5a), only contains the transverse component E_z^{d} of the electric field and the components H_x^{d} and H_y^{d} of the magnetic field. The same remark applies to the complementary set, but this time with the transverse component H_z^{d} of the magnetic field and the components E_x^{d} and E_y^{d} of the electric field. We deduce that the general problem can be decomposed into two elementary mathematical problems, called TE (transverse electric) or TM (transverse magnetic) according to whether they deal with E_z^{d} or H_z^{d}. Equations (2.3a) and (2.3b) for the TE problem and (2.4a) and (2.4b) for the TM problem provide a simple means of deducing the non-transverse components from E_z^{d} and H_z^{d}. So, introducing these values in the remaining equations enables us to derive two independent sets of equations, each set containing only one function.

In the TE problem, (2.3a), (2.3b) and (2.4c) show that E_z^{d} satisfies a Helmholtz equation:

$$\nabla^2 E_z^{\mathrm{d}} + k^2 E_z^{\mathrm{d}} = 0 \qquad \text{in } R_+. \tag{2.6a}$$

In addition, (2.5a) gives the boundary condition:

$$E_z^{\mathrm{d}} = -E_z^{\mathrm{i}} = -a_z \exp(\mathrm{i}\alpha x - \mathrm{i}\beta f(x)) \qquad \text{on } \mathscr{P}. \tag{2.7a}$$

Similarly, for the TM problem, (2.3c), (2.4a) and (2.4b) yield the Helmholtz equation

$$\nabla^2 H_z^{\mathrm{d}} + k^2 H_z^{\mathrm{d}} = 0 \qquad \text{in } R_+, \tag{2.6b}$$

and the boundary condition on $\mathscr{P}$ is derived from (2.4a), (2.4b) and (2.5b):

$$\begin{aligned} \mathrm{d}H_z^{\mathrm{d}}/\mathrm{d}n &= -\mathrm{d}H_z^{\mathrm{i}}/\mathrm{d}n \\ &= \mathrm{i}b_z(\beta n_y - \alpha n_x)\exp(\mathrm{i}\alpha x - \mathrm{i}\beta f(x)) \qquad \text{on } \mathscr{P}, \end{aligned} \tag{2.7b}$$

where $\mathrm{d}/\mathrm{d}n$ denotes the normal derivative on $\mathscr{P}$. Now, the question which arises is whether the two couples (2.6a), (2.7a) or (2.6b), (2.7b) are sufficient to determine the transverse components of the fields. To this question, the mathematician gives a negative answer since these equations do not include conditions at $y = \infty$. Therefore, to complete these equations, we must add the so-called radiation condition (or outgoing wave condition), which seems to be obvious for the physicist: when y tends to infinity, E_z^{d} and H_z^{d} must remain bounded and can be described as a superposition of outgoing plane waves propagating upward. Since (2.6a)–(2.7b) have been obtained using the elementary laws of electromagnetism, the physicist is tempted to deduce that the solution of these equations satisfying the radiation condition exists and is unique. However, it appears that its existence has never been established by mathematicians. Yet, this question cannot be called useless: physical evidence is not a proof of mathematical existence! On the other hand, the uniqueness of the TE problem ((2.6a), (2.7a) and the radiation condition) can be proved assuming the local square integrability of the field (CADILHAC [1980]), but the generalization of this theorem to the TM problem is not straighforward. Perhaps, the specialist of Helmholtz equations and boundary value problems will be surprised to read that existence and uniqueness theorems, which have been demonstrated for diffracting structures apparently very close to the grating, have not been generalized to gratings (WILCOX [1975]). In fact, the difficulty for the grating problem lies in the fact that both region R_+ and boundary $\mathscr{P}$ are unbounded. In spite of this lack of demonstration, the existence and uniqueness of the TE and TM problems will be assumed in the following. More generally, we assume these properties of existence and uniqueness when the right-hand member of (2.7a) or (2.7b) is replaced by a locally square integrable function of x.

2.3. THE TWO FUNDAMENTAL (TM and TE) PROBLEMS

Let us define the fundamental TE case of polarization (FTE) where a plane wave of unit amplitude strikes the grating ($a_z = 1$, $a_x = a_y = b_z = 0$). For this particular case, the TE problem is described by (2.6a) and (2.7a) with $a_z = 1$. On the other hand, the TM problem of this FTE case is described by (2.6b) and (2.7b) and is homogeneous, since $b_z = 0$. Thanks to the uniqueness of the solution, we can deduce that the solution H_z^{d} of the TM problem for the FTE case vanishes. *So, we conclude that, in the FTE case, the diffracted electric field remains parallel to* Oz *and is given by* (2.6a), (2.7a) *and the radiation condition.* The same considerations hold for a TM polarized incident wave where the fundamental TM case of polarization (FTM) can be defined ($b_z = 1$, $b_x = b_y = a_z = 0$). *Thus, for each of these two fundamental cases of polarization, the problem is scalar.* Let us denote by F^{i} and F the amplitude of E_z^{i} and E_z^{d} (for the FTE case) or the amplitude of H_z^{i} and H_z^{d} (for the FTM case). From the equations of TE and TM problems, we derive that F is the solution of Dirichlet (FTE case) or Neumann (FTM case) boundary value problems:

$$\nabla^2 F + k^2 F = 0 \quad \text{in } R_+, \tag{2.8}$$

$$\text{on } \mathscr{P} \begin{cases} F = -F^{\mathrm{i}} = -\exp(\mathrm{i}\alpha x - \mathrm{i}\beta f(x)) & \text{for the FTE case,} \quad (2.9\mathrm{a}) \\ \mathrm{d}F/\mathrm{d}n = -\mathrm{d}F^{\mathrm{i}}/\mathrm{d}n = \mathrm{i}(\beta n_y - \alpha n_x)\exp(\mathrm{i}\alpha x - \mathrm{i}\beta f(x)) & \\ & \text{for the FTM case,} \quad (2.9\mathrm{b}) \end{cases}$$

$$F \text{ satisfies a radiation condition when } y \to \infty, \tag{2.10}$$

the incident field being given by:

$$F^{\mathrm{i}} = \exp(\mathrm{i}\alpha x - \mathrm{i}\beta y). \tag{2.11}$$

When the FTE and the FTM problems have been solved, the solution of a general problem corresponding to an arbitrary polarization can be derived very simply: it suffices to multiply by a_z the solution of the FTE problem and by b_z the solution of the FTM problem to obtain E_z^{d} and H_z^{d}. So, it will suffice to study these two fundamental cases and first, we show that (2.8)–(2.10) allow us to derive well known elementary properties of the diffracted field F.

2.4. THE PSEUDO-PERIODICITY OF THE FIELD

In this section, it is shown that $F(x, y)\exp(-\mathrm{i}\alpha x)$ is a periodic function. To this aim, it is useful to consider the function G defined by:

$$G(x, y) = F(x + d, y)\exp(-\mathrm{i}\alpha d).$$

To prove the periodicity of $F(x, y) \exp(-\mathrm{i}\alpha x)$ it suffices to show that $G(x, y) = F(x, y)$. Owing to the uniqueness of the solution, this equality will be satisfied if $G(x, y)$ obeys eqs. (2.8)–(2.10). Bearing in mind that $F(x, y)$ satisfies (2.8) and (2.10), it is clear that $G(x, y)$ satisfies these equations also. It is not difficult to see that G satisfies (2.9a) or (2.9b) if the periodicity d of $f(x)$ has been remembered.

In conclusion, it has been proved that $F(x, y)$ can be expressed in the form:

$$F(x, y) = \exp(\mathrm{i}\alpha x)\, T(x, y),$$

where $T(x, y)$ has a period d in x. Such a function will be called a pseudo-periodic function, with a coefficient of pseudo-periodicity α and pseudo-period d.

For a given grating, we shall say that two functions F and F' have the same pseudo-periodicity when $\alpha - \alpha' = pK$, p being an integer, the pseudo-periods d and d' being identical and equal to $2\pi/K$.

2.5. THE RAYLEIGH EXPANSION ABOVE THE GROOVES

Let us develop $T(x, y)$ in a Fourier series:

$$T(x, y) = \sum_n t_n(y) \exp(\mathrm{i}nKx),$$

$\sum_n$ denoting a sum from $-\infty$ to $+\infty$. The two equations above imply:

$$F(x, y) = \sum_n t_n(y) \exp(\mathrm{i}\alpha_n x),$$

with

$$\alpha_n = \alpha + nK = k \sin(\theta) + nK. \tag{2.12a}$$

So, a pseudo-periodic function can be developed in a pseudo-Fourier series of $\exp(\mathrm{i}\alpha_n x)$.

Now, we have to find the form of $t_n(y)$. Introducing the Fourier expansion of F in (2.8), then multiplying by $\exp(-\mathrm{i}\alpha x)$ yields:

$$\sum_n \left[\frac{\mathrm{d}^2 t_n}{\mathrm{d}y^2} + (k^2 - \alpha_n^2)t_n\right] \exp(\mathrm{i}nKx) = 0 \qquad \text{in } R_+.$$

If $y > y_{\mathrm{M}}$, the above equation is valid for any value of x and each term of the Fourier series must be null:

$$\frac{\mathrm{d}^2 t_n}{\mathrm{d}y^2} + (k^2 - \alpha_n^2)t_n = 0 \qquad \text{if } y > y_{\mathrm{M}}.$$

Of course, the physicist is very well acquainted with the above equation. To express $t_n(y)$, we denote by U the finite set of integers for which $k^2 - \alpha_n^2$ is positive and we define:

$$\beta_n = (k^2 - \alpha_n^2)^{1/2} \qquad \text{if } n \in U$$
$$= \mathrm{i}(\alpha_n^2 - k^2)^{1/2} \qquad \text{if } n \notin U. \tag{2.12b}$$

So, the expression of $t_n(y)$ is given by:

$$t_n(y) = A_n \exp(-\mathrm{i}\beta_n y) + B_n \exp(\mathrm{i}\beta_n y) \qquad \text{if } y > y_{\mathrm{M}}.$$

The first term, of coefficient A_n, becomes infinite when $y \to \infty$ if $n \notin U$, or represents an incoming plane wave propagating toward the grating surface if $n \in U$. Thus, the radiation condition means that the A_n vanish and the diffracted field $F(x, y)$ can be expressed as a series $R(x, y)$ of plane waves $R_n(x, y)$:

$$F(x, y) = R(x, y) = \sum_n R_n(x, y) = \sum_n B_n \Phi_n(x, y) \qquad \text{if } y > y_{\mathrm{M}} \tag{2.13}$$

with

$$\Phi_n(x, y) = \exp(\mathrm{i}\alpha_n x + \mathrm{i}\beta_n y). \tag{2.14}$$

The above result is very simple: *above the top of the grooves, the diffracted field can be represented by the so-called "Rayleigh expansion"* (RE) used by RAYLEIGH [1907], i.e. a series of outgoing plane waves having the same pseudo-periodicity. It should be borne in mind that these plane waves can be divided into two categories. If $n \notin U$, β_n is imaginary and $R_n(x, y)$ represents an evanescent wave propagating in the x direction. On the other hand, for the finite number of $n \in U$, $R_n(x, y)$ is an homogeneous plane wave propagating upward with a wave-vector $\boldsymbol{k}_n(\alpha_n, \beta_n)$. Denoting by θ_n the angle of $\boldsymbol{k}_n$ with the Oy axis (measured anticlockwise this time), it can be stated:

$$\sin\theta_n = \alpha_n/(\alpha_n^2 + \beta_n^2)^{1/2} = \alpha_n/k, \tag{2.12c}$$

which, after replacing α_n by its value given by (2.12a) yields the famous grating formula:

$$\sin\theta_n = \sin\theta + n\lambda/d. \tag{2.15}$$

The complicated convention for θ and θ_n is justified by the fact that the above formula takes exactly the same form as the one classically demonstrated for transmission gratings, using optical path considerations. For the opticist, the evanescent waves have no interest (except in integrated optics) and the only interesting quantities are the finite number of B_n, for $n \in U$. Indeed, the efficiency e_n in the nth diffracted order, viz. the nth order diffracted energy to incident energy ratio, is very easily deduced from the B_n. It suffices to calculate the flux of Poynting's vector of the incident wave and of the nth order diffracted wave through a plane surface parallel to the xz plane at $y > y_M$, of unit length on the Oz axis and of length d on the Ox axis, to deduce:

$$e_n = B_n \overline{B_n} \cos\theta_n / \cos\theta, \qquad (2.16)$$

$\overline{B_n}$ denoting the conjugate of B_n.

The validity of the RE above the top of the grooves, which is rigorous, should not be confused with the so-called "Rayleigh hypothesis" (RH) which assumes that the plane wave expansion remains valid inside the groove, and whose theoretical validity will be studied further.

2.6. A LEMMA AND ITS APPLICATIONS: ENERGY BALANCE CRITERION, RECIPROCITY AND EXPRESSION OF THE B_n

Here, we give a mathematical lemma, inspired by some previous papers (URETSKY [1965], PETIT [1967, 1980], MAYSTRE and MC PHEDRAN [1974], CADILHAC [1980]), which enables us to establish some important properties of the coefficients B_n. To this end, we consider a given grating and we define a certain class $\mathscr{F}_0$ of functions which contains in particular the two functions $F(x, y)$ corresponding to the two fundamental cases of polarization for a given incidence θ. More precisely, any function $u(x, y)$ of $\mathscr{F}_0$ will satisfy the following conditions:

- It is pseudo-periodic with coefficient of pseudo-periodicity α.
- It satisfies (2.8) (Helmholtz equation).
- It is bounded for $y \to \infty$.

It will be assumed that $u(x, y)$ also satisfies a local square integrability in x and y and that the boundary values on $\mathscr{P}$ for u and $\mathrm{d}u/\mathrm{d}n$ are square-integrable too. Obviously, $\mathscr{F}_0$ contains $F(x, y)$, whatever the fundamental case of polarization, but we do not impose a boundary condition on the grating, or a radiation condition. As a consequence, $\mathscr{F}_0$ contains the incident field F^{i} too. Using the same arguments as in § 2.5, it can be shown that such an arbitrary function $u(x, y)$ can be represented by a sum of plane waves above the grooves:

$$u(x, y) = \sum_{n \in U} A_n \exp(\mathrm{i}\alpha_n x - \mathrm{i}\beta_n y) + \sum_n B_n \exp(\mathrm{i}\alpha_n x + \mathrm{i}\beta_n y). \qquad (2.17)$$

The above expansion differs slightly from the expansion of F (see (2.13)) since it contains incident homogeneous plane waves of amplitudes A_n, but it cannot contain incident terms corresponding to $n \notin U$ since it is bounded at infinity.

Considering a second function $v(x, y)$ of the same set, with amplitudes A'_n and B'_n, we will establish the following lemma:

$$\frac{1}{2id}\int_{\mathscr{P}}\left(u\frac{d\bar{v}}{dn}-\bar{v}\frac{du}{dn}\right)ds=\sum_{n\in U}^{n}\beta_n(A_n\overline{A'_n}-B_n\overline{B'_n}), \tag{2.18}$$

where $\int_{\mathscr{P}}$ denotes a curvinilear integral on a period of the profile $\mathscr{P}$, ds being the differential of the curvilinear abscissa s on $\mathscr{P}$. The demonstration of this lemma lies in the fact that u and v (thus $\bar{v}$) satisfy the same Helmholtz equation:

$$\nabla^2 u + k^2 u = 0, \qquad \nabla^2\bar{v} + k^2\bar{v} = 0.$$

Multiplying the first equation by $\bar{v}$, the second equation by u and subtracting the first from the second yield:

$$u\nabla^2\bar{v} - \bar{v}\nabla^2 u = 0 \qquad \text{in } R_+ .$$

By integrating the above equation in the hatched area of Fig. 2.2 and applying the second Green identity, it can be deduced:

$$\int_{\mathscr{C}}\left(u\frac{d\bar{v}}{dn}-\bar{v}\frac{du}{dn}\right)d\ell=0,$$

where $\mathscr{C}$ is the boundary of the hatched area and ℓ its curvilinear abscissa. Since $u\,d\bar{v}/dn$ and $\bar{v}\,du/dn$ are periodic, the contributions of QR and SP are opposite and cancel each other. Moreover the normal to RS is parallel to Oy, thus the above equation yields:

$$\int_{\mathscr{P}}\left(u\frac{d\bar{v}}{dn}-\bar{v}\frac{du}{dn}\right)ds=\int_{\mathrm{SR}}\left(u\frac{d\bar{v}}{dy}-\bar{v}\frac{du}{dy}\right)dx.$$

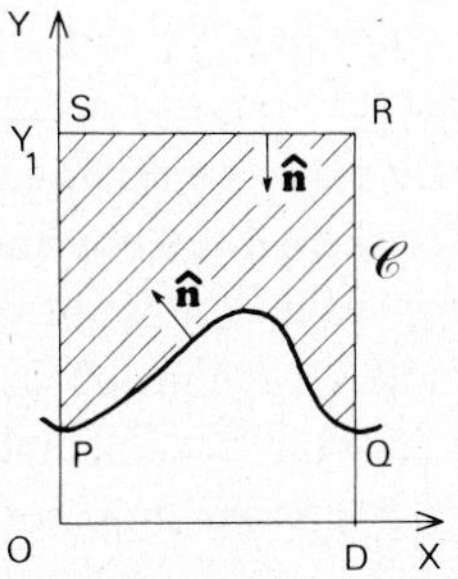

Fig. 2.2. Demonstration of a lemma. $\mathscr{C}$ is the boundary PQRS of the hatched area and ℓ denotes the curvilinear abscissa on $\mathscr{C}$.

It is a cumbersome, but not difficult, exercise to transform the right-hand member of the above equation in order to get (2.18). The reader may reach this goal by introducing expansions (2.17) for u and v in this right-hand member, separating the terms $n \notin U$ from the others, and bearing in mind that $\int_0^d \exp(inKx)\,\mathrm{d}x = d \cdot \delta_n$, where $\delta_n = 0$ if $n \neq 0$ and $\delta_0 = 1$.

The energy balance criterion is obtained by taking $v = u = F + F^{\mathrm{i}}$ in (2.18). Since $F + F^{\mathrm{i}}$ or $\mathrm{d}(F + F^{\mathrm{i}})/\mathrm{d}n$ vanishes on $\mathscr{P}$, the left-hand member is null and we deduce:

$$\sum_{n \in U} B_n \overline{B_n} \beta_n = \beta_0,$$

or in other terms, using (2.12b), (2.12c) and (2.16):

$$\sum_{n \in U} e_n = 1. \tag{2.19}$$

It is important to notice that the evanescent orders are not involved in this relation, called the energy balance criterion, which expresses a result with an obvious physical meaning: *the diffracted energy is equal to the incident energy.*

The lemma will enable us to establish another result, familiar to opticists: the reciprocity relation. Here we take $u = F + F^{\mathrm{i}}$, as before (case 1), for a given fundamental case of polarization. Now, we consider the pth order of diffraction ($p \in U$) of diffraction angle θ_p. Let us call θ''_p, α''_p, β''_p and B''_p the parameters of the Rayleigh expansion of the field v'' obtained for the same fundamental case of polarization, but this time with the angle of incidence $\theta'' = -\theta_p$ (case 2). Bearing in mind our conventions for incident and diffracted angles, the incident wave in this second case has a direction of propagation which is just the opposite of that of the pth diffracted order in the first case. From the grating formula, a straightforward calculation shows that $\theta''_p = -\theta$, $\alpha''_p = -\alpha_0$ and $\beta''_p = \beta_0$, which means that the pth diffracted order in case 2 and the incident wave in case 1 have opposite directions, a fact which will probably seem obvious to opticists familiar with the reversion theorem (Fig. 2.3). Function u and $v = \overline{v''}$ wholly satisfy the conditions of the lemma since the grating formula shows that they have the same pseudo-periodicity. On the other hand, u and v are very different since v includes all the possible incident waves ($n \in U$) but only one propagating diffracted wave, due to the conjugation which transforms an incident wave into a diffracted wave propagating in the opposite direction, and conversely. Proceeding carefully to calculate the right-hand member of (2.18) and remarking that the left-hand member of this equation vanishes, it can be found:

$$\beta_0 \overline{A'_0} = \beta''_p B''_p = \beta_p B_p. \tag{2.20}$$

The King's Library

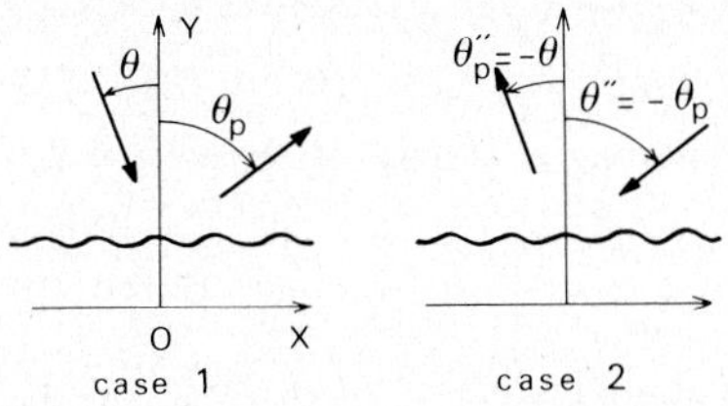

Fig. 2.3. The reciprocity theorem: the efficiency of the pth order is the same in the two cases.

Multiplying each member by its conjugate, and dividing by $\beta_p \beta_0''$ yield, after simplification:

$$B_p'' \overline{B_p''} \beta_p'' / \beta_0'' = B_p \overline{B_p} \beta_p / \beta_0, \tag{2.21}$$

a result which can be easily outlined: *The efficiency in the* pth *order is the same in the two cases of* Fig. 2.3.

After the famous reciprocity relation, let us obtain a last result, which generally appears as the consequence of much more complicated considerations: the coefficients B_n may be given by a curvilinear integral on $\mathscr{P}$. Once more, the value of u will be the function $F + F^{\mathrm{i}}$, corresponding to a fundamental case of polarization. The function v will be very simple:

$$v = \Phi_p(x, y) = \exp(\mathrm{i}\alpha_p x + \mathrm{i}\beta_p y), \qquad n \in U.$$

In other words, v will be a single diffracted wave corresponding to the nth order. So, the A_m' and all the B_m' except $B_n' = 1$ are nil. It follows that the right-hand member of (2.18) will be equal to $-\beta_n B_n$. Evaluating the left-hand member requires the separation of FTE and FTM problems. Since either u, or $\mathrm{d}u/\mathrm{d}n$ vanishes on $\mathscr{P}$, it follows immediately that:

$$B_n = \frac{1}{2\mathrm{i}d\beta_n} \int_{\mathscr{P}} \frac{\mathrm{d}u}{\mathrm{d}n} \overline{\Phi}_n(x, f(x))\, \mathrm{d}s \qquad \text{for the FTE case,}$$

$$= -\frac{1}{2\mathrm{i}d\beta_n} \int_{\mathscr{P}} u \frac{\mathrm{d}\overline{\Phi}_n}{\mathrm{d}n}(x, f(x))\, \mathrm{d}s \qquad \text{for the FTM case.}$$

By replacing $\mathrm{d}s$ by $(1 + f'(x)^2)^{1/2}\, \mathrm{d}x$ and using the value of $\Phi_n(x, y)$, it can be derived that

$$B_n = \frac{1}{2\mathrm{i}d\beta_n} \int_0^d \exp(-\mathrm{i}\alpha_n x - \mathrm{i}\beta_n f(x)) \phi(x)\, \mathrm{d}x \tag{2.22a}$$

for the FTE case,

$$B_n = \frac{1}{2d}\int_0^d \left(1 - \frac{\alpha_n}{\beta_n} f'(x)\right) \exp(-\mathrm{i}\alpha_n x - \mathrm{i}\beta_n f(x))\psi(x)\,\mathrm{d}x \quad (2.22\mathrm{b})$$

for the FTM case. Thus

$$\phi(x) = (1 + f'(x)^2)^{1/2}\,\mathrm{d}(F + F^{\mathrm{i}})/\mathrm{d}n, \qquad (2.22\mathrm{c})$$

$$\psi(x) = F(x, f(x)) + F^{\mathrm{i}}(x, f(x)). \qquad (2.22\mathrm{d})$$

This result is very interesting. *The coefficients* B_n, *and thus the efficiencies, are known from a simple integral as soon as the normal derivative of the total field* (for the FTE case) *or the field on the grating* (for the FTM case) *is given.* A simple physical interpretation of this result may be given by noticing that functions ϕ and ψ are closely linked to the surface current density $\boldsymbol{j}_{\mathscr{P}}$ on the grating surface. Indeed, $\boldsymbol{j}_{\mathscr{P}} = \hat{\boldsymbol{n}} \times \boldsymbol{H}$ and, from (2.3a) and (2.3b), we derive:

$$\phi(x) = -\mathrm{i}\omega\mu_0(1 + f'(x)^2)^{1/2} j_{\mathscr{P}}(x) \qquad \text{for the FTE case,} \quad (2.23\mathrm{a})$$

$$\psi(x) = j_{\mathscr{P}}(x) \qquad \text{for the FTM case,} \quad (2.23\mathrm{b})$$

the scalar function $j_{\mathscr{P}}$ being the z component of $\boldsymbol{j}_{\mathscr{P}}$ for the FTE case and the projection $\boldsymbol{j}_{\mathscr{P}} \cdot \boldsymbol{t}$ of $\boldsymbol{j}_{\mathscr{P}}$ of the unit tangent vector $\hat{\boldsymbol{t}} = \hat{\boldsymbol{n}} \times \hat{\boldsymbol{z}}$ for the FTM case. It is not surprising to see that the B_n can be derived from the surface current density since $\boldsymbol{j}_{\mathscr{P}}$ actually generates the diffracted field. It is worth noting that (2.22a) and (2.22b), demonstrated for $n \in U$, hold for $n \notin U$, as we shall see further. Similar formulae can be obtained for the incident amplitudes A_n, by setting $v = \exp(\mathrm{i}\alpha_n x - \mathrm{i}\beta_n y)$. Since $A_n = \delta_n$ for FTE and FTM cases:

$$\delta_n = -\frac{1}{2\mathrm{i}d\beta_n}\int_0^d \exp(-\mathrm{i}\alpha_n x + \mathrm{i}\beta_n f(x))\phi(x)\,\mathrm{d}x$$

$$\text{for the FTE case,} \qquad (2.24\mathrm{a})$$

$$\delta_n = \frac{1}{2d}\int_0^d \left(1 + \frac{\alpha_n}{\beta_n} f'(x)\right) \exp(-\mathrm{i}\alpha_n x + \mathrm{i}\beta_n f(x))\psi(x)\,\mathrm{d}x$$

$$\text{for the FTM case,} \qquad (2.24\mathrm{b})$$

and these relations hold when $n \notin U$, as we shall see later.

2.7. WHY RIGOROUS VECTOR METHODS?

For a long time, it has been difficult to convince some grating users of the usefulness of rigorous grating theories and, from my own experience, it seems that even today, some opticists are not aware of the troubles they may encounter using approximate theories. So, we have chosen a theory, one of the most familiar to opticists, based on the Kirchhoff approximation (BECKMANN and SPIZZICHINO [1963], BECKMANN [1967]), in order to compare its numerical results with exact values. Though the basic approximation is the same, the demonstration developed here to obtain the approximate formula will be rather different from the demonstration made by BECKMANN, and the resultant formula will appear to be slightly different.

Let us consider the TE fundamental case of polarization where, according to (2.22a), the B_n are known as soon as the function ϕ (closely linked with the normal derivative of the total field) is known. The approximate value ϕ_B of ϕ will be obtained in the following manner: we shall assume that the normal derivative of the field at a certain point of $\mathscr{P}$ is the same as the normal derivative of the field which would be obtained by replacing $\mathscr{P}$ by a plane mirror tangent to $\mathscr{P}$ at the same point. So the "local" value of $F^i + F$ in the vicinity of $\mathscr{P}$ will be the sum of the incident wave and the wave reflected by the fictitious mirror. An effortless calculation shows that:

$$\phi_B(x) = -2i(\alpha f'(x) + \beta) \exp(i\alpha x - i\beta f(x)). \tag{2.25}$$

Introducing $\phi_B(x)$ in (2.22a) and integrating by parts to remove the term $f'(x)$ gives the following formula:

$$B_n = -\frac{1}{d} \frac{1 + \cos(\theta + \theta_n)}{\cos\theta_n(\cos\theta + \cos\theta_n)} \int_0^d \exp[-inKx - i(\beta + \beta_n)f(x)]\, dx, \tag{2.26}$$

a formula which differs from the BECKMANN formula by the presence of the factor $\cos\theta_n$ (instead of $\cos\theta$) in the denominator, but we have to bear in mind that the quantity given by BECKMANN is not B_n. After calculation, the value of B_n obtained in the FTM case is identical, except for the sign $(-)$ in front of the right-hand member, which must be eliminated. Though the expression of B_n given by (2.26) obviously satisfies the reciprocity theorem stated in (2.20), we must notice that it finally gives a scalar result, which does not predict any polarization effect, at least for the efficiency of perfectly conducting gratings. In the particular case of the sinusoidal grating:

$$f(x) = \frac{H}{2} \cos Kx, \tag{2.27}$$

the integral can be evaluated in closed form using classical formulae

(ABRAMOWITZ and STEGUN [1970]), in such a way that, for the FTE case:

$$B_n = -(-\mathrm{i})^n \frac{1 + \cos(\theta + \theta_n)}{\cos\theta_n(\cos\theta + \cos\theta_n)} J_n\left[k\frac{H}{2}(\cos\theta + \cos\theta_n)\right]. \tag{2.28}$$

The corresponding efficiency e_{-1} has been drawn in Fig. 2.4 for $H/d = 0.2$ in Littrow mount. In this mounting, very familiar to grating users, the -1 order diffracted wave and the incident wave are propagating in opposite directions, so $\theta_{-1} = -\theta$, and from the grating formula we deduce that in this mounting, the wavelength, groove spacing and incidence are linked by a very simple relation:

$$\lambda/d = 2\sin\theta. \tag{2.29}$$

Furthermore, from (2.16) and (2.28) the efficiency e_{-1} is given by:

$$e_{-1} = J_1^2(kH\cos\theta)/\cos^4\theta.$$

Figure 2.4 also shows the efficiency obtained from a computer code developed by the present author, using a rigorous integral theory, which will be described in § 5. It will be shown further that the accuracy of the "rigorous" results is better than 1 per cent and it is to be hoped that the reader will accept this

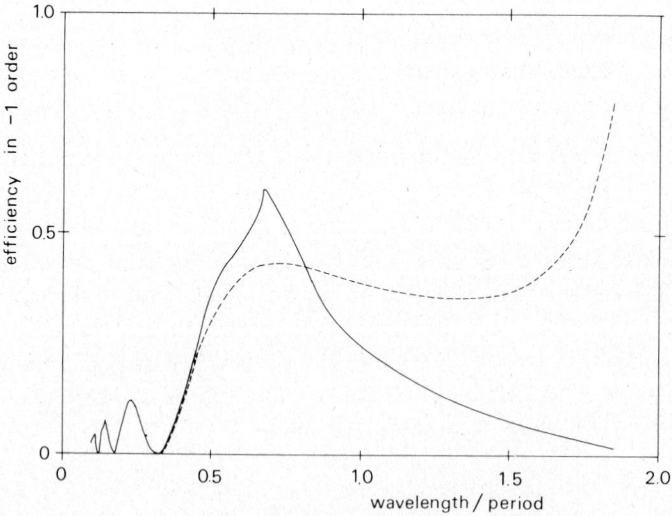

Fig. 2.4. Comparison of the numerical results obtained from a rigorous theory and a non-rigorous one. Solid line: rigorous result from the integral theory; dashed line: result from the BECKMANN theory.

assertion at this point. Obviously, the agreement between the two theories is very good as long as λ/d is lower than 0.4 approximately: the results are nearly identical. Unfortunately, the agreement does not hold any longer in the complementary region since the efficiency predicted by BECKMANN's formula tends to infinity when λ/d tends to 2 (this corresponds to a grazing incidence, as shown by (2.29)). The limit of the region of validity of BECKMANN's theory varies with H, and decreases when H increases. Since many commercial gratings have groove depth to groove spacing ratios higher than 0.2, we must conclude that *it is dangerous to trust in such a theory in the resonance region.* However, *it must be emphasized that such a theory is very valuable when λ/d is low*: it gives accurately the properties of gratings without the help of a big computer and moreover, the rigorous computation becomes more and more difficult and expansive when λ/d decreases. So, there does not exist any challenge between rigorous and non-rigorous theories, but rather a complementarity. We hope only to convince the reader that using non-rigorous theories has a prerequisite: the thorough knowledge of its domain of validity, in order to avoid disasters. Of course, this conclusion still holds for non-periodic rough surfaces.

§ 3. The Rayleigh method

We have seen in the previous section that the RE (Rayleigh Expansion, see (2.13)) is valid above the top of the grooves. Rayleigh made the assumption that this expansion was a valid representation of the diffracted field F not only for $y > y_M$ but throughout the region R_+, including the grooves (RAYLEIGH [1907, 1945]). The first objections about the RH (Rayleigh Hypothesis) were formulated thirty years ago (DERIUGIN [1952], LIPPMANN [1953]) and until now, this topic has raised considerable controversy. At present, there is no doubt that the RH is neither always valid, nor always invalid. The main theoretical achievements in this topic will be presented in a synthetic manner. Due to the difficulty of some demonstrations based on the properties of complex functions of complex variables, we thought it useful to give at the beginning a non-rigorous but much more simple demonstration founded on BECKMANN's approximation, which nevertheless leads to an exact result. The problem we are dealing with in the last part of this section is the main subject of the actual controversies: *why and in what conditions can the RH be used to obtain accurate numerical results, even in conditions where this RH theoretically fails*?

Now, to begin this discussion, we intend to convince the reader that the question of the validity of the RH is not academic: the RH simplifies considerably the work of the theoretician and provides a very simple means of calculating the grating efficiencies.

3.1. BASIC EQUATION AND NUMERICAL IMPLEMENTATION OF THE RAYLEIGH METHOD

After Rayleigh, the Rayleigh hypothesis has been used by many authors (FANO [1941], MEECHAM [1956], STROKE [1960], BOUSQUET [1963a, 1963b], PETIT [1963], DELEUIL [1963], JANOT and HADNI [1963], YAKOVLEV [1965], PETIT and CADILHAC [1966], WIRGIN [1967], JIRACEK [1973]). Some of them implemented this theory numerically when sufficiently powerful computers became available. Here, we describe three ways of implementing this method. They have in common the basic equation, obtained by writing that the RE described by the series $R(x, y)$ (see 2.13)), which obviously satisfies the Helmholtz equation and radiation condition stated in (2.8) and (2.10), must also satisfy the boundary condition on the grating surface expressed by (2.9a) or (2.9b). For example, in the FTE case, we obtain:

$$\sum_n B_n \Phi_n(x, f(x)) + F^{\mathrm{i}}(x, f(x)) = 0, \qquad \forall x. \tag{3.1a}$$

In this equation, the only unknown is the set of complex coefficients B_n which appear to be the coefficients of the expansion of $-F^{\mathrm{i}}(x, f(x))$ into a series of functions $\Phi_n(x, f(x))$. All the numerical methods described here reduce (3.1a) to a system of $(2N+1)$ linear equations with $(2N+1)$ unknowns: the coefficients B_n for $-N \leqslant n \leqslant +N$. So, these methods neglect all the B_n corresponding to $|n| > N$ and replace (3.1a) by:

$$s_N(x) = \sum_{n=-N}^{+N} B_n^{(N)} \Phi_n(x, f(x)) + F^{\mathrm{i}}(x, f(x)) = 0. \tag{3.1b}$$

Of course, it will be necessary to check the convergence of the $B_n^{(N)}$, approximations of B_n obtained for a size $2N+1$ of the linear system, when N increases.

In the first method, called the point-matching method (PMM), (3.1b) is written at $2N+1$ points of abscissae x_m placed on one period of the grating:

$$\sum_{n=-N}^{+N} B_n^{(N)} \Phi_n(x_m, f(x_m)) = -F^{\mathrm{i}}(x_m, f(x_m)), \ \forall m \in (1, 2N+1). \tag{3.2}$$

The linear system of $2N+1$ equations with $2N+1$ unknowns can be solved on the computer using classical methods. Of course, the choice of the x_m is fundamental. In practice, the x_m are chosen equidistant.

In the second numerical method, called the Fourier Series Method (FSM), (3.1b) is projected on the $2N+1$ functions $\exp(\mathrm{i}\alpha_m x)$ $(-N \leqslant m \leqslant +N)$ in order to obtain a linear system of equations:

$$\sum_{n=-N}^{+N} B_n^{(N)} \langle \exp(\mathrm{i}\alpha_m x)\, \Phi_n(x, f(x))\rangle = -\langle \exp(\mathrm{i}\alpha_m x), F^{\mathrm{i}}(x, f(x))\rangle,$$
$$\forall m \in (-N, +N), \qquad (3.3)$$

with
$$\langle p(x), q(x)\rangle = \int_0^d \bar{p}(x) q(x)\,\mathrm{d}x.$$

To justify this method, it suffices to notice that the left-hand member of (3.1b) is the product of a periodic function and $\exp(\mathrm{i}\alpha x)$. In practice, two cases are observed: either the scalar products can be calculated in closed form or they require a numerical integration.

In the last method, described in this section, the functional

$$T_N = \int_0^d |s_N(x)|^2\,\mathrm{d}x$$

is minimized. To obtain a linear system, it must be noticed that

$$T_N = \langle s_N(x), s_N(x)\rangle,$$

and that T_N, considered as a function of the $B_n^{(N)}$, must be stationary:

$$\delta T_N = 2 \sum_{n=-N}^{+N} \mathrm{Re}\{\delta \bar{B}_n^{(N)} \langle \Phi_n, s_N(x)\rangle\} = 0,$$

$\mathrm{Re}\{p(x)\}$ denoting the real part of $p(x)$. Therefore, this Fourier variational method (FVM) leads to the following equations, obtained by equating to zero all the coefficients of $\delta \bar{B}_n^{(N)}$:

$$\sum_{m=-N}^{+N} B_n^{(N)} \langle \Phi_n(x, f(x)), \Phi_m(x, f(x))\rangle + \langle \Phi_n(x, f(x)), F^{\mathrm{i}}(x, f(x))\rangle = 0,$$
$$\forall n \in (-N, +N). \qquad (3.4)$$

This time, (3.1b) is not projected on the Fourier basis but on the set of functions $\Phi_n(x, f(x))$.

When they were implemented on a computer, between 1960 and 1970, these methods were applied, in the first instance, to ruled echelette gratings, viz. gratings having triangular grooves, and also to sinusoidal gratings. In these two cases, the coefficients of the linear system can be evaluated in closed form for the FSM and for the FVM: exponentials for echelette gratings and Bessel functions for sinusoidal gratings. For shallow gratings the efficiencies obtained from these methods satisfied the energy balance criterion and the reciprocity theorem with very high accuracy. They also converged when N was increased. On the other hand, the Rayleigh methods completely failed for deeper grooves. The region of convergence of the FSM and the FVM was significantly larger than that of the PMM. For instance, calculations using the FSM for symmetrical echelette gratings (Fig. 3.1) of groove spacing $d = 1.25$ μm used in normal incidence with $\lambda = 0.546$ μm gave very good results until $\tan b = 0.15$. These results were less precise but acceptable for deeper grooves until $\tan b = 0.3$, for which the energy balance criterion was satisfied to within 2%. For greater values of $\tan b$, the results were unprecise and for $\tan b = 1$, a divergence was observed (PETIT [1963]). For the same grating, the numerical domain of convergence is dependent on the wavelength to groove spacing ratio (MC PHEDRAN [1973]). Due to this numerical failure, many authors questioned the validity of the Rayleigh hypothesis and remembered the former warnings of DERIUGIN and LIPPMANN. For these authors, the RE is not valid at a point P inside the grooves since it does not contain waves propagating towards $y = -\infty$, which are generated by the surface currents at the points of the grating surface situated above the point of observation. We shall see that the serious objections to the RH apparently have no connection with the remarks of DERIUGIN and LIPPMANN.

3.2. CONCERNING THE VALIDITY OF THE RAYLEIGH HYPOTHESIS

3.2.1. *Preliminary results*

From a theoretical point of view, it is interesting to consider the behavior of the Rayleigh expansion $R(x, y)$ inside the grooves, i.e. in the region where it

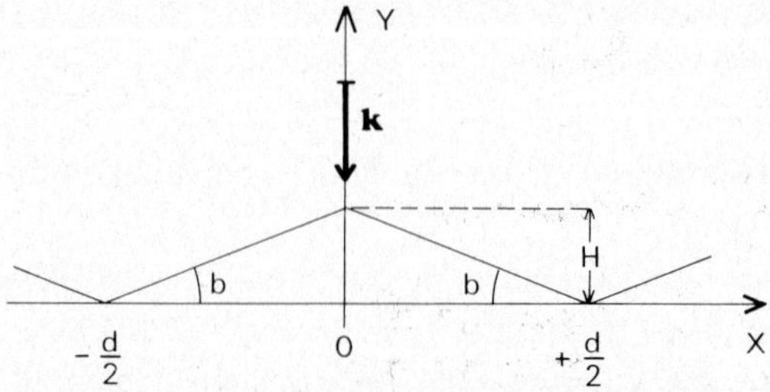

Fig. 3.1. The symmetrical echelette grating used in normal incidence.

cannot be used a priori to describe the field. First, let us demonstrate that, if the RE, which represents the diffracted field above the top of the grooves, converges at any point inside the grooves, it actually represents the diffracted field F at this point. To this end, let us study the asymptotic behavior of the nth term $R_n(x, y)$ of the RE when $n \to \pm\infty$. We denote by p the positive integer $p = |n|$ and we call $\tilde{\beta}_{\pm p}$, $\tilde{\Phi}_{\pm p}(x, y)$, $\tilde{R}_{\pm p}(x, y)$ asymptotic expressions of $\beta_{\pm p}$, $\Phi_{\pm p}(x, y)$ and $R_{\pm p}(x, y)$, respectively. From (2.12a), (2.12b) and (2.14), we derive:

$$\tilde{\beta}_{\pm p} = \mathrm{i}pK \pm \mathrm{i}\alpha, \tag{3.5}$$

$$\tilde{\Phi}_p = u(x, y)\, w(x, y)^p, \tag{3.6a}$$

$$\tilde{\Phi}_{-p} = \bar{u}(x, y)^{-1}\, \overline{w}(x, y)^p, \tag{3.6b}$$

with

$$u(x, y) = \exp(\mathrm{i}\alpha x - \alpha y), \tag{3.7}$$

$$w(x, y) = \exp(\mathrm{i}Kx - Ky). \tag{3.8}$$

The first step of the demonstration is described in Appendix A. Let us state the conclusion:

If the RE converges at a certain point of ordinate y_0, it converges absolutely and uniformly to an analytic function of the two real variables x and y in the half plane $y \geqslant y_1$, whatever the value $y_1 > y_0$.

An obvious consequence of this fundamental property is that the RE diverges in the half plane $y < y_3$ if it diverges for a certain point of ordinate y_2, whatever the value $y_3 < y_2$. So, we are led to define an "ordinate of convergence" y_c such that:

- the RE converge for $y \geqslant y_1$, whatever $y_1 > y_c$,
- the RE diverges for $y \leqslant y_3$, whatever $y_3 < y_c$.

Of course, since the RE represents the field $F(x, y)$ and converges for $y \geqslant y_M$, it can be inferred that $y_c \leqslant y_M$ and that $F(x, y)$ is an analytic function of the real variables x and y when $y > y_M$. In fact, $F(x, y)$ is an analytic function of x and y in the entire region R_+, because of the following mathematical property of the Helmholtz equation:

Since the Helmholtz equation $\nabla^2 u + k^2(x, y)u = 0$ is a linear elliptic partial differential equation, if k is constant (or analytical in x and y) in a certain domain Ω_1 of R^2, every continuous solution $u(x, y)$ is an analytic function of the two real variables x and y in this domain (SOMMERFELD [1949], COURANT and HILBERT [1962], MIRANDA [1970]). We deduce that $F(x, y)$ is analytic in R_+.

Now, let us summarize the above theoretical results:

$R(x, y)$ is an analytic function of x and y for $y > y_1 > y_c$; $F(x, y)$ is an analytic function of x and y in R_+; $F(x, y) \equiv R(x, y)$ for $y > y_M$.

So, in the domain Ω_1 of R_+ situated above $y_1 > y_c$, $v(x, y) = F(x, y) - R(x, y)$ is an analytic function of x and y. Since $v(x, y)$ vanishes in the subdomain Ω_2 of Ω_1 defined by $y > y_M$, it vanishes in the entire domain Ω_1. *So, the Rayleigh expansion $R(x, y)$ well represents the diffracted field $F(x, y)$ in Ω_1, i.e. above the ordinate of convergence y_c.* Now, if $y_c < y_m$, the RE converges and identifies to the diffracted field throughout R_+, therefore the RH is valid. This conclusion does not hold any longer if $y_c > y_m$, since the RE, which diverges in a certain part of R_+, cannot be used to express the field at any point of the profile $\mathscr{P}$, in order to write the boundary conditions (2.9a) or (2.9b). Finally, the study of the validity of the RH reduces to the study of the ordinate of convergence y_c, which will be compared with the ordinate of the bottom of the grooves y_m. In the following, the value of y_c will be deduced from the asymptotic behavior of $R_n(x, y)$ when $n \to +\infty$. It is worth noting that the value of $R(x, y)$ in the region of R_- situated above $y = y_c$ is the analytic continuation of the diffracted field $F(x, y)$ below $\mathscr{P}$.

3.2.2. *A simple but non-rigorous demonstration*

The expression of B_n obtained from Beckmann's theory (cf. (2.26)) may be used to evaluate the asymptotic behavior of the RE, provided the integral can be calculated in closed form. Though the Beckmann approximation becomes more and more precise when $\lambda \to 0$, this obviously cannot be considered as a rigorous investigation but rather an attempt at suggesting a fundamental result from elementary mathematics. For the sinusoidal grating whose profile is described by (2.27), introducing the expression of $\tilde{B}_{\pm p}$ given by (2.28) and remarking that $k \cos\theta_{\pm p} = \beta_{\pm p} \to ipK$ when $p \to \infty$ we find, for the FTE case:

$$\tilde{B}_{\pm p} = c_{\pm} I_p(ph)/p,$$

with $h = \pi H/d$, $c_{\pm}$ being two complex coefficients and I_p the modified Bessel function (ABRAMOWITZ and STEGUN [1970]). From the uniform asymptotic behavior of $I_p(pz)$ which can be found in the same reference, we derive:

$$\tilde{B}_{\pm p} = c'_{\pm} p^{-3/2} \exp\{p[(1 + h^2)^{1/2} - \operatorname{arcsinh}(1/h)]\},$$

$c'_{\pm}$ being independent of p, which yields, taking into account (3.6a) and (3.6b),

$$\tilde{R}_{\pm p}(x, y) = \tilde{B}_{\pm p} \tilde{\Phi}_{\pm p} = c''_{\pm} p^{-3/2} \exp[pK(y_c - y)],$$

with

$$K y_c = (1 + h^2)^{1/2} - \operatorname{arcsinh}(1/h). \tag{3.9}$$

So, the general condition of validity of the RH, i.e. $y_c < y_m = -H/2$, becomes $h < h_c$, h_c being given by the transcendental equation:

$$h_c + (1 + h_c^2)^{1/2} - \operatorname{arcsinh}(1/h_c) = 0, \tag{3.10}$$

which finally gives:

$$h_c = 0.447743, \tag{3.11}$$

i.e. the value first given by PETIT and CADILHAC [1966] from a counterexample described in the next section.

3.2.3. *A counterexample*

Now, we briefly outline the first rigorous theoretical demonstration of the non-validity of the RH (PETIT and CADILHAC [1966]). It applies to sinusoidal gratings, whose profile is given by (2.27).

Let us assume that the RE is valid at any point of the profile $\mathscr{P}$. Thus, it satisfies the boundary condition (3.1a) which, for sinusoidal gratings, becomes:

$$\sum_n B_n \exp\left(\mathrm{i}\alpha_n x + \mathrm{i}\beta_n \frac{H}{2} \cos Kx\right) + \exp\left(\mathrm{i}\alpha x - \mathrm{i}\beta \frac{H}{2} \cos Kx\right) = 0, \ \forall x. \tag{3.12}$$

It will be shown that the above equation cannot be satisfied, at least above the critical value H_c of H deduced from (3.11).

First, we must envisage mathematical considerations which perhaps will appear strange to the physicist: the behavior of (3.12) for complex values of x. In fact, we have already investigated in Appendix A some properties of the Rayleigh expansion $R(x, y)$ for complex values of x and y. These properties apply as well to the first term of (3.12) which is simply $R(x, \frac{1}{2}H \cos Kx)$, a function of a single complex variable x. If (3.12) is satisfied for any value of x, it applies in particular to $Kx = -\pi$, i.e. at the bottom of a groove. Of course, this requires the convergence of $R(-\pi/K, -H/2)$. A direct consequence of Appendix A is that $R(x, \frac{1}{2}H \cos Kx)$ will converge to an analytic function of the complex variable x in the region $\Omega_x \in C$ defined by:

$$\left| w\left(x, \frac{H}{2} \cos Kx\right) \right| \leqslant \left| w\left(x_1, \frac{H}{2} \cos Kx_1\right) \right|,$$

$$\left| \overline{w}\left(x, \frac{H}{2} \cos Kx\right) \right| \leqslant \left| w\left(x_1, \frac{H}{2} \cos Kx_1\right) \right|,$$

x_1 being a real number different from $-\pi/K$ (and therefore whose ordinate on $\mathscr{P}$ is greater than y_m) but which can be very close to it.

We have drawn in Fig. 3.2 the contour lines of $|w(x, \frac{1}{2}H\cos Kx)|$ for $H/d = 0.175$, i.e. $h = 0.55$. The dashed line shows the contour line passing through a point of abscissa $\xi_1 = Kx_1$ close to $-\pi$. Taking into account the contour lines of $\overline{w}(x, \frac{1}{2}H\cos Kx)$, which are symmetrical to those of Fig. 3.2 with respect to the Oξ axis, Ω_x is the hatched region of Fig. 3.3. So, the left hand member of (3.12), which is analytic in Ω_x and vanishes on the real axis AE, must vanish throughout Ω_x, and in particular on the line $\mathscr{D}$ of positive abscissa $\xi_0 < \pi/2$. Now, a careful examination of (3.12) shows that, on this line, the series converges and even tends to 0 when $\eta \to -\infty$, whereas the second term F^{i} tends to infinity. Thus, (3.12) cannot be verified on $\mathscr{D} \in \Omega_x$, a fact which is in contradiction with the definition of Ω_x. So, our hypothesis (the convergence of the RE on the entire profile) is false. The demonstration holds as long as the countour line passing through the saddle point S of Fig. 3.2 crosses the vertical $\xi = -\pi$ above point A. Otherwise, the two dashed lines meet and cut the Oη axis, the domain Ω_x is bounded and the demonstration cannot be achieved. The limit value h_{c}' is obtained by writing that the contour line passing through A

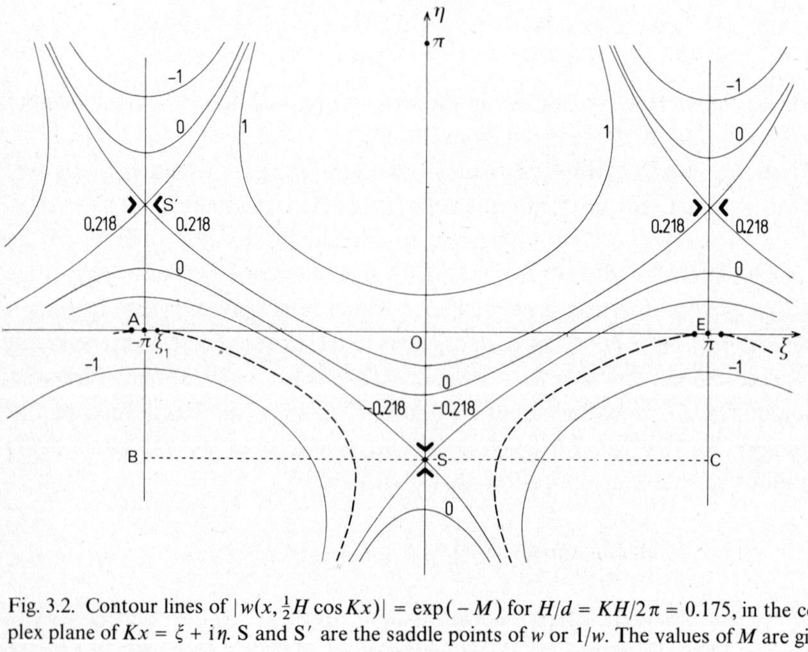

Fig. 3.2. Contour lines of $|w(x, \frac{1}{2}H\cos Kx)| = \exp(-M)$ for $H/d = KH/2\pi = 0.175$, in the complex plane of $Kx = \xi + \mathrm{i}\eta$. S and S′ are the saddle points of w or $1/w$. The values of M are given on the curves.

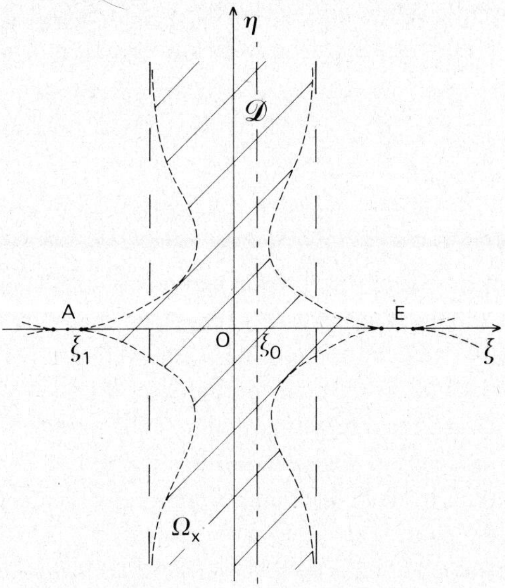

Fig. 3.3. Had the Rayleigh expansion been convergent at the point A, it would have been convergent throughout the finite hatched region Ω_x.

contains S. To find the value of x_s corresponding to the saddle point, we must look for the solution of $\mathrm{d}w(x, f(x))/\mathrm{d}x = 0$, i.e.

$$f'(x_s) = \mathrm{i}, \tag{3.13}$$

and for the sinusoidal grating

$$Kx_s = -\mathrm{i}\,\operatorname{arcsinh}(1/h), \tag{3.14}$$

the corresponding value M_s of M being

$$M_s = (1 + h^2)^{1/2} - \operatorname{arcsinh}(1/h). \tag{3.15}$$

Remarking that the value of M at the point A is equal to $-h$, we derive h_c', satisfying (3.10), and thus $h_c' = h_c$. So, this rigorous demonstration shows that the RH is untenable if $h > h_c$. MILLAR [1971] showed that the RH is valid below h_c.

3.2.4. *A general investigation*

Now, our aim is to investigate the validity of the RH for gratings whose profiles are given by analytic functions $f(x)$, for both the FTE and the FTM cases. The asymptotic values $\tilde{B}_{\pm p}$ will be derived from (2.22a) and (2.22b). Let us consider the FTE case. By using (3.5) and (3.6a), it can be shown that:

$$\tilde{B}_p = -\frac{1}{4\pi p}\int_0^d \phi(x)u(x, f(x))^{-1}w(x, f(x))^{-p}\,\mathrm{d}x, \tag{3.16a}$$

$$= -\frac{1}{4\pi p}\int_0^d \exp[p(-\mathrm{i}Kx + Kf(x))]\,\phi(x)u(x, f(x))^{-1}\,\mathrm{d}x. \tag{3.16b}$$

The asymptotic behavior of this integral can be obtained using a saddle point method (DENNERY and KRZYWICKI [1967]), as made by HILL and CELLI [1978]. Since $f(x)$ is analytic and may be extended to complex values of x, and assuming that $\phi(x)$ satisfies the same property (cf. MILLAR [1969] in the case of sinusoidal gratings), the integration path may be modified in the x complex plane according to Cauchy's theorem and the asymptotic behavior is deduced from a path passing through the saddle point. For instance, for sinusoidal gratings, the saddle point S of $1/w$ is shown in Fig. 3.2. The initial path AE is replaced by ABSCE, and finally by BSC since the contributions of AB and CE cancel each other. It turns out finally:

$$\tilde{B}_p = T_+ p^{-3/2}\exp[pKf(x_\mathrm{s}) - \mathrm{i}pKx_\mathrm{s}], \tag{3.17}$$

T_+ being a complex coefficient.

For the FTM case, though the coefficient in $1/\beta_n$ is removed in (2.22b), (3.17) still holds if we notice that $(1 - f'(x)\alpha_p/\beta_p)$ asymptotically tends to 0 at the saddle point. From (3.6a) and (3.17), we derive an asymptotic value $\tilde{R}_p$ of the RE:

$$\tilde{R}_p(x, y) = T_+\, u(x, y)p^{-3/2}\exp\{\mathrm{i}pK[(x - x_\mathrm{s}) + \mathrm{i}(y - f(x_\mathrm{s}))]\}. \tag{3.18}$$

So, for real or complex values of x and y, a necessary condition for the convergence of the RE is:

$$\mathrm{Im}\{x - x_\mathrm{s} + \mathrm{i}(y - f(x_\mathrm{s}))\} > 0. \tag{3.19a}$$

A similar calculation for the terms R_{-p} leads to a second necessary condition:

$$\mathrm{Im}\{\bar{x} - x_\mathrm{s} + \mathrm{i}(\bar{y} - f(x_\mathrm{s}))\} > 0. \tag{3.19b}$$

It emerges that the domain of convergence $\Omega \in C^2$ of the RE is the intersection of the two regions defined by (3.19a) and (3.19b).

For real values of x and y, (3.19a) and (3.19b) show that the RE converges above the horizontal line passing through the point J of real coordinates $(x_\mathrm{J}, y_\mathrm{J})$ such that:

$$x_\mathrm{J} + \mathrm{i}y_\mathrm{J} = x_\mathrm{s} + \mathrm{i}f(x_\mathrm{s}), \tag{3.20}$$

a fundamental result previously found by other authors using steepest descent methods (HILL and CELLI [1978], DE SANTO [1981]) or an elementary conformal mapping (VAN DEN BERG and FOKKEMA [1979]).

To interpret (3.20), it is necessary to bear in mind that x_s, given by (3.13) is complex, whilst x_J and y_J are real. In its region of convergence, the RE identifies with $F(x, y)$, viz. the diffracted field or its analytic continuation below $\mathscr{P}$. Therefore, it may be shown that the point J is a singularity of $F(x, y)$. Since F is analytic in R_+, we deduce that $J \in R_-$.

From the fundamental result stated in (3.20) it can be expected that, in general:

$$y_J = y_c. \tag{3.21}$$

However, this is not a general rule. For instance, CADILHAC recently suggested the case of the following grating (private communication):

$$f(x) = \frac{1}{K} \operatorname{arcsinh}(a \cos Kx),$$

which is close to the sinusoidal grating for low values of a. For this grating, there is no saddle point since (3.13) has no solution and $y_J = -\infty$. However, a branch point limits the domain of convergence of the RE, and y_c is finite.

From another remark of Cadilhac, let us state an original result which could permit the opticists to find a physical meaning for the point J. The equation of the tangent to $\mathscr{P}$ at a point $(x_0, f(x_0))$ is:

$$y - f(x_0) - f'(x_0)(x - x_0) = 0,$$

and at the saddle point $x = x_s$ of $\mathscr{P}$, it becomes, according to (3.13):

$$y - f(x_s) = \mathrm{i}(x - x_s),$$

which is the equation of the first isotropic line tangent to $\mathscr{P}$. It can be shown easily that the point J, which obviously belongs to this first isotropic line (IYANAGA and KAWADA [1980]) tangent to $\mathscr{P}$, according to (3.20), belongs to the second isotropic line as well. *Thus, J is the intersection of the two isotropic lines tangent to $\mathscr{P}$. For conics, this is one of the definitions of the focus. So, we are led to extend this definition of the mathematical focus to an arbitrary curve given by an analytic function $f(x)$. The focus J of the profile is a singularity of the analytic continuation of the diffracted field in region R_- and the RE does not converge below it. So, the RH is not valid when the focus is situated above the bottom of the grooves. It may be expected in general that the RH is valid when the focus is situated below the bottom of the grooves.*

This fundamental remark leads us to generalize the result obtained in this section. For instance, if the profile $\mathscr{P}$ is given by parametric equations:

$$x = x(t), \qquad y = y(t),$$

(2.22a) and (2.22b) can be used as well to investigate the asymptotic behavior of B_n, using the saddle point method in the t complex plane. The result is the same: the singularity of the RE is the focus of the curve. It will be found by writing that the tangent to $\mathscr{P}$ at the point t_0,

$$-y'(t_0)(x - x_0) + x'(t_0)(y - y_0) = 0, \tag{3.22}$$

is an isotropic line which satisfies

$$y'(t_0) = \mathrm{i}x'(t_0), \tag{3.23}$$

and by looking for the real coordinates x_J and x_J, such that

$$x_J + \mathrm{i}y_J = x(t_0) + \mathrm{i}y(t_0).$$

The same remark applies to a grating for which $\mathscr{P}$ is given by $\chi(x, y) = 0$ where the focus will be found by solving:

$$\frac{\partial \chi}{\partial y}(x_0, y_0) - \mathrm{i}\frac{\partial \chi}{\partial x}(x_0, y_0) = 0, \tag{3.24}$$

$$\chi(x_0, y_0) = 0, \tag{3.25}$$

and by writing that $x_J + \mathrm{i}y_J = x_0 + \mathrm{i}y_0$.

For the opticist, it is not surprising to find a singularity of the diffracted field at a focus. Unfortunately, this is not the optical focus of geometrical optics, which cannot be defined for an arbitrary profile $\mathscr{P}$. It is the mathematical focus, which may be considered as the "asymptotic optical focus of the evanescent waves" and which is independent of the wavelength and of the incidence. This author verified an interesting property in the case of sinusoidal gratings: the mathematical focus J is always close to the optical focus obtained from geometrical optics by replacing the top of $\mathscr{P}$ by circles having the same radius of curvature at this point. Thanks to the definition of the focus, all the above conclusions can be generalized to any two-dimensional diffracting subject: *the focus is a singularity of the solution of the Dirichlet and Neumann problems of diffraction.*

The reader interested in calculating the limit of validity of the RH for a particular grating can use the above equations. Many examples of calculation of y_c have been given by NEVIERE and CADILHAC [1970], HILL and CELLI

[1978], VAN DEN BERG and FOKKEMA [1979], DE SANTO [1981]. For non-analytic profiles, it has been shown by MILLAR [1969], NEVIERE and CADILHAC [1970], that in general, the RH fails as soon as the upper point of $\mathscr{P}$ is an edge. This phenomenon can be easily explained by the fact that the field F is singular on the edge (a singularity is a point where the field is not analytic), a fact which implies the divergence of the RE on the edge. This rule is not general since counterexamples have been given for echelette gratings in the FTM case (MARECHAL and STROKE [1959]), echelette and lamellar gratings for the FTE and FTM cases (MAYSTRE [1974, 1980b]), in which the diffracted field is described by a small number of plane waves and has no singularities.

It is amazing to notice that one of the main results of this section was implicitly stated by SOMMERFELD [1949] in his theory of "characteristics". He showed that the necessary condition for finding the analytic continuation of the field at the vicinity of the profile was that the characteristic lines (which are nothing but the isotropic lines for the Helmholtz equations) must not be tangent to the profile. A simple generalization of this remark to complex values of x and y shows that singularities of $F(x, y)$ may be expected in the vicinity of the point $(x_s, f(x_s))$.

3.3. A PARADOXICAL RESULT: THE SYSTEM OF RAYLEIGH FUNCTIONS IS A TOPOLOGICAL BASIS

Apparently, the theoretical results about the validity of the RH are the explanation for the numerical difficulties observed in its first implementations. This explains why rigorous theories have been developed further. Nevertheless, since 1971, a series of theoretical and numerical studies have raised a new interest in the Rayleigh method. Exploiting an analysis of VEKUA [1953] for other diffracting structures, some authors (YASUURA and IKUNO [1971], IKUNO and YASUURA [1973]) claimed the completeness of the set of $\Phi_n(x, f(x))$ in the Hilbert space of square integrable functions (cf. MILLAR [1973] or CADILHAC [1980]). Let us outline this demonstration for the interested reader. Assuming that the pseudo-periodic, locally square integrable function $\Gamma(x)$ is orthogonal to the $\Phi_n(x, f(x))$, i.e. that $\langle \Gamma(x), \Phi_n(x, f(x)) \rangle = 0$, $\forall n$, we must prove that $\Gamma(x) \equiv 0$. With this aim we use some results established in Appendix B. The functions $f(x)$ and $\Gamma(x)$ being given, we consider the profile $\mathscr{P}'$ given by $y = f(-x)$, separating R'_+ and R'_- and a continuous function $U(x, y)$ satisfying (B.8) and (B.9), i.e. Helmholtz equations in both R'_+ and R'_-, and radiation conditions for $y \rightarrow \pm\infty$. If we give the jump $\bar{\Gamma}(-x)(1+(\mathrm{d}f(-x)/\mathrm{d}x)^2)^{-1/2}$ of its normal derivative, then $U(x, y)$ is given by (B.10) and this expression reduces, for $y < y_{\mathrm{m}}$, to a sum of plane waves (see (B.16)) whose coefficients D_n^- are given by (B.17). Here, since U is continuous, $\tau = 0$ and moreover, $\eta = \bar{\Gamma}(-x)$, according to (B.13) and (B.14). We deduce that

$$D_n^- = [s_n^-, \bar{\Gamma}(-x)] = \frac{1}{2\mathrm{i}\beta_n} \langle \Gamma(-x), \Phi_n(-x, f(-x)) \rangle$$

$$= \frac{1}{2\mathrm{i}\beta_n} \langle \Gamma(x), \Phi_n(x, f(x)) \rangle = 0.$$

So, $U(x, y)$ vanishes for $y < y_{\mathrm{m}}$ and since U is analytic in R'_-, it vanishes throughout R'_-. By continuity, its boundary value above $\mathscr{P}$ is null and the uniqueness of the solution for the FTE case allows us to state that U vanishes throughout R'_+. Finally, the jump of the normal derivative of U is null and $\Gamma(x) \equiv 0$.

Considering (3.1a), the basic equation of the Rayleigh method, the completeness of the Rayleigh basis seems to justify the RH and to be in contradiction with our former theoretical study since the incident field $F^{\mathrm{i}}(x, f(x))$ is a locally square integrable function. To understand the paradox, we must recall that the $\Phi_n(x, f(x))$ are not orthogonal. In fact, they are members of a topological basis. Physicists are very well acquainted with this notion for orthogonal systems but the situation is much more difficult in our problem. Indeed, due to the non-orthogonality of the Φ_n, the completeness of the topological basis means that a square integrable function, for instance $-F^{\mathrm{i}}(x, f(x))$, can be represented as a limit of linear combinations of the $\Phi_n(x, f(x))$:

$$-F^{\mathrm{i}}(x, f(x)) = \lim_{N\to\infty} S_N(x), \tag{3.26}$$

with

$$S_N(x) = \sum_{n=-N}^{+N} B_n^{(N)} \Phi_n(x, f(x)) = T_N(x, f(x)), \tag{3.27}$$

$$T_N(x, y) = \sum_{n=-N}^{+N} B_n^{(N)} \Phi_n(x, y),$$

and we shall prove that:

$$\lim_{N\to\infty} B_n^{(N)} = B_n, \tag{3.28}$$

where B_n is the actual Rayleigh coefficient. Two fundamental remarks must be made about these equations. First, (3.26) must be understood in the sense of the least squares approximation. In other words, the $B_n^{(N)}$ must be chosen in order to minimize I_N such that:

$$I_N = \int_0^d |S_N(x) + F^i(x, f(x))|^2 \, dx.$$

Equation (3.26) *only means that* $I_N \to 0$ *when* $N \to \infty$, *and does not mean that the two members are equal for any value of* x. The second fundamental remark is that the $B_n^{(N)}$ of (3.27) depend on N. *So, the convergence of* $S_N(x)$ *when* $N \to \infty$ *does not imply the convergence of the RE, even though* $B_n^{(N)} \to B_n$. To cast light on this rather subtle fact, let us give a simple numerical example of such an apparent paradox. S_N is now, in this simple example, defined by:

$$S_N = \sum_{n=0}^{N} r_n^{(N)} \qquad \text{with} \qquad r_n^{(N)} = 1 - 2n/N.$$

Obviously, $S_N = 0$ for any value of N since $r_{N-n}^{(N)} = -r_n^{(N)}$ and, of course, S_N converges to O when $N \to \infty$. On the other hand, remarking that $\lim_{N \to \infty} r_n^{(N)} = 1$ when n is fixed, the series

$$R = \sum_{n=0}^{\infty} \lim_{N \to \infty} r_n^{(N)}$$

diverges. This is exactly what happens in our case: the convergence of the $S_N(x)$ does not imply the convergence of the RE, $R(x, f(x))$, since in this series, the $B_n^{(N)}$ are replaced by their asymptotic values B_n.

Of course, these new theoretical considerations have raised a new interest in the RVM described previously, which appears to be justified from a theoretical point of view. However, recent numerical experiments have proved that, even though the RVM can be used for some non-shallow gratings, it converges slowly and does not converge for groove depths higher than $H_c = 0.54d$ in the case of sinusoidal gratings (WIRGIN [1979a], HUGONIN, PETIT and CADILHAC [1981]).

It has been shown (see for instance the paper by MILLAR [1973]) that the set of $S_N(x)$ uniformly converges to the diffracted field F in any closed subset above $\mathscr{P}$, assuming the existence of a certain Green function vanishing on the grating surface $\mathscr{P}$. Here, we use a more general demonstration to show that S_N uniformly converges to F on $\mathscr{P}$ provided the profile is analytic.

Let us define, for a given profile $\mathscr{P}$, a second periodic curve $\mathscr{P}^*$, given by $y = f^*(x) \leqslant f(x)$. We denote by R_+^* the region $y > f^*(x)$. We assume that the analytic continuation of F (also denoted by F) has no singularities in R_+^* and is locally square integrable on $\mathscr{P}^*$, properties always satisfied if $\mathscr{P}^* \equiv \mathscr{P}$.

If we denote by $G_{\mathscr{P}^*}(x, y, x', y')$ the pseudo-periodic Green function vanishing when $(x, y) \in \mathscr{P}^*$ and satisfying a radiation condition at $y = \infty$, we can write:

$$F(x, y) = \int_{\mathscr{P}^*} F(x, f^*(x)) \frac{\mathrm{d}G_{\mathscr{P}^*}}{\mathrm{d}n'}(x, y, x', f^*(x'))\, \mathrm{d}s^*, \tag{3.29}$$

s^* being the curvilinear abscissa of the point of abscissa x' on $\mathscr{P}^*$, and x, y the coordinates of a point P of R^*_+ (cf., for example, MORSE and FESHBACH [1953], for the definition and properties of a similar Green's function obtained when $\mathscr{P}^*$ is replaced by a closed surface). Now, let us call $S^*_N(x)$ the set given by the sum described in (3.27), the B^N_n being chosen, this time, in order to minimize I^*_N:

$$I^*_N = \int_0^d |S^*_N(x) + F(x, f^*(x))|^2\, \mathrm{d}x.$$

The completeness of the $\Phi_n(x, f^*(x))$ implies the convergence of I^*_N to 0 when $N \to \infty$. Since each term of the sum representing $S^*_N(x)$ can be represented as in (3.29), we derive:

$$F(x, y) - T^*_N(x, y) = \int_{\mathscr{P}^*} [F(x, f^*(x)) - S^*_N(x)] \frac{\mathrm{d}G_{\mathscr{P}^*}}{\mathrm{d}n'}(x, y, x', f^*(x'))\, \mathrm{d}s^*$$

and the Schwartz inequality yields:

$$\lim_{N\to\infty} |F(x, y) - T^*_N(x, y)| \leqslant \left[\int_{\mathscr{P}^*} |F(x, f^*(x)) - S^*_N(x)|^2\, \mathrm{d}s^*\right]^{1/2}$$

$$\times \left[\int_{\mathscr{P}^*} \left|\frac{\mathrm{dG}_{\mathscr{P}^*}}{\mathrm{dn}'}(\mathrm{x}, \mathrm{y}, \mathrm{x}', \mathrm{f}^*(\mathrm{x}'))\right|^2 \mathrm{ds}^*\right]^{1/2}.$$

Since the first term of the right-hand member tends to zero and the second is finite:

$$\lim_{N\to\infty} |F(x, y) - T^*_N(x, y)| = 0 \tag{3.30}$$

in any closed subset of R^*_+.

The interpretation of this result differs according to whether $\mathscr{P}$ is analytic or not. In the second case, we can expect a singularity of F on $\mathscr{P}$. So, $\mathscr{P}^*$ and $\mathscr{P}$ necessarily have some points in common and (3.30) is not valid on the entire profile $\mathscr{P}$. Nevertheless, taking $\mathscr{P}^* \equiv \mathscr{P}$ shows that *the "Rayleigh set" $T^*_N(x, y)$ uniformly converges to $F(x, y)$ above the profile.* Since this property is true in particular for $y > y_M$, where F is given by the RE, it is easy to deduce that $B_n^{(N)} \to B_n$ when $N \to \infty$. If we notice that the $B_n^{(N)}$ so defined are nothing but the coefficients determined by using the RVM (cf. § 3.1), we are led to the conclusion, stated above without a complete demonstration, that *this method is theoretically valid for any kind of grating, even when the RH fails*. Of course, all these conclusions hold when $\mathscr{P}$ is analytic, but in addition, we can conjecture that $\mathscr{P}^*$ can have no point in common with $\mathscr{P}$, because of the absence of singularity of F on $\mathscr{P}$. So, R_+^* includes $\mathscr{P}$ and from (3.30) we derive that:

$$\lim_{N\to\infty} \sum_{n=-N}^{+N} B_n^{(N)} \Phi_n(x, f(x)) + F^{\mathrm{i}}(x, f(x)) = 0, \tag{3.31}$$

in the sense of uniform convergence. The existence of such $B_n^{(N)}$ is the proof of the theoretical validity of the PMM and FSM (cf. § 3.1) for such profiles. In this case, the FSM seems to be better and simpler than the FVM, as shown numerically by WIRGIN [1979b]: for sinusoidal gratings, the first method converges below $H/d = 0.7$, a groove depth five times greater than the theoretical limit of the RH. This success of the FSM probably has to be qualified for non-sinusoidal analytic profiles, where the coefficients of the linear system of equations are not analytic. This remark is not without substance: it can be conjectured that the progress of the numerical implementation of the FSM in the last fifteen years is also linked to the progress in the computation of the Bessel functions on computers.

So, we are led to an unexpected conclusion: *the numerical limits of the Rayleigh methods are not directly linked to the theoretical failure of the RH, but to purely numerical reasons* at least for analytic profiles. For instance, in the PMM, these reasons are obvious. According to (3.2), (3.6a), (3.6b) and (3.8), the n, m coefficients $\Phi_n(x_m, f(x_m))$ of the linear system of equations to be solved are very different in magnitude. If, for simplicity, we assume that the ordinate of the top of the grooves $y_M = 0$, the coefficient at the corresponding abscissa will tend to 1 in the modulus, while at the other points x_m, it will tend to 0 in the modulus when $n \to \infty$. So, it suffices to adopt a sufficiently large value of N to obtain identical lines for high values of n, at least for computer purposes, and the matrix becomes singular. This phenomenon arises for any value of H, including those inferior to H_c, and this proves that the PMM is ill-conditioned. The same kind of limitation appears for FSM and for FVM but for higher groove depths. The interested reader can find theoretical investigations, thorough comparisons of these methods and comparison with other rigorous methods in the papers by BATES [1975], KALHOR [1976], NAMIOKA, HARADA and YASUURA [1979], WIRGIN [1979a, 1979b, 1979c, 1980a, 1980b,

1981], VAN DEN BERG [1980, 1981], HUGONIN, PETIT and CADILHAC [1981].

In conclusion, the possibilities of the Rayleigh theory were underestimated fifteen years ago, as stressed by WIRGIN, and the Rayleigh methods may be valuable in the study of some types of gratings, in particular the FSM. However, because of the theoretical and numerical limitations, these methods are far from being a near universal tool such as the integral method described in § 5, at least for the present.

§ 4. The Waterman method

The basic idea of the WATERMAN method, for the FTE case, is to use the completeness of the set of functions $\Phi_n(x, -f(x))$ to deduce the function $\phi(x)$ from (2.24a), then to calculate the amplitudes B_n by introducing $\phi(x)$ in (2.22a). As regards the FTM case, the same procedure may be applied using (2.24b) and (2.22b). It is worth noting that the fundamental equations (2.24a) and (2.24b) are sometimes presented as consequences of the so-called "extinction theorem", well known in Optics and Electromagnetism. As for Rayleigh's method, there exist various ways of implementing (2.24a). For instance, a point matching method can be used by representing $\phi(x)$ by its values at $2N+1$ points x_m of the interval $(0, d)$ and by writing (2.24a) for $n \in (-N, +N)$. If a trapezoidal rule is used, the coefficients of the linear system of equations to be inverted are proportional to $\Phi_n(x_m, f(x_m))$. Obviously, the remarks of § 3.3 about the ill-conditioning of the system for the PMM in the Rayleigh theory apply to the PMM in the WATERMAN theory as well. Therefore, the numerical limitations may be expected to be close to each other. A similar remark can be made about the FSM where $\phi(x)$ is represented as a series of $\exp(\mathrm{i}\alpha_m x)$. The coefficients, proportional to $\langle \exp -\mathrm{i}\alpha_m x, \Phi_n(x, -f(x)) \rangle$, will be closely linked with those obtained in the Rayleigh method, according to (3.3), and the numerical limitation may be expected to be of the same order (cf. for instance WIRGIN [1979a, 1979b]).

This method, or methods relying on the same basic ideas, has been used by various authors (WATERMAN [1975], URETSKY [1965], DE SANTO [1975], WHITMAN and SCHWERING [1977], WIRGIN [1978, 1979a, 1979b], WHITMAN, LESKIV and SCHWERING [1980]. An interesting feature of the DE SANTO method, which has been numerically implemented (JORDAN and LANG [1979], LANG and JORDAN [1980]) lies in the representation of $\phi(x)$ by the product of a Fourier series by the physical optics approximation (viz. the function ϕ_B

given in (2.25)). As may be expected from Fig. 2.4, this feature enhances the accuracy of the numerical results (at least when λ/d is low), as announced by WHITMAN, LESKIV and SCHWERING [1980] who developed the same formalism independently. In conclusion, the main quality of the Waterman methods rests on its simplicity, equivalent to that of the Rayleigh method. Unfortunately, the numerical limitations of these two methods are also very close to each other.

§ 5. The integral method

5.1. HISTORICAL SURVEY AND INTUITIVE APPROACH

To our knowledge, the integral method has been the most widely used of the rigorous theories of gratings. Until now, it has been the most versatile and the most accurate tool for investigating grating properties. On the other hand, its main shortcoming lies in its mathematical complexity and in the toughness of numerical implementation.

After the pioneering studies of PETIT and CADILHAC [1964], WIRGIN [1964, 1967], URETSKY [1965], PETIT [1965, 1966a], PAVAGEAU, EIDO and KOBEISSE [1967], the various ways of implementing this theory have been widely investigated, at least for perfectly conducting gratings. Thus, many integral treatments of the grating problem are now available (GREEN [1970], DUMERY and FILIPPI [1970], MAYSTRE and PETIT [1970a, 1970b], MAYSTRE [1974, 1980a], VAN DEN BERG [1971a, 1971b], KALHOR and NEUREUTHER [1971], MC PHEDRAN [1973], BOTTEN [1978a, 1978b], GARCIA and CABRERA [1978]). Many of these methods have been inspired by the ideas of MAUE [1949]. MC CLELLAN and STROKE [1966] also published an integral equation for the FTM case but no numerical application supported the theory, which seems to be questionable since it requires the continuity of the tangential component of the magnetic field on the grating surface.

In spite of its complexity, the integral formalism lies on a very simple and intuitive background as we shall briefly describe.

Due to the incident field, there exists on $\mathscr{P}$ a surface current density $j_{\mathscr{P}}(x)$. Unfortunately, finding $j_{\mathscr{P}}(x)$ is not straightforward: the existence of $j_{\mathscr{P}}(x)$ at a given point M of $\mathscr{P}$ is not only the effect of the reflection of the incident wave at this point, as assumed in the BECKMANN theory; in fact, the surface current densities $j_{\mathscr{P}}(x')$ at the other points M′ of $\mathscr{P}$ act like secondary sources which radiate a field in all directions, and in particular toward M. *So,* $j_{\mathscr{P}}(x)$ *depends not only on the characteristics of the incident field at M, but also on the value of* $j_{\mathscr{P}}(x')$ *at the other points M′ of* $\mathscr{P}$.

Now, it is felt intuitively that the diffracted field is nothing but the field generated by this surface current; therefore it may be expected that the knowledge of the function $j_{\mathscr{P}}(x)$ will enable us to express the diffracted field F at any point of region R_+. This intuitive fact is fully confirmed by mathematics: after defining the field generated at the point P of region R_+ by a single elementary current placed at $\mathrm{M}' \in \mathscr{P}$ (the so-called Green function), we obtain $F(\mathrm{P})$ by integrating on $\mathscr{P}$ the effects of all these elementary currents. So, we obtain an integral expression of $F(\mathrm{P})$ containing the function $j_{\mathscr{P}}(x)$. Even though $j_{\mathscr{P}}(x)$ is unknown, the first step of the integral formalism is achieved: *the problem of finding a function* $F(x, y)$ *of two variables reduces now to the finding of a function* $j_{\mathscr{P}}(x)$ *of one variable.*

It is very interesting that the integral expression of $F(\mathrm{P})$ in terms of $j_{\mathscr{P}}(x)$ satisfies both the Helmholtz equation (2.8) and the radiation condition (2.10), *even though an arbitrary function* $a(x)$ *is introduced within it, instead of* $j_{\mathscr{P}}(x)$. Of course, the actual value of $j_{\mathscr{P}}(x)$ will be found by writing that the integral expression satisfies the boundary conditions (2.9a) or (2.9b) on $\mathscr{P}$: the condition so obtained is an integral equation. The remarkable mathematical feature of this equation is that it concentrates all the conditions of the boundary value problem.

5.2. THE BASIC MATHEMATICAL FORMULAE

The basic mathematical formulae are demonstrated in Appendix B. For the reader not interested in the mathematical demonstrations, let us summarize the results in the following theorem, called below FTIM (Fundamental Theorem of Integral Method).

If a pseudo-periodic function $U(x, y)$, *continuous everywhere except on* $\mathscr{P}$, *satisfies* (B.8) *and* (B.9), *i.e. Helmholtz equations in* R_+ *and* R_-, *and radiation conditions for* $y \to \pm\infty$, *it can be expressed by* (B.10) *at any point of space in terms of the jumps of* U *and* $\mathrm{d}U/\mathrm{d}n$ *on* $\mathscr{P}$. *Furthermore, the limits* $U^{\pm}$ *of* U *on both sides of* $\mathscr{P}$ *are given in terms of these jumps by* (B.21) (*or* (B.24) *in an operator form*) *and* (B.25) *provides the values* $\mathrm{d}U^{\pm}/\mathrm{d}n$ *of the normal derivative in the particular case where* U *is continuous on* $\mathscr{P}$. *Outside the grooves,* $U(x, y)$ *can be represented as series of plane waves* (B.15) *and* (B.16) *whose coefficients are given by* (B.17), (B.18) *and* (B.19).

It is worth noting that Appendix B uses “Dirac functions” $\delta(x)$, easy to handle for the physicist but incorrect for the mathematician. The interested reader may find rigorous demonstrations of the above theorem using the Schwartz theory of distributions (SCHWARTZ [1966a, 1966b]) in a paper by the present author (MAYSTRE [1980a]).

5.3. THE FTE CASE

The FTE case is described by the Dirichlet boundary value problem stated by (2.8), (2.9a), (2.10). According to § 5.1, the first step is to express the diffracted field F in terms of the surface current density. We define the "generalized diffracted field" $U(x, y)$ *at any point of space* by the difference between the total field (which vanishes in region R_-) and the incident field:

$$\begin{aligned} U(x, y) &= F(x, y) && \text{in } R_+, \\ &= -F^{\mathrm{i}}(x, y) && \text{in } R_-. \end{aligned} \tag{5.1}$$

Obviously, $U(x, y)$ fulfills the conditions of the FTIM and can be expressed in terms of $\eta = (1 + f'(x)^2)^{1/2} (\mathrm{d}U^+/\mathrm{d}n(x) - \mathrm{d}U^-/\mathrm{d}n(x))$ and $\tau = U^+(x) - U^-(x)$. Bearing in mind the definition of U stated in (5.1), the boundary condition (2.9a) clearly shows that $\tau = 0$, while (2.22c) and (2.23a) indicate that $\eta = \phi$, i.e. a function closely linked with the surface current density $j_{\mathscr{P}}$. So, the FTIM enables us to show that the intuitive remark stated above is true: the diffracted field can be expressed in terms of $j_{\mathscr{P}}$. From (B.10) we deduce:

$$U(x, y) = \int_0^d \mathscr{A}(x, y, x')\, \phi(x')\, \mathrm{d}x', \tag{5.2}$$

$\mathscr{A}(x, y, x')$ being given by (B.11).

The FTIM also allows us to state that above the grooves, the diffracted field can be represented by a Rayleigh expansion, as previously shown in § 2.5. This enables us to find again the grating formula (cf. (2.15)) and to express the coefficients B_n of (2.13) using (B.17):

$$B_n = [s_n^+, \phi], \tag{5.3}$$

s_n^+ being given by (B.18). This formula is identical to (2.22a) but has been demonstrated in Appendix B for any value of n. Equation (2.24a) may be derived from (B.17) as well, by writing that U identifies with $-E^{\mathrm{i}}$ in region R_-.

To find $\phi(x)$, we remember that the right-hand member of (5.2) satisfies both (2.8) and (2.10), even though an arbitrary function is introduced in the integral instead of $\phi(x')$. To fulfill the last condition (2.9a), we write:

$$U^+(x) = U^-(x) = -F^{\mathrm{i}}(x, f(x)). \tag{5.4}$$

Replacing the left-hand member by its value given by (B.21) yields:

$$G\{\phi\} = -\psi_0, \tag{5.5}$$

with

$$\psi_0(x) = F^{\mathrm{i}}(x, f(x)), \tag{5.6}$$

which is an integral equation of the first kind (PETIT and CADILHAC [1964]) since the unknown function ϕ is not present outside the integral contained in the left-hand member. The kernel $G(x, x')$ is given by (B.22) and (B.11). It is straightforward to see that introducing the solution of (5.5) in the right-hand member of (5.2) provides an expression for U (thus an expression of F in R_+) which satisfies all the conditions of the FTE case boundary value problem.

Another equivalent way to write the boundary condition for F is:

$$\frac{\mathrm{d}U^-}{\mathrm{d}n}(x) = -\frac{\mathrm{d}F^{\mathrm{i}}}{\mathrm{d}n}(x, f(x)). \tag{5.7}$$

Indeed, U given by (5.2) satisfies the Helmholtz equation in region R_- and radiation condition for $y \to -\infty$. So, if this expression satisfies (5.7), it must identify with $-F^{\mathrm{i}}(x, y)$ in region R_-, since $-F^{\mathrm{i}}$ is an obvious solution of this Neumann boundary value problem whose solution has been assumed to be unique in § 2.2. Now, since U given by (5.2) is continuous, we derive that (5.4) is satisfied. Thus, (5.7) implies (5.4). Finally, introducing the value $\mathrm{d}U^-/\mathrm{d}n$ given by (B.25) in (5.7), then multiplying the two members by $(1 + f'(x)^2)^{1/2}$ yields:

$$\phi/2 = \phi_0 + N\{\phi\}, \tag{5.8}$$

with

$$\phi_0(x) = -\mathrm{i}(\beta + \alpha f'(x)) \exp(\mathrm{i}\alpha x - \mathrm{i}\beta f(x)), \tag{5.9}$$

$N(x, x')$ being given by (B.26), which is an integral equation of the second kind (PAVAGEAU, EIDO and KOBEISSE [1967], PAVAGEAU and BOUSQUET [1970]). Either (5.5) or (5.8) may be used to determine ϕ.

5.4. THE FTM CASE

In the FTM case, we have to solve the Neumann boundary value problem stated by (2.8), (2.9b) and (2.10). After using (5.1) to define a "generalized diffracted field" which satisfies the conditions of the FTIM, $U(x, y)$ can be expressed by (B.10) as a function of its jump, owing to the continuity of its normal derivative stated in (2.9b):

$$U(x, y) = \int_0^d \mathscr{B}(x, y, x')\, \psi(x')\, \mathrm{d}x', \tag{5.10}$$

$\mathscr{B}(x, y, x')$ being defined by (B.11) and $\psi(x')$ being the surface current density, according to (2.22d) and (2.23b). As in the preceding section, $U(x, y)$ has been expressed as a function of the surface current density, which satisfies (2.8) and (2.10), whatever the function introduced in (5.10) instead of $\psi(x')$.

Due to the limitation of the FTIM for non-continuous functions $U(x, y)$, it is not possible to express directly the boundary condition (2.9b). Let us show that the very simple condition:

$$U^-(x) = -F^{\mathrm{i}}(x, f(x)) \tag{5.11}$$

is equivalent to this boundary condition.

Indeed, the uniqueness of the solution of the Dirichlet boundary value problem allows us to assert that U given by (5.10), which satisfies the Helmholtz equation in region R_- and the radiation condition for $y \to \infty$, is equal to the obvious solution $-F^{\mathrm{i}}(x, y)$ in region R_- if (5.11) is satisfied. So $\mathrm{d}U^-/\mathrm{d}n(x) = -\mathrm{d}F^{\mathrm{i}}/\mathrm{d}n(x, f(x))$ and since the expression (5.10) has a continuous normal derivative on $\mathscr{P}$, $\mathrm{d}U^+/\mathrm{d}n = -\mathrm{d}F^{\mathrm{i}}/\mathrm{d}n$ and (2.9b) is satisfied. Using (B.24) to express the left-hand member of (5.11) yields:

$$\psi/2 = \psi_0 + N'\{\psi\}, \tag{5.12}$$

N' being given by (B.23) and (B.12).

After solving this integral equation of the second kind, the amplitudes B_n of the Rayleigh expansion of F above the grooves are deduced from (B.17):

$$B_n = [t_n^+, \psi], \tag{5.13}$$

t_n^+ being given by (B.19). This formula extends (2.22b) to any value of F_n.

It is interesting to notice that other integral equations may be obtained for FTE and for FTM cases, using the FTIM (MAYSTRE [1980a]).

5.5. NUMERICAL CONSIDERATIONS

In the first attemps at solving the above integral equations, a Fourier Series Method was used: the unknown function was represented as a series of $\exp(\mathrm{i}\alpha_m x)$. Truncating this series from $m = -N$ to $m = +N$ leads to a system of $2N + 1$ linear equations with $2N + 1$ unknowns: the coefficients of the series. This method is interesting for echelette or trapezoidal gratings since the coefficients of the linear system can be calculated in closed form (PETIT and CADILHAC [1964], MAYSTRE and PETIT [1970a, 1970b]). Unfortunately, it does not supply great precision and is difficult to generalize to other profiles.

This explains why, to our knowledge, this method has been abandoned and replaced by a point matching method.

In (5.5), (5.8) and (5.12), we have to solve an equation of the form:

$$\sigma\gamma(x) = \gamma_0(x) + \int_0^d L(x, x')\,\gamma(x')\,\mathrm{d}x', \tag{5.14}$$

where γ_0 is a known function, L being the kernel, given by a series, σ a coefficient (equal to 0 for (5.5)) and γ the unknown. Equation (5.14) is written at M points x_i of the period from $x = 0$ to $x_M < d$:

$$\sigma\gamma(x_i) = \gamma_0(x_i) + \int_0^d L(x_i, x')\,\gamma(x')\,\mathrm{d}x', \qquad i \in (1, M). \tag{5.15}$$

It is now sufficient to express the integral in terms of the M unknowns $\gamma(x_i)$. With this aim, we set:

$$\int_0^d L(x_i, x')\,\gamma(x')\,\mathrm{d}x' = \sum_{j=1}^{M} L_{ij}\,\gamma(x_j). \tag{5.16}$$

Introducing the right-hand member of (5.16) in (5.15) yields a linear system of M equations with M unknowns. So, the main work is to find "good values" of L_{ij}. Indeed, these values cannot be exact from a mathematical point of view, at least if M is finite: we know that the integral of the left-hand member of (5.16) cannot be rigorously achieved with a finite number of sample points. So, the problem is to choose the best method of integration. When the kernel is regular, this is no great undertaking: because of the periodicity of the integrand, the trapezoidal rule is very precise (MAYSTRE [1980a]), and the value of L_{ij} will be given by:

$$L_{ij} = (x_{j+1} - x_{j-1})\cdot L(x_i, x_j)/2, \qquad \text{if } j \neq 1 \text{ or } M,$$

x_{j-1} being replaced by $x_M - d$ for $j = 1$, and x_{j+1} being replaced by d for $j = M$.

The kernels N and N' of (5.8) and (5.12) are regular, even when $x' \to x$, contrary to what is expected at first glance (PAVAGEAU, EIDO and KOBEISSE [1967], MAYSTRE [1974, 1980a]). This fortunate feature no longer holds good for (5.5), where the kernel has a logarithmic singularity when $x' \to x$. Here, efficient methods may be employed to remove the singularity and perform the computations of L_{ij} (MAYSTRE [1980a]). The number M of sample points on $\mathscr{P}$ may be estimated using the following empirical rule: about six points per wavelength must be placed on $\mathscr{P}$ if a precision of about 1% is required. In all cases, great attention must be paid to the summation of the kernels: a rough truncation of the series leads to imprecise numerical results (MAYSTRE [1980a]). The vital importance of the care which must be devoted to these numerical problems of summation and integration cannot be over-emphasized. From our own experience, we are *able to warn the non-specialist, that a rough numerical application of the integral formalism cannot furnish reliable numerical results.*

Finally, because of the presence of $f'(x)$ or $f'(x')$ in the expression of the kernels N and N' (cf. (B.26) and (B.23)), the question which arises is: what happens when the profile has edges? In fact, the problem of edges is a serious and general problem in electromagnetism, which cannot be treated in this chapter, and we must refer to the studies of MEIXNER [1972]. For our specific problem, it will be noticed that the numerical difficulty disappears as soon as the function $f(x)$ is replaced by its truncated Fourier series, whose corresponding profile has no edges. Numerical experience shows that in general the efficiencies of the new gratings so obtained tend rapidly toward those of the grating having edges, when the number of Fourier coefficients increases. When the profile cannot be given by a function $y = f(x)$ (for instance when a vertical $x = x_0$ cuts $\mathscr{P}$ at several points), the integral formalism applies as well, provided s is used instead of x in the integrals.

§ 6. Other methods

6.1. DIFFERENTIAL METHODS

Differential methods have played a valuable role in the field of grating theories, but mainly for dielectric and metallic gratings, and that is why one of these methods will be described in detail in the next paragraph. For perfectly conducting gratings, two kinds of differential methods have been used. In the first method, the Maxwell equations are projected onto the x, y and z axes and directly written at a finite number of points placed inside the groove between $y = y_M$ and $y = f(x)$. This is a striking example of a very simple theory, which unfortunately can lead to strong numerical instabilities and long computer time (MOAVENI, KALHOR and AFRASHTEH [1975]).

More sophisticated are methods of the second kind, whose basic idea is to replace the x, y coordinates by new coordinates X, Y which map the grating surface $\mathscr{P}$ onto the plane $Y = 0$. The expression $F(X, Y)$ of the diffracted field, in the new coordinate system, no longer satisfies the initial Helmholtz equation. However, the more complicated equation so obtained leads, after projection of

F on the $\exp(i\alpha_m X)$, to a system of coupled differential equations which can be solved on the computer. Thus, in outline, *the basic idea is to replace the complicated boundary condition of* $\mathscr{P}$ *by a simpler condition, on* $Y = 0$. *Unfortunately, there exists a disadvantage: the Helmholtz equation is replaced by a more complicated one.*

The earliest study in this domain employs a conformal mapping technique (NEVIERE, CERUTTI-MAORI and CADILHAC [1971], NEVIERE, CADILHAC and PETIT [1973], VINCENT [1980b]). It leads to short computer times and good accuracy provided the adequate conformal mapping is found. A second study, using the simpler transformation $Y = y - f(x)$ (CHANDEZON, MAYSTRE and RAOULT [1980]) appeared more recently. This work has been extended to finite conductivity grating covered with dielectric layers (CHANDEZON, DUPUIS, CORNET and MAYSTRE [1982]) and the resulting computer code can deal with very deep gratings.

6.2. MODAL METHODS

Modal methods are of great interest and have in common three fundamental features: *they are based on very simple mathematics*; *they are very effective and can deal with highly modulated gratings with low computer time*; *they are restricted to special groove geometries*. More precisely, they can apply as soon as the field inside the grooves can be expressed in the form of a modal expansion, similar to the modal expansions encountered in the problems of wave-guides. In this case, the grating problem may be solved by matching the modal expansion to the Rayleigh expansion on the line $y = y_M$. This leads to the inversion of a linear system of equations whose unknowns are the coefficients of the modal expansion. This method is particularly useful for lamellar gratings, i.e. gratings whose grooves are rectangles, the sides of which are parallel to Ox and Oy (MAYSTRE and PETIT [1972b], HESSEL, SCHMOYS and TSENG [1975], ANDREWARTHA, FOX and WILSON [1979a, 1979b]). Though the theory is presented in quite a different manner, we can class in this group also the methods used for lamellar gratings by WIRGIN [1967, 1969] and WIRGIN and DELEUIL [1969]. The modal theory has been generalized by ROUMIGUIERES, MAYSTRE and PETIT [1974, 1975] to perfectly conducting rectangular rods lying on a dielectric stack. Recently, a new modal theory has been applied by BOTTEN, CRAIG, MC PHEDRAN, ADAMS and ANDREWARTHA [1981a, 1981b] to rectangular groove dielectric or metallic gratings.

JOVICEVIC and SESNIC [1972] used a modal theory to investigate the properties of echelette gratings, but their method relied on a questionable assumption (PETIT [1980]), contrary to the formalism of ITOH and MITTRA [1969] for the same gratings, which is quite rigorous. This kind of formalism is also well adapted to the study of gratings with semi-circular grooves (ANDREWARTHA, DERRICK and MC PHEDRAN [1981a]). A generalization of the modal theory to gratings with arbitrary profiles has been proposed by FOX [1980], and also by ANDREWARTHA, DERRICK and MC PHEDRAN [1981b] but, in our view, the modal method loses its main advantages over other classical methods for such arbitrary profiles.

6.3. MISCELLANEOUS

MARCUSE [1976] proposed for dielectric gratings an original method in which the field inside the grooves is represented by a double Fourier series, with a period d in x and a period $d' > H$ in y. He applied this method to echelette gratings, though some numerical difficulties appeared for large groove depth.

The specialist of boundary value problems would be astonished by the absence of a variational method for the grating problem. A recent study by BREIDNE and MAYSTRE [1982] investigated the possibilities of this method, very well known in electromagnetism (VAN BLADEL [1964]) and in many other domains of physics. Comparisons with results obtained from the integral formalism show that this method gives accurate results for medium-depth gratings ($H/d \lesssim 0.6$) and is particularly well adapted to the study of ghosts, viz. directions of diffractions not predicted by the grating formula and due to irregularities of period nd on the grating profile.

Finally, let us point out the original work of FACQ [1976, 1977] who investigated some properties of finite gratings using an integral method. Though it is not easy to renounce the periodicity of the grating problem, FACQ was able to investigate the properties of gratings having a small number of grooves (less than fifteen) using the inversion algorithm of AKAIKE and ROBIN for large systems of linear equations (AKAIKE [1973]).

It is difficult to give some idea of the computation time required by all the methods described in this paragraph and in previous ones. They are generally very low, less than one second on a CDC 7600 computer. The reader interested in more precise details may refer to a recent work for comparison of computation times of the Rayleigh methods with that of a computer code written by the present author using an integral formalism (HUGONIN, PETIT and CADILHAC [1981]). As regards the precision of the results, the experience of our laboratory, where practically all these methods have been numerically implemented, allows the statement that the results of the rigorous methods agree to within 1% in the resonance region, provided the numerical implementation is carefully carried out. As far as the integral, differential and modal methods are concerned, an agreement better than 10^{-5} in relative value on the efficiencies is reached in general. Perhaps, the reader will judge such a precision to be quite impractical. This opinion was justified until recent years but at present, this precision is sometimes needed for important investigations, for example the use of gratings as beam samplers for high power lasers (MAYSTRE, NEVIERE and PETIT [1980]).

§ 7. Metallic and dielectric gratings

Nowadays, the computer codes devoted to real metallic gratings are in most widespread use, even though they are more complicated and time-consuming than those devoted to perfectly conducting gratings. This is due to theoretical (PETIT, MAYSTRE and NEVIERE [1972], MAYSTRE [1973, 1974]) and experimental (HUTLEY and BIRD [1973], LOEWEN, MAYSTRE, MC PHEDRAN and WILSON [1974]) evidence that the model of perfectly conducting gratings fails in the visible and near infrared regions, at least for the FTM case. This phenomenon is not intuitive since the reflectivity of metals like aluminium, silver or gold employed for the construction of commercial gratings exceeds 90% in these regions.

7.1. BOUNDARY VALUE PROBLEM AND ELEMENTARY PROPERTIES

In this section, R_- is filled with a dielectric or a metal of complex index ν. Proceeding as in § 2.2, but this time with the Maxwell equations in both regions R_+ and R_-, it can be shown that the total field does not depend on z and remains TE (or TM) polarized when the incident field is TE (or TM). So, for both polarizations, we shall deal with the projections of the field on Oz. The diffracted field $F(x, y)$ is defined as the difference between the total field and the incident field in R_+ and identifies with the total field in R_-. So, it is possible to define the FTE (or FTM) cases, where the incident electric (or magnetic) field F^{i} is given by (2.11). In these fundamental cases, the diffracted field F obeys the following boundary value problem:

$$\nabla^2 F + k^2 F = 0 \qquad \text{in } R_+, \tag{7.1}$$

$$\nabla^2 F + k^2 \nu^2 F = 0 \qquad \text{in } R_-, \tag{7.2}$$

$$F^+(x) + F^{\mathrm{i}}(x, f(x)) = F^-(x), \tag{7.3}$$

$$\mathrm{d}F^+/\mathrm{d}n(x) + \mathrm{d}F^{\mathrm{i}}/\mathrm{d}n(x, f(x)) = C_{\mathrm{p}} \frac{\mathrm{d}F^-}{\mathrm{d}n}(x), \tag{7.4}$$

$$F \text{ satisfies a radiation condition for } y \to \pm\infty, \tag{7.5}$$

with

$$C_{\mathrm{p}} = 1 \qquad \text{for FTE}, \tag{7.6}$$

$$C_{\mathrm{p}} = 1/\nu^2 \qquad \text{for FTM}.$$

It is worth noting that, for the FTE case, (7.3), (7.4) and (7.6) mean that the total field and its normal derivative are continuous on $\mathscr{P}$. This continuity holds for the total field but not for its normal derivative for the FTM case since $C_{\mathrm{p}} \neq 1$. The uniqueness of the solution of the FTE problem has been shown by CADILHAC [1980], at least when ν is real. As in § 2.4 and § 2.5, it can be established from the above equation that F is pseudo-periodic in x, and is described above the top of the grooves by the Rayleigh expansion (2.13) and below the bottom of the grooves by another Rayleigh expansion:

$$F(x, y) = \textstyle\sum_n B_n^{(\nu)} \exp(\mathrm{i}\alpha_n x - \mathrm{i}\beta_n^{(\nu)} y) \qquad \text{if } y < y_{\mathrm{m}}, \tag{7.7}$$

with

$$\beta_n^{(\nu)} = (k^2 \nu^2 - \alpha_n^2)^{1/2}. \tag{7.8}$$

The above definition of $\beta_n^{(\nu)}$ is ambiguous since ν may be complex. Its determination will be fixed by the following inequality:

$$\mathrm{Re}\{\beta_n^{(\nu)}\} + \mathrm{Im}\{\beta_{\mathrm{n}}^{(\nu)}\} \geqslant 0, \tag{7.9}$$

which generalizes the convention for β_n stated in (2.12b). So, the classical grating formula for reflected orders is unchanged but must be completed for dielectric gratings (ν real) by the grating formula for transmitted orders, which deduces easily from (2.12a) and (7.8):

$$\nu \sin\theta_n^{(\nu)} = \sin\theta + n\lambda/d, \tag{7.10}$$

where the diffraction angle $\theta_n^{(\nu)}$ of the nth transmitted order is measured anticlockwise like the incidence angle. The set $U^{(\nu)}$ of propagating orders in R_- may be defined easily when ν is real but, more generally, $U^{(\nu)}$ will be defined for an arbitrary value of ν by:

$$\mathrm{Re}\{\beta_n^{(\nu)}\} - \mathrm{Im}\{\beta_n^{(\nu)}\} \geqslant 0 \qquad \text{if } \mathrm{Im}\{\nu\} \geqslant 0,$$

$$U^{(\bar{\nu})} = U^{(\nu)} \qquad \text{if } \mathrm{Im}\{\nu\} < 0.$$

In order to state some fundamental properties of such gratings, it is interesting to generalize the lemma of § 2.6. We consider two pseudo-periodic functions u and v containing incoming and outgoing waves on both sides and satisfying:

$$\nabla^2 u + k^2 u = 0, \qquad \nabla^2 v + k^2 v = 0, \qquad \text{in } R_+,$$

$$\nabla^2 u + k^2 \nu^2 u = 0, \qquad \nabla^2 v + k^2 \bar{\nu}^2 v = 0, \qquad \text{in } R_-,$$

and the boundary conditions (7.3) and (7.4) on $\mathscr{P}$, ν having a positive imaginary part. So, u may be written in the form:

$$u = \sum_{n \in U} A_n \exp(\mathrm{i}\alpha_n x - \mathrm{i}\beta_n y) + \sum_n B_n \exp(\mathrm{i}\alpha_n x + \mathrm{i}\beta_n y) \qquad \text{in } R_+,$$

$$u = \sum_{n \in U^{(\nu)}} A_n^{(\nu)} \exp(\mathrm{i}\alpha_n x + \mathrm{i}\beta_n^{(\nu)} y) + + \sum_n B_n^{(\nu)} \exp(\mathrm{i}\alpha_n x - \mathrm{i}\beta_n^{(\nu)}) \qquad \text{in } R_-. \tag{7.11}$$

By definition, the expression for v will be derived from the above expressions for u by replacing A_n, B_n, $A_n^{(\nu)}$ and $B_n^{(\nu)}$ by A_n', B_n', $A_n^{(\nu)\prime}$ and $B_n^{(\nu)\prime}$, $\beta_n^{(\nu)}$ being replaced by $\beta_n^{(\bar{\nu})} = \overline{\beta_n^{(\nu)}}$, $U^{(\bar{\nu})}$ being equal to $U^{(\nu)}$. Proceeding as in § 2.6 for R_+, then R_- yields:

$$\frac{1}{2\mathrm{i}d} \int_{\mathscr{P}} \left(u^+ \frac{\mathrm{d}\overline{v^+}}{\mathrm{d}n} - \overline{v^+} \frac{\mathrm{d}u^+}{\mathrm{d}n} \right) \mathrm{d}s = \sum_{n \in U} \beta_n (A_n \overline{A_n'} - B_n \overline{B_n'}), \tag{7.12}$$

$$\frac{1}{2\mathrm{i}d} \int_{\mathscr{P}} \left(u^- \frac{\mathrm{d}\overline{v^-}}{\mathrm{d}n} - \overline{v^-} \frac{\mathrm{d}u^-}{\mathrm{d}n} \right) \mathrm{d}s = \sum_{n \in U^{(\nu)}} \beta_n^{(\nu)} (B_n^{(\nu)} \overline{B_n^{(\bar{\nu})\prime}} - A_n^{(\nu)} \overline{A_n^{(\bar{\nu})\prime}}), \tag{7.13}$$

and from the boundary conditions (7.3) and (7.4), we can eliminate the left-hand members of the above equations:

$$\sum_{n \in U} \beta_n (A_n \overline{A_n'} - B_n \overline{B_n'}) = C_{\mathrm{p}} \sum_{n \in U^{(\nu)}} \beta_n^{(\nu)} (B_n^{(\nu)} \overline{B_n^{(\bar{\nu})\prime}} - A_n^{(\nu)} \overline{A_n^{(\bar{\nu})\prime}}). \tag{7.14}$$

Using the same functions u and v as in § 2.6 for the reciprocity theorem, the "reciprocity by reflection" stated in (2.20) remains unchanged, whatever the index ν of R_- (MAYSTRE and MC PHEDRAN [1974]).

For the "reciprocity by transmission" (NEVIERE and VINCENT [1976]), the function v is defined as the conjugate of the total field v'' of Rayleigh coefficients B_n'' above $\mathscr{P}$, which exists when an incident wave $\exp(\mathrm{i}\alpha'' x + \mathrm{i}\beta'' y)$ propagates upward in R_-, with $\alpha'' = -\alpha_n$ and $\beta'' = \beta_n^{(\nu)}$. The result is that, when $n \in U$:

$$\beta B_n'' = C_{\mathrm{p}} \beta_n^{(\nu)} B_n^{(\nu)}. \tag{7.15}$$

When ν is real, the energy balance criterion can be derived from (7.14) by setting $u = v$, which yields:

$$\sum_{n\in U} \beta_n|A_n|^2 + C_p \sum_{n\in U^{(\nu)}} \beta_n^{(\nu)}|A_n^{(\nu)}|^2 = \sum_{n\in U} \beta_n|B_n|^2 + C_p \sum_{n\in U^{(\nu)}} \beta_n^{(\nu)}|B_n^{(\nu)}|^2. \tag{7.16}$$

Here, also, the use of Poynting's theorem shows that the quantities in (7.16) are no other than the flux of energy of incident waves in R_+ (first term), or in R_- (second term), and the flux of energy of diffracted waves in R^+ (third term) or R^- (last term). So, the physical meaning is very simple: the incident energy is equal to the diffracted energy. It is interesting to note that, when ν is real, (7.15) reveals that the efficiency is the same in the two cases: there exists a "reciprocity by transmission".

7.2. THE INTEGRAL FORMALISM

The first rigorous theoretical studies on this topic proposed the solving of two coupled integral equations (WIRGIN [1968], NEUREUTHER and ZAKI [1969], VAN DEN BERG [1971a, 1971b]). Indeed, contrary to what happens with perfectly conducting gratings, where $j_{\mathscr{P}}$ is a relevant function to express the diffracted field, here the physical meaning cannot suggest a comparable function defined on the profile $\mathscr{P}$. However, it turns out that the field $F^+(x)$ on $\mathscr{P}$ and its normal derivative $\mathrm{d}F^+/\mathrm{d}n(x)$ are sufficient to express the diffracted field in R_+ and R_- using (B.10). Indeed, (B.10) may be used in the first instance to express a function of U and $\mathrm{d}U/\mathrm{d}n$ being equal to F^+ and $\mathrm{d}F^+/\mathrm{d}n$. Likewise, the field F in R_- may be expressed by applying (B.10) in region R_- (viz. replacing $\mathscr{A}$ and $\mathscr{B}$ by $\mathscr{A}^{(\nu)}$ and $\mathscr{B}^{(\nu)}$ deduced from (B.11) and (B.12) by replacing β_n by $\beta_n^{(\nu)}$) to a function equal to 0 in R_+ and to F in R_-. Of course, two integral equations are necessary to find the two unknown functions: they are obtained by using the boundary conditions (7.3) and (7.4). Unfortunately, these two integral equations are coupled, a fact which leads to serious numerical limitations.

Now, we shall present the theory developed by the present author, which leads to a single integral equation (MAYSTRE [1972, 1973, 1974, 1980a]). From an intuitive point of view, this formalism relies upon the following idea: *a fictitious surface current density $j_{\mathscr{P}}(x)$ exists which, placed on $\mathscr{P}$, would generate in free space (the material filling region R_- being replaced by air) a diffracted field F^- identical in R_+ to the actual diffracted field F.* Of course, the field F' diffracted by this fictitious surface current density in R_- has no physical meaning. If we assume the existence of this function $j_{\mathscr{P}}(x)$, we must notice that it enables us to express F' at any point in space and thus, to express F in region R_+. So, Appendix B can be used to express the limits F^+ and $\mathrm{d}F^+/\mathrm{d}n$ above $\mathscr{P}$ and then the limits F^- and $\mathrm{d}F^-/\mathrm{d}n$ will be deduced from (7.3) and (7.4). Now, the field in R_- may be derived from F^- and $\mathrm{d}F^-/\mathrm{d}n$ using (B.10) for a material of index ν. Finally, we have expressed the field at any point of space in terms of $j_{\mathscr{P}}(x)$ and obviously, a single integral equation is enough to determine this single unknown function. It is worth noting that, from this point of view, the theory leading to a pair of integral equations is redundant: in fact, the two unknown functions F^+ and $\mathrm{d}F^+/\mathrm{d}n$ may be deduced from a single relevant function $j_{\mathscr{P}}(x)$.

More precisely, the mathematical steps of this method are outlined in Fig. 7.1. We first define a function $U(x, y)$ by:

$$U = F \qquad \text{in } R_+, \tag{7.17}$$

$$\nabla^2 U + k^2 U = 0 \qquad \text{in } R_-, \tag{7.18}$$

$$U^-(x) = U^+(x), \tag{7.19}$$

U satisfies a radiation condition for $y \to -\infty$. (7.20)

The function U, which is defined simply by (7.17) in R_+, is also well defined in R_- by (7.18)–(7.20), due to the existence and uniqueness of the Dirichlet problem assumed in § 2 (since $U^+ = F^+$ is fixed, U^- is fixed too, according to (7.19)). Moreover, U obviously satisfies the conditions of the FTIM stated in § 5.2. So, using (B.10), $U(x, y)$ which is continuous on $\mathscr{P}$ (cf. (7.19)) may be expressed at any point of space from the jump $\phi(x)/(1 + f'(x)^2)^{1/2}$ of its normal derivative:

$$U(x, y) = \int_0^d \mathscr{A}(x, y, x')\,\phi(x')\,\mathrm{d}x' \tag{7.21}$$

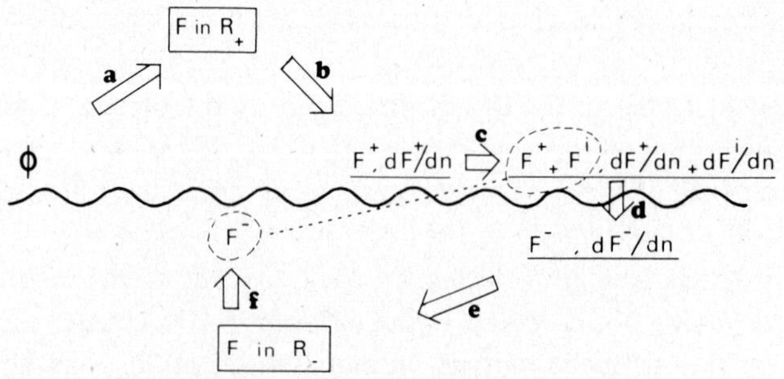

Fig. 7.1. An integral method for finite conductivity gratings in outline.

It is interesting to notice that the above formula, quite similar to (5.2), gives the expression of the field U generated in free space by a surface current density $j_{\mathscr{P}}(x) = \mathrm{i}\phi(x)(1 + f'(x)^2)^{-1/2}/\omega\mu_0$. So, we now have the expression of F in R_+ (arrow **a** of Fig. 7.1) and (B.17) allows us to express the coefficients B_n of the diffracted field in R_+:

$$B_n = [s_n^+, \phi]. \tag{7.22}$$

Furthermore, (B.24) and (B.25) give the limits F^+ and $\mathrm{d}F^+/\mathrm{d}n$ of U above $\mathscr{P}$ (arrow **b**):

$$F^+ = G\{\phi\}, \tag{7.23}$$

$$(1 + f'(x)^2)^{1/2}\,\mathrm{d}F^+/\mathrm{d}n = \phi/2 + N\{\phi\}. \tag{7.24}$$

The limits F^- and $\mathrm{d}F^-/\mathrm{d}n$ can be easily deduced from (7.23) and (7.24) by adding the incident field to F (arrow **c**) and using the boundary conditions (7.3) and (7.4) (arrow **d**):

$$F^- = G\{\phi\} + \psi_0, \tag{7.25}$$

$$(1 + f'(x)^2)^{1/2}\,\mathrm{d}F^-/\mathrm{d}n = [\phi/2 + N\{\phi\} + \phi_0]/C_\mathrm{p}, \tag{7.26}$$

ψ_0 and ϕ_0 being given by (5.6) and (5.9).

Now, to express the total field F in R_- in terms of ϕ, it suffices to use (B.10) for a function $U'(x, y)$ satisfying the conditions of the FTIM (but this time with a Helmholtz equation of wave-number $k\nu$):

$$U'(x, y) = 0 \qquad \text{in } R_+,$$

$$U'(x, y) = F(x, y) \qquad \text{in } R_-,$$

and (B.10) yields (arrow **e**):

$$U'(x, y) = -\int_0^d \mathscr{A}^{(\nu)}(x, y, x')\frac{\mathrm{d}F^-}{\mathrm{d}n}(x')(1 + f'(x')^2)^{1/2}\,\mathrm{d}x'$$

$$-\int_0^d \mathscr{B}^{(\nu)}(x, y, x')\,F^-(x')\,\mathrm{d}x'. \tag{7.27}$$

Equation (B.17) permits us to evaluate the amplitude $B_n^{(\nu)}$ of the Rayleigh expansion of F (i.e. U') in R_- from (7.25) and (7.26):

$$B_n^{(\nu)} = -\frac{1}{C_\mathrm{p}}[s_n^{(\nu)-}, \phi/2 + N\{\phi\} + \phi_0] - [t_n^{(\nu)-}, G\{\phi\} + \psi_0], \tag{7.28}$$

$s_n^{(\nu)-}$ and $t_n^{(\nu)-}$ being obtained from s_n^- and t_n^- by replacing β_n by $\beta_n^{(\nu)}$. Similarly, putting (7.25) and (7.26) in (B.24) and bearing in mind that $U' = 0$ above $\mathscr{P}$ allows us to express the limit F^- of U' below $\mathscr{P}$ (arrow **f**):

$$F^- = (G\{\phi\} + \psi_0)/2 - G^{(\nu)}\{\phi/2 + N\{\phi\} + \phi_0\}/C_p - N^{(\nu)\prime}\{G\{\phi\} + \psi_0\}. \tag{7.29}$$

The integral equation is obtained by writing that the above value is linked to $F^+ + F^i$, by (7.3). Using F^+ given by (7.23) yields:

$$(G\{\phi\} + G^{(\nu)}\{\phi\}/C_p)/2 + N^{(\nu)\prime}\{G\{\phi\}\} + G^{(\nu)}\{N\{\phi\}\}/C_p$$
$$= -(\psi_0/2 + N^{(\nu)\prime}\{\psi_0\} + G^{(\nu)}\{\phi_0\}/C_p). \tag{7.30}$$

It can be proved that the diffracted field F derived from (7.21) and (7.27) well satisfies the boundary value problem stated in § 7.1, provided (7.30) is verified (MAYSTRE [1980a]).

This integral equation is very well adapted to highly conducting gratings: when $|\nu| \to \infty$, it tends toward integral equations devoted to perfectly conducting gratings. The above formalism has been generalized to more complicated gratings of great practical interest, for instance metallic or dielectric gratings covered with a stack of dielectric layers (MAYSTRE [1978a, 1978b], BOTTEN [1978a, 1978b]) or bimetallic gratings, i.e. gratings whose surface is composed of two alternating metals (BOTTEN [1980]). Another interest of this formalism lies in the fact that the integral equation is practically the same for the FTE and FTM cases: it is very interesting to treat these two cases simultaneously. The computation time is of the order of a second on a CDC 7600, if one is content with a precision of 1%.

7.3. DIFFERENTIAL METHOD

We have seen in § 6.1 that differential methods permit us to solve the problem of perfectly conducting gratings. In this section, we shall describe a particular differential formalism, which had great importance in the development of grating theories, not only because of its particular possibilities, but also since it permitted fruitful comparisons with the results obtained from the integral formalism of finite conductivity gratings described in the previous paragraph.

A feature of this particular differential formalism is that it cannot be applied to perfectly conducting gratings. On the other hand, it can deal with periodic structures like arrays of identical dielectric (or metallic) cylinders, or even 'generalized gratings' described by a non piecewise continuous index $\nu(x, y)$

periodic in x, for which domains R_+ and R_- cannot be defined. For the sake of simplicity, the description of the formalism will be restricted to the simplest case of Fig. 2.1, where $v(x, y)$ is equal to 1 in R_+ and to v in R_-.

The method was first proposed by PETIT [1966b] and implemented by CERUTTI-MAORI, PETIT and CADILHAC [1969]. Unfortunately, due to the algorithm used in the computer code for solving the system of coupled differential equations (finite difference method), strong difficulties for conducting gratings emerged. The breakthrough in the differential method was realized by NEVIERE [1975] and NEVIERE, VINCENT and PETIT [1974], who not only improved the formalism but also used reliable and accurate numerical methods which revolutionized the domain of application of this formalism (see also CHANG, SHAH and TAMIR [1980]). However, the curse of this method, to which no solution has so far been found, lies in its inability to provide reliable results for highly conducting gratings, such as aluminum gratings in the visible region, even though some progress has been made recently (VINCENT [1980a]).

We deal with the FTE case, and denoting by $\chi(x, y)$ a function equal to k^2 in R_+ and $k^2 v^2$ in R_-, we deduce that the total field $F^{\mathrm{t}}(x, y)$ satisfies a Helmholtz equation obtained by putting (7.1) and (7.2) together and bearing in mind that F^{i} satisfies (7.1) in R_+:

$$\nabla^2 F^{\mathrm{t}}(x, y) + \chi(x, y)\, F^{\mathrm{t}}(x, y) = 0. \tag{7.31}$$

Of course, the diffracted field F satisfies (7.31), except on $\mathscr{P}$ where F is discontinuous. On the other hand, (7.31) is valid at any point of space, even on $\mathscr{P}$, due to the continuity of F^{t} and its normal derivative $\mathrm{d}F^{\mathrm{t}}/\mathrm{d}n$ on $\mathscr{P}$ stated in (7.3) and (7.4). One can say that (7.31) *is valid in the sense of distributions*, which means, in this particular case, that the definition of the left-hand member of (7.31) does not cause any problem for those who are not acquainted with the theory of distributions. Now, we can transform (7.31) by expanding F^{t} and χ in series:

$$F^{\mathrm{t}}(x, y) = \sum_n F_n(y) \exp(\mathrm{i}\alpha_n x), \tag{7.32}$$

$$\chi(x, y) = \sum_n \chi_n(y) \exp(\mathrm{i}nKx), \tag{7.33}$$

and we derive easily:

$$\sum_n \Big(F_n'' - \alpha_n^2 F_n + \sum_q \chi_{n-q} F_q\Big) \exp(\mathrm{i}\alpha_n x) = 0. \tag{7.34}$$

Since (7.34) is valid at any point of space, including $\mathscr{P}$, (7.34) is satisfied for any value of x and y, and so any term of the series must vanish:

$$F_n''(y) - \sum_q J_{nq}(y) F_q(y) = 0, \tag{7.35}$$

with
$$J_{nq}(y) = \alpha_n^2 \delta_{n-q} - \chi_{n-q}(y). \tag{7.36}$$

Furthermore, (7.31) and the continuity of F^t and dF^t/dn (and thus of grad F^t) on $\mathscr{P}$ imply the continuity of $F_n(y)$ and $F_n'(y)$ for any value of y. Above y_M and below y_m, $J_{nq}(y)$ is constant since χ remains constant; thus, we find again that F^t is described by plane wave expansions which can be deduced from (2.13) and (7.7) by adding the incident field to F in R_+. From (7.32), we derive the expressions for $F_n(y)$ imposed by the radiation conditions:

$$F_n(y) = \exp(-i\beta y)\delta_n + \sum_n B_n \exp(i\beta_n y) \qquad \text{for } y \geqslant y_M, \tag{7.37}$$

$$F_n(y) = \sum_n B_n^{(v)} \exp(-i\beta_n^{(v)} y) \qquad \text{for } y \leqslant y_m. \tag{7.38}$$

Obviously, the values of the Rayleigh coefficients B_n and $B_n^{(v)}$ may be deduced from $F_n(y_M)$ and $F_n(y_m)$. Now, to determine the two sets of $F_n(y_M)$ and $F_n(y_m)$ we must recall that $F_n(y)$ satisfies between y_m and y_M a system of coupled differential equations (7.35). From a mathematical point of view, this is not sufficient since boundary conditions for $F_n(y)$ at y_m and y_M are necessary. These boundary conditions may be inferred from the special form of $F_n(y)$ at the limits y_m and y_M, given by (7.37) and (7.38). Indeed, it is easy to verify on these equations that:

$$F_n'(y_m) + i\beta_n^{(v)} F_n(y_m) = 0, \tag{7.39}$$

$$F_n'(y_M) - i\beta_n F_n(y_M) = -2i\beta \exp(-i\beta y_M)\delta_n. \tag{7.40}$$

So, the grating problem has been reduced to the solving of the system of coupled differential equations (7.35) with boundary conditions (7.39) and (7.40) on the boundaries y_m and y_M.

The numerical implementation is based on the "shooting method". First, we have to truncate the infinite differential system. A system of $2N + 1$ coupled differential equations with $2N + 1$ unknowns $F_n(y)$ is obtained by eliminating the F_n having a number n outside $(-N, +N)$. The first step of the "shooting method" is to construct a computer code numerically able to provide the $F_n(y_M)$ as soon as the $F_n(y_m)$ are known from $-N$ to $+N$. Let us show briefly that this is not a difficult problem. Indeed, as soon as the $F_n(y_m)$ are known, the $F_n'(y_m)$ can be derived from (7.39) in a linear manner. Now, standard algorithms like Numerov, Runge–Kutta or Adams–Moulton algorithms, enable the calculation of $F_n(y)$ and $F_n'(y)$ up to y_M (VINCENT [1980b]). Intuitively, it is easy to see that as soon as the $F_n(y)$ and $F_n'(y)$ are known all the derivatives $F_n^{(m)}(y)$ may be deduced from (7.35) or by differentiating (7.35). So, we can compute with the desired precision the values $F_n(y + \Delta y)$ and $F'_n(y + \Delta y)$ at a slightly higher ordinate using, for instance, a Taylor expansion for $F_n(y)$. Proceeding in like manner at M points between y_{m} and y_{M} finally gives $F_n(y_{\mathrm{M}})$ and $F'_n(y_{\mathrm{M}})$. In fact, the numerical algorithms used in practice are much simpler and very efficient. From $F_n(y_{\mathrm{M}})$ and $F'_n(y_{\mathrm{M}})$, we may compute the set of:

$$U_n = F'_n(y_{\mathrm{M}}) - \mathrm{i}\beta_n F_n(y_{\mathrm{M}}). \tag{7.41}$$

Of course, if the exact values of $F_n(y_{\mathrm{m}})$ corresponding to the physical problem have been employed at the starting point of the integration, U_n must be equal to $-2\mathrm{i}\beta \exp(-\mathrm{i}\beta y_{\mathrm{M}})\delta_n$, according to (7.40), if the integration has been done with infinite precision. But the exact $F_n(y_{\mathrm{m}})$ are unknown. To understand the "shooting method", it must be envisaged that the $F_n(y_{\mathrm{m}})$ are replaced by arbitrary values c_n in (7.39), and it must be noted that the corresponding U_n numerically obtained by using (7.39), (7.35) and (7.41) are linearly linked to these c_n arbitrarily chosen at the starting point of integration:

$$U_n = \sum_{q=-N}^{+N} T_{nq} c_q. \tag{7.42}$$

By adopting a set of $c_q = \delta_{q-r}$, we shall find out the set of $U_n = T_{nr}$, which permits us to reconstitute the rth column of the matrix T of size $2N + 1$. Thus, $2N + 1$ successive integrations corresponding to $2N + 1$ different values of r permit the calculation of the entire matrix T.

Now, it suffices to use (7.42) to derive the actual coefficients $F_n(y_{\mathrm{m}})$ by solving a system of $2N + 1$ linear equations with $2N + 1$ unknowns:

$$\sum_{q=-N}^{+N} T_{nq} F_q(y_{\mathrm{m}}) = -2\mathrm{i}\beta \exp(-\mathrm{i}\beta y_{\mathrm{M}})\delta_n. \tag{7.43}$$

The field between y_{m} and y_{M} and the Rayleigh coefficients B_n and $B_n^{(\nu)}$ may be deduced easily from $F_m(y_{\mathrm{m}})$, as seen before.

The FTM case of polarization may be treated by the differential method but is more complicated because (7.31) is not true in the sense of distributions.

One of the origins of the instabilities observed using the differential formalism for highly conducting metals may be understood from an intuitive point of view. When the conductivity increases, the skin depth decreases. So, for a given grating of groove depth H, the function $F^{\mathrm{t}}(x, y_{\mathrm{m}})$ tends to a "delta function" when the skin depth to groove depth ratio becomes very weak. So, the truncation of (7.35) will entail a large error, whatever the value of N. It seems that problems of stability of algorithms have at least the same importance in this failure.

It is interesting to notice that the integration of the system of differential equations can be performed in closed form in the particular case of the lamellar grating, where the χ_n are independent of y. This feature permits the representation of the field in the form of "modes" and the numerical application leads to very efficient computer codes (KNOP [1978]).

§ 8. The grating in conical diffraction

8.1. DEFINITIONS

The term conical diffraction is used as soon as the incident wave-vector $\boldsymbol{k}$ does not lie in the $x\mathrm{O}y$ plane. All the problems posed by this new kind of mounting are solved by a theorem of invariance (PETIT and MAYSTRE [1972], MAYSTRE [1980a]), at least for perfectly conducting gratings. We shall give this theorem without demonstration. The FTE (or FTM) cases of polarization may be defined in this more complicated case as well. In the FTE (or FTM) case, the incident magnetic field (or electric field) is perpendicular to $\mathrm{O}z$ and it can be shown that the corresponding diffracted fields satisfy the same property provided the grating is perfectly conducting. To state the "invariance theorem", let us associate an equivalent classical mounting (i.e. the usual mounting described in Fig. 2.1) to an arbitrary diffraction mounting by projecting the incident wave-vector $\boldsymbol{k}$ and the diffracted wave-vectors $\boldsymbol{k}_n$ on the $\mathrm{O}z$ axis and the $x\mathrm{O}y$ plane:

$$\boldsymbol{k} = \boldsymbol{k}^{xy} + k^z\hat{z},$$

$$\boldsymbol{k}_n = \boldsymbol{k}_n^{xy} + k_n^z\hat{z},$$

$\hat{z}$ being the unit vector along $\mathrm{O}z$.

The equivalent classical mounting is defined by an incident wave-vector equal to $\boldsymbol{k}^{xy}$. This not only gives the incidence θ^{xy} but also the wavelength $\lambda^{xy} = 2\pi/|\boldsymbol{k}^{xy}| = \lambda/\cos\Lambda$, Λ being the angle between $\boldsymbol{k}$ and the $x\mathrm{O}y$ plane.

8.2. THEOREM OF INVARIANCE

When a perfectly conducting grating is used in conical diffraction, the projection k_n^z *of the wave-vector of the* n*th order diffracted wave on the* $\mathrm{O}z$ *axis is equal to* k^z *and the direction of* $\mathbf{k}_n^{xy}$ *in the* $x\mathrm{O}y$ *plane is identical to the direction of the* n*th diffracted order in the equivalent classical mounting. The efficiency in the* n*th* *diffracted order is equal to the efficiency in the* nth *diffracted order for the equivalent classical mounting, in the same fundamental case of polarization.*

So, the directions of diffraction $\boldsymbol{k}_n$ lie on a cone containing $\boldsymbol{k}$ (this is the origin of the name "conical diffraction") and the projections of the directions of diffraction on the xOy plane are deduced from θ^{xy} by the classical grating formula, using a fictitious wavelength λ^{xy}. The computer codes devoted to the usual mounting can be used to determine the efficiencies from the parameters θ^{xy} and λ^{xy}. It is worth noting that the invariance theorem enables the realization of a mounting with a constant efficiency (MAYSTRE and PETIT [1972a]): it suffices to keep θ^{xy} constant and to vary λ and Λ in order to keep $\lambda/\cos(\Lambda)$ constant.

The part of the theorem devoted to the directions of diffraction still holds for metallic and dielectric gratings. On the other hand, the efficiency of such gratings does not obey the theorem since the vector boundary value problem cannot be separated into two scalar boundary value problems: a TE (or TM) incident wave in general does not generate a TE (or TM) diffracted wave. However, rigorous integral (MAYSTRE and PETIT [1974]) and differential (VINCENT, NEVIERE and MAYSTRE [1978]) methods show that the invariance theorem can be a good empirical rule for metallic gratings, even in the extreme UV region, provided the variation of reflectivity of the metal with angle Λ is taken into account (MAYSTRE, NEVIERE and PETIT [1980]).

§ 9. Conclusion

Some recent developments in the field of rigorous grating theories are not described in this chapter. The reader interested in the very complicated problem of diffraction by "crossed gratings" (viz. gratings which are not invariant in z but periodic in x and z with groove spacings d and d') may refer to the review of MC PHEDRAN, DERRICK and BOTTEN [1980]. Though the present computer codes devoted to these crossed gratings are time-consuming and do not yield exceptional performance, they have the merit of showing the high interest, as filters in infrared or solar selective elements, of these new gratings which can be constructed holographically.

The fascinating field of inverse scattering for gratings has been investigated in the last few years. Here, the unknown is the profile of the grating, which must be deduced from an efficiency curve of the grating, for instance, in Littrow mount. Though the mathematical problems encountered in the first investigations are tremendously difficult to handle, computer codes have been constructed (ROGER and MAYSTRE [1981]) and the first theoretical reconstitution of a grating profile from actual experimental data has been achieved by ROGER and BREIDNE [1980].

At present, one can consider that rigorous vector theories of gratings are able to solve almost all the classical problems of gratings encountered by manufacturers and users. Moreover, the numerous computer codes provide for the theoretician the opportunity to evaluate the advantages and shortcomings of each rigorous theory, as well as the precise limits of non-rigorous theories. This information is invaluable for investigating the somewhat tougher and very important problems of optics, such as diffraction by non-periodic rough surfaces. Finally, these accurate computer codes have often played a role comparable to that of accurate experimental devices: numerical exploitation of them has given valuable clues in phenomenological studies and has enabled the precise verification of purely theoretical predictions. This explains why some specialists of grating theories have dubbed themselves "numerical experimenters" to show the importance they attach to this research.

Appendix A

Let us assume, by hypothesis, that the Rayleigh expansion $R(x, y)$ converges for $x = x_0$ and $y = y_0$. We derive that the modulus of $R_n(x, y)$ tends to 0 when $n \to \pm\infty$:

$$|B_n \cdot \Phi_n(x_0, y_0)| \to 0 \qquad \text{if } n \to \pm\infty. \tag{A.1}$$

Now, it is useful to consider the behavior of $R(x, y)$ in $\mathbf{C}^2$, i.e. for complex values of x and y. What we intend to show in the first place is that the RE converges uniformly and absolutely in the region $\Omega \in \mathbf{C}^2$ defined by:

$$|w(x, y)| \leqslant |w(x_1, y_1)|, \tag{A.2}$$

$$|\overline{w}(x, y)| \leqslant |w(x_1, y_1)|, \tag{A.3}$$

provided x_1 and y_1 are real, with $y_1 > y_0$, which implies from (3.8):

$$|w(x_1, y_1)/w(x_0, y_0)| = q < 1. \tag{A.4}$$

To clarify (A.3), it must be recalled that $\overline{w}(x, y)$ denotes the complex function $\overline{w(\overline{x}, \overline{y})}$ which is analytic in $\mathbf{C}^2$ (contrary to $w(\overline{x}, \overline{y})$ or $\overline{w(x, y)}$). The exact limits of Ω may be derived easily from (3.8) but it suffices to note from this equation and (A.2) that the subset of Ω, corresponding to real values of x and y, is the half plane defined by $y > y_1$.

Assuming, for simplicity, that the incidence angle $\theta = 0$ (which implies $\alpha = 0$ and $u(x, y) = 1$, according to (3.7)), it is interesting to express $|R_n(x, y)| = |B_n \Phi_n(x, y)|$ in the form:

$$|R_n(x, y)| = |B_n \Phi_n(x_0, y_0)| \cdot \left|\frac{w_n(x_0, y_0)^n}{\Phi_n(x_0, y_0)}\right| \cdot \left|\frac{\Phi_n(x, y)}{w_n(x, y)^n}\right| \cdot \left|\frac{w_n(x, y)^n}{w_n(x_0, y_0)^n}\right| \tag{A.5}$$

Let us examine the four terms of the right-hand member of (A.5) when $n \to +\infty$:

– the first term tends to 0, according to (A.1);

– the second and third terms tend to 1, according to (3.6a);

– the last term is bounded by q^n, according to (A.2) and (A.4).

So, $|R_n|$ is bounded, for sufficiently large values of n, by the nth term q^n of a convergent geometrical series independent of x and y. The same conclusion may be drawn, for $n \to -\infty$, by replacing $w(x, y)$ by $\overline{w}(x, y)$ in (A.5). This shows that $R(x, y)$ converges absolutely and uniformly in Ω. Since each term $R_n(x, y)$ is an analytic function of the two complex variables x and y, the series $R(x, y)$ uniformly converges in Ω to an analytic function of $\mathbb{C}^2$. Of course, the restriction of $R(x, y)$ to real values of x and y is an analytic function of the two real variables x and y in the part of Ω which belongs to $\mathbb{R}^2$, i.e. the half plane $y \geqslant y_1$.

Appendix B

Let U be a function satisfying in R_+ the following conditions:

$$\nabla^2 U(x, y) + k^2 U(x, y) = 0,$$

$$\text{a radiation condition for } y \to \infty, \tag{B.1}$$

k^2 being a real positive number or eventually a complex number with a positive imaginary part. Let us obtain an integral expression of U in R_+ from the values of U and its normal derivative $\mathrm{d}U/\mathrm{d}n$ on $\mathcal{P}$.

To this end, we first define the elementary solution $\mathcal{G}(x, y)$ of the Helmholtz equation (B.1), pseudo-periodic in x and satisfying a radiation condition for $y \to \pm\infty$:

$$\nabla^2 \mathcal{G} + k^2 \mathcal{G} = \delta(y) \sum_n \exp(\mathrm{i}n\alpha d)\, \delta(x - nd), \tag{B.2}$$

where δ denotes the Dirac function. Since $\mathcal{G}$ is pseudo-periodic, it can be expressed as a series of $\exp(\mathrm{i}\alpha_n x)$:

$$\mathcal{G}(x, y) = \sum_n \mathcal{G}_n(y) \exp(\mathrm{i}\alpha_n x).$$

Introducing the above expression in (B.2), multiplying by $\exp(-\mathrm{i}\alpha x)$ and identifying the nth terms of the Fourier series of both members yield:

$$\frac{\mathrm{d}^2 \mathcal{G}_n}{\mathrm{d}y^2} + \beta_n^2 \mathcal{G}_n = \delta(y)/d. \tag{B.3}$$

Remarking that the right-hand member vanishes for $y \neq 0$, it turns out that $\mathcal{G}_\mathrm{n}$ can be expressed in the following form:

$$\begin{aligned}\mathcal{G}_n(y) &= a \exp(\mathrm{i}\beta_n y) + a' \exp(-\mathrm{i}\beta_n y) \qquad \text{if } y > 0,\\ &= \mathrm{b} \exp(i\beta_n y) + b' \exp(-\mathrm{i}\beta_n y) \qquad \text{if } y < 0.\end{aligned}$$

So, $\mathcal{G}(x, y)$ is expressed for both regions $y > 0$ and $y < 0$ in the form of plane waves, and the radiation condition allows us to eliminate the incoming waves in such a way that $a' = b = 0$. The determination of a and b' is achieved by observing that (B.3) implies the continuity of $\mathcal{G}_n$ at $y = 0$ and a jump of $1/d$ of its derivative in y, which entails $a = b' = 1/2\mathrm{i}d\beta_n$. Finally, we obtain:

$$\mathcal{G}(x, y) = \frac{1}{2\mathrm{i}d} \sum_n \frac{1}{\beta_n} \exp(\mathrm{i}\alpha_n x + \mathrm{i}\beta_n |y|). \tag{B.4}$$

This elementary solution of the Helmholtz equation is no other than the field diffracted at the point P of coordinates x and y by an infinite set of elementary currents $j_n = \exp(\mathrm{i}n\alpha d)$ in wires placed at the points $x = nd$ of the Ox axis.

To express $U(x, y)$, we consider Green's function, so called, $\mathcal{G}(x - x', y - y')$ which satisfies:

$$\begin{aligned}&\nabla^2 \mathcal{G}(x - x', y - y') + k^2 \mathcal{G}(x - x', y - y')\\ &\qquad = \delta(y - y') \sum_n \exp(\mathrm{i}n\alpha d)\, \delta(x - x' - nd),\end{aligned} \tag{B.5}$$

the differentiation being made either in x and y or in x' and y'.

Now, let us outline a classical calculation, which enables us to express U in an integral form. A first equation is obtained from (B.5) by multiplying the two members by $U(x', y')$. The second one is obtained by replacing x and y by x' and y' in (B.1), then multiplying the left-hand member by $\mathcal{G}(x - x', y - y')$. Subtracting the second equation from the first, integrating in the hatched domain of Fig. 2.2 with contour $\mathscr{C}$ (the segment SR being at $y = \infty$) and using the second Green identity yield:

$$\int_{\mathscr{C}} \mathscr{G}(x-x', y-y') \frac{\mathrm{d}U^+}{\mathrm{d}n} \mathrm{d}\ell' - \int_{\mathscr{C}} \frac{\mathrm{d}\mathscr{G}}{\mathrm{d}n'}(x-x', y-y')\, U^+ \,\mathrm{d}\ell'$$
$$\begin{aligned} &= U(x, y) \qquad \text{in } R_+, \\ &= 0 \qquad \text{in } R_-, \end{aligned} \tag{B.6}$$

$\mathrm{d}\mathscr{G}/\mathrm{d}n'$ being the normal derivative of $\mathscr{G}(x-x', y-y')$ considered as a function of x' and y', U^+ and $\mathrm{d}U^+/\mathrm{d}n$ the limit values of U and $\mathrm{d}U/\mathrm{d}n$ on $\mathscr{C}$, and ℓ' the curvilinear abscissa on $\mathscr{C}$. It is worth noting that the signs in the left-hand member of (B.6), unusual in the Green theorem, are imposed by the sense of the normal $\hat{\boldsymbol{n}}$, which is oriented toward the interior of R_+. Since the integrand of (B.6) is periodic in x', the contributions of QR and SP cancel each other. The contribution of SR vanishes, as shown by the lemma of § 2.6, by setting $u = U(x', y')$, $v = \mathscr{G}(x-x', y-y')$, and remarking that u only contains outgoing waves whereas v is incoming. So,

$$\begin{aligned} U(x, y) = &\int_{\mathscr{P}} \mathscr{G}(x-x', y-f(x')) \frac{\mathrm{d}U^+}{\mathrm{d}n}(x')\, \mathrm{d}s' \\ &- \int_{\mathscr{P}} \frac{\mathrm{d}\mathscr{G}}{\mathrm{d}n'}(x-x', y-f(x'))\, U^+(x')\, \mathrm{d}s', \end{aligned} \tag{B.7a}$$

the right-hand member vanishing in R_-.

An equation similar to (B.7a) can be written for a function $U(x, y)$ satisfying a Helmholtz equation in R_- and a radiation condition for $y \to -\infty$:

$$\begin{aligned} U(x, y) = &\int_{\mathscr{P}} \frac{\mathrm{d}\mathscr{G}}{\mathrm{d}n'}(x-x', y-f(x'))\, U^-(x')\, \mathrm{d}s' \\ &- \int_{\mathscr{P}} \mathscr{G}(x-x', y-f(x')) \frac{\mathrm{d}U^-}{\mathrm{d}n}(x')\, \mathrm{d}s', \end{aligned} \tag{B.7b}$$

the right-hand member vanishing in R_+.
This time, the sign of the right-hand member is the usual one since the normal is oriented toward the exterior of R_-, to keep $\hat{\boldsymbol{n}}$ unchanged on $\mathscr{P}$.

Finally, let us generalize (B.7a) and (B.7b) by considering a function U satisfying:

$$\nabla^2 U + k^2 U = 0 \qquad \text{in } R_+ \text{ and } R_-, \tag{B.8}$$

$$\text{a radiation condition for } y \to \pm\infty. \tag{B.9}$$

Such a function can be considered as the sum of two functions U_1 and U_2, U_1 being identical to U in R_+ and null in R_-, U_2 being null in R_+ and equal to U in R_-. Using (B.7a) and (B.7b) to express U_1 and U_2 in all the space, then adding the two expressions to derive $U = U_1 + U_2$ and finally replacing $\mathrm{d}s'$ by $(1 + f(x')^2)^{1/2}\,\mathrm{d}x'$ yield:

$$U(x, y) = \int_0^d \mathscr{A}(x, y, x')\, \eta(x')\, \mathrm{d}x' + \int_0^d \mathscr{B}(x, y, x')\, \tau(x')\, \mathrm{d}x' \qquad \text{(B.10)}$$

with:

$$\mathscr{A}(x, y, x') = \frac{1}{2\mathrm{i}d} \sum_n \frac{1}{\beta_n} \exp[\mathrm{i}\alpha_n(x - x') + \mathrm{i}\beta_n |y - f(x')|], \qquad \text{(B.11)}$$

$$\mathscr{B}(x, y, x') = \frac{1}{2d} \sum_n \left[\mathrm{sign}(y - f(x')) - \frac{\alpha_n}{\beta_n} f'(x')\right]$$
$$\times \exp[\mathrm{i}\alpha_n(x - x') + \mathrm{i}\beta_n |y - f(x')|], \qquad \text{(B.12)}$$

$$\eta(x') = (1 + f'(x')^2)^{1/2}\, [\mathrm{d}U^+/\mathrm{d}n(x') - \mathrm{d}U^-/\mathrm{d}n(x')], \qquad \text{(B.13)}$$

$$\tau(x') = U^+(x') - U^-(x'). \qquad \text{(B.14)}$$

When $y > y_\mathrm{M}$ or $y < y_\mathrm{m}$, the sign of $y - f(x')$ is equal to $+1$ or -1 and remains constant, so, the symbol $|\quad|$ in (B.11) and (B.12) may be replaced by $\pm(\quad)$, the terms in x and y may be removed from the integrand of (B.10) and we derive:

$$U(x, y) = \sum_n D_n^+ \exp(\mathrm{i}\alpha_n x + \mathrm{i}\beta_n y) \qquad \text{if } y > y_\mathrm{M}, \qquad \text{(B.15)}$$

$$U(x, y) = \sum_n D_n^- \exp(\mathrm{i}\alpha_n x - \mathrm{i}\beta_n y) \qquad \text{if } y < y_\mathrm{m}, \qquad \text{(B.16)}$$

with
$$D_n^\pm = [s_n^\pm, \eta] + [t_n^\pm, \tau], \qquad \text{(B.17)}$$

$$s_n^\pm = \frac{1}{2\mathrm{i}d\beta_n} \exp(-\mathrm{i}\alpha_n x \mp \mathrm{i}\beta_n f(x)), \qquad \text{(B.18)}$$

$$t_n^{\pm} = \frac{1}{2d}\left(\pm 1 - \frac{\alpha_n}{\beta_n} f'(x)\right) \exp(-\mathrm{i}\alpha_n x \mp \mathrm{i}\beta_n f(x)), \tag{B.19}$$

$$[u, v] = \langle \bar{u}, v \rangle = \int_0^d u(x)\, v(x)\, \mathrm{d}x.$$

So, (B.10)–(B.14) enable us to express U satisfying (B.8) and (B.9) at any point of space, once the limit values of U and $\mathrm{d}U/\mathrm{d}n$ are known on both sides of $\mathscr{P}$. These equations are generally considered as the fundamental equations of the integral theory. Nevertheless, we believe that it is very important to supply an answer to a second problem, which will cast light on the integral approach. Let us imagine that the functions $\eta(x')$ and $\tau(x')$ of (B.10) are not derived from (B.13) and (B.14), i.e. from an actual function U, but are pseudo-periodic functions arbitrarily chosen. Under these circumstances, what is the function $U(x, y)$ obtained after integration of (B.10)? It is not difficult to show that such a function satisfies (B.8) and (B.9) and thus the unknown limit values of U and $\mathrm{d}U/\mathrm{d}n$ on both sides of $\mathscr{P}$ satisfy (B.13) and (B.14). Unfortunately, these two equations cannot provide the four limit values U^+, U^-, $\mathrm{d}U^+/\mathrm{d}n$, $\mathrm{d}U^-/\mathrm{d}n$, and two supplementary equations must be found for this purpose. To this end, it can be noticed that, for given y, the right-hand member of (B.10) is the product of a continuous function and a Fourier series of x. Hence, the value $U(x, f(x))$ of the right-hand member of (B.10) on $\mathscr{P}$ is the mean value of the two limits on both sides:

$$U^+(x) + U^-(x) = 2U(x, f(x)). \tag{B.20}$$

Eliminating U^+ or U^- from (B.14) and (B.20) and replacing $U(x, f(x))$ by its expression yield:

$$U^{\pm}(x) = \pm\frac{\tau}{2} + \int_0^d G(x, x')\, \eta(x')\, \mathrm{d}x' + \int_0^d N'(x, x')\, \tau(x')\, \mathrm{d}x', \tag{B.21}$$

with

$$G(x, x') = \mathscr{A}(x, f(x), x'), \tag{B.22}$$

$$N'(x, x') = \mathscr{B}(x, f(x), x'). \tag{B.23}$$

For simplicity, (B.21) will be written in a symbolic operator form:

$$U^{\pm} = \pm\frac{\tau}{2} + G\{\eta\} + N'\{\tau\}. \tag{B.24}$$

The same considerations about discontinuities of Fourier series, applied to grad U, allow us to write, in the particular case where $\tau(x) = 0$:

$$[1 + f'(x')^2]^{1/2}\frac{\mathrm{d}U^{\pm}}{\mathrm{d}n} = \pm\frac{\eta}{2} + N\{\eta\}, \tag{B.25}$$

with

$$N(x, x') = \frac{1}{2d}\sum_n [\operatorname{sign}(f(x) - f(x')) - \frac{\alpha_n}{\beta_n} f'(x)] \times \exp[\mathrm{i}\alpha_n(x - x') + \mathrm{i}\beta_n|f(x) - f(x')|]. \tag{B.26}$$

Equations (B.21) and (B.25), which express $U^{\pm}$ and $\mathrm{d}U^{\pm}/\mathrm{d}n$ as functions of η and τ, appear as being reciprocal of (B.13) and (B.14). They can be considered as the two fundamental equations of the integral formalism.

References

ABRAMOWITZ, M. and A. STEGUN, 1970, Handbook of Mathematical functions (Dover Publ., New York).

AKAIKE, H., 1973, SIAM J. Appl. Math. **24,** 234.

ANDREWARTHA, J. R., G. H. DERRICK and R. C. MC PHEDRAN, 1981a, Opt. Acta **28,** 1177.

ANDREWARTHA, J. R., G. H. DERRICK and R. C. MC PHEDRAN, 1981b, Opt. Acta **28,** 1501.

ANDREWARTHA, J. R., J. R. FOX and I. J. WILSON, 1979a, Opt. Acta **26,** 69.

ANDREWARTHA, J. R., J. R. FOX and I. J. WILSON, 1979b, Opt. Acta **26,** 197.

BATES, R. H., 1975, IEEE Trans. Microwave Theory & Tech. **23,** 605.

BECKMANN, P., 1967, in: Progress in Optics, ed. E. Wolf (North-Holland, Amsterdam) vol. VI, p. 53.

BECKMANN, P. and A. SPIZZICHINO, 1963, The Scattering of Electromagnetic Waves from Rough Surfaces (Pergamon, Macmillan, London, New York).

BORN, M. and E. WOLF, 1965, Principles of Optics (Pergamon Press, Oxford) p. 401.

BOTTEN, L. C., 1978a, Opt. Acta **25,** 481.

BOTTEN, L. C., 1978b, Ph. D. Thesis (Univ. of Tasmania, Hobart).

BOTTEN, L. C., 1980, J. Opt. **11,** 161.

BOTTEN, L. C., M.S. CRAIG, R. C. MC PHEDRAN, J. L. ADAMS and J. R. ANDREWARTHA, 1981a, Opt. Acta **28,** 413.

BOTTEN, L. C., M.S. CRAIG, R. C. MC PHEDRAN, J. L. ADAMS and J. R. ANDREWARTHA, 1981b, Opt. Acta **28,** 1087.

BOUSQUET, P., 1963a, C. R. Acad. Sci. Paris **256,** 3422.

BOUSQUET, P., 1963b, C. R. Acad. Sci. Paris **257,** 80.

BREIDNE, M. and D. MAYSTRE, 1982, J. Opt. Soc. Amer. **72,** 499.

CADILHAC, M., 1980, in: Electromagnetic Theory of Gratings, ed. R. Petit (Springer, Berlin) p. 53.
CERUTTI-MAORI, G., R. PETIT and M. CADILHAC, 1969, C. R. Acad. Sci. Paris **B8,** 1060.
CHANDEZON, J., M. T. DUPUIS, G. CORNET and D. MAYSTRE, 1982, J. Opt. Soc. Amer. **72,** 839.
CHANDEZON, J., D. MAYSTRE and G. RAOULT, 1980, J. Opt. **11,** 235.
CHANG, K. C., V. SHAH and T. TAMIR, 1980, J. Opt. Soc. Amer. **70,** 804.
COURANT, R. and D. HILBERT, 1962, Methods of Mathematical Physics, vol. 2 (Interscience, New York).
DELEUIL, R., 1963, C. R. Acad. Sci. Paris **258,** 506.
DENNERY, P. and A. KRZYWICKI, 1967, Mathematics for Physicists (Harper & Row, New York, Evanston, London, and John Weatherhill, Tokyo) p. 87.
DERIUGIN, L. N., 1952, Dokl. Akad. Nauk. SSSR **87**(6), 913.
DE SANTO, J. A., 1975, J. Acoust. Soc. Amer. **57,** 1195.
DE SANTO, J. A., 1981, Radio Sci. **16,** 1315.
DUMERY, G. and P. FILIPPI, 1970, C. R. Acad. Sci. Paris **270,** 137.
FACQ, P., 1976, Ann. Telecommun. **31,** 99.
FACQ, P., 1977, Thesis (Univ. of Limoges, France).
FANO, U., 1941, J. Opt. Soc. Amer. **31,** 213.
FOX, J. R., 1980, Opt. Acta **27,** 289.
FRAUNHOFER, J., 1821, Denkschr. Akad. Wiss. München **8,** 1.
GARCIA, N. and N. CABRERA, 1978, Phys. Rev. **B18,** 576.
GREEN, R., 1970, IEEE Trans. Microwave Theory & Tech. **18,** 313.
HESSEL, A., J. SCHMOYS and D. Y. TSENG, 1975, J. Opt. Soc. Amer. **65,** 380.
HILL, N. R. and V. CELLI, 1978, Phys. Rev. **B17,** 2478.
HUGONIN, J. P., R. PETIT and M. CADILHAC, 1981, J. Opt. Soc. Amer. **71,** 593.
HUTLEY, M. C. and V. M. BIRD, 1973, Opt. Acta **20,** 771.
IKUNO, H. and K. YASUURA, 1973, IEEE Trans. Antennas & Propag. **2,** 657.
ITOH, T. and R. MITTRA, 1969, IEEE Trans. Microwave Theory & Tech. **17,** 319.
IYANAGA, S. and Y. KAWADA, 1980, Encyclopedic Dictionary of Mathematics (MIT Press, Cambridge, MA) p. 344E.
JANOT, C. and A. HADNI, 1963, J. Phys. **24,** 1073.
JIRACEK, G. R., 1973, IEEE Trans. Antennas & Propag. **21,** 393.
JORDAN, A. K. and R. H. LANG, 1979, Radio Sci. **14,** 1077.
JOVICEVIC, S. and S. SESNIC, 1972, J. Opt. Soc. Amer. **62,** 865.
KALHOR, H., 1976, IEEE Trans. Antennas & Propag. **24,** 884.
KALHOR, H. and A. NEUREUTHER, 1971, J. Opt. Soc. Amer. **61,** 43.
KNOP, K., 1978, J. Opt. Soc. Amer. **68,** 1206.
LABEYRIE, A. and J. FLAMAND, 1969, Opt. Commun. **1,** 9.
LANG, R. H. and A. K. JORDAN, 1980, NRL Memorandum Rept. 4188, March 17, Naval Research Laboratory, Washington, DC 20375.
LIPPMANN, B. A., 1953, J. Opt. Soc. Amer. **43,** 408.
LOEWEN, E. G., D. MAYSTRE, R. C. MC PHEDRAN and I. J. WILSON, 1974, Jpn. J. Appl. Phys., suppl. **14-1,** 143.
MADDEN, R.P. and J. STRONG, 1958, Concepts of Classical Optics (Freeman & Co, San Francisco).
MARCUSE, D., 1976, Bell Syst. Tech. J. **55,** 1295.
MARECHAL, A. and G. W. STROKE, 1959, C. R. Acad. Sci. **249,** 2042.
MAUE, A. W., 1949, Z. Phys. **126,** 601.
MAYSTRE, D., 1972, Opt. Commun. **6,** 50.
MAYSTRE, D., 1973, Opt. Commun. **8,** 216.

MAYSTRE, D., 1974, Thesis (Univ. of Aix-Marseille 3, France).
MAYSTRE, D., 1978a, J. Opt. Soc. Amer. **68,** 490.
MAYSTRE, D., 1978b, Opt. Commun. **26,** 127.
MAYSTRE, D., 1980a, in: Electromagnetic Theory of Gratings, ed. R. Petit (Springer, Berlin) p. 63.
MAYSTRE, D., 1980b, Proc. Int. U.R.S.I. Symp. 1980 (Munich, August 26–29) p. 144 B/1.
MAYSTRE, D. and R. C. MC PHEDRAN, 1974, Opt. Commun. **12,** 164.
MAYSTRE, D., M. NEVIERE and R. PETIT, 1980, in: Electromagnetic Theory of Gratings, ed. R. Petit (Springer, Berlin) p. 159.
MAYSTRE, D. and R. PETIT, 1970a, C. R. Acad. Sci. Paris **271,** 400.
MAYSTRE, D. and R. PETIT, 1970b, Opt. Commun. **2,** 309.
MAYSTRE, D. and R. PETIT, 1972a, Opt. Commun. **5,** 35.
MAYSTRE, D. and R. PETIT, 1972b, Opt. Commun. **5,** 90.
MAYSTRE, D. and R. PETIT, 1974, J. Spectr. Soc. Jpn, **23** suppl., 61.
MC CLELLAN, R. C. and G. W. STROKE, 1966, J. Math. Phys. **45,** 383.
MC PHEDRAN, R. C., 1973, Ph. D. Dissertation (Univ. of Tasmania, Hobart).
MC PHEDRAN, R. C., G. H. DERRICK and L. C. BOTTEN, 1980, in: Electromagnetic Theory of Gratings, ed. R. Petit (Springer, Berlin) p. 227.
MEECHAM, W. C., 1956, J. Appl. Phys. **27,** 361.
MEIXNER, J., 1972, IEEE Trans. Antennas & Propag. **20,** 442.
MILLAR, R. F., 1969, Proc. Camb. Phil. Soc. **65,** 773.
MILLAR, R. F., 1971, Proc. Camb. Phil. Soc. **69,** 217.
MILLAR, R. F., 1973, Radio Sci. **8,** 785.
MIRANDA, C., 1970, Partial Differential Equations (Springer, Berlin).
MOAVENI, M. K., H. A. KALHOR and A. AFRASHTEH, 1975, Comput. & Electr. Eng. **2,** 265.
MORSE, P. M. and H. FESHBACH, 1953, Methods of Theoretical Physics, Part I (Mc Graw-Hill, New York) p. 804.
NAMIOKA, T., T. HARADA and K. YASUURA, 1979, Opt. Acta **26,** 1009.
NEUREUTHER, A. and K. ZAKI, 1969, Alta Freq. **38,** 282.
NEVIERE, M., 1975, Thesis (Univ. of Aix Marseille 3, France).
NEVIERE, M. and M. CADILHAC, 1970, Opt. Commun. **2,** 235.
NEVIERE, M., M. CADILHAC and R. PETIT, 1973, IEEE Trans. Antennas & Propag. **21,** 37.
NEVIERE, M., G. CERUTTI-MAORI and M. CADILHAC, 1971, Opt. Commun. **3,** 48.
NEVIERE, M. and P. VINCENT, 1976, Opt. Acta **23,** 557.
NEVIERE, M., P. VINCENT and R. PETIT, 1974, Nouv. Rev. Opt. **5,** 65.
PAVAGEAU, J. and J. BOUSQUET, 1970, Opt. Acta **17,** 469.
PAVAGEAU, J., R. EIDO and H. KOBEISSE, 1967, C. R. Acad. Sci. Paris **264,** 424.
PETIT, R., 1963, Rev. Opt. **42,** 263.
PETIT, R., 1965, C. R. Acad. Sci. Paris **260,** 4454.
PETIT, R., 1966a, Rev. Opt. **45,** 249.
PETIT, R., 1966b, Rev. Opt. **8,** 353.
PETIT, R., 1967, Opt. Acta **14,** 301.
PETIT, R., 1980, in: Electromagnetic Theory of Gratings, ed. R. Petit (Springer, Berlin) p. 1.
PETIT, R. and M. CADILHAC, 1964, C. R. Acad. Sci. Paris **259,** 2077.
PETIT, R. and M. CADILHAC, 1966, C. R. Acad. Sci. AB **262,** 468.
PETIT, R. and D. MAYSTRE, 1972, Rev. Phys. Appl. **7,** 427.
PETIT, R., D. MAYSTRE and M. NEVIERE, 1972, Communication to the Ninth International Congress of Optics, Santa-Monica, CA (see Space Opt., 1974, Proc. of the ninth ICO, p. 667).
RAYLEIGH, Lord, 1907, Proc. Roy. Soc. **A79,** 399.
RAYLEIGH, Lord, 1945, The Theory of Sound, vol. 2 (Dover, New York) p. 89.
RITTENHOUSE, D., 1786, Trans. Amer. Phil. Soc. **2,** 201.

ROGER, A. and M. BREIDNE, 1980, Opt. Commun. **35,** 299.
ROGER, A. and D. MAYSTRE, 1981, J. Opt. Soc. Amer. **70,** 1483.
ROUMIGUIERES, J. L., D. MAYSTRE and R. PETIT, 1974, Proc. Fifth Colloquium on Microwave Communications, Budapest, ET 305.
ROUMIGUIERES, J. L., D. MAYSTRE and R. PETIT, 1975, Opt. Commun. **7,** 402.
ROWLAND, H. A., 1882, Phil. Mag. **13,** 469.
RUDOLPH, D. and G. SCHMAHL, 1967, Umschau Wiss. Tech. **67,** 225.
SCHWARTZ, L., 1966a, Mathematics for the Physical Sciences (Addison-Wesley, Reading, MA).
SCHWARTZ, L., 1966b, Théorie des Distributions (Hermann, Paris).
SOMMERFELD, A., 1949, Partial Differential Equations (Academic, New York).
STROKE, G. W., 1960, Rev. Opt. **39,** 350.
URETSKY, J. L., 1965, Ann. Phys. **33,** 400.
VAN BLADEL, J., 1964, Electromagnetic Fields (Mc Graw-Hill, New York).
VAN DEN BERG, P. M., 1971a, Appl. Sci. Res. **24,** 261.
VAN DEN BERG, P. M., 1971b, Thesis (Delft, The Netherlands).
VAN DEN BERG, P. M., 1980, Proc. Int. URSI Symp. 1980 (Munich, August 26–29) p. 211 B/1.
VAN DEN BERG, P. M., 1981, J. Opt. Soc. Amer. **71,** 1224.
VAN DEN BERG, P. M. and J. T. FOKKEMA, 1979, J. Opt. Soc. Amer. **69,** 27.
VEKUA, I. N., 1953, Dokl. Akad. Nauk. SSSR **90,** 715.
VINCENT, P., 1980a, SPIE Proc., 240.
VINCENT, P., 1980b, in: Electromagnetic Theory of Gratings, ed. R. Petit (Springer, Berlin) p. 101.
VINCENT, P., M. NEVIERE and D. MAYSTRE, 1978, Nucl. Instrum. and Meth. **152,** 123.
WATERMAN, P. C., 1975, J. Acoust. Soc. Amer. **57,** 791.
WHITMAN, G., D. M. LESKIV and F. SCHWERING, 1980, J. Opt. Soc. Amer. **70,** 1495.
WHITMAN, G. and F. SCHWERING, 1977, IEEE Trans. Antennas & Propag. **25,** 869.
WILCOX, C. H., 1975, Scattering Theory for the d'Alembert Equation in Exterior Domains (Springer, Berlin).
WIRGIN, A., 1964, Rev. Opt. **9,** 449.
WIRGIN, A., 1967, Thesis (Univ. of Paris, France).
WIRGIN, A., 1968, Rev. Cethedec **5,** 131.
WIRGIN, A., 1969, Alta Frequenza, Special Issue from URSI Symposium **38,** 327.
WIRGIN, A., 1978, Opt. Commun., **27,** 189.
WIRGIN, A., 1979a, C. R. Acad. Sci. Paris, **A289,** 259.
WIRGIN, A., 1979b, C. R. Acad. Sci. Paris, **B289,** 273.
WIRGIN, A., 1979c, C. R. Acad. Sci. Paris, **B288,** 179.
WIRGIN, A., 1980a, J. Acoust. Soc. Amer. **68,** 692.
WIRGIN, A., 1980b, Opt. Acta **27,** 1671.
WIRGIN, A., 1981, Opt. Acta **28,** 1377.
WIRGIN, A. and R. DELEUIL, 1969, J. Opt. Soc. Amer. **59,** 1348.
YAKOVLEV, E. A., 1965, Opt. Spektrosk. **19,** 417.
YASUURA, K. and H. IKUNO, 1971, Proc. Int. Symp. on Antennas and Propagation, Sendai, Japan, 173.

E. WOLF, PROGRESS IN OPTICS XXI
© ELSEVIER SCIENCE PUBLISHERS B.V. 1984

II

THEORY OF OPTICAL BISTABILITY

BY

LUIGI A. LUGIATO

Istituto di Fisica dell' Università,
Milan, Italy

CONTENTS

This paper is dedicated to the memory of Mario Gronchi, beloved friend and respected collaborator whose outstanding contributions to the theory of optical bistability are well known in the field. An absurd destiny has prevented us from enjoying further his humanity and intelligence.

§ 1. Introduction

Bistable operation in optical systems has been theoretically predicted and sometimes experimentally observed in several situations, for instance with two-mode lasers (LAMB [1965]), Zeeman lasers (HEER and GRAFT [1965], SARGENT, LAMB and FORK [1967]), lasers with a saturable absorber (KAZANTSEV, RAUTIAN and SURDUTOVICH [1968], LUGIATO, MANDEL, DEMBINSKI and KOSSAKOWSKI [1978], RUSHIN and BAUER [1979]), parametric oscillators (WOO and LANDAUER [1971], LUGOVOI [1979], DRUMMOND, MCNEIL and WALLS [1980a]), dye lasers (BACZYNSKI, KOSSAKOWSKI and MARSZALEK [1976], SCHAEFER and WILLIS [1976], HAAG, MUNZ and MAROWSKY [1981]), reflection by a nonlinear interface (KAPLAN [1977], SMITH, TOMLINSON, MALONEY and HERMANN [1981]), and bidirectional ring cavity (L. MANDEL, ROY and SINGH [1981]). However, the name "optical bistability" has been given specifically to the bistable behavior which arises in the class of systems described below.

1.1. WHAT IS OPTICAL BISTABILITY

Let us consider a cw laser beam injected into an optical cavity (for instance a Fabry–Perot or a ring cavity), tuned or nearly tuned to the incident light. In general, the incident field is partially transmitted, partially reflected and partially absorbed. When the cavity is empty, the transmitted power P_T is proportional to the incident power P_I, and the proportionality constant depends on the cavity detuning and on the finesse of the cavity. The interesting case is when the cavity is filled with absorbing material resonant or nearly resonant with the incident field. In this case, P_T becomes a nonlinear function of P_I. The behavior of the system is determined by the ratio of the absorption parameter αL and the mirror transmissivity T, where α is the unsaturated absorption coefficient per unit length on resonance and L the length of the sample. Increasing $\alpha L/T$, one finds that the steady state curve of transmitted versus incident power develops first a portion with differential gain $\mathrm{d}P_T/\mathrm{d}P_I$ larger than unity. In this condition, if one slowly modulates the incident intensity, the modulation is transferred to the transmitted field via the nonlinear steady state relation $P_T = P_T(P_I)$ and turns out to be *amplified*. Thus, the system works as an *optical transistor*. If one further increases the ratio $\alpha L/T$ the steady state curve $P_T = P_T(P_I)$ becomes S-shaped. The segment with negative slope is unstable; hence, there is a definite range of values of P_I in correspondence with which the system is *bistable*. If we slowly sweep the incident power from zero to a value beyond the bistable region, and then sweep it back, we obtain a hysteresis cycle with a low and a high transmission branch. The bistable behavior arises from the interplay of the nonlinearity of the atom-field interaction with the feedback action of the mirrors. The threshold value of $\alpha L/T$ for which one yields bistability, depends on several parameters such as the cavity mistuning, the atomic detuning, the inhomogeneous linewidth, the type of cavity and so on. When the incident field is in perfect resonance with the atomic line, dispersion does not play any role, so that one has *purely absorptive bistability*. Otherwise, one has the general mixed absorptive and dispersive case. When the atomic detuning is so large that absorption becomes negligible one has *purely dispersive bistability*.

The systems which show this behavior are usually called *all-optical (or intrinsic) bistable systems*. We also consider the so-called *hybrid electro-optical systems*, which have been devised in many variants. A typical device of this type is obtained by replacing the absorber by an electro-optic crystal, which is monitored by the output field and produces changes in the refraction index proportional to the output power.

Already from this brief description it is evident that these systems (both all-optical and electro-optical) have great potential as devices, because they can work as optical transistors, memory elements, or pulse shapers, which eliminate the noisy parts of the imput light (clippers, discriminators and limiters). Also, we shall see later, they can work as converters of cw light into pulsed light. Therefore, there is presently a big effort towards the construction of practical, miniaturized and fast operating optical devices of this kind.

Optical bistability has also aroused a wide theoretical interest, renewing in part the enthusiasms that in the sixties were devoted to the laser. In fact, optical bistability is a remarkable example of cooperative behavior in an open system far from thermal equilibrium, and therefore is naturally a chapter of Haken's *Synergetics* (HAKEN [1977]) and of Prigogine's *theory of dissipative structures* (NICOLIS and PRIGOGINE [1977]). First of all, it is immediately evident that optical bistability is an example of *non-equilibrium steady state behavior*, analogous to first-order phase transitions in equilibrium systems. Furthermore, as we shall see later, by controlling the external parameters one can induce either the emergence of *spontaneous pulsations* in the system (self-pulsing behavior) or the appearance of *chaotic* (turbulent) behavior. Clearly, the theoretical interest of hybrid systems is reduced with respect to all-optical systems.

1.2. A HISTORICAL SKETCH

Let us give a brief description of the history of optical bistability, concerning the intrinsic bistability in all-optical systems and the theoretical aspects. Absorptive optical bistability was theoretically predicted by SZÖKE, DANEU, GOLDHAR and KURNIT [1969]. The same problem was considered also by SEIDEL [1971], AUSTIN and DESHAZER [1971] and SPILLER [1972]. Some years later MCCALL [1974] proved that under suitable conditions the same system can show differential gain with transistor action, and also treated absorptive optical bistability in a Fabry–Perot cavity by numerical analysis of the Maxwell–Bloch equations. This work led to the experiments of GIBBS, MCCALL and VENKATESAN (1976) in sodium, in which both transistor operation and bistability were observed. The analysis of the data showed that the observed bistability was of the dispersive type, with few exceptions. The mechanism which produces dispersive optical bistability was explained with the help of a simple phenomenological cubic model.

These results stimulated theoretical and experimental activity; in particular, three papers that were crucial for the following developments are those of FELBER and MARBURGER [1976], BONIFACIO and LUGIATO [1976] and SMITH and TURNER [1977]. Successively, other experiments on optical bistability have been performed by VENKATESAN and MCCALL [1977], BISCHOFBERGER and SHEN [1978, 1979], GRISCHKOWSKI [1978], MCCALL and GIBBS [1978], GARMIRE, MARBURGER, ALLEN and WINFUL [1979], GIBBS, MCCALL, VENKATESAN, GOSSARD, PASSNER and WIEGMANN [1979], MILLER, SMITH and JOHNSTON [1979], MILLER and SMITH [1979], MILLER, SMITH and SEATON [1981], GRYNBERG, GIACOBINO, DEVAUD and BIRABEN [1980], SANDLE and GALLAGHER [1981], ARIMONDO, GOZZINI, LOVITCH and PISTELLI [1981], WEYER, WIEDENMANN, RATEIKE, MCGILLIVRAY, MEYSTRE and WALTHER [1981], ARECCHI, GIUSFREDI, PETRIELLA and SALIERI [1982], GRANT and KIMBLE [1982] and MEIER, HOLZNER, DERIGHETTI and BRUN [1982].

The first *exact analytical* theory of optical bistability, fully including propagation effects and saturation, was given by BONIFACIO and LUGIATO [1978a], in the case of a unidirectional ring cavity in the purely absorptive situation. In particular, this paper showed the crucial importance of the double limit $\alpha L \to 0$, $T \to 0$, with $\alpha L/T$ constant (mean-field limit) to obtain a simplified theory. In fact, in this limit one recovers exactly the previously formulated "mean field theory" (BONIFACIO and LUGIATO [1976, 1977], which produced quite a number of predictions also concerning transient behavior and quantum statistical effects. The exact analytical solution of BONIFACIO and LUGIATO [1978a] was extended to the mixed absorptive–dispersive case independently by IKEDA [1979], BONIFACIO, LUGIATO and GRONCHI [1979] and ROY and ZUBAIRY [1980a]. The generalization to the case of the bidirectional ring cavity has been given in AGRAWAL [1981], ASQUINI and CASAGRANDE [1981] and KAPLAN and MEYSTRE [1982].

The mean field model for a ring cavity has been generalized, both at semiclassical and quantum statistical level, to the case of mixed absorptive and dispersive bistability by BONIFACIO and LUGIATO [1978c]. The general bistability conditions have been worked out in HASSAN, DRUMMOND and WALLS [1978], BONIFACIO, GRONCHI and LUGIATO [1979a] and AGRAWAL and CARMICHAEL [1979]. Further analyses of dispersive bistability are given in SCHWENDIMANN [1979], BOWDEN and SUNG [1979], WILLIS and DAY [1979], TEWARI [1979], MILLER [1981], GRAHAM and SCHENZLE [1981] and BOWDEN [1981].

At the semiclassical level, the mean field model has been used to investigate various aspects of the transient behavior, in particular with regard to switching characteristics (BONIFACIO and LUGIATO [1978d], BONIFACIO and MEYSTRE [1978, 1979], MEYSTRE and HOPF [1979], BENZA and LUGIATO [1979a], HOPF, MEYSTRE, DRUMMOND and WALLS [1979], LUGIATO, MILANI and MEYSTRE [1982]).

At the quantum statistical level, the mean field model was used to describe the spectrum of transmitted and fluorescent light (AGARWAL, NARDUCCI, FENG and GILMORE [1977], AGARWAL, NARDUCCI, GILMORE and FENG [1978a,b, 1979], CARMICHAEL and WALLS [1977], BONIFACIO and LUGIATO [1978d,e], LUGIATO [1979], CASAGRANDE and LUGIATO [1980]) and the photon statistics of the transmitted light in the good cavity case. In BONIFACIO, GRONCHI and LUGIATO [1978] the bimodal character of the distribution function in the bistability region was discussed describing the behavior of the mean value and the fluctuations of the transmitted light. This behavior completes the analogy between optical bistability and first-order phase transitions, showing, on the other hand, the nonthermodynamic character of the transition, which stems from the fact that the diffusion coefficient of the Fokker–Planck equation is intensity-dependent. These results have been further developed in WILLIS [1978, 1981], SCHENZLE and BRAND [1978, 1979], GRAGG, SCHIEVE and BULSARA [1978, 1979], ARECCHI and POLITI [1979], DRUMMOND and WALLS [1980, 1981), ZARDECKI [1980], HANGGI, BULSARA and JANDA [1980], LUGIATO, FARINA and NARDUCCI [1980], LUGIATO, CASAGRANDE and PIZZUTO [1982], BONIFACIO, LUGIATO, FARINA and NARDUCCI [1981] and ENGLUND, SCHIEVE, ZUREK and GRAGG [1981].

The stability analysis of the exact stationary solution in a ring cavity showed that, under suitable conditions, a part of the high transmission branch becomes unstable (BONIFACIO and LUGIATO [1978b], LUGIATO [1980a]). In this situation, the system works as a converter of cw into pulsed light (BONIFACIO, GRONCHI and LUGIATO [1979b], GRONCHI, BENZA, LUGIATO, MEYSTRE and SARGENT [1981]). For proper values of the parameters the sequence of pulses is chaotic (IKEDA [1979], IKEDA, DAIDO and AKIMOTO [1980]). An analytical treatment of self-pulsing has been given in BENZA and LUGIATO [1979b, 1981, 1982], BENZA, LUGIATO and MEYSTRE [1980] and LUGIATO, BENZA, NARDUCCI and FARINA [1981, 1982]. Self-pulsing and chaotic behavior have been observed in hybrid systems (MCCALL [1978] and GIBBS, HOPF, KAPLAN and SHOEMAKER [1981], respectively).

The analysis of BONIFACIO and LUGIATO [1978a] has also been generalized to the case of a Fabry–Perot cavity. MEYSTRE [1978] first analyzed the mean field limit in this framework and the deviations from the mean field theory that one finds when T is not small enough. To treat this problem he used some equations obtained in BONIFACIO and LUGIATO [1978d] by suitably truncating the infinite hierarchy of equations derived by FLECK [1968]. The analysis of MEYSTRE [1978] has been extended by ABRAHAM, BULLOUGH and HASSAN [1979], ABRAHAM, HASSAN and BULLOUGH [1980], ABRAHAM and HASSAN [1980] and by ROY and ZUBAIRY [1980b]. In particular, ABRAHAM, BULLOUGH and HASSAN [1979] analyzed the generation of hysteresis cycles by pulses. On the other hand, CARMICHAEL [1980] and HERMANN [1980] solved both numerically and analytically the steady state Maxwell–Bloch equations derived by MCCALL [1974]. These equations include the standing wave effects more completely than the truncated hierarchy equations, producing corrections on the order of 15%. This analysis was extended to the mixed absorptive–dispersive case in CARMICHAEL and HERMANN [1980], who have shown that in the mean field limit one obtains for a Fabry–Perot a state equation which differs from that valid for a ring cavity (again quantitatively the difference is on the order of 15%). The state equation for a Fabry–Perot coincides, apart from a change of sign to convert the absorber into an amplifier, with that previously obtained by SPENCER and LAMB [1972] for a laser with injected signal. MCCALL and GIBBS [1980] derived from this state equation the general bistability conditions for a Fabry–Perot.

The effects of the radial shape of the electric field injected into the cavity have been analyzed in MARBURGER and FELBER [1978], BALLAGH, COOPER, HAMILTON, SANDLE and WARRINGTON [1981], ARIMONDO, GOZZINI, LOVITCH and PISTELLI [1981], DRUMMOND [1981], ROSANOV and SEMENOV [1981], FIRTH and WRIGHT [1982] and MOLONEY, BELIC and GIBBS [1982].

Considerable attention has been devoted also to two-photon bistability or multistability (ARECCHI and POLITI [1978], AGRAWAL and FLYTZANIS [1980], HERMANN and THOMPSON [1980, 1981]) which has been also experimentally observed (GRYNBERG, GIACOBINO, DEVAUD and BIRABEN [1980]).

Recently, the possibility of bistable or multistable operation with three-level atoms (WALLS, ZOLLER and STEYN-ROSS [1981]), degenerate two-level atoms (KITANO, YABUZAKI and OGAWA [1981], HAMILTON, BALLAGH and SANDLE [1982]), or optical pumping (ARECCHI, GIUSFREDI, PETRIELLA and SALIERI [1982], GOZZINI [1982]) has been pointed out.

Another problem to which a remarkable theoretical interest is presently devoted is the description of bistability in semiconductors. This interest is stimulated by the recent progress in constructing miniaturized all-optical bistable devices using GaAs (GIBBS, MCCALL, VENKATESAN, GOSSARD, PASSNER and WIEGMANN [1979]), InSb (MILLER, SMITH and JOHNSTON [1979]) or tellurium (STAUPENDAHL and SCHINDLER [1980]). The first approaches have been made by the experimentalists themselves in order to obtain an overall picture of the phenomenon (GIBBS, MCCALL, VENKATESAN, GOSSARD, PASSNER and WIEGMANN [1979], MILLER, SMITH and SEATON [1981], MILLER [1981]). These authors essentially adapt the two-level description to the situation in semiconductors. First-principle approaches, including the details of the dynamics in semiconductors, have been developed recently (GOLL and HAKEN [1980], STEYN-ROSS and GARDINER [1983]).

1.3. AIM OF THE ARTICLE

Let us now outline the scope of this chapter. In recent years, a huge amount of literature on optical bistability has accumulated. A large number of variants has been considered; see for instance the recent works of AGRAWAL and FLYTZANIS [1981], LUGOVOI [1981], BJORKHOLM, SMITH and TOMLINSON

[1981] and SARID [1981]. The aim of this article is not to give a review of all the papers on optical bistability, but to select a few topics which illustrate in the simplest way the main physical principles of optical bistability and related phenomena. The treatment is theoretical, but we systematically discuss many of the points which are relevant for the experiments. Furthermore, the analysis is restricted to intrinsic all-optical systems.

For all the points which are not discussed in this paper, we advise the reader to consult one of the following general references: the Proceedings of the Asheville Conference, edited by BOWDEN, CIFTAN and ROBL [1981], the special issue of the IEEE Journal of Quantum Electronics edited by SMITH [1981], the book on Dissipative Systems in Quantum Optics, edited by BONIFACIO [1982], and the recent review by ABRAHAM and SMITH [1982b]. In particular, the last article complements the present paper. Other, shorter reviews can be found in GIBBS, MCCALL and VENKATESAN [1979], GIBBS, MCCALL and VENKATESAN [1980], COLLINS and WASMUNDT [1980] and ABRAHAM and SMITH [1982a].

Our selection is dictated by the general criterium of maximum simplicity. Thus, we restrict ourselves to the case of unidirectional ring cavity, plane wave approximation, two-level atoms, homogeneous broadening, and one-photon transitions. Both the absorptive and the dispersive cases will be treated. The starting point of our analysis is the set of coupled Maxwell–Bloch equations, obtained in the slowly varying envelope and rotating wave approximations. The connections of this treatment with the mode description of optical bistability are studied in detail. At a quantum statistical level, we base ourselves on a suitable many-mode master equation, which allows a deep analysis of the fluctuations in the system. Starting from a comprehensive description of the steady state behavior, the treatment will be developed with particular emphasis on the cooperative effects, on the nonstationary behavior (transient effects and self-pulsing) and on the quantum effects.

Section 2 is devoted to the semiclassical treatment, and § 3 to the quantum statistical theory. In particular, § 2.1 concerns the steady state behavior, §§ 2.2 and 2.3 the transient effects, and §§ 2.4 and 2.5 the self pulsing behavior.In the quantum statistical part, §§ 3.1 and 3.2 discuss the basic equations, §§ 3.3 and 3.4 analyze the spectrum of transmitted and fluorescent light, while § 3.5 describes the photon statistics of the transmitted light. Section 3.6 studies the transient behavior in its quantum statistical aspects, and finally, § 3.7 gives a brief discussion on the observability of the quantum effects.

§ 2. Semiclassical treatment

In order to describe theoretically the phenomenon of optical bistability, it is easier to consider a unidirectional *ring cavity* (Fig. 1) than a Fabry–Perot, because in this cavity one has to deal with propagation only in one direction, thus avoiding standing wave difficulties. For simplicity we assume that mirrors 3 and 4 have 100% reflectivity. We call R and T (with $R + T = 1$) the reflection and transmission coefficients of mirrors 1 and 2. We describe the dynamics of the coupled system atoms plus radiation field by the well known, one-sided Maxwell–Bloch equations, which incorporate the *plane wave approximation.*

In the case of a *homogeneously broadened system* of N *two level atoms*, the Maxwell–Bloch equations read (Arecchi and Bonifacio [1965], Haken [1970], Sargent, Scully and Lamb [1974], Allen and Eberly [1975])

$$\frac{\partial E}{\partial t} + c\frac{\partial E}{\partial z} = -gP, \tag{1a}$$

$$\frac{\partial P}{\partial t} = \frac{\mu}{\hbar} ED - [\gamma_\perp + i(\omega_a - \omega_0)]P, \tag{1b}$$

$$\frac{\partial D}{\partial t} = -\frac{\mu}{2\hbar}(EP^* + E^*P) - \gamma_\parallel (D - N/2). \tag{1c}$$

E is the slowly varying envelope of the electric field, P is the macroscopic atomic polarization and D is one half the difference between the populations of the lower and of the upper level. The equations for E^* and P^* are the complex conjugates of eqs. (1a) and (1b) respectively. μ is the modulus of the dipole

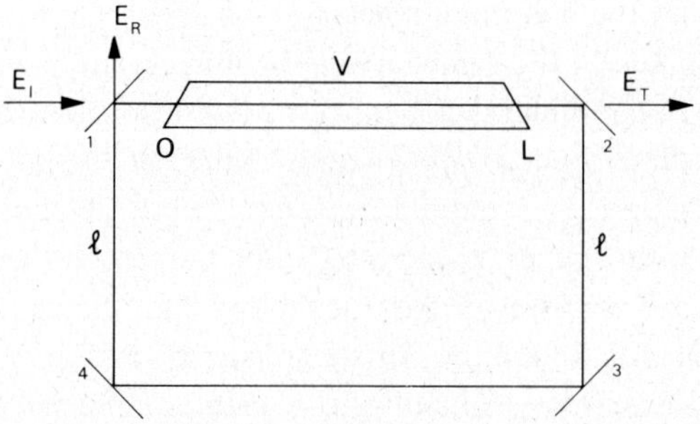

Fig. 1. Ring Cavity. E_I, E_T and E_R are the incident, transmitted and reflected fields respectively.

moment of the atoms and g is a coupling constant given by

$$g = \frac{4\pi\omega_0\mu}{V}, \tag{2}$$

where ω_0 is the frequency of the incident field, and V the volume of the atomic sample. $\gamma_\parallel$ and $\gamma_\perp$ are the inverse of the atomic relaxation times T_1 and T_2 respectively, and ω_a is the transition frequency of the atoms.

The coherent cw field E_I enters into the cavity from the left (Fig. 1) and drives the atoms. E_I is taken as real and positive for definiteness. The cavity imposes two relations between E_I, the transmitted field amplitude E_T, and the fields $E(0, t)$ and $E(L, t)$ (Bonifacio and Lugiato [1978a])

$$E_T(t) = \sqrt{T}\, E(L, t), \tag{3a}$$

$$E(0, t) = \sqrt{T}\, E_I + R \exp(-i\delta_0)\, E(L, t - \Delta t), \tag{3b}$$

where L is the length of the atomic sample and $\Delta t = (2l + L)/c$ is the time the light takes to travel from mirror 2 to mirror 1. δ_0 is the cavity detuning:

$$\delta_0 = \frac{\omega_c - \omega_0}{c/\mathscr{L}}, \tag{3c}$$

where ω_c is the frequency of the cavity that is nearest to resonance with the incident field, and $\mathscr{L} = 2(L + l)$ is the total length of the cavity. In particular, eq. (3b) is a boundary condition characteristic of the ring cavity. The second contribution on the right hand side describes a feedback mechanism due to the mirrors, which is essential to give rise to bistability.

2.1. STEADY STATE BEHAVIOR

Let us first consider the steady state ($\partial E/\partial t = \partial P/\partial t = \partial D/\partial t = 0$). The stationary equation for the field in general has the form

$$\frac{dE}{dz} = -\chi(|E|^2)E, \tag{4}$$

where

$$\chi = \chi_a + i\chi_d \tag{5}$$

is the complex dielectric susceptibility, with an absorptive (dispersive) component $\chi_a(\chi_d)$. In the particular case of eqs. (1), χ has the expression

$$\chi = \alpha(1 - i\Delta)[1 + \Delta^2 + |E|^2/I_s]^{-1}, \tag{6}$$

where α is the unsaturated absorption coefficient on resonance,

$$\alpha = \frac{\mu g N}{2\hbar c \gamma_\perp}, \tag{7a}$$

Δ is the atomic detuning parameter

$$\Delta = (\omega_a - \omega_0)/\gamma_\perp \tag{7b}$$

and I_s is the saturation intensity

$$I_s = \frac{\hbar^2 \gamma_\perp \gamma_\parallel}{\mu^2}. \tag{7c}$$

However, eq. (4) is more general than eqs. (1); for instance it holds for inhomogeneously broadened systems or for Kerr media, with suitable expressions of the susceptibility χ.

2.1.1. *The case of perfect resonance*

Let us consider first the case in which the incident field, the atoms and the cavity are perfectly in resonance; i.e. $\omega_0 = \omega_a = \omega_c$. This is a particular case of purely absorptive bistability. From eqs. (4) and (6) we have

$$\frac{dF}{dz} = -\alpha \frac{F}{1 + F^2}, \tag{8}$$

where F is the normalized adimensional electric field

$$F = \frac{E}{\sqrt{I_s}} = \frac{\mu E}{\hbar(\gamma_\perp \gamma_\parallel)^{1/2}} \tag{9}$$

and the field has been taken as real. Together with F, it is suitable to consider the normalized incident and transmitted amplitudes

$$y = \frac{\mu E_I}{\hbar(\gamma_\perp \gamma_\parallel T)^{1/2}} = \frac{E_I}{(I_s T)^{1/2}}, \qquad x = \frac{\mu E_T}{\hbar(\gamma_\perp \gamma_\parallel T)^{1/2}} = \frac{E_T}{(I_s T)^{1/2}}, \tag{10}$$

so that eqs. (3) become, at steady state,

$$x = F(L), \tag{11a}$$

$$F(0) = Ty + Rx. \tag{11b}$$

Equation (8) can be solved immediately and gives

$$\ln \frac{F(0)}{x} + \tfrac{1}{2}[F^2(0) - x^2] = \alpha L. \tag{12}$$

By combining eqs. (11b) and (12) we finally obtain (BONIFACIO and LUGIATO [1978a])

$$\ln\left[1 + T\left(\frac{y}{x} - 1\right)\right] + \frac{x^2}{2}\left\{\left[1 + T\left(\frac{y}{x} - 1\right)\right]^2 - 1\right\} = \alpha L. \tag{13}$$

This equation gives an exact relation between the transmitted field x and the incident field y. It depends on the two parameters αL and T. The structure of eq. (13) can be intuitively understood by a graphical representation of eqs. (11b) and (12) as shown in Fig. 2. The steady state values of x are the intersections

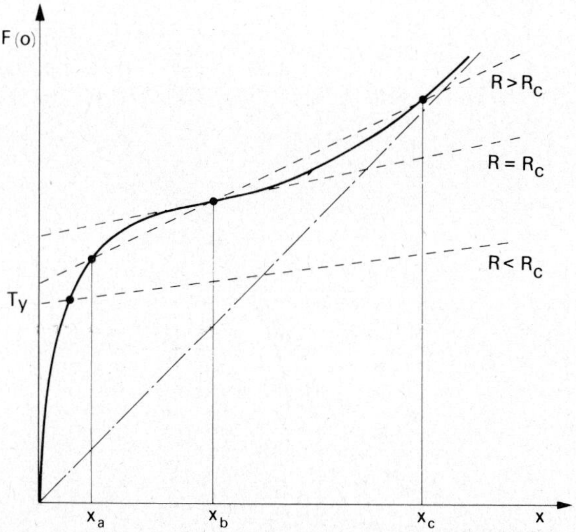

Fig. 2. Qualitative graph of the normalized field $F(0)$ at $z = 0$ as a function of the field $F(L) \equiv x$ at $z = L$ (transfer function of the atomic medium at steady state). For $R = 0$ one has $F(0) = y$. Quantities x and y are proportional to the transmitted and incident fields E_T and E_I respectively (see eq. (10)). For a generic R, the function $x = x(y)$ is obtained by intersecting the curve with the straight line $F(0) = RF(L) + Ty$.

of the straight line (11b) with the curve (12). The first one is the boundary condition of the cavity. The second is the transfer function of the medium, which expresses the field at $z = 0$ as a function of the field at $z = L$ (and vice versa). It has neither maxima nor minima (see Fig. 2), but it has an inflection point. The angular coefficient R_c of the tangent at the inflection point is such that $0 < R_c < 1$. R_c depends only on αL. For $R < R_c$ there is only one intersection point for all values of y. For $R > R_c$ there is a range of values of y in correspondence with which one finds three intersection points $x_a < x_b < x_c$. Points x_b are unstable, hence, this is a bistable situation. If we plot the steady state solutions x as a function of the incident field y, we obtain an S-shaped curve (Fig. 3) which leads to a hysteresis cycle.

From this analysis we see that bistability arises from the combined action of the *nonlinear* transfer of the medium (eq. (12)) and of the feedback from the mirrors (eq. (11b)). This feedback action is essential, because as one sees from Figs. 2 and 3 there is no bistability for $R = 0$.

If we add a linear loss term (representing, for example, diffraction) $-\zeta E$ to the right hand side of eq. (1a), it is easy to verify that it reduces the bistable region. This effect was first pointed out by SPILLER [1972].

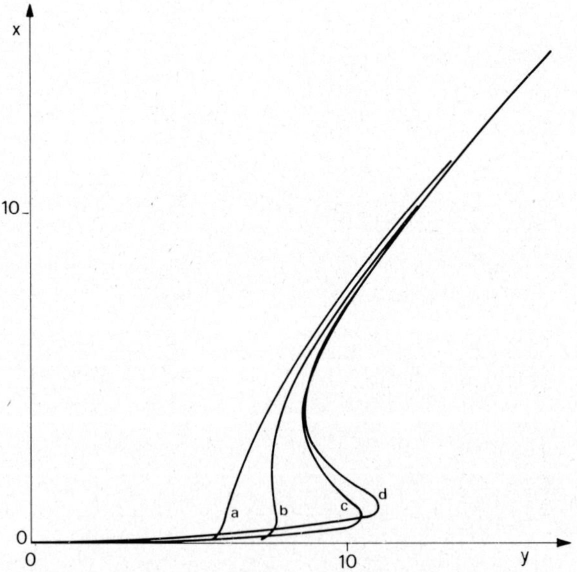

Fig. 3. Plot of transmitted light versus incident light at steady state for $C = \alpha L/2T$ fixed equal to 10, and different values of αL and T. For $\alpha L \to 0$ one approaches the behavior predicted by the mean field theory (see eq. (32)): (a) $\alpha L = 20$, $T = 1$; (b) $\alpha L = 10$, $T = 0.5$; (c) $\alpha L = 2$, $T = 0.1$; (d) mean field, $C = 10$.

2.1.2. *The general case*

Let us now consider the general case, in which E is complex at steady state. By writing

$$E(z) = \rho(z) \exp[\mathrm{i}\varphi(z)], \tag{14}$$

we obtain from eq. (4)

$$\frac{\mathrm{d}\rho}{\mathrm{d}z} = -\chi_{\mathrm{a}}(\rho^2)\rho, \tag{15a}$$

$$\frac{\mathrm{d}\varphi}{\mathrm{d}z} = -\chi_{\mathrm{d}}(\rho^2). \tag{15b}$$

On the other hand, indicating by $P_{\mathrm{I}} = E_{\mathrm{I}}^2$ and $P_{\mathrm{T}} = |E_{\mathrm{T}}|^2$ the incident and transmitted intensities respectively, using eqs. (3a) and (3b) we obtain, for the transmissivity of the cavity,

$$\mathscr{T} = \frac{P_{\mathrm{T}}}{P_{\mathrm{I}}} = \frac{T^2}{(\eta - R)^2 + 4R\eta \sin^2\{\frac{1}{2}[\varphi(L) - \varphi(0) - \delta_0]\}}, \tag{16}$$

with

$$\eta = \frac{\rho(0)}{\rho(L)} \quad \Rightarrow \quad \eta \geqslant 1. \tag{17}$$

Let us now consider a few cases separately.

a) Empty cavity. In this case $\chi_{\mathrm{a}} = \chi_{\mathrm{b}} = 0$, so that $\eta = 1$, and $\varphi(L) = \varphi(0)$. Hence, eq. (16) reduces to the usual expression of the transmissivity as a function of the cavity detuning (Born and Wolf [1970])

$$\mathscr{T} = \frac{1}{1 + 4R \sin^2 \frac{\delta_0}{2} \Big/ T^2}. \tag{18}$$

b) Kerr medium. (Felber and Marburger [1976], Marburger and Felber [1978].) In this case one has

$$\chi_{\mathrm{a}} = 0, \qquad \chi_{\mathrm{d}} = c_1 + c_2\rho^2, \tag{19}$$

where c_1 and c_2 are constants. Hence, $\eta = 1$ and

$$\varphi(L) - \varphi(0) = -L(c_1 + c_2 P_{\mathrm{T}}/T)$$

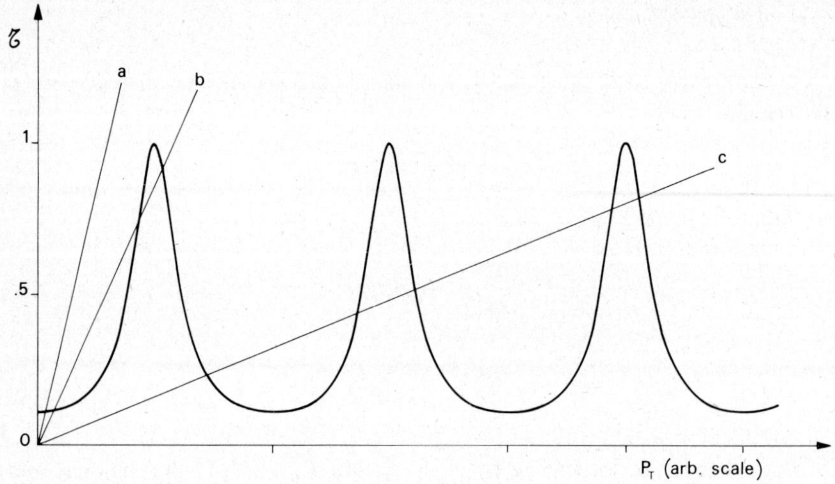

Fig. 4. The steady state transmission $\mathcal{T}$ of the cavity, filled with a Kerr medium, is graphed as a function of the transmitted power P_T. **a**, **b** and **c**, show the straight line $\mathcal{T} = P_T/P_I$ for different values of the incident power P_I. The intersections with the curve correspond to the (stable or unstable) stationary states of the system for a fixed value of P_I.

so that eq. (16) becomes

$$\mathcal{T} = \frac{1}{1 + 4R\sin^2\{\frac{1}{2}(\delta + \beta P_T)\}/T^2}, \tag{20}$$

with $\delta = \delta_0 + Lc_1$, $\beta = Lc_2/T$. Therefore the plot of $\mathcal{T}$ versus P_T has the same shape as the usual graph of transmissivity as a function of detuning (FELBER and MARBURGER [1976]) (Fig. 4). Using the procedure of FELBER and MARBURGER, the steady state solutions can be found by intersecting the line (20) with the straight line $\mathcal{T} = P_T/P_I$ in the plane $(\mathcal{T}, P_T)$. The value of P_I controls the angular coefficient of the straight line. For small P_I one has only one intersection (line a). By increasing P_I, the intersections become three (line b), with the middle one being unstable. Thus, we obtain bistability and the plot of P_T versus P_I shows a hysteresis cycle. This bistability is of a purely dispersive type. For larger values of P_I one obtains multiple solutions (line *c*), which leads to *multistability* and multiple hysteresis cycles. When T approaches unity, the curve (20) flattens and bistability disappears as usual.

c) *Two-level, homogeneously broadened atomic system in the general absorptive plus dispersive case* (BONIFACIO, LUGIATO and GRONCHI [1979]). Let us introduce the normalized incident and transmitted intensities

$$Y = \left(\frac{\mu E_{\mathrm{I}}}{\hbar}\right)^2 \frac{1}{\gamma_\perp \gamma_\parallel T} = y^2 = \frac{P_{\mathrm{I}}}{I_{\mathrm{s}} T},$$

$$X = \left(\frac{\mu |E_{\mathrm{T}}|}{\hbar}\right)^2 \frac{1}{\gamma_\perp \gamma_\parallel T} = |x|^2 = |F(L)|^2 = \frac{P_{\mathrm{T}}}{I_{\mathrm{s}} T}, \tag{21}$$

From eqs. (15) with eqs. (3a), (5), (6) and (9) we obtain

$$X = \frac{2}{\eta^2 - 1} [\alpha L - (1 + \Delta^2) \ln \eta], \tag{22}$$

$$\varphi(L) - \varphi(0) = \Delta \ln \eta. \tag{23}$$

By solving eq. (22) with respect to η, we obtain the function $\eta = \eta(X)$. Hence, by inserting eq. (22) into eq. (16) we find the expression of the transmissivity as a function of the normalized transmitted intensity X:

$$\mathscr{T} = \frac{X}{Y} = \frac{T^2}{[\eta(X) - R]^2 + 4R\,\eta(X) \sin^2\{\frac{1}{2}[\Delta \ln \eta(X) - \delta_0]\}} \tag{24}$$

Therefore in this case the shape of the function $\mathscr{T}(X)$ is governed by the dependence of η on X. In particular, for large X η approaches unity, so that the transmissivity becomes constant and equal to the empty cavity value (18). In general, the curve $\mathscr{T}(X)$ is quite different from the Kerr medium curve (20). However, when dispersion is dominant, under suitable conditions one or a few resonances survive (see Fig. 5).

The possibility of multistability (Fig. 6) depends on the number of oscillations that the function $\sin\{\frac{1}{2}[\Delta \ln \eta(X) - \delta_0]\}$ undergoes (IKEDA [1979]). As one sees from eq. (22), η is a monotonically decreasing function of X which varies from unity to $\exp[\alpha L/(1 + \Delta^2)]$. Hence, the quantity $\Delta \ln \eta - \delta_0$ varies from $-\delta_0$ to $\alpha L \Delta/(1 + \Delta^2) - \delta_0$. Therefore the number of oscillations of the sine function is determined by the parameter $\alpha L \Delta/(1 + \Delta^2)$.

Equation (24) can be rephrased as follows:

$$Y = X(\eta) \frac{1}{T^2} \{(\eta - R)^2 + 4R\,\eta \sin^2[\tfrac{1}{2}(\Delta \ln \eta - \delta_0)]\}. \tag{25}$$

Equations (22) and (25) together give a parametric representation $X = X(\eta)$, $Y = Y(\eta)$ of the function $X(Y)$ of transmitted versus incident intensity.

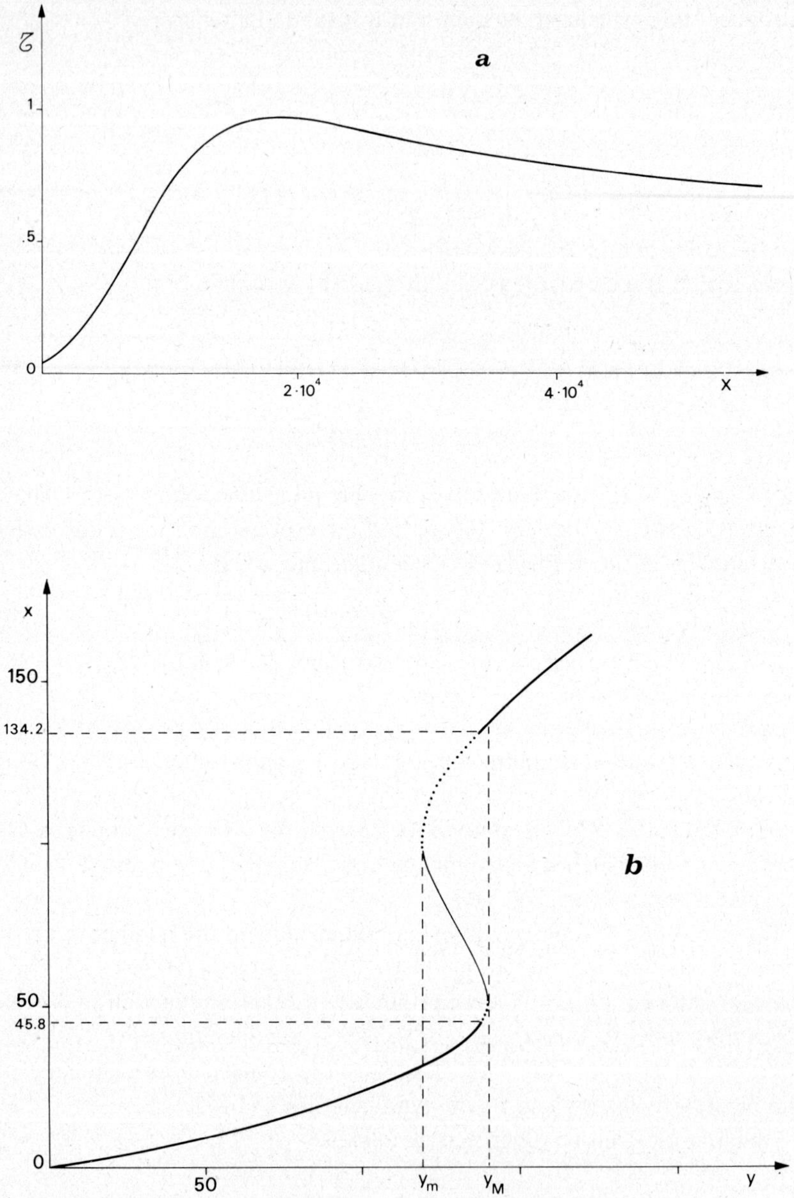

Fig. 5. (a) Same as Fig. 4, but for a cavity filled with a two-level medium; X is the normalized transmitted intensity. $\alpha L = 3.6$, $T = 0.1$ (hence $C = 180$), $\Delta = 60$, $\theta = \delta_0/T = 1$. (b) Plot of transmitted field $x(=\sqrt{X})$ versus incident field y for the same values of the parameters of Fig. 5a. The part of the curve with negative slope is unstable. The approach to the states in the solid positive slope part of the curve is oscillatory, and that in the dotted part is monotonic.

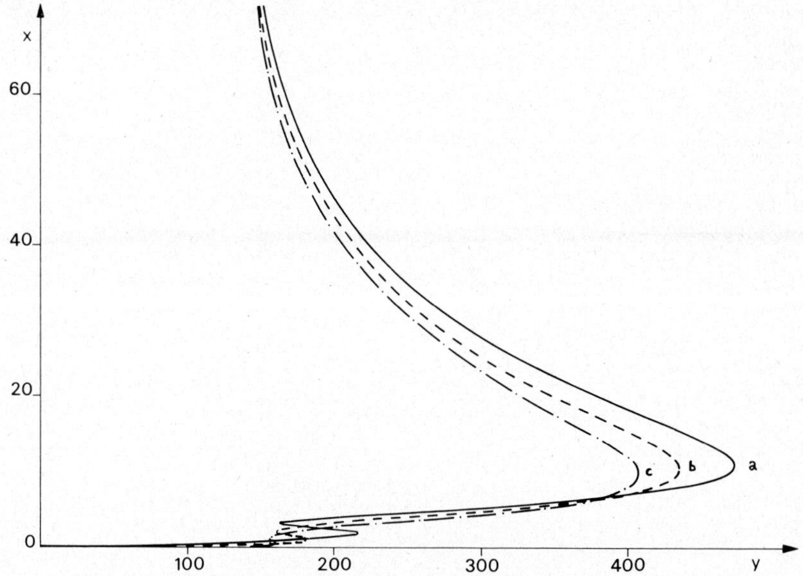

Fig. 6. Optical multistability. $x = \sqrt{X}$ is the normalized transmitted field. In all curves $C = 900$, $\Delta = 5$, $\theta = 0.05$. Curve **a** corresponds to the case of homogeneous broadening ($T_2^* = \infty$). In **b**, $\gamma_\perp T_2^* = 1$; in **c**, $\gamma_\perp T_2^* = 0.5$. Clearly, multistability disappears with increased inhomogeneous broadening.

Examples of this curve are given in Fig. 7. The generalization of eqs. (22) and (25) to the case of Lorentzian inhomogeneous broadening has been given in GRONCHI and LUGIATO [1980]; for the Fabry–Perot case see CARMICHAEL and AGRAWAL [1980].

2.1.3. *The stationary solution in the mean field limit*

From now on, we consider exclusively the case of a two-level atomic system. In general, the susceptibility has the structure

$$\chi(|E|^2) = \alpha\tilde{\chi}(|F|^2), \tag{26}$$

where α and F are defined in eqs. (7a) and (9) respectively. Therefore the steady state solution becomes particularly simple in the case $\alpha L \ll 1$, because, as one sees from eqs. (15), the field becomes practically uniform in space. More precisely, we shall perform the multiple limit (BONIFACIO and LUGIATO [1978a,c])

$$\alpha L \to 0, \qquad T \to 0, \qquad \delta_0 \to 0$$

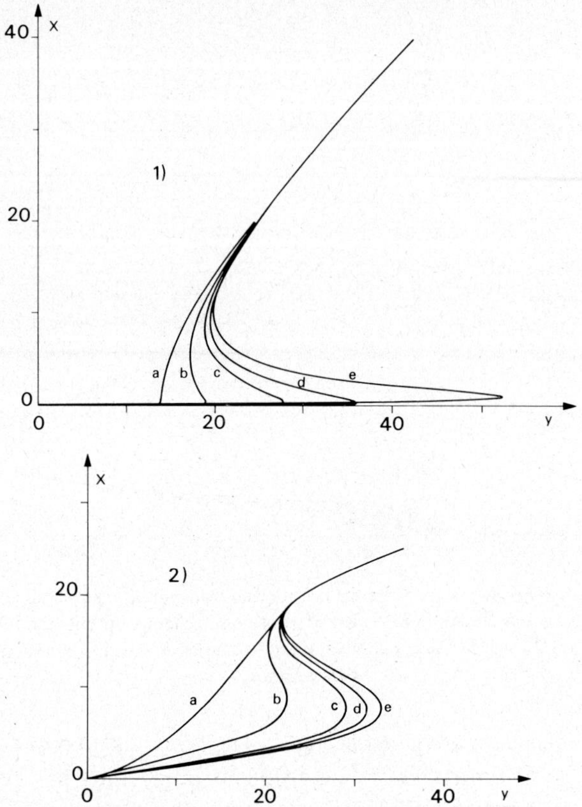

Fig. 7. Plot of the transmitted amplitude $x = \sqrt{X}$ as a function of the incident amplitude y in the homogeneously broadened case. In both Figs. 7.1 and 7.2 curves **a**, **b**, **c** and **d** show the exact stationary solution (eqs. (22) and (25)); curve **e** is the mean field result (eq. (31)). In Fig. 7.1 $C = 50$, $\Delta = \theta = 0$; in Fig. 7.2 $C = 50$, $\Delta = 10$, $\theta = 2.25$. Curves **a**: $\alpha L = 100$, $T = 1$; curves **b**: $\alpha L = 50$, $T = 0.5$; curves **c**: $\alpha L = 20$, $T = 0.2$; curves **d**: $\alpha L = 10$, $T = 0.1$.

with
$$C \equiv \frac{\alpha L}{2T} \equiv \text{constant}$$
$$\theta \equiv \frac{\delta_0}{T} = \frac{\omega_c - \omega_0}{cT/\mathscr{L}} \equiv \text{constant}. \tag{27}$$

It is easy to derive the steady state solution in the limit (27). In fact, from eqs. (15a,b) and (17), (27) we have to first order in αL

$$\eta = 1 + \alpha L \tilde{\chi}_a(|F(L)|^2 = 1 + \alpha L \tilde{\chi}_a(X),$$
$$\varphi(L) - \varphi(0) = \alpha L \tilde{\chi}_d(X). \tag{28}$$

By inserting eq. (28) into eq. (16) we obtain, in the limit (27),

$$\mathscr{T} = \{[1 + 2C\tilde{\chi}_a(X)]^2 + [\theta - 2C\tilde{\chi}_d(X)]^2\}^{-1} \tag{29}$$

and because $\mathscr{T} = Y/X$ we have

$$Y = X\{[1 + 2C\tilde{\chi}_a(X)]^2 + [\theta - 2C\tilde{\chi}_d(X)]^2\}. \tag{30}$$

Equation (30) coincides with the state equation of optical bistability in a ring cavity, derived from the mean field model of optical bistability (see § 2.2.1). For this reason, the limit (27) is called "mean field limit".

The explicit expression of eq. (30), in the case of Lorentzian inhomogeneous broadening, is given in BONIFACIO and LUGIATO [1978c]. For a homogeneously broadened system, from eq. (6) we have (BONIFACIO and LUGIATO [1978c], HASSAN, DRUMMOND and WALLS [1978])

$$Y = X\left\{\left(1 + \frac{2C}{1 + \Delta^2 + X}\right)^2 + \left(\theta - \frac{2C\Delta}{1 + \Delta^2 + X}\right)^2\right\}. \tag{31}$$

In the particular case $\Delta = \theta = 0$, eq. (31) can be written in terms of amplitudes instead of intensities as follows (BONIFACIO and LUGIATO [1976]):

$$y = x + \frac{2Cx}{1 + x^2}. \tag{32}$$

Equations (31) and (32) can be also recovered directly from the exact solutions (22) + (25) and (13), respectively, by performing the limit (27).

Relations of the type (31) between incident and transmitted intensity were given in SZÖKE, DANEU, GOLDHAR and KURNIT [1969] and GIBBS, MCCALL and VENKATESAN [1976] on the basis of phenomenological arguments. Our approach derives this formula from first principles as an analytical solution of the Maxwell–Bloch equations with boundary condition, pointing out its limit of validity, which is $\alpha L \ll 1$, $T \ll 1$, $\delta_0 \ll 1$.

Let us now briefly comment on the physical meaning of the limit (27). First, $\alpha L \to 0$ (i.e. $\alpha \to 0$) is the weak coupling limit in the interaction between the electric field and the atoms. However, if we only let $\alpha L \to 0$ but keep T finite, C vanishes and therefore we obtain the empty cavity solution $Y = X(1 + \theta^2)$. On the contrary, if we also let $T \to 0$, the parameter C is arbitrary and we obtain the nonlinear terms in eq. (30), which produce all the interesting phenomena. The physical meaning of the limit $T \to 0$ is that the mean lifetime $\mathscr{L}/cT$ of the photons in the cavity becomes infinite, so that the photons can experience the interaction with the atoms even when this becomes vanishingly small. Finally, the limit

$$\delta_0 = \frac{\omega_c - \omega_0}{c/\mathscr{L}} \to 0, \quad \theta = \frac{\omega_c - \omega_0}{cT/\mathscr{L}} \quad \text{finite}$$

means that the cavity detuning must be smaller than the free spectral range, but of the same order of magnitude as the cavity linewidth k, given by

$$k = cT/\mathscr{L}. \tag{33}$$

This limit implies that the system operates only with the cavity mode resonant with the incident field.

Note that, despite the limit $\alpha L \to 0$, eqs. (30) and (31) do not give a weak coupling theory. In fact, the limit (27) is peculiar because the internal field E becomes infinite, in such a way that the normalized variables F, X and Y remain finite in the limit. Since X is proportional to μ^2, one sees that the coupling constant appears in eq. (31) at all orders.

Figure 7 shows how the curve (31) is approached in the limit (27) (see also Fig. 3). The parts of the curves with negative slope are unstable, as we shall see later, so that one finds a hysteresis cycle. Curve (e) in Fig. 7.1 is obtained from eq. (31) for $C = 50$, $\Delta = \theta = 0$ (purely absorptive case); curve (e) in Fig. 7.2 comes from eq. (31) for $C = 50$, $\Delta = 10$, and $\theta = 2.25$ (dispersive case). In both Figs. 7.1 and 7.2 the curves a, b, c and d show the exact solution (22) and (25) for different values of αL and of the transmissivity, chosen in such a way that $C = \alpha L/2T$ is a constant, equal to 50. For large values of αL and T, as in curve (a), there is no bistability whereas the bistable behavior increases by decreasing αL and T. In this way one approaches the mean field result (31) which is already a good approximation for $\alpha L \approx 1$. With C and T fixed the mean field curve is a better approximation in the dispersive case (Fig. 7.2) than in the absorptive one (Fig. 7.1). This is due to the fact that absorption is reduced in the dispersive case so that the variation of the field in space is not strong even when αL is large. This can be understood on the basis of the fact that in the dispersive case the mean field condition $\alpha L \ll 1$ can be replaced by $\bar{\alpha} L \ll 1$, where $\bar{\alpha}$ is the unsaturated absorption coefficient off resonance $\bar{\alpha} = \alpha/(1 + \Delta^2)$.

In the following two subsections we shall analyze the mean field state equation (31), which expresses the incident intensity as a function of the transmitted intensity. It depends on three parameters, the cooperativity parameter C, the atomic detuning Δ and the cavity mistuning θ. In comparing eq. (31) with experimental data, the definition (27) of C must be changed into

$$C = \alpha L \mathscr{F} / 2\pi, \tag{34}$$

where $\mathscr{F}$ is the effective finesse of the cavity.

A general property of eq. (31) is that, contrary to eqs. (22) and (25), it can never produce multistability, but at most can lead to bistability.

2.1.4. *Bistability conditions in the resonant case* (*mean field limit*) (BONIFACIO and LUGIATO [1976])

The field internal to the cavity is in general quite different from the incident field, because there is a reaction field, cooperatively produced by the atoms, which counteracts the incident one.

In the purely absorptive, resonant case $\Delta = \theta = 0$, the steady state behavior is described by eq. (32). The nonlinear term $2Cx/(1 + x^2)$ arises from the reaction field and hence, from atomic cooperation, which is "measured" by the parameter C. For very large x eq. (32) reduces to the empty cavity solution $x = y$ (i.e. $E_T = E_I$). The atomic system is saturated, so that the medium is bleached. In this situation each atom interacts with the incident field as if the other atoms were not there: this is the noncooperative situation, and in fact the quantum statistical treatment shows that atom–atom correlations are negligible. On the other hand, for small X eq. (32) reduces to $y = (2C + 1)x$. Here the linearity arises simply from the fact that, for a small external field, the response of the system is linear. In this situation the atomic system is unsaturated; for large C the atomic cooperation is dominant and one has strong atom–atom correlations (LUGIATO [1979].

The curves $y(x)$, obtained by varying C, are analogous to the Van der Waals curves for the liquid–vapor phase transition, with y, x and C playing the role of pressure, volume and temperature respectively. For $C < 4$, y is a monotonic function of x, so that one has no bistability (Fig. 8). However, in part of the curve the differential gain dx/dy is larger than unity, so that in this situation one has the possibility of transistor operation. In fact, if the incident intensity is adiabatically modulated around a value of P_I, such that $dP_T/dP_I = (x/y)dx/dy > 1$, the modulation is amplified in the transmitted light (MCCALL [1974]).

For $C = 4$ (critical curve) the graph has an inflection point with horizontal tangent. Finally for $C > 4$ the curve develops a maximum and a minimum, which for $C \gg 1$ correspond to $(x_M \approx 1,\ y_M \approx C)$ and $(x_m \approx \sqrt{2C},$

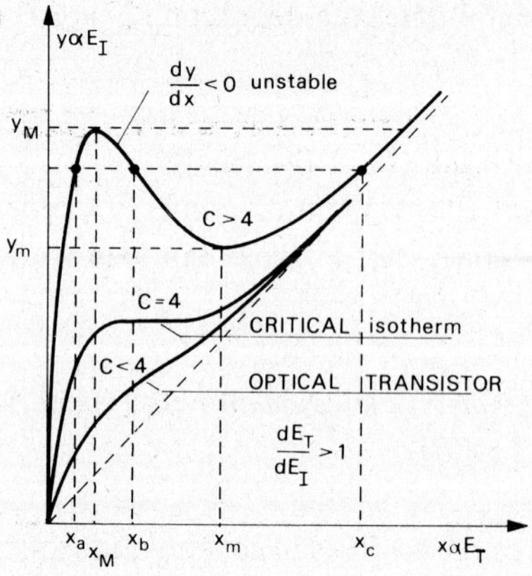

Fig. 8. Plot of the mean field state equation (32) for purely absorptive bistability with $\theta = 0$, for different values of the bistability parameter C.

$y_m \approx \sqrt{8C}$). Hence, for $y_m < y < y_M$ one finds three stationary solutions $x_a < x_b < x_c$. As we shall show in § 2.4.2, solutions x_b on the part of the curve with negative slope are unstable. Therefore we have a bistable situation, and by exchanging the axes x and y we immediately obtain the hysteresis cycle of transmitted versus incident light. Since atomic cooperation is dominant in the states x_a, and negligible in the states x_c, we shall call x_a "cooperative stationary state" and x_c "one-atom stationary state".

2.1.5. *Bistability conditions in the general case* (*mean field limit*)

Let us now consider eq. (31) for general values of Δ and θ (HASSAN, DRUMMOND and WALLS [1978], BONIFACIO, GRONCHI and LUGIATO [1979a], AGRAWAL and CARMICHAEL [1979]). We assume that $\Delta\theta > 0$, because for $\Delta\theta < 0$ it is more difficult to obtain bistability. For definiteness, we take Δ, $\theta \geqslant 0$; however the situation is symmetrical with respect to a simultaneous sign change in Δ and θ. The function $Y(X)$, defined by eq. (31) always has a single inflection point at

$$X_{\text{inf}} = \frac{2C - \Delta\theta + 1}{C + \Delta\theta - 1}(\Delta^2 + 1). \tag{35}$$

In order to have bistability the obvious conditions are

$$X_{\text{inf}} > 0, \qquad \left.\frac{\mathrm{d}Y}{\mathrm{d}X}\right|_{X_{\text{inf}}} < 0. \tag{36}$$

The first condition guarantees that the inflection point is within the physical region $X > 0$, while the second one identifies the values of the parameters for which the curve $Y(X)$ has a maximum and a minimum. For $\Delta\theta > 0$, the first condition in eq. (36) reads

$$2C > \Delta\theta - 1. \tag{37}$$

The second condition gives

$$(2C - \Delta\theta + 1)^2(C + 4\Delta\theta - 4) > 27C(\Delta + \theta)^2. \tag{38}$$

The analysis of eqs. (37) and (38) leads us (BONIFACIO, GRONCHI and LUGIATO [1979a]) to conclude that

(i) bistability is impossible for $C < 4$.

(ii) for a fixed value of $C > 4$, the largest hysteresis cycle is obtained for $\Delta = \theta = 0$ and bistability exists only in a finite domain of the plane $\{\Delta, \theta\}$ around the origin.

(iii) if we keep C and Δ fixed and C satisfies condition (38) for $\theta = 0$, by increasing θ the size of the hysteresis cycle increases until it reaches a maximum and then decreases. Finally the cycle vanishes in correspondence to a value of θ smaller than $(2C + 1)/\Delta$ (see eq. (37)).

(iv) if we keep $C > 4$ fixed and increase Δ and θ, simultaneously from zero, with the ratio Δ/θ kept fixed, the hysteresis cycle of the curve $X(Y)$ shifts to the left and decreases in size, until it disappears.

Therefore in homogeneously broadened, two-level systems, when absorptive bistability for $\theta = 0$ is not possible, dispersive bistability for general values of Δ and θ is also impossible. This is no longer true in the case of inhomogeneously broadened systems ($T_2^* < \infty$). For fixed Δ, θ and T_2^*, one obtains bistability, provided that C is larger than a suitable value $C_{\min}$, which depends on Δ, θ, T_2^*. $C_{\min}$ increases rapidly with $(T_2^*)^{-1}$. The important point is that, for $(\gamma_\perp T_2^*)^{-1} \gg 1$, one finds values of C such that the system is not bistable for $\Delta = \theta = 0$, but becomes bistable when Δ and θ are large enough. In other words, for these values of T_2^* and C one does not find absorptive bistability, but only dispersive bistability. (BONIFACIO and LUGIATO [1978c], HASSAN, DRUMMOND and WALLS [1978]).

So far, we have only considered hysteresis cycles obtained by varying the incident field intensity and keeping the parameters C, Δ, θ fixed. Of course one can also consider cycles obtained by keeping Y fixed and varying C, Δ, or θ, or some of these parameters simultaneously (AGRAWAL and CARMICHAEL [1979]). For instance, one can perform an experiment in which one adiabatically sweeps the incident field frequency, thereby varying Δ and θ together. This procedure is followed in SANDLE and GALLAGHER [1981], and ARIMONDO, GOZZINI, LOVITCH and PISTELLI [1981].

By exploiting optical pumping one can reduce the switching power and obtain a wide tuning range (ARECCHI, GIUSFREDI, PETRIELLA and SALIERI [1982], GOZZINI [1982]).

In the case of Fabry–Perot, the bistability threshold C_{min} is somewhat higher than for a ring cavity. For example, in the purely absorptive case with homogeneous broadening the bistability condition is $C > 4.96$ (CARMICHAEL [1980], HERMANN[1980], MCCALL and GIBBS [1980]).

The effects of the radial shape of the electric field have been studied in BALLAGH, COOPER, HAMILTON, SANDLE and WARRINGTON [1981], ARIMONDO, GOZZINI, LOVITCH and PISTELLI [1981], DRUMMOND [1981], ROSANOV and SEMENOV [1981], FIRTH and WRIGHT [1982] and MOLONEY, BELIC and GIBBS [1982]. These treatments predict that the bistability threshold C_{min} and the switching intensity are raised by the Gaussian profile of the incident field. This effect, which is found both for a ring cavity and for a Fabry–Perot, is largest in the purely absorptive case (in which C_{min} increases by more than a factor two), whereas it is small in the purely dispersive case. With the exception of ROSANOV and SEMENOV [1981], FIRTH and WRIGHT [1982] and MOLONEY, BELIC and GIBBS [1982], these treatments introduce ab initio the hypothesis that when the incident field corresponds to a TEM_{00} mode of the cavity the internal field also has the same radial shape. This assumption remains to be substantiated by further analysis.

MARBURGER and FELBER [1978] predicted that under proper conditions the self-focusing can lower the threshold.

2.1.6. *Absorptive versus dispersive bistability*

As we have said, we have purely absorptive optical bistability when the atomic detuning Δ vanishes. On the contrary, we have purely dispersive optical bistability when Δ is so large that the absorptive part of the nonlinear susceptibility is negligible; hence, the state equation (30) reduces to

$$Y = X\{1 + [\theta - 2C\tilde{\chi}_{d}(X)]^2\}. \tag{39}$$

In the case of a homogeneously broadened, two-level system eq. (39) is a good approximation of eq. (30) when (BONIFACIO, GRONCHI and LUGIATO [1979a]

$$\Delta^2 \gg 1, \qquad \Delta\theta \gg 1, \qquad \Delta \gg \theta. \tag{40}$$

Moreover, if the condition

$$\frac{2C}{\Delta\theta} - 1 \ll 1 \tag{41}$$

is also satisfied, eq. (39) reduces in turn to the relation

$$Y = X\left\{1 + \left[\theta - \frac{2C}{\Delta} + \frac{2C}{\Delta^3}X\right]^2\right\}. \tag{42}$$

Equation (42) predicts bistability for $(2C/\Delta) - \theta > \sqrt{3}$. The condition (41) guarantees that $X \ll \Delta^2$ along the hysteresis cycle, which in turn implies that the atoms remain in the ground state (i.e., no absorption) in the whole bistable domain (see eq. (50), below).

Equation (42) is a particular example of the cubic model of purely dispersive optical bistability, that was first considered by GIBBS, MCCALL and VENKATESAN [1976]:

$$Y = X\{1 + [B - AX]^2\}, \tag{43}$$

which gives bistability for $B > \sqrt{3}$. Equation (43) describes optical bistability in several materials, for instance in a Kerr medium in the limit of small $\delta + \beta P_T$ (see eq. (20)).

Let us now describe the physical mechanisms that give rise to hysteresis in absorptive and dispersive optical bistability. In the absorptive case, let us consider for simplicity the resonant situation $\theta = 0$. In the cooperative (i.e. lower transmission) branch the transmission is small, because the presence of the saturable absorber drastically decreases the quality factor Q of the cavity. Most of the incident light is reflected from the cavity. Increasing the incident field causes the absorber to saturate, which allows the Q to increase. This in turn increases the internal field which again increases the saturation and so on, until the absorber is bleached, so that $P_T \approx P_I$. On the other hand, when the system is in the one-atom (i.e. higher transmission) branch and the incident intensity is decreased, the field internal to the cavity is already strong enough to maintain the absorber saturated, and therefore the transmitted light switches "off" at an incident power lower than that necessary to switch "on", thereby producing hysteresis.

In the case of purely dispersive optical bistability the mechanism is quite different, and was first pointed out in GIBBS, MCCALL and VENKATESAN [1976]. In the lower branch the transmission is low because the empty cavity frequency ω_c is detuned from the incident frequency ω_0. If the atomic and cavity detunings have the same sign, by increasing the incident field the nonlinear refractive index changes the effective optical length of the cavity towards resonance. This in turn increases the internal field, which further drives the effective cavity frequency $\omega_c' = \omega_c - k\chi_d(X)$ towards the incident field frequency and so on, until resonance is reached, so that $P_T \approx P_I$. On the other hand, when the system is in the higher transmission branch and the incident intensity is decreased, the internal field is already strong enough to maintain resonance, which again produces hysteresis.

In order to complete the discussion of the steady state behavior, let us illustrate the relative advantages of absorptive and dispersive optical bistability. First of all, it is clear that dispersive optical bistability is "easier" mainly for two reasons:

a) it does not require saturation of the medium as it appears from the cubic model (43);

b) in absorptive optical bistability the resonance condition between the incident field and the atoms cannot be easily maintained for a time long enough to allow the system to reach steady state, owing to the jitter in laser frequency. This problem emerges from the experiment of WEYER, WIEDENMANN, RATEIKE, MCGILLIVRAY, MEYSTRE and WALTHER [1981].

As we have seen in the previous section, in the case of homogeneous broadening absorptive optical bistability has the advantage of exibiting the largest hysteresis cycle for fixed C, when $\Delta = \theta = 0$. However, this is no longer true in the case of inhomogeneous broadening. Furthermore, even in the case of homogeneous broadening, the switching from the low transmission to the high transmission branch occurs for lower values of the imput field when Δ and θ are different from zero (see § 2.1.5, point (iv)). This is an important advantage, also because the presence of too intense a field in the absorber might produce undesirable effects, for instance heating of the medium.

For these reasons most experiments on optical bistability are in dispersive conditions. Absorptive optical bistability has been first observed by GIBBS, MCCALL and VENKATESAN [1976] and SANDLE, BALLAGH and GALLAGHER [1981], and later in greater detail by WEYER, WIEDENMANN, RATEIKE, MCGILLIVRAY, MEYSTRE and WALTHER [1981] and GRANT and KIMBLE [1982] (see Fig. 32, below) at optical frequencies, by GOZZINI, LONGO and MACCARRONE [1982] at microwave frequencies, and by MEIER, HOLZNER, DERIGHETTI and BRUN [1982] at radio frequencies.

On the other hand, from the viewpoint of theory and hence, of the comparison between experiment and theory, absorptive optical bistability with $\theta = 0$ is certainly much easier to deal with, because in the Maxwell–Bloch equations (1) all the fields can be safely assumed as real. This is the reason why most theoretical papers treat the absorptive case, and this also explains why some parts of this article are exclusively devoted to the problem of absorptive optical bistability. This occurs especially in the treatment of the most sophisticated problems, such as self-pulsing or quantum effects. The generalization to the case Δ, $\theta \neq 0$ involves quite cumbersome calculations and would reduce the pedagogical impact of this paper.

A final remark concerns the nomenclature. We called the bistability absorptive or dispersive according to whether it is produced by the absorptive or by the dispersive part of the susceptibility. This, however, does not mean that absorption is dominant in the purely absorptive case and necessarily absent in purely dispersive optical bistability. In general, the incident energy is in part transmitted, in part reflected, and in part absorbed by the atomic sample. By "absorbed" we mean both the energy diffused as fluorescent light and the energy dissipated in the medium. The absorbed energy is proportional to the population of the upper level. In the case of purely absorptive optical bistability, when the system is in the higher transmission branch most energy is transmitted, except in a neighborhood of the lower bistability threshold x_m (Fig. 8), where the absorbed and the transmitted energies have the same order of magnitude (see § 3.4.2). In the lower transmission branch absorption is dominated by reflection. On the other hand, in the two-level system there are situations that fulfil conditions (40) of purely dispersive optical bistability, in which, however, the population of the upper level is not negligible in the bistable region. An example is $C = 180$, $\Delta = 60$, and $\theta = 1$ (BONIFACIO, GRONCHI and LUGIATO [1979a]); see also Fig. 5. Only when condition (41) is satisfied is absorption really absent in the range of values of the incident field for which the system is bistable.

2.1.7. *The cooperation parameter C*

We stress that the parameter C is crucial not only in optical bistability, but also in many other cooperative phenomena in quantum optics. For instance, in superfluorescence (see the book edited by BONIFACIO [1982]), that is, in cooperative emission from a pencilshaped, mirrorless sample of two-level atoms prepared in a state of complete inversion, the necessary condition for the rise of cooperativity is $\alpha L \gg 1$, i.e. (since $T = 1$) $C \gg 1$.

Another interesting example is that of the laser with injected signal (DEGIORGIO and SCULLY [1970], SPENCER and LAMB [1972], CHOW, SCULLY and VAN STRYLAND [1975] and LUGIATO [1978]). The configuration of this system is the same as for optical bistability, but with an amplifier instead of an absorber. That is, the atoms are continuously pumped to yield a positive unsaturated inversion per atom σ. The Maxwell–Bloch equations for this system are identical to eqs. (1), with N replaced by $-\sigma N$. Hence, all the formulas derived up to now remain valid for the laser with injected signal, provided that one replaces α by $-\sigma\alpha$ and therefore defines C as:

$$C = -\frac{\sigma\alpha L}{2T}.$$

With this definition, the state equation for the laser with injected signal in the perfect resonant case is still eq. (32), but now C is negative.

For $y = 0$, eq. (32) reduces to the steady state equation of the normal laser (i.e., without injected signal), which has the two solutions $x = 0$ (nonlasing system) and $x = (2|C| - 1)^{1/2}$ (oscillating laser). Since the trivial solution is unstable for $|C| > 1/2$, the graph of the stationary solution x versus the pump parameter C shows a second-order phase transition in correspondence to the threshold $|C| = 1/2$ (DEGIORGIO and SCULLY [1970], GRAHAM and HAKEN [1970]).

In the case $y \neq 0$, for $|C| > 1/2$ the plot of transmitted light versus incident light obtained from eq. (32) shows multiple solutions (LUGIATO [1978]) (Fig. 9). However, one does not have a bistable situation, because all the part of the curve where $x < 0$ is unstable due to phase fluctuations. However, using the procedure of BÖSIGER, BRUN and MEIER [1981] one can maintain a fixed difference of 180° between the phases of the transmitted and incident fields, so that the system actually becomes bistable. In such a way bistability was observed in a laser with injected signal based on the dynamics of nuclear spin systems. These experiments correspond to "bad cavity" conditions with $\gamma_{\parallel} \ll \gamma_{\perp}$ (MEIER, HOLZNER, DERIGHETTI and BRUN [1982]).

2.2. SIMPLIFIED TREATMENT OF OPTICAL BISTABILITY IN THE LIMIT OF SMALL ABSORPTION, TRANSMISSION AND DETUNING

As we have seen in § 2.1, the simplest situation corresponds to (see eq. (27))

$$\alpha L \ll 1, \qquad T \ll 1, \qquad \delta_0 \ll 1 \tag{44}$$

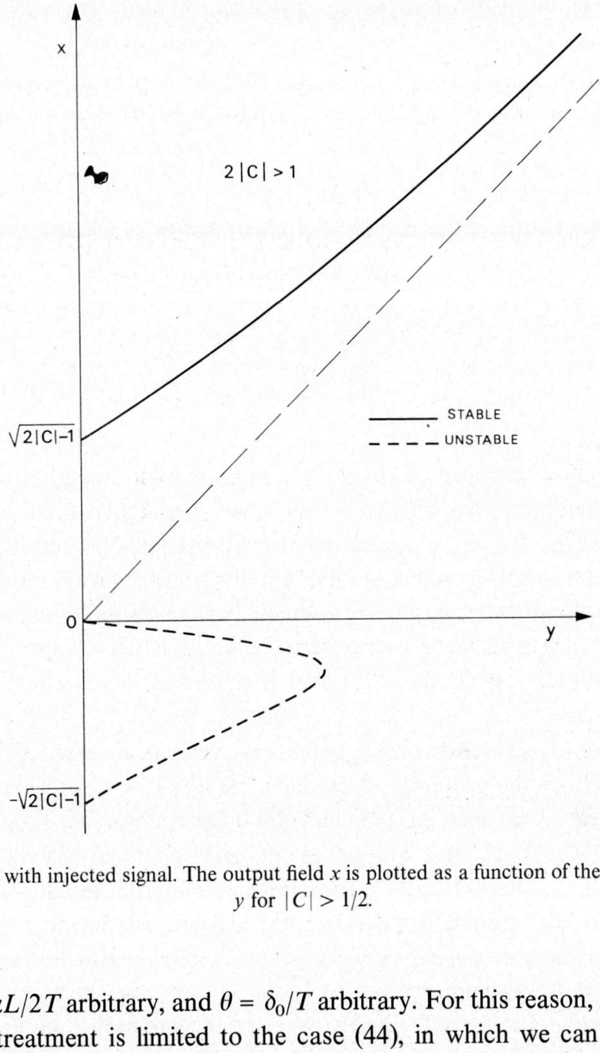

Fig. 9. Laser with injected signal. The output field x is plotted as a function of the incident field y for $|C| > 1/2$.

with $C = \alpha L/2T$ arbitrary, and $\theta = \delta_0/T$ arbitrary. For this reason, most of the following treatment is limited to the case (44), in which we can obtain the maximum physical insight and the maximal amount of analytical results. However, it is important to stress that conditions (44) may not be ideal from a practical viewpoint. In fact, a very small transmissivity renders the empty cavity response time $\mathscr{L}/cT$ long, and this is opposite to the requirement of a fast operating device.

2.2.1. *The Maxwell–Bloch equations in the limit $\alpha L \ll 1$, $T \ll 1$, $\delta_0 \ll 1$*

As shown in Appendix A, in the limit (44) the Maxwell equation (1a) and the boundary condition (3b) can be reformulated as (LUGIATO [1981])

$$\frac{\partial E}{\partial t'} + c\frac{L}{\mathscr{L}}\frac{\partial E}{\partial z} = -\mathrm{i}k\theta E - k\left(E - \frac{E_\mathrm{I}}{\sqrt{T}}\right) - g\frac{L}{\mathscr{L}}P, \tag{45}$$

$$E(0, t') = E(L, t'), \tag{46}$$

where the variable t' is defined by

$$t' = t + \Delta t\,\frac{z}{L}. \tag{47}$$

The time t' is introduced in order to eliminate the retardation Δt from the boundary condition (3b), which then becomes the standard periodicity condition in space (46). In a sense, the transformation (47) has the effect of "bending" the atomic sample onto itself, transforming it into a ring. Note that t' can be replaced by the real time t whenever the evolution occurs on a time scale much longer than the cavity transit time $\mathscr{L}/c$. This is generally true in the limit (44) for a wide class of initial conditions, except when self-pulsing behavior arises.

Equation (45) describes propagation with velocity $c' = cL/\mathscr{L}$. The factor $L/\mathscr{L}$ arises from the geometry of the cavity (Fig. 1).

The difference between eq. (45) and the original Maxwell–Bloch equation (1a) arises from the fact that most of the information contained in the boundary condition (3b) has been transferred into the time evolution equation for the field (45). Namely, the external field E_I, the transmissivity coefficient, and the cavity mistuning parameter appear now in eq. (45), which also incorporates the retardation Δt via the time t'.

From now on, it is suitable to express the time evolution equations in terms of the normalized quantities F defined by eq. (9) and

$$\tilde{P} = \left[\frac{N}{2}\left(\frac{\gamma_\parallel}{\gamma_\perp}\right)^{1/2}\right]^{-1} P, \qquad \tilde{D} = \left(\frac{N}{2}\right)^{-1} D. \tag{48}$$

Thus, eqs. (45) and (1b,c) become

$$\frac{\partial F}{\partial t'} + c\frac{L}{\mathscr{L}}\frac{\partial F}{\partial z} = k[-\mathrm{i}\theta F - (F - y) - 2C\tilde{P}], \tag{49a}$$

$$\frac{\partial \tilde{P}}{\partial t'} = \gamma_{\perp}[F\tilde{D} - (1 + \mathrm{i}\Delta)\tilde{P}], \tag{49b}$$

$$\frac{\partial \tilde{D}}{\partial t'} = -\gamma_{\parallel}[\tfrac{1}{2}(F\tilde{P}^* + F^*\tilde{P}) + \tilde{D} - 1]. \tag{49c}$$

It is straightforward to calculate the stationary solutions from eqs. (49). In fact, in the limit (44) the steady state fields are uniform in space, so that one must set not only $\partial F/\partial t' = \partial \tilde{P}/\partial t' = \partial \tilde{D}/\partial t' = 0$ but also $\partial F/\partial z = 0$. In such a way, using eqs. (10), (11b), and (21), one obtains

$$\tilde{P}_{\mathrm{st}} = \frac{(1 - \mathrm{i}\Delta)x_{\mathrm{st}}}{1 + \Delta^2 + X_{\mathrm{st}}}, \qquad \tilde{D}_{\mathrm{st}} = \frac{1 + \Delta^2}{1 + \Delta^2 + X_{\mathrm{st}}} \tag{50}$$

and the mean field state equation (31) for X_{st}.

2.2.2. *Formulation in terms of equations for the modes. The mean field model of optical bistability*

It may be suitable to re-express eqs. (49) in terms of mode variables for the field and the atomic system. This procedure is quite common in quantum optics (Haken [1970], Sargent, Scully and Lamb [1974]). The cavity frequencies are $\omega_{\mathrm{c}} + \alpha_n$, where ω_{c} is the frequency that is nearest to the incident frequency, and

$$\alpha_n = \frac{2\pi c}{\mathcal{L}} n, \qquad n = 0, \pm 1, \ldots. \tag{51}$$

Let us consider the following expansions

$$\begin{Bmatrix} F(z, t') \\ \tilde{P}(z, t') \\ \tilde{D}(z, t') \end{Bmatrix} = \sum_n \mathrm{e}^{\mathrm{i}k_n z} \begin{Bmatrix} f_n(t') \\ p_n(t') \\ d_n(t') \end{Bmatrix}, \tag{52}$$

with

$$k_n = \frac{\alpha_n}{cL/\mathcal{L}} = \frac{2\pi}{L} n, \qquad n = 0, \pm 1, \ldots. \tag{53}$$

The functions $\exp(\mathrm{i}k_n z)$ obey the periodicity boundary condition (46). By substituting eqs. (52) into eqs. (49), and taking into account the orthogonality of the functions $\exp(\mathrm{i}k_n z)$ in the internal $0 \leqslant z \leqslant L$, one finds the following equations for the mode amplitudes f_n, p_n and d_n:

$$f_n = -\mathrm{i}\alpha_n f_n + k[-\mathrm{i}\theta f_n - (f_n - y\delta_{n,0}) - 2Cp_n], \tag{54a}$$

$$\dot{p}_n = \gamma_\perp \left[\sum_{n'} f_{n'} d_{n-n'} - (1 + \mathrm{i}\Delta) p_n \right], \tag{54b}$$

$$\dot{d}_n = -\gamma_\| \left[\tfrac{1}{2} \sum_{n'} (f_{n'} p^*_{n'-n} + f^*_{n'} p_{n'+n}) + d_n - \delta_{n,0} \right]. \tag{54c}$$

where the dot means derivative with respect to time t'. When the fields F, $\tilde{P}$ and $\tilde{D}$ are uniform in space only the amplitudes f_0, p_0, d_0, corresponding to the resonant mode, are different from zero. Note from eqs. (54) that if the fields are initially (i.e. at $t' = 0$) uniform, they remain uniform during the whole evolution. This is a consequence of the assumption (44). Hence, in this case, eqs. (54) reduce to

$$k^{-1} \dot{x} = -\mathrm{i}\theta x - (x - y) - 2Cp_0, \tag{55a}$$

$$\gamma_\perp^{-1} \dot{p}_0 = x d_0 - (1 + \mathrm{i}\Delta) p_0, \tag{55b}$$

$$\gamma_\|^{-1} \dot{d}_0 = -\tfrac{1}{2}(x p_0^* + x^* p_0) - d_0 + 1, \tag{55c}$$

where we have taken into account that in this case $x = f_0$. Equations (55) were first introduced in BONIFACIO and LUGIATO [1976, 1978d], and are usually called the "mean field model" of optical bistability. Here the words "mean field" refer to the fact that the fields are uniform in space. Therefore, they have quite a different meaning, for instance, from the one used in the theory of equilibrium phase transitions. The mean field model (55) holds in the limit (44), provided that the initial condition is perfectly uniform. As we shall see in § 2.4.3, in order to describe self-pulsing we shall have to consider initial conditions that are at least slightly nonuniform in space.

Hence, the mean-field theory of optical bistability is a *one-mode theory*. As one immediately verifies, at steady state ($\dot{f}_0 = \dot{p}_0 = \dot{d}_0 = 0$) one obtains, for p_0 and d_0, the expressions (50), and one immediately derives the state equation (31). In fact, at steady state, only the resonant mode has a nonvanishing amplitude in the limit (44).

2.3. TRANSIENT BEHAVIOR

For the sake of simplicity, we shall describe the transient behavior in the framework of the mean field model (55). However, most of these results also remain valid out of the range of parameters specified by (44).

We shall first analyze the regression of the system to steady state, after it has been slightly displaced from it. In §§ 2.3.3 and 2.3.4 we shall describe the complete transient approach to steady state, emphasizing a few points that are relevant for the switching properties of the system.

2.3.1. *Regression to steady state and critical slowing down*

Let us consider a stable stationary state of the system. If the system is slightly shifted from this stationary state, the regression to the steady state is ruled by the set of equations obtained by linearizing eqs. (55) around steady state. By calling $\delta x = x - x_{st}$ etc. one obtains

$$k^{-1}\,\delta\dot{x} = -(1 + i\theta)x - 2C\delta p_0, \tag{56a}$$

$$\gamma_{\perp}^{-1}\,\delta\dot{p}_0 = x_{st}\,\delta d_0 + d_{0,st}\,\delta x - (1 + i\Delta)\delta p_0, \tag{56b}$$

$$\gamma_{\parallel}^{-1}\,\delta\dot{d}_0 = \tfrac{1}{2}(x_{st}\,\delta p_0^* + p_{0,st}^*\,\delta x + x_{st}^*\,\delta p_0 + p_{0,st}\,\delta x^*) - \delta d_0. \tag{56c}$$

The equations for δx^* and δp_0^* are the complex conjugates of eqs. (56a) and (56b) respectively. $p_{0,st}$ and $d_{0,st}$ are given by eq. (50). To be more specific, we consider the following experiment. Let us assume that the system is initially in a steady state corresponding to some value E_I of the incident field. If E_I is rapidly changed into $E_I + \delta E_I$ ($|\delta E_I| \ll E_I$) the system approaches the new, slightly different, steady state corresponding to $E_I + \delta E_I$. This approach is described by a solution of the linearized equations (56) and can be experimentally observed by looking at the transient behavior of the transmitted light. The solutions of the linearized equations are linear combinations of five exponentials $\exp(\lambda_i t)$, $i = 1, \ldots, 5$. When the decay constants $-\operatorname{Re}\lambda_i$ are well separated, the approach to the stationary situation is mainly characterized by the decay constant $\bar{\lambda}$, which is equal to the smallest among the quantities $-\operatorname{Re}\lambda_i$. We prove in general the following results:

1) Let us consider a point (x, y) on the cooperative branch, very near to the upper discontinuity point ($x = x_M$, $y = y_M$) (see Fig. 8). The approach to the steady state (x, y) is very slow, and becomes slower the nearer y is to y_M. Hence, there is a *critical slowing down* in correspondence to the discontinuity point y_M.

This critical slowing down is similar to that which one finds in tunnel diodes; see LANDAUER and WOO [1973]. More specifically, one finds that $\bar{\lambda} \to 0$ as $(y_M - y)^{1/2}$ when y approaches y_M from below. Hence, $\bar{\lambda}$ corresponds with a *soft mode* of the system.

2) A similar critical slowing down is found in correspondence to the lower discontinuity point ($x = x_m$, $y = y_m$). In fact, let us consider a point (x, y) on the one-atom branch, very near to ($x = x_m$, $y = y_m$). The damping constant $\bar{\lambda}$, which characterizes the approach to (x, y), tends to zero as $(y - y_m)^{1/2}$ when y approaches y_m from above.

This critical slowing down at the boundaries of the steady state hysteresis cycle was first pointed out in BONIFACIO and LUGIATO [1976], and is an absolutely general feature of optical bistability, that is, it holds for whatever values of the important parameters in play, including αL, T and δ_θ, for any kind of cavity and absorbing material.

The other features of the behavior of $\bar{\lambda}$ when y is varied depend on whether Δ and θ vanish or not, and on the relative order of magnitude of the constants k, $\gamma_\perp$, $\gamma_\parallel$. Two typical situations, which we shall consider in the following, are (BONIFACIO and LUGIATO [1978d]):

a) $k \ll \gamma_\perp, \gamma_\parallel$. In this situation, which is usual in laser amplifiers, the empty cavity width is much smaller than the atomic linewidth. We shall call this case "good (quality) cavity case".

b) $k \gg \gamma_\perp, \gamma_\parallel$. We shall call this situation "bad (quality) cavity case" even if this name is somewhat improper when $T \ll 1$, because in this case the finesse of the cavity is large.

For $T \ll 1$ the condition $k \gg \gamma_\perp, \gamma_\parallel$ can be achieved only with a very short cavity or with an atomic material with long relaxation times.

In case a), two of the roots λ_i have a much smaller (in modulus) real part than all the others. They are the solution of the equation

$$\lambda^2 + 2k\left[1 + \frac{2C(1 + \Delta^2)}{(1 + \Delta^2 + X_{st})^2}\right]\lambda + k^2 \left.\frac{dY}{dX}\right|_{X_{st}} = 0, \tag{57}$$

where the function $Y(X)$ is defined by eq. (31). When Δ and θ are different from zero, the solutions of eq. (57) are complex conjugate over a large part of the steady state curve, with the exception of two segments including the boundaries of the cycle (see Fig. 5b). For instance, for $X_{st} \ll 1$ the solutions are

$$\lambda = -k\{[1 + 2C/(1 + \Delta^2)] \pm i[\theta - 2C\Delta/(1 + \Delta^2)]\}$$

and for $X_{st} \to \infty$ they are $\lambda = -k(1 \pm i\theta)$. Hence, in these cases the approach to steady state is oscillatory.

For $\Delta = \theta = 0$ the solutions are always real:

$$\lambda_1 = -k \left.\frac{\mathrm{d}y}{\mathrm{d}x}\right|_{x_{\mathrm{st}}} = -k\left(1 + 2C\frac{1 - x_{\mathrm{st}}^2}{(1 + x_{\mathrm{st}}^2)^2}\right),$$

$$\lambda_2 = -k\frac{y}{x_{\mathrm{st}}} = -k\left(1 + \frac{2C}{1 + x_{\mathrm{st}}^2}\right), \tag{58}$$

where the function $y(x)$ is defined in eq. (32). Hence, in this case the approach to steady state is always monotonic.

In case b), we limit ourselves to considering the situation $\Delta = \theta = 0$. When (x, y) lies on the cooperative branch the approach is monotonic. For $C \gg 1$ one has approximately

$$\bar{\lambda} = 2\gamma_{\parallel}\left(1 - \frac{y^2}{C^2}\right)^{1/2} \Bigg/ \left[1 + \left(1 - \frac{y^2}{C^2}\right)^{1/2}\right]. \tag{59}$$

When (x, y) lies on the one-atom branch the approach is oscillatory except when (x, y) is very near to the lower discontinuity point $(x_{\mathrm{m}}, y_{\mathrm{m}})$ or when the ratio $\gamma_{\parallel}/\gamma_{\perp}$ is too small. In particular, for $y \gtrsim y_{\mathrm{M}}$ (BONIFACIO and LUGIATO [1976])

$$\bar{\lambda} \approx (\gamma_{\perp} + \gamma_{\parallel})/2 \tag{60}$$

and the oscillation frequency is the Rabi frequency of the incident field,

$$\Omega_{\mathrm{I}} = \mu E_{\mathrm{I}}/\hbar\sqrt{T}. \tag{61}$$

Rabi oscillations in the bad cavity case have been reported in ROHART and MACKE [1980]. Observation of transients in optical bistability was reported in GRYNBERG, BIRABEN and GIACOBINO [1981].

Of course, situations different from $k \ll \gamma_{\perp}, \gamma_{\parallel}$ or $k \gg \gamma_{\perp}, \gamma_{\parallel}$ can be considered. For instance, DRUMMOND [1982] studied the case $\gamma_{\parallel} \ll k, \gamma_{\perp}$ which is close to the situation one finds in GaAs (GIBBS, MCCALL and VENKATESAN [1979]). A general study of the mean field model for absorptive optical bistability in the limit of very large C has been done by MANDEL and ERNEUX [1982].

2.3.2. *Adiabatic elimination of the atomic variables in the good cavity case*

Let us now consider in detail the good cavity case $k \ll \gamma_{\perp}, \gamma_{\parallel}$. In this situation, the atomic variables vary in time much more rapidly than the field variables. After a short transient of the order of T_1, T_2, the atomic variables attain a stationary situation, i.e. we can put $\dot{p}_0 = \dot{d}_0 = 0$ in eqs. (55). Hence, the atomic variables p_0 and d_0 follow without retardation the motion of the field variable x, or using Haken's language are *slaved* by the field variable. In fact, the expressions for $p_0(t)$ and $d_0(t)$ for times much larger than T_1, T_2 are given by eq. (50) with x_{st} replaced by $x(t)$ and X_{st} by $|x|^2(t)$. By substituting the expression for p_0 into eq. (55a) one obtains a closed equation for the field variable which reads (BONIFACIO and LUGIATO [1978c])

$$k^{-1}\dot{x} = y - x\left[\left(1 + \frac{2C}{1 + \Delta^2 + |x|^2}\right) + \mathrm{i}\left(\theta - \frac{2C\Delta}{1 + \Delta^2 + |x^2|}\right)\right]. \quad (62)$$

This procedure is called *adiabatic elimination of the atomic variables* (HAKEN [1977]).

The following two subsections illustrate some properties of the transient that are relevant for the switching behavior of the system.

2.3.3. *Good cavity, case of perfect resonance: complete transient approach to steady state*

Let us first consider the simplest setting, that is the case of perfect resonance $\Delta = \theta = 0$. In this situation, by writing $x = x_1 + \mathrm{i}x_2$ eq. (62) becomes

$$k^{-1}\dot{x}_1 = -\frac{\partial \tilde{V}_y}{\partial x_1}, \qquad k^{-1}\dot{x}_2 = -\frac{\partial \tilde{V}_y}{\partial x_2},$$

$$\tilde{V}_y(x_1, x_2) = -yx_1 + \tfrac{1}{2}(x_1^2 + x_2^2) + C\ln(1 + x_1^2 + x_2^2), \quad (63)$$

where $\tilde{V}_y$ plays the role of a mechanical potential in the overdamped motion of a particle. Hence, if $x_2(0) = 0$, the imaginary part vanishes at all times. In this case, the dynamics are ruled by the equation

$$k^{-1}\dot{x} = -\frac{\partial V_y}{\partial x} = y - x - \frac{2Cx}{1 + x^2}, \qquad V_y(x) = \tilde{V}_y(x, 0), \quad (64)$$

where we have written x instead of x_1. We shall use eq. (64) to discuss the following problem. Let us assume that initially the system is at steady state, with a vanishingly small external field ($y \approx 0$), so that $x(0) \equiv x_0 \approx 0$. At this point, we abruptly switch the incident light on to some operating value y_{op}, larger than the upper bistability threshold y_M (Fig. 8). Hence, the transmitted light approaches the stationary value $\bar{x}$ in the high transmission branch corresponding to the value y_{op} of the incident field, that is the solution of eq. (32) for $y = y_{op}$. Equation (64), which describes the transient approach to the steady state $\bar{x}$, can be easily solved. For $y = y_{op} > y_M$, $x(0) = x_0$ we obtain (BENZA and LUGIATO [1979a])

$$\begin{aligned} k t' &= f(x) - f(x_0), \\ f(x) &= A_1 \ln|x - \bar{x}| + \tfrac{1}{2}A_2 \ln\left[x^2 + (\bar{x} - y_{op})x + \frac{y_{op}}{\bar{x}}\right] \\ &\quad + \frac{1}{\sqrt{A_3}}[2A_4 - A_2(\bar{x} - y_{op})] \tan^{-1} \frac{2x + \bar{x} - y_{op}}{\sqrt{A_3}}, \\ A_1 &= \frac{1 + \bar{x}^2}{\bar{x}y_{op} - 2\bar{x}^2 - y_{op}/\bar{x}}, \qquad A_2 = \frac{\bar{x}^2 - 1 - \bar{x}y_{op} + y_{op}/\bar{x}}{\bar{x}y_{op} - 2\bar{x}^2 - y_{op}/\bar{x}}, \\ A_3 &= 4\frac{y_{op}}{\bar{x}} - (\bar{x} - y_{op})^2 > 0, \qquad A_4 = \frac{2(y_{op} - \bar{x})}{\bar{x}y_{op} - 2\bar{x}^2 - y_{op}/\bar{x}}. \end{aligned} \tag{65}$$

The plot of x versus $\tau \equiv kt'$ is obtained from the graph of τ versus x by simply exchanging the axes. Figure 10a shows several plots of x versus τ for $C = 20$ and different values of y_{op}. Clearly, the approach shows a kind of "lethargy", and the time the system takes to reach steady state becomes longer and longer as y_{op} approaches y_M from above (BONIFACIO and MEYSTRE [1979], BENZA and LUGIATO [1979a]). This behavior is another facet of the *critical slowing down* mentioned in § 2.3.1. Let us consider eq. (64). The potential $V_y(x)$ has two minima in correspondence with the two stable stationary solutions $x = x_a$ and $x = x_c$ (Fig. 8), and one maximum in correspondence to the unstable state $x = x_b$. For $y_M - \varepsilon < y < y_M$, with $\varepsilon \ll y_M$, the left minimum is very shallow (Fig. 11a); hence, if we slightly displace the system from the minimum the system takes a long time to return to the steady state x_a. This lengthening is the critical slowing down discussed in § 2.3.1. For $y = y_M$ the left minimum becomes an inflection point with horizontal tangent (Fig. 11b). Finally for $y_M < y < y_M + \varepsilon$ there is only one stationary state $\bar{x}$, but the potential has a very flat part, hence, the system takes a long time to reach the steady state $\bar{x}$ (Fig. 11c).

This critical slowing down behavior in the approach to the steady state in the high transmission branch has been experimentally confirmed by GARMIRE, MARBURGER, ALLEN and WINFUL [1979] in a hybrid system, and by

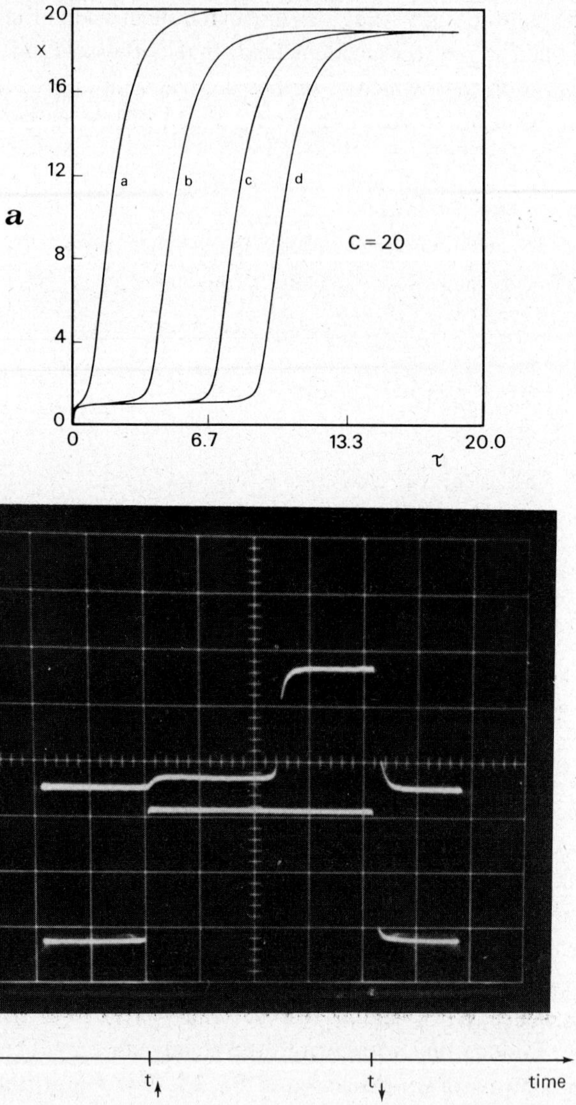

Fig. 10. (a) Time evolution of the transmitted field in the good cavity case for C = 20 and: **a**, y_{op} = 22; **b**, y_{op} = 21.1; **c**, y_{op} = 21.05; **d**, y_{op} = 21.04. Time is in units of k^{-1}. One has y_M = 21.0264 (see Fig. 8). (b) Observation of critical slowing down in microwave absorptive bistability in ammonia (from BARBARINO, GOZZINI, LONGO, MACCARRONE and STAMPACCHIA [1982]). Upper trace: transmitted power, lower trace: incident power. At $t = t_\uparrow$ the incident power is switched on to a value slightly larger than the bistability threshold y_M (see Fig. 8). At $t = t_\downarrow$ the incident power is switched off. The time scale is 10 ms/div. In this experiment T_1 and T_2 are of the order of a microsecond, and the cavity buildup time is ≈60 ns. The gas pressure is on the order of a millitorr and the mirror transmissivity is 0.01.

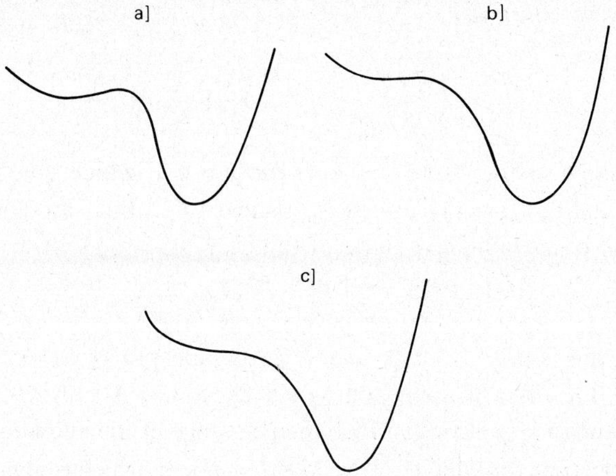

Fig. 11. Qualitative shape of the potential $V_y(x)$ (see eqs. (63) and (64)), when (a) y is slightly smaller than y_{M}, (b) $y = y_{\mathrm{M}}$, and (c) y is slightly larger than y_{M}.

BARBARINO, GOZZINI, LONGO, MACCARRONE and STAMPACCHIA [1982] in an all-optical system (Fig. 10b). Needless to say, a similar behavior arises if we start with the system at steady state for $y \gg y_{\mathrm{M}}$ and abruptly decrease the incident field to a value y_{op} smaller than y_{m} (see Fig. 8).

Coming back now to Fig. 10a, we can easily explain the main features of the time evolution for $C \gg 1$, $y_{\mathrm{op}} = y_{\mathrm{M}} + \varepsilon$, which are the following: (i) the curve $x(\tau)$ exhibits a plateau with a very slow evolution, followed by a steep rise to the stationary value, and (ii) as y_{op} becomes nearer and nearer to y_{M}, the length of the plateau increases very rapidly, whereas the slope of the steep part remains practically unchanged.

In order to understand these points, let us consider the inflection points of the curve $\tau = \tau(x)$. From eq. (64) one finds easily that this curve has two inflection points which correspond with the maximum and minimum x_{M} and x_{m} (Fig. 8), of the function $y = y(x)$, defined by eq. (32), that are given by

$$x_{\substack{\mathrm{M}\\ \mathrm{m}}} = [C - 1 \mp (C^2 - 4C)^{1/2}]^{1/2}. \tag{66}$$

Precisely, the point $x = x_{\mathrm{M}}$, $\tau = \tau(x_{\mathrm{M}})$ lies in the plateau of the curve $x = x(\tau)$, whereas the point $x = x_{\mathrm{m}}$, $\tau = \tau(x_{\mathrm{m}})$ lies in the steep part. Taking into account that

$$x_{\substack{\mathrm{M}\\ \mathrm{m}}} - \frac{2Cx_{\substack{\mathrm{M}\\ \mathrm{m}}}}{1 + x^2_{\substack{\mathrm{M}\\ \mathrm{m}}}} = y_{\substack{\mathrm{M}\\ \mathrm{m}}}$$

we obtain, from eq. (64),

$$\left(\frac{d\tau}{dx}\right)_{x_{\mathrm{m}}^{\mathrm{M}}} = (y_{\mathrm{op}} - y_{\mathrm{m}}^{\mathrm{M}})^{-1}. \tag{67}$$

Hence, the slope of the curve $x = x(\tau)$ for $x = x_{\mathrm{M}}$, which characterizes the plateau, is equal to $y_{\mathrm{op}} - y_{\mathrm{M}}$. When y_{op} is near to y_{M} this slope becomes very small. On the other hand the slope of the curve $x = x(\tau)$ for $x = x_{\mathrm{m}}$, which characterizes the steep part, is equal to $y_{\mathrm{op}} - y_{\mathrm{m}}$. For $C \gg 1$ one has $y_{\mathrm{op}} - y_{\mathrm{m}} \approx C \gg 1$, so that the slope is very large. When y_{op} becomes nearer and nearer to y_{M}, the slope $y_{\mathrm{op}} - y_{\mathrm{m}} \approx y_{\mathrm{M}} - y_{\mathrm{m}}$ remains practically unchanged. On the contrary, the time of approach to steady state $\bar{t}$ is very sensitive to the difference between y_{op} and y_{M}. Since the last stage of the evolution is fast, we identify $\bar{t}$ as the time for which $x = x_{\mathrm{m}}$, i.e. $\bar{t} = \tau(x_{\mathrm{m}})$, which is given by eq. (65) with $x_0 = 0$. One sees that $\bar{t}$ diverges for $y_{\mathrm{op}} \to y_{\mathrm{M}}$ from above, because $A_3 \to 0$. The plot of $\bar{t}$ as a function of y_{op} is shown in Fig. 12.

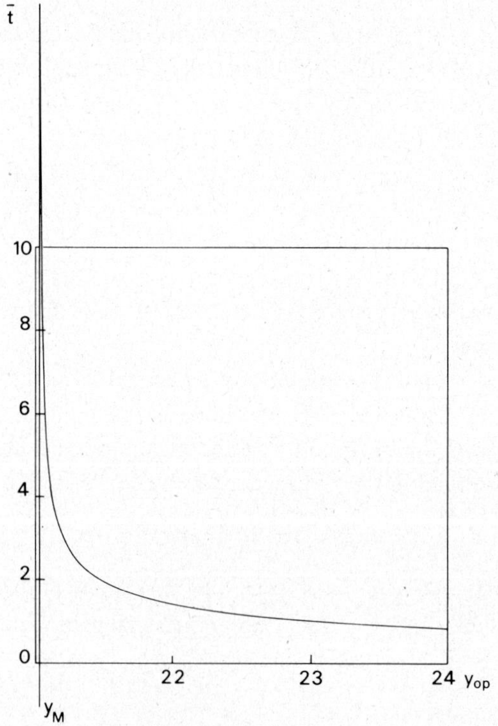

Fig. 12. Plot of the switching time $\bar{t}$ (in units of k^{-1}) as a function of the operating value y_{op} of the incident field for $C = 20$ and good cavity limit. $\bar{t}$ diverges for $y_{\mathrm{op}} \to y_{\mathrm{M}}$.

The critical slowing down is also important in connection with the following question. In order to obtain the hysteresis cycle of transmitted versus incident light, we must sweep the incident intensity back and forth. What is the maximum sweeping velocity such that the cycle is practically identical to the steady state cycle? A naïve answer is that in the good cavity case the sweeping time must be much larger than the buildup time of the field in the empty cavity. That is, in the resonant case $\theta = 0$, much larger than k^{-1}. Actually, the non-linearity of the dynamics gives a much more stringent condition. This feature emerges clearly from the experiment by WEYER, WIEDENMANN, RATEIKE, MCGILLIVRAY, MEYSTRE and WALTHER [1981]. In fact when the incident field y is varied, the potential V_y gets continuously deformed. In particular, when y is near to y_M, the potential changes from the configuration of Fig. 11a to that of Figs. 11b and 11c. During this stage, the system lies in a flat part of the potential, so that the time evolution is quite slow. For this reason the switching time from the low to the high transmission branch is on the order of 10^2–$10^3\ k^{-1}$ and the sweeping time must be larger than this lower bound in order to obtain a static hysteresis cycle. Otherwise, the hysteresis cycle is rounded by transient effects. The generation of hysteresis cycles by pulses was first analyzed by BISHOFBERGER and SHEN [1978] and ABRAHAM, BULLOUGH and HASSAN [1979].

2.3.4. *Good cavity, purely dispersive case: anomalous switching*

Let us come back to eq. (62) with Δ, $\theta \neq 0$, and consider the following problem. As in § 2.3.3, we start with the system initially at steady state, with a vanishingly small external field $y \approx 0$. Abruptly, we change the incident field to a value y_{op} in the bistability region, that is $y_m < y_{op} < y_M$ (Fig. 8). In the absorptive case, the system always approaches the lower transmission steady state corresponding to the value y_{op} of the incident field, and only when y_{op} is made larger than y_M can the high transmission branch be reached. In the dispersive case this is no longer true in general. One achieves switching as soon as y_{op} exceeds a threshold value y_{th} which lies between y_m and y_M and whose value depends on the parameters in play. This phenomenon, which was first pointed out in the framework of a cubic model of dispersive optical bistability by HOPF, MEYSTRE, DRUMMOND and WALLS [1979], is called *anomalous switching*, and has been experimentally observed in hybrid systems (GOLDSTONE, HO and GARMIRE [1981]).

The difference in the switching behavior between the cases $\Delta = \theta = 0$ and Δ, $\theta \neq 0$ arises from the fact that in the first one the electric field can be treated as a real variable, whereas in dispersive optical bistability it is necessarily a complex quantity which evolves in a two-dimensional phase space. Furthermore, the approach to the steady state for $k \ll \gamma_\perp, \gamma_\parallel$ is always monotonic for $\Delta = \theta = 0$, whereas it can be oscillatory for $\Delta, \theta \neq 0$ (see Fig. 5b). This fact can explain the overshoot found in Kerr media when the system switches from the low to the high transmission branch (BISHOFBERGER and SHEN [1979]).

In order to obtain physical insight into the phenomenon of anomalous switching we follow the procedure of LUGIATO, MILANI and MEYSTRE [1982]. Together with eq. (62) let us consider the equation obtained by dropping the damping terms:

$$\dot{x} = y - \mathrm{i}x\left(\theta - \frac{2C\Delta}{1 + \Delta^2 + |x|^2}\right). \tag{68}$$

The advantage of eq. (68) with respect to eq. (62) is that eq. (68) is analytically soluble. We note that eqs. (62) and (68) describe systems which are not only quantitatively but also qualitatively different, since in the case of eq. (62) the system approaches one of the two stationary solutions, whereas in the case of eq. (68), and due to the lack of dissipation, the motion is periodic in time. That is, the trajectory is a *closed* curve in the phase plane. However, our aim is only to determine the threshold of anomalous switching. For this purpose, it is enough to ensure that the solution of eq. (68), which starts from the origin, is a good approximation to the exact trajectory during that part of the time evolution which is relevant for the onset of anomalous switching. The relevant part of the trajectory involves only values of x smaller than (or at most on the order of) Δ.

For the sake of argument, let us consider the conditions (40) of purely dispersive optical bistability. Furthermore, we want the dispersive terms to be dominant during the relevant stage of the time evolution, so that eq. (62) is well approximated by eq. (68). This is achieved by imposing the condition

$$2C/\Delta \gg 1. \tag{69}$$

Let us consider the solutions of eq. (68). The trajectories in the phase plane $x_1 = \mathrm{Re}\, x$, $x_2 = \mathrm{Im}\, x$ are ruled by the equation

$$\frac{\mathrm{d}x_1}{\mathrm{d}x_2} = -\frac{y + x_2 f(x_1, x_2)}{x_1 f(x_1, x_2)}, \tag{70}$$

where

$$f(x_1, x_2) = \theta - \frac{2C\Delta}{1 + \Delta^2 + x_1^2 + x_2^2}.$$

The solution of eq. (70) that starts from the origin obeys the equation

$$\tfrac{1}{2}\theta(x_1^2 + x_2^2) + yx_2 - C\Delta \ln \frac{1 + \Delta^2 + x_1^2 + x_2^2}{1 + \Delta^2} = 0. \tag{71}$$

The behavior of the trajectory (71) as a function of y is illustrated in Fig. 13. For y smaller than a critical value $\bar{y}$, the curve (71) is disconnected into two circle-like parts. The inner part, that includes the origin, has a small radius, so that the intensity (proportional to $|x|^2$) remains small during the whole time evolution. The outer part, which corresponds to a large intensity, cannot be

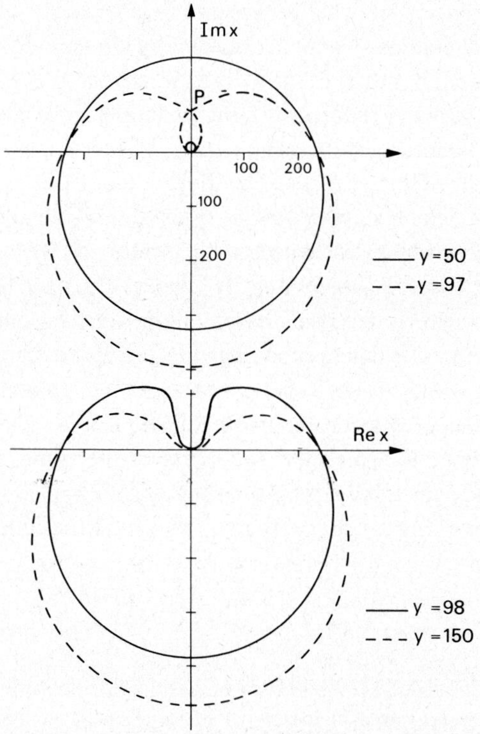

Fig. 13. The curve given by eq. (71) is shown for $C = 180$, $\Delta = 60$, $\theta = 1$ and four different values of y. For $y < \bar{y}$ the curve is disconnected. For $y = \bar{y} = 97$ the two parts of the curve coalesce at their north poles. For $y > \bar{y}$ the curve is connected.

TABLE 1
Threshold of anomalous switching

C	$\bar{y}$	y_{th}	$\lvert y - y_{\text{th}}\rvert / y_{\text{th}}$
1000	748	790	5.3×10^{-2}
2000	1551	1610	3.6×10^{-2}
4000	3219	3250	9.5×10^{-3}

reached for $x(0) = 0$. For $y = \bar{y}$ the two parts touch each other, so that the curve becomes connected (and remains connected for $y > \bar{y}$). This means that for $y \geqslant \bar{y}$ the intensity can reach large values even though it vanishes initially.

We now argue that the transition point $\bar{y}$, obtained from eq. (68), can be substantially identified with the anomalous threshold y_{th} of eq. (62). For definiteness, let us take $\Delta = 60$ and $\theta = 1$. In table 1 we compare for various values of C the value $\bar{y}$, for which the curve (71) becomes connected, to the threshold y_{th} of anomalous switching obtained by numerically solving eq. (62). Clearly the agreement between $\bar{y}$ and y_{th} is excellent for $C \gtrsim 500$. In LUGIATO, MILANI and MEYSTRE [1982] one finds a simple graphical procedure to determine $\bar{y}$, and hence, y_{th}.

The anomalous switching is caused by the fact that for $y = y_{\text{th}}$ the trajectory in the phase plane, which starts from the origin, "hits" the unstable stationary solution of eq. (62). Hence, the trajectory is "scattered" by the unstable state, which is a repeller, towards larger intensity regions. The orbit for $y = y_{\text{th}}$, which starts from the origin, is covered in an infinite time because it contains a stationary state. Hence, when approaching the anomalous threshold one again finds a "critical slowing down" effect, similar to that found in the previous subsection. The same behavior occurs in eq. (68) for $y = \bar{y}$. In fact, the point P, in correspondence to which the two parts of the orbit touch each other (Fig. 13), is just the unstable stationary solution of eq. (68).

Finally, it is interesting to observe that if we perform the cubic approximation

$$\frac{1}{1 + \Delta^2 + |x|^2} \approx \frac{1}{1 + \Delta^2}\left(1 - \frac{|x|^2}{1 + \Delta^2}\right)$$

eq. (68) becomes formally identical to the time evolution equation discussed in the context of the one-electron theory of the free electron laser formulated in BONIFACIO, CASAGRANDE and LUGIATO [1981].

2.4. INSTABILITIES IN OPTICAL BISTABILITY

The search for instabilities is the crucial point in the study of cooperative phenomena in an open system far from thermodynamic equilibrium (HAKEN [1977], NICOLIS and PRIGOGINE [1977]). In fact, let us consider in general a parameter y that measures the strength of the interaction of an open system with the external world (in our case y is the incident field amplitude). When y is small, the system is in a steady state which is the direct continuation of the thermodynamic equilibrium state (quasi-equilibrium state). On the other hand, when y increases, the system becomes more and more unbalanced. In correspondence to a suitable threshold value y_c (in our case y_M in Fig. 8), the quasi-equilibrium state becomes unstable, and the system can show, roughly speaking, three different types of behavior:

a) It can perform a transition to a new steady state (in our case, the higher transmission branch). If this new state arises in a continuous way (or, as one usually says in the jargon of instability theory, "bifurcates") from the quasi-equilibrium state, the behavior is analogous to second-order phase transitions in equilibrium systems. Otherwise one has an analogy with first-order phase transitions, and one usually finds bistability and hysteresis.

b) It can approach a nonstationary situation, that is, one which is periodic in time. In other words, the state of the system in the long time limit is not represented by a fixed point in the phase space (steady state), but by a limit cycle. In this case the system shows a pulsing behavior, or more precisely a "self-pulsing" behavior, because it does not arise from external manipulation but is spontaneously generated by the self-organization of the system.

c) It can approach a nonperiodic behavior that does not exhibit any kind of regularity in time, so that it is called *chaotic*.

These three types of transition can also appear in succession when y is varied, as a result of successive bifurcations.

In the following subsections we shall show that in optical bistability we find not only a behavior of type a), but also self-pulsing and chaotic behavior.

With the exception of § 2.4.4, we shall always consider the situation (44), and furthermore we shall assume the good quality cavity condition $k \ll \gamma_\perp, \gamma_\parallel$. Therefore, we start from eqs. (49). First, we consider a stationary solution $F_{st} = x_{st}$, $\tilde{P}_{st}$, $\tilde{D}_{st}$, and introduce the deviations from steady state

$$F(z, t') = x_{st} + \delta F(z, t'),$$

$$\tilde{P}(z, t') = \tilde{P}_{st} + \delta P(z, t'),$$

$$\tilde{D}(z, t') = \tilde{D}_{st} + \delta D(z, t'). \tag{72}$$

In the following analysis of instabilities in optical bistability, we shall always write x instead of x_{st}. By inserting eqs. (72) into (49) we obtain the following equations for the deviations

$$\frac{\partial \delta F}{\partial t'} + c\frac{L}{\mathscr{L}}\frac{\partial \delta F}{\partial z} = -k[(1+\mathrm{i}\theta)\delta F + 2C\delta P], \tag{73a}$$

$$\frac{\partial \delta P}{\partial t'} = \gamma_\perp[x\delta D + \tilde{D}_{st}\delta F - (1+\mathrm{i}\Delta)\delta P] + \gamma_\perp \delta F\cdot\delta D, \tag{73b}$$

$$\frac{\partial \delta D}{\partial t'} = -\gamma_\parallel[\tfrac{1}{2}(x\delta P^* + x^*\delta P + \tilde{P}^*_{st}\delta F + \tilde{P}_{st}\delta F^*) + \delta D]$$

$$-\frac{\gamma_\parallel}{2}(\delta F\delta P^* + \delta F^*\delta P). \tag{73c}$$

The equations for δF and δP^* are the complex conjugates of eqs. (73a) and (73b), respectively.

2.4.1. *Linear stability analysis: eigenvalues of the linearized problem*

When the deviations δF, etc., are small, we can neglect the nonlinear terms in eqs. (73). Thus we obtain a linear set of equations, which are the basis of the stability analysis of the system. We recall that this linear stability analysis has a local character. A global type of stability analysis will be considered in the quantum statistical treatment (see § 3.5).

Let us now consider the five-component vector

$$\boldsymbol{q}(z,t') = \begin{pmatrix} \delta F(z,t') \\ \delta F^*(z,t') \\ \delta P(z,t') \\ \delta P^*(z,t') \\ \delta D(z,t') \end{pmatrix}. \tag{74}$$

Next, we introduce the ansatz

$$\boldsymbol{q}(z,t') = \exp(\lambda t' + \mathrm{i}k_n z)\boldsymbol{q}^0, \tag{75}$$

where λ is a complex number and k_n is defined by eq. (53), and we insert eq. (75) into the linearized equations. We obtain the eigenvalue equation

$$\hat{\mathscr{L}}_n \boldsymbol{q}^0 = \lambda \boldsymbol{q}^0, \tag{76}$$

where $\hat{\mathscr{L}}_n$ is the matrix

$$\hat{\mathscr{L}}_n = \begin{pmatrix} -k(1+\mathrm{i}\theta+\mathrm{i}\alpha_n) & 0 & -2Ck & 0 & 0 \\ 0 & -k(1-\mathrm{i}\theta+\mathrm{i}\alpha_n) & 0 & -2Ck & 0 \\ \gamma_\perp \tilde{D}_{\mathrm{st}} & 0 & -\gamma_\perp(1+\mathrm{i}\Delta) & 0 & \gamma_\perp x \\ 0 & \gamma_\perp \tilde{D}_{\mathrm{st}} & 0 & -\gamma_\perp(1-\mathrm{i}\Delta) & \gamma_\perp x^* \\ -\frac{\gamma_\parallel}{2}\tilde{P}^*_{\mathrm{st}} & -\frac{\gamma_\parallel}{2}\tilde{P}_{\mathrm{st}} & -\frac{\gamma_\parallel}{2}x^* & -\frac{\gamma_\parallel}{2}x & -\gamma_\parallel \end{pmatrix}, \tag{77}$$

and α_n is defined by eq. (51). The characteristic equation $\det(\mathscr{L}_n - \lambda) = 0$ has five roots. Hence, the eigenvalues λ_{nj} are labeled by two indices, the frequency index $n = 0, \pm 1, \ldots$ and the index $j = 1, \ldots, 5$. In particular, $n = 0$ corresponds to the resonant frequency. The steady state is stable if, and only if, Re $\lambda_{nj} \leqslant 0$ for all n and j. It is easy to solve the characteristic equation to lowest order in the ratios $k/\gamma_\perp$, $k/\gamma_\parallel$. After substituting the expressions (50) of $\tilde{P}_{\mathrm{st}}$ and $\tilde{D}_{\mathrm{st}}$, one obtains (LUGIATO [1980a])

$$\lambda_{n1} = -\mathrm{i}\alpha_n + k\left\{\lambda_n^{(+)} + O\left(\frac{k}{\gamma_\perp}, \frac{k}{\gamma_\parallel}\right)\right\},$$

$$\lambda_{n4} = -\mathrm{i}\alpha_n + k\left\{\lambda_n^{(-)} + O\left(\frac{k}{\gamma_\perp}, \frac{k}{\gamma_\parallel}\right)\right\}, \tag{78}$$

where

$$\lambda_n^{(\pm)} = -1 - \{[(1-\mathrm{i}\tilde{\alpha}_n)^2 + \Delta^2](\bar{d} - \mathrm{i}\tilde{\alpha}_n) + \bar{d}X(1-\mathrm{i}\tilde{\alpha}_n)\}^{-1}$$

$$\times \frac{2C(1+\Delta^2)}{1+\Delta^2+X}\left[-\tilde{\alpha}_n^2 - \mathrm{i}\tilde{\alpha}_n\left(1+\bar{d}-\frac{1}{2}\frac{\bar{d}X}{1+\Delta^2}\right)+\bar{d}\right] \pm \mathscr{H}^{1/2}, \tag{79}$$

with

$$\tilde{\alpha}_n = \alpha_n/\gamma_\perp, \qquad \bar{d} = \gamma_\parallel/\gamma_\perp, \tag{80}$$

$$\mathscr{H} = -\theta^2 + \{[(1 - \mathrm{i}\tilde{\alpha}_n)^2 + \Delta^2](\bar{d} - \mathrm{i}\tilde{\alpha}_n) + \bar{d}X(1 - \mathrm{i}\tilde{\alpha}_n)\}^{-2}$$

$$\times \left\{ \left[\frac{2C(1 + \Delta^2)}{1 + \Delta^2 + X}\right]^2 \left\{ \tilde{\alpha}_n^2 \left[\Delta^2 \left(1 + \frac{\bar{d}X}{1 + \Delta^2}\right) - \left(\frac{1}{2} \frac{\bar{d}X}{1 + \Delta^2}\right)^2\right] \right.\right.$$

$$\left. + \mathrm{i}\bar{d}\tilde{\alpha}_n \left(2\Delta^2 + \bar{d}X \frac{\Delta^2 - X}{1 + \Delta^2}\right) + \bar{d}^2\Delta^2 \left(\frac{X}{1 + \Delta^2} - 1\right) + \bar{d}^2 \frac{X(X - \Delta^2)}{1 + \Delta^2} \right\}$$

$$+ 2C\theta\Delta \frac{1 + \Delta^2}{1 + \Delta^2 + X} \left\{ \tilde{\alpha}_n^4 \left(2 + \frac{\bar{d}X}{1 + \Delta^2}\right) \right.$$

$$+ \mathrm{i}\tilde{\alpha}_n^3 \left[2\bar{d} + \left(2 + \frac{\bar{d}X}{1 + \Delta^2}\right)(\bar{d} + 2)\right]$$

$$- \tilde{\alpha}_n^2 \left[2\bar{d}(\bar{d} + 2) + \left(2 + \frac{\bar{d}X}{1 + \Delta^2}\right)(2\bar{d} + 1 + \Delta^2 + \bar{d}X)\right]$$

$$- \mathrm{i}\tilde{\alpha}_n \left[4\bar{d}^2 + 2\bar{d}(1 + \Delta^2) + 2\bar{d}^2X + \bar{d}\left(2 + \frac{\bar{d}X}{1 + \Delta^2}\right)(1 + \Delta^2 + X)\right]$$

$$\left.\left. + 2\bar{d}^2(1 + \Delta^2 + X) \right\} \right\}. \tag{81}$$

On the other hand, for $j = 2, 3$, and 5 we have

$$\lambda_{nj} = \lambda_{nj}^{(0)} \left[1 + O\left(\frac{k}{\gamma_\perp}, \frac{k}{\gamma_\parallel}\right)\right], \tag{78'}$$

where $\lambda_{nj}^{(0)}$ are the solutions of the cubic equation

$$[(\lambda + \gamma_\perp)^2 + \gamma_\perp^2 \Delta^2](\lambda + \gamma_\parallel) + \gamma_\perp \gamma_\parallel X(\lambda + \gamma_\perp) = 0, \tag{82}$$

such that, for $\Delta = 0$,

$$\lambda_{n\frac{2}{3}}^{(0)} = -\tfrac{1}{2}\{\gamma_\perp + \gamma_\parallel \pm [(\gamma_\perp - \gamma_\parallel)^{1/2} - 4\gamma_\perp \gamma_\parallel X]^{1/2}\},$$

$$\lambda_{n5}^{(0)} = -\gamma_\perp. \tag{83}$$

For $\Delta = \theta = 0$ the expression for $\lambda_n^{(\pm)}$ simplifies as follows

$$\lambda_n^{(+)} = -1 - \frac{2C}{1 + x^2} \frac{\bar{d}(1 - x^2) - \mathrm{i}\tilde{\alpha}_n}{(1 - \mathrm{i}\tilde{\alpha}_n)(\bar{d} - \mathrm{i}\tilde{\alpha}_n) + \bar{d}x^2}, \tag{84a}$$

$$\lambda_n^{(-)} = -1 - \frac{2C}{1 + x^2} \frac{1}{1 - \mathrm{i}\tilde{\alpha}_n}. \tag{84b}$$

2.4.2. *Instability conditions*

Clearly, the real part of λ_{n2}, λ_{n3} and λ_{n5} is always negative. On the contrary, under suitable conditions one finds that Re $\lambda_{n1} > 0$ or Re $\lambda_{n4} > 0$, or both for some n, so that the steady state is unstable. Hence, the *instability condition* is

$$\text{Re}\,\lambda_{n1}^{(+)} > 0 \qquad \textit{and/or} \qquad \text{Re}\,\lambda_{n4}^{(-)} > 0. \tag{85}$$

Let us first consider the case $n = 0$, that is the resonant mode. The eigenvalues λ_{0j} coincide with the constants λ_i discussed in § 2.3.1. In fact, in the mean field theory, we consider only the resonant mode. Using eqs. (79–81) we can easily verify that λ_{01} and λ_{04} coincide with the two solutions of eq. (57), which was derived in the good cavity limit. For $\mathrm{d}Y/\mathrm{d}X > 0$ both solutions of eq. (57) have negative real part. Instead, when $\mathrm{d}Y/\mathrm{d}X < 0$, the eigenvalue λ_{01} has positive real part. This proves what we anticipated, that is, that the part of the curve $X(Y)$ with negative slope is unstable. In fact, in correspondence to these stationary solutions the resonant frequency is unstable.

Next, let us consider the off-resonant frequencies $n \neq 0$. First, we analyze the case $\Delta = \theta = 0$ in which $\lambda_n^{(\pm)}$ are given by eq. (84). From eq. (84b) we see that, in this situation, Re $\lambda_n^{(-)} < 0$ for all n. On the other hand, the analysis of the instability condition Re $\lambda_n^{(+)} > 0$ leads to a biquadratic equation, the discussion of which yields the following conclusions (BONIFACIO and LUGIATO (1978b]): The stationary state is unstable when the following two conditions are simultaneously satisfied:

$$\bar{R} \geqslant 0, \qquad \bar{S} + \bar{R}^{\,1/2} \geqslant 0,$$

$$\bar{R} = \gamma_\perp^2 \gamma_\parallel^2 x^4 \left(1 - \frac{y}{x}\right)^2 + \left(\gamma_\parallel^2 - \gamma_\perp^2 \frac{y}{x}\right)^2$$

$$- 2\gamma_\perp \gamma_\parallel x^2 \left[3\gamma_\parallel^2 + 4\gamma_\perp \gamma_\parallel + \frac{y}{x}(3\gamma_\perp^2 - \gamma_\parallel^2) + \gamma_\perp^2 \frac{y^2}{x^2}\right],$$

$$\bar{S} = \gamma_\parallel (3\gamma_\perp x^2 - \gamma_\parallel) - \frac{y}{x}\gamma_\perp(\gamma_\perp + \gamma_\parallel x^2), \tag{86}$$

provided that at least one of the discrete values α_n lies in the interval $\alpha_{\min} < \alpha_n < \alpha_{\max}$, where

$$\alpha_{\substack{\max\\ \min}} = \frac{1}{\sqrt{2}}(\bar{S} \pm \sqrt{\bar{R}})^{1/2}. \tag{87}$$

Under suitable conditions it happens that, in correspondence to a part of the plot $x(y)$ *with positive slope*, some off-resonance frequencies are unstable. The unstable frequencies are those such that $\alpha_{\min} < |\alpha_n| < \alpha_{\max}$. As $\lambda_{-n}^{(+)} = (\lambda_n^{(+)})^*$, the frequencies become unstable in pairs, symmetrically with respect to the resonant frequency.

For the sake of definiteness, let us analyze the case $\gamma_\perp = \gamma_\parallel \equiv \gamma$. The resulting picture is as follows:

1) For $C < 2(1 + \sqrt{2})$ all the points of the plot $x = x(y)$, which lie on the part with positive slope, are stable.

2) For $C > 2(1 + \sqrt{2})$ the points in the one-atom branch (high transmission branch), such that $x < C/2$ (see Fig. 14), are unstable *provided* that at least one of the discrete values α_n lies in the range $\alpha_{\min}(x) < \alpha_n < \alpha_{\max}(x)$, where (see Fig. 15)

$$\alpha_{\substack{\max\\ \min}}(x) = \gamma[x^2 - C - 1 \pm (C^2 - 4x^2)^{1/2}]^{1/2}. \tag{88}$$

On the contrary, for $\gamma_\parallel \ll \gamma_\perp$ (strong elastic collisions) no positive slope instability arises.

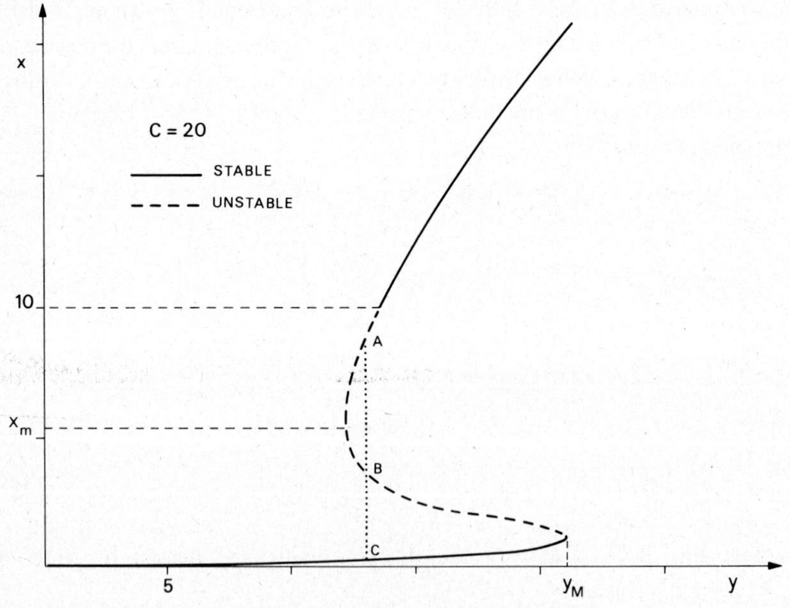

Fig. 14. Purely absorptive bistability S-curve with indication of the stable and unstable states for $\alpha L \to 0, T \to 0, C = \alpha L/2T = 20, \gamma_\perp = \gamma_\parallel = \gamma$. In the broken-line segment of the part with positive slope, the points are unstable, provided that at least one cavity frequency α_n lies in the range $\alpha_{\min}(x) < \alpha_n < \alpha_{\max}(x)$, where $\alpha_{\min}(x)$ and $\alpha_{\max}(x)$ are shown in Fig. 15.

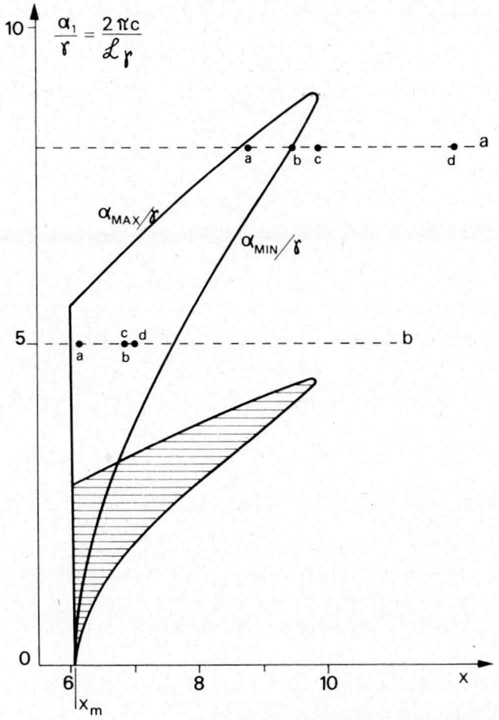

Fig. 15. Instability region in the plane of the variables x and α_1/γ for $\alpha L \to 0$, $T \to 0$, $C = \alpha L/2T = 20$, $\gamma_\perp = \gamma_\parallel = \gamma$. x is the normalized transmitted field in the high transmission branch, $\alpha_1 = 2\pi c/\mathscr{L}$ is the difference between the frequencies of the adjacent mode and of the resonant mode. The curves $\alpha_{max}(x)$, $\alpha_{min}(x)$ obey eq. (88). The shaded region indicates the points in correspondence to which the frequencies $n = \pm 2$ are unstable. The points on the lines **a** and **b** indicate the values of α_1/γ and x in correspondence to Figs. 19a–d and 21a–d respectively.

Let us now consider the general case Δ, $\theta \neq 0$. In this case, both Re $\lambda_n^{(+)}$ and Re $\lambda_n^{(-)}$ can become positive. For $\gamma_\perp = \gamma_\parallel = \gamma$ (i.e. $\bar{d} = 1$), we draw the following conclusions (LUGIATO [1980a]):

a) When θ is at most of order unity and $\Delta \gg 1$, the instability domain in the part of the curve $X = X(y)$ with positive slope practically vanishes. Note from eq. (40) that this is precisely the situation of purely dispersive optical bistability. This might explain why self-pulsing instabilities have never been observed in the experiments on dispersive optical bistability.

b) When $\Delta \lesssim 1$, $\theta > 1$ and C is large enough, part of the high transmission branch is unstable. The interesting feature is that the instability domain can be much larger than the bistability domain, or even can exist in the absence of

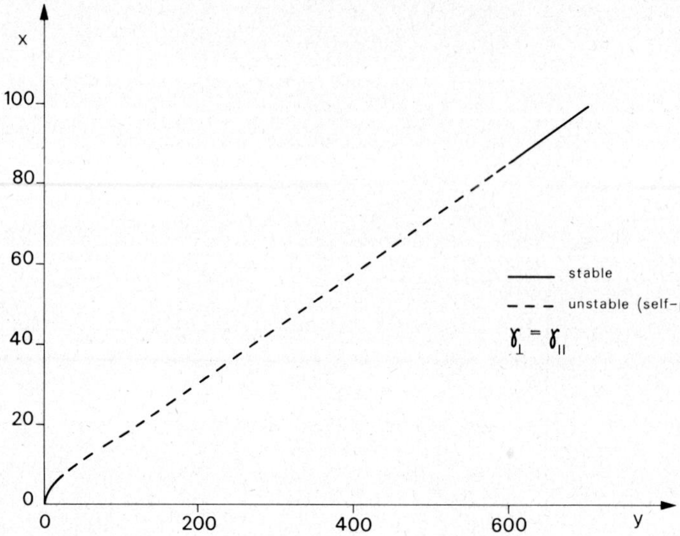

Fig. 16. Dispersive case. Stable and unstable stationary states in the limit $\alpha L \to 0$, $T \to 0$, with $C = \alpha L/2T = 30$, $\gamma_\perp = \gamma_\parallel = \gamma$. One has $\Delta = \theta = 7$. $x = \sqrt{X}$ and $y = \sqrt{Y}$ are the normalized transmitted and incident field amplitudes respectively.

bistability. Note that this instability occurs also for $\Delta = 0$ but $\theta \neq 0$. In this case the situation is purely absorptive.

c) For $\theta \approx \Delta > 1$ and C large enough, the instability domain turns out to be quite extended (Fig. 16).

The stability analysis in the case of a Fabry–Perot cavity has been performed in CASAGRANDE, LUGIATO and ASQUINI [1980]. (See also SARGENT [1980]).

2.4.3. *Self-pulsing and precipitation*

At this point, one asks what happens when the stationary state in the higher transmission branch becomes unstable. The most straightforward procedure to answer this question is offered by computer solutions of the Maxwell–Bloch equations (1) (BONIFACIO, GRONCHI and LUGIATO [1979b], GRONCHI, BENZA, LUGIATO, MEYSTRE and SARGENT [1981]). One finds that the system either precipitates to the lower transmission state, which corresponds to the same value of the incident field, or approaches a self-pulsing behavior. These two types of behavior are shown in Fig. 17, which exhibits the envelope of the time evolution of the transmitted field, that oscillates between the upper and the lower envelope. The initial condition of the evolution shown in Fig. 17 is such that the system is slightly displaced from the unstable steady state in the

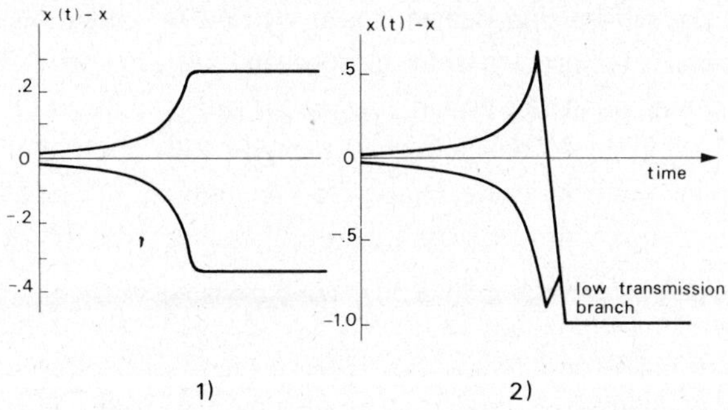

Fig. 17. Envelope of the time evolution of the transmitted field for $C = 20$, $\gamma_\perp = \gamma_\parallel = \gamma$, $\mathscr{L} = 5L$, $T = 0.1$, obtained by numerically solving the Maxwell–Bloch equations. The transmitted field oscillates between the upper and the lower envelope with a period roughly equal to the cavity transit time $\mathscr{L}/c$. (1) $x = 9$, $\alpha_1/\gamma = 8$. In the long-time limit, the system approaches a steady self-pulsing regime in which the envelope is perfectly flat. (2) $x = 6.15$, $\alpha_1/\gamma = 3.6$. For long times, the oscillations vanish and the system precipitates to the low transmission branch.

higher transmission branch. The oscillations are first exponentially amplified, during the stage of the time evolution that is governed by the linearized equations. In the case of Fig. 17.2, the oscillations finally vanish because the system precipitates to the lower transmission branch. In the case of Fig. 17.1, in due time the system approaches a regular self-pulsing regime, in which the transmitted light is given by a periodic sequence of short pulses, with a period on the order of the cavity transit time $\mathscr{L}/c$. The shape of the pulses depends on the number of unstable modes. When only the two modes adjacent to the resonant one are unstable the pulses are sinusoidal. This behavior is interesting also from the practical viewpoint, because it suggests an *all-optical device to convert* cw light into pulsed light.

Another type of converter, based on a completely different mechanism, was suggested by McCall [1978], who showed that a bistable device can pulsate when the nonlinearity has two contributions of opposite sign and different time constants. In the same paper the author reports the observation of this behavior in a hybrid electro-optical device, and suggests that the same phenomenon can be observed in an intrinsic all-optical bistable system in which switching is due to a fast electronic effect, but a slower thermal effect prevents either state from being stable. Observation of this behavior in a micron-sized GaAs etalon at 80 K has been recently reported (Jewell, Gibbs, Tarng, Gossard and Wiegmann [1982]). In this system, the period of the pulsations is of the order of several microseconds, whereas in the self-pulsing behavior we described before the period is of the order of nanoseconds.

Note that the self-pulsing behavior that we find in optical bistability has a completely different character from the so-called passive Q-switching that one finds in lasers with saturable absorbers (see for instance POWELL and WOLGA [1971]). In fact, passive Q-switching is a single-mode phenomenon, in which the field remains uniform in the cavity. On the contrary, self-pulsing in optical bistability is a many-mode phenomenon, because the frequencies which become unstable are different from the resonant frequency. In this case, there is a pulse which propagates in the cavity. On the other hand, the pulsed behavior in optical bistability is quite analogous to the self-pulsing in ring lasers beyond the so-called second threshold, predicted by RISKEN and NUMMEDAL [1968], and GRAHAM and HAKEN [1968].

In order to understand the mechanism which underlies the self-pulsing behavior, let us come back to the equations for the mode amplitudes (54) and let us make the following steps:

a) We consider the case $\theta = \Delta = 0$.

b) We perform the transformation of variables

$$\bar{f}_n(t') = \exp(\mathrm{i}\alpha_n t')f_n(t'), \qquad \bar{p}_n(t') = \exp(\mathrm{i}\alpha_n t')p_n(t'), \tag{89}$$

c) We take into account that $\gamma_\perp, \gamma_\parallel \gg k$, and therefore we adiabatically eliminate (see § 2.3.2) the atomic variables, by setting $\dot{\bar{p}}_n = \dot{d}_n = 0$.

d) We subdivide $\bar{f}_0$ into steady value and deviation: $\bar{f}_0 = x + \delta f_0$. Remember in this connection that for $n \neq 0$ the amplitudes f_n vanish at steady state.

e) We subdivide $\bar{f}_n$ for $n \neq 0$ into real and imaginary parts:

$$\bar{f}_n(t') = f_n'(t') + \mathrm{i}f_n''(t'). \tag{90}$$

As a consequence, we obtain the following equations for the *off-resonant* modes $n \neq 0$:

$$\begin{aligned} \dot{f}_n' &= k\lambda_n^{(+)}f_n' + \text{nonlinear terms}, \\ \dot{f}_n'' &= k\lambda_n^{(-)}f_n'' + \text{nonlinear terms}. \end{aligned} \tag{91}$$

When $\mathrm{Re}\,\lambda_n^{(+)} > 0$ the nth mode experiences amplification and the steady state is unstable. Now, using eq. (84a), $k\,\mathrm{Re}\,\lambda_n^{(+)}$ can be rewritten as

$$k\,\mathrm{Re}\,\lambda_n^{(+)} = \mathscr{G}_n - k,$$

$$\mathscr{G}_n = \frac{2Ck}{1+x^2}\,\frac{\bar{d}(1-x^2)[\bar{d}(1+x^2)-\tilde{\alpha}_n^2] + \tilde{\alpha}_n^2(1+\bar{d})}{[\bar{d}(1+x^2)-\tilde{\alpha}_n^2]^2 + \tilde{\alpha}_n^2(1+\bar{d})^2}. \tag{92}$$

When $\mathscr{G}_n$ is positive, eq. (92) is a gain-minus-loss form. The instability arises when, for at least one off-resonant mode, the gain exceeds the loss k. Note that, contrary to the loss term which is the same as for the usual laser, the gain which produces the off-resonance frequency buildup is quite different from the gain of the laser. In fact, in the laser the gain arises from population inversion. In the case of optical bistability, we deal with a purely passive system with negative population inversion. The gain in optical bistability requires the presence in the medium of a mode with a large amplitude (the resonant mode, which arises from the incident field). Via mode–mode interaction, the resonant mode induces gain for the side-modes. Hence, with respect to the unstable modes the system works as a novel type of laser without population inversion (GRONCHI, BENZA, LUGIATO, MEYSTRE and SARGENT [1981]). It is important to observe that $\mathscr{G}_n$ coincides with the gain one finds in saturation spectroscopy (see SARGENT [1978]), that is with the gain experienced by a weak probe field traveling through an atomic passive medium, saturated by a strong field detuned from the probe field. $\mathscr{G}_n$ also coincides with the differential gain of MCCALL [1974].

We also note that, as one sees from eq. (91), in order to allow the unstable modes to build up it is necessary that at least one of these modes has a nonvanishing initial amplitude. Hence, at least a small initial inhomogeneity of the field is necessary to trigger the instability. Fluctuations can easily achieve this goal.

Let us now consider, still for $\Delta = \theta = 0$, the case that only the two frequencies immediately and symmetrically adjacent to the resonant frequency are unstable. In the steady self-pulsing regime, the field inside the cavity is described to excellent approximation by an expression of the type

$$F(z, t') = x + A \cos\left[\alpha_1\left(t' - \frac{z\mathscr{L}}{cL}\right) + \psi_0\right], \tag{93}$$

where the phase ψ_0 depends on the initial conditions. Hence, using the relation of the electric field $\mathscr{E}$ with its envelope E,

$$\mathscr{E}(z, t) = E(z, t) \exp\left[-\mathrm{i}\omega_0\left(t - \frac{z}{c}\right)\right] + \mathrm{c.c.}$$

and taking into account eqs. (9) and (47), and the relation $\Delta t = (\mathscr{L} - L)/c$, we obtain

$$\mathscr{E}(z, t) = \bar{E} \cos\left[\omega_0\left(t - \frac{z}{c}\right)\right]$$

$$+ \mathscr{A} \left\{ \cos\left[(\omega_0 + \alpha_1)\left(t - \frac{z}{c}\right) + \psi_0 \right] \right.$$

$$\left. + \cos\left[(\omega_0 - \alpha_1)\left(t - \frac{z}{c}\right) - \psi_0 \right] \right\}, \tag{94}$$

where we have introduced the symbols $\bar{E}$, $\mathscr{A}$ with obvious meaning. Hence, part of the incident light is transferred from the resonant mode, of frequency ω_0, to the adjacent modes, of frequencies $\omega_0 \pm \alpha_1$, which experience gain and build up. This gives rise to the undamped spiking behavior.

When the system is in the instability region, the dynamics involve a competition between the resonant mode and the unstable modes: when the unstable modes prevail the system approaches the self-pulsing regime; when the resonant mode dominates the dynamics, the system precipitates to the low transmission branch.

2.4.4. *Case of αL large. Chaotic self-pulsing*

Up to now, in studying the stability of the steady states, we have always assumed that αL is small. In this subsection, we shall drop assumption (44). Actually, the first paper which predicted a positive slope instability (Bonifacio and Lugiato [1978b]) treated the case of absorptive optical bistability for general values of αL and T. Gronchi, Benza, Lugiato, Meystre and Sargent [1981] analyze the effects of increasing αL and T on the self-pulsing behavior in absorptive optical bistability.

In this subsection, we shall treat the absorptive + dispersive case. We start from the general Maxwell–Bloch equations (1), linearized around steady state. From eq. (3b), and taking into account the fact that the steady state field $E_{st}(z)$ obeys the boundary condition (3b) itself, we have that the boundary condition for the deviation $\delta E(z, t)$ from steady state is

$$\delta E(0, t) = R \exp(-\mathrm{i}\delta_0)\, \delta E(L, t - \Delta t). \tag{95}$$

The linearized Maxwell–Bloch equations admit solutions of exponential type, namely $\delta E(z, t)$, $\delta P(z, t)$, $\delta D(z, t) \propto \exp(\lambda t)$. By inserting this ansatz into the linearized Maxwell–Bloch equations and using the boundary condition (95), one obtains an equation that determines the eigenvalues. In the limits (44), and $k \ll \gamma_\perp, \gamma_\parallel$ this equation reduces to the one discussed in § 2.4.1.

For the sake of simplicity, let us now assume that the cavity transit time $\mathscr{L}/c$

is much longer than all atomic relaxation times, i.e.

$$\gamma_\perp \mathscr{L}/c \gg 1, \qquad \gamma_\parallel \mathscr{L}/c \gg 1. \tag{96}$$

In this situation $\gamma_\perp$ and $\gamma_\parallel$ are much larger than both the real and the imaginary part of many of the relevant eigenvalues λ_{n1} and λ_{n4}. In fact, Re $\lambda_{n{1 \atop 4}}$ is proportional to the cavity linewidth $k = cT/\mathscr{L}$, while Im $\lambda_{n{1 \atop 4}}$ is on the order of $\alpha_n = 2\pi nc/\mathscr{L}$, $n = 0, \pm 1, \ldots$. In the limit

$$\gamma_\perp \mathscr{L}/c \to \infty, \qquad \gamma_\parallel \mathscr{L}/c \to \infty, \tag{96$'$}$$

$\gamma_\perp$ and $\gamma_\parallel$ are much larger than both the real and the imaginary parts of *all* the eigenvalues λ_{n1} and λ_{n4}. Hence, in the linearized equations for $\delta P(z, t)$ and $\delta D(z, t)$ one can neglect λ with respect to $\gamma_\perp$ and $\gamma_\parallel$, which leads to a simplified eigenvalue equation. A more straightforward procedure to derive the same equation is the following: on the basis of assumption (96′) we can adiabatically eliminate the atomic variables in the Maxwell–Bloch equations, that is we can set $\partial P/\partial t = \partial D/\partial t = 0$ in eqs. (1b) and (1c). Thus, using the variables ρ_F and φ defined by eq. (14) and by $\rho_F = \mu\rho/\hbar(\gamma_\perp \gamma_\parallel)^{1/2}$, we obtain the equations

$$\frac{\partial \rho_F}{\partial z} + \frac{1}{c}\frac{\partial \rho_F}{\partial t} = -\alpha \frac{\rho_F}{1 + \Delta^2 + \rho_F^2}, \tag{97a}$$

$$\frac{\partial \varphi}{\partial z} + \frac{1}{c}\frac{\partial \varphi}{\partial t} = \frac{\alpha \Delta}{1 + \Delta^2 + \rho_F^2}. \tag{97b}$$

Next, we linearize these equations around the steady state. By setting

$$\delta\rho_F(z, t) = \rho_F(z, t) - \rho_{F,st}(z), \qquad \delta\varphi(z, t) = \varphi(z, t) - \varphi_{st}(z)$$

we have

$$\frac{\partial \delta\rho_F}{\partial z} + \frac{1}{c}\frac{\partial \delta\rho_F}{\partial t} = -\alpha \frac{1 + \Delta^2 - \rho_{F,st}^2(z)}{[1 + \Delta^2 + \rho_{F,st}^2(z)]^2}\,\delta\rho_F, \tag{98a}$$

$$\frac{\partial \delta\varphi}{\partial z} + \frac{1}{c}\frac{\partial \delta\varphi}{\partial t} = -\alpha\Delta \frac{2\rho_{F,st}(z)}{[1 + \Delta^2 + \rho_{F,st}^2(z)]^2}\,\delta\varphi. \tag{98b}$$

By combining eq. (98b) with the stationary equation

$$\partial\rho_{F,st}/\partial z = -\alpha\rho_{F,st}(z)/[1 + \Delta^2 + \rho_{F,st}^2(z)]$$

(see eqs. (15a), (5), and (6)) and defining $\delta\overline{\varphi}(z, t) = \rho_{F,st}(z)\,\delta\varphi(z, t)$, we obtain

$$\frac{\partial\delta\overline{\varphi}}{\partial z} + \frac{1}{c}\frac{\partial\delta\overline{\varphi}}{\partial t} = -\frac{\alpha}{1 + \Delta^2 + \rho^2_{F,st}(z)}\delta\overline{\varphi} - \frac{2\alpha\Delta\rho^2_{st}(z)}{[1 + \Delta^2 + \rho^2_{F,st}(z)]^2}\delta\rho_F. \quad (98c)$$

The boundary condition (95) can be rephrased as

$$\delta\rho_F(0, t) = R\left[(\cos\Phi)\,\delta\rho_F\left(L, t - \frac{\mathscr{L} - L}{c}\right) - (\sin\Phi)\,\delta\overline{\varphi}\left(L, t - \frac{\mathscr{L} - L}{c}\right)\right], \quad (99a)$$

$$\delta\overline{\varphi}(0, t) = R\left[(\sin\Phi)\,\delta\rho_F\left(L, t - \frac{\mathscr{L} - L}{c}\right) + (\cos\Phi)\,\delta\overline{\varphi}\left(L, t - \frac{\mathscr{L} - L}{c}\right)\right], \quad (99b)$$

where

$$\Phi = \varphi_{st}(L) - \varphi_{st}(0) - \delta_0. \quad (100)$$

Two relations which will be useful in the sequel and which follow from eqs. (15), (5), and (6) are

$$\rho_{F,st}(z) = \rho_{F,st}(0)\exp\{-\overline{\alpha}z\overline{D}_{st}(z)\}, \quad (101a)$$

$$\varphi_{st}(z) - \varphi_{st}(0) = \overline{\alpha}z\Delta\overline{D}_{st}(z), \quad (101b)$$

where $\overline{D}_{st}(z)$ is the space average of the population difference at steady state from 0 to z:

$$\overline{D}_{st}(z) = \frac{1}{z}\int_0^z dz'\,D_{st}(z') = \frac{1}{z}\int_0^z dz'\,\frac{1 + \Delta^2}{1 + \Delta^2 + \rho^2_{F,st}(z')}, \quad (102)$$

and $\overline{\alpha} = \alpha/(1 + \Delta^2)$ is the absorption coefficient off resonance. Equation (101a) generalizes the well known Beer absorption law to the case of a nonlinear medium. Note that $\exp[\overline{\alpha}L\overline{D}(L)]$ coincides with the quantity η defined by eq. (17).

From now on, we follow the treatment of BONIFACIO and LUGIATO [1978b]. First, we introduce the ansatz

$$\begin{Bmatrix}\delta\rho_F(z, t)\\ \delta\overline{\varphi}(z, t)\end{Bmatrix} = e^{\lambda t}\begin{Bmatrix}\delta\rho_F(z)\\ \delta\overline{\varphi}(z)\end{Bmatrix} + \text{c.c.} \quad (103)$$

and integrate eqs. (98a) and (98c), obtaining

$$\delta\rho_{\mathrm{F}}(z) = C_1 \exp\left[-\frac{\lambda}{c}z - \bar{\alpha}z\bar{D}_{\mathrm{st}}(z)\right]\frac{1+\Delta^2+\rho^2_{\mathrm{F,st}}(0)}{1+\Delta^2+\rho_{\mathrm{F,st}}(z)}, \tag{104a}$$

$$\delta\bar{\varphi}(z) = \exp\left[-\frac{\lambda}{c}z - \bar{\alpha}z\bar{D}_{\mathrm{st}}(z)\right]\left[C_2 + C_1\Delta\frac{\rho^2_{\mathrm{F,st}}(z)-\rho^2_{\mathrm{F,st}}(0)}{1+\Delta^2+\rho^2_{\mathrm{F,st}}(z)}\right], \tag{104b}$$

where C_1 and C_2 are arbitrary constants. Next, we insert eqs. (103) and (104) into eqs. (99a) and (99b), thus obtaining a homogeneous system of two equations for C_1 and C_2. The condition that this system admits nontrivial solutions gives the simplified eigenvalue equation. By setting

$$v = R\exp\left\{-\frac{\lambda\mathscr{L}}{c} - \bar{\alpha}L\bar{D}_{\mathrm{st}}(L)\right\}, \tag{105a}$$

$$B = \left(\frac{1+\Delta^2+\rho^2_{\mathrm{F,st}}(0)}{1+\Delta^2+\rho^2_{\mathrm{F,st}}(L)}\right)^{1/2}, \tag{105b}$$

$$\begin{aligned}\mathscr{S} = {} & [(1+\Delta^2+\rho^2_{\mathrm{F,st}}(0))(1+\Delta^2+\rho^2_{\mathrm{F,st}}(L))]^{-1/2} \\ & \times\left[\frac{\Delta}{2}(\rho^2_{\mathrm{F,st}}(0)-\rho^2_{\mathrm{F,st}}(L))\sin\Phi \right. \\ & \left. + \left(1+\Delta^2+\frac{\rho^2_{\mathrm{F,st}}(0)+\rho^2_{\mathrm{F,st}}(L)}{2}\right)\cos\Phi\right],\end{aligned} \tag{105c}$$

the equation reads

$$Bv^2 - 2\mathscr{S}v + B^{-1} = 0. \tag{106}$$

Hence, the eigenvalues are given by

$$\lambda_{n,\pm} = -\frac{c}{\mathscr{L}}\ln\frac{\mathscr{S}\pm(\mathscr{S}^2-1)^{1/2}}{BR\exp(-\bar{\alpha}L\bar{D}(L))} + \mathrm{i}\alpha_n, \qquad n = 0, \pm 1, \ldots. \tag{107}$$

Using eqs. (100) and (101), the eigenvalues are expressed as functions of $\bar{\alpha}$, Δ, δ_0, $\rho^2_{\mathrm{st}}(0)$ and $\bar{D}_{\mathrm{st}}(L)$. From these expressions, one sees (LUGIATO, ASQUINI and NARDUCCI [1982]) that eq. (107) coincides with the expression of the eigenvalues given in SNAPP, CARMICHAEL and SCHIEVE [1981] and CARMICHAEL, SNAPP and SCHIEVE [1982] obtained from the treatment of optical bistability in a ring cavity given by IKEDA [1979]. In fact, roughly one year after the work of BONIFACIO and LUGIATO [1978a,b] on the absorptive case, Ikeda reconsidered the same model but in the general absorptive and dispersive case. By adiabatically eliminating the atomic polarization and population difference under conditions (96′), he transformed the Maxwell–Bloch equations with boundary conditions into a system of finite difference equations, with a time step equal to the cavity transit time. By linearization of the system, we obtain the eigenvalues (107). Note that all the eigenvalues λ_{n+} have the same real part and the same holds for the eigenvalues λ_{n-}. Hence, all the frequencies become simultaneously unstable when Re $\lambda_{n\pm} > 0$. This is a consequence of the limit (96′). On the other hand, if one does not perform the limit (96′), only a finite set of frequencies become unstable as we have seen in § 2.4.1.

From eq. (107), one has that the steady state is unstable when

$$2\bar{B}|\mathscr{S}| > 1 + \bar{B}^2, \qquad \bar{B} = BR \exp[-\bar{\alpha}L\bar{D}(L)] \tag{108}$$

For $\mathscr{S} > 0$ the condition $2\bar{B}\mathscr{S} > 1 + \bar{B}^2$ is equivalent to the condition $\mathrm{d}Y/\mathrm{d}X < 0$, as one easily verifies using eqs. (22), (25) and (17). On the other hand, for $\mathscr{S} < 0$ the condition $2\bar{B}\mathscr{S} < -(1 + \bar{B}^2)$, which can be satisfied only for Δ and θ different from zero, leads to instabilities in the parts of the curve $X(Y)$ with positive slope. Under suitable conditions (IKEDA [1979]) several-segments of the curve $X(Y)$ are unstable, and one can find ranges of values of the incident intensity in correspondence to which no stable steady state exists. In this situation the system shows a self-pulsing behavior, which can be either a regular self-pulsing with a period roughly equal to an even multiple of the cavity transit time and square-wave type pulses (Fig. 18b), or a *chaotic self-pulsing* (Fig. 18c). As it has been shown in SNAPP, CARMICHAEL and SCHIEVE [1981], this period doubling bifurcation behavior, which finally leads to chaos, is in agreement with the general theory of nonlinear maps developed by FEIGENBAUM [1978, 1979]. Following the suggestions of IKEDA, DAIDO and AKIMOTO [1980], GIBBS, HOPF, KAPLAN and SHOEMAKER [1981] built a hybrid electro-optical device which reproduced this type of behavior (Fig. 18, see also OKADA and TAKIZAWA [1981]). The problem of the possibility of instabilities of this type has been studied in FIRTH [1981] and ABRAHAM, FIRTH and WRIGHT [1982] for Fabry–Perot cavities and by WINFUL and COOPERMAN [1982] for distributed feedback bistable optical devices. KITANO, YABUZAKI and OGAWA [1981b] predicted a magnetically induced self-pulsing in optical tristability.

Since the instability predicted in BONIFACIO and LUGIATO [1978b] was derived in a context of differential equations, whereas the instability found in

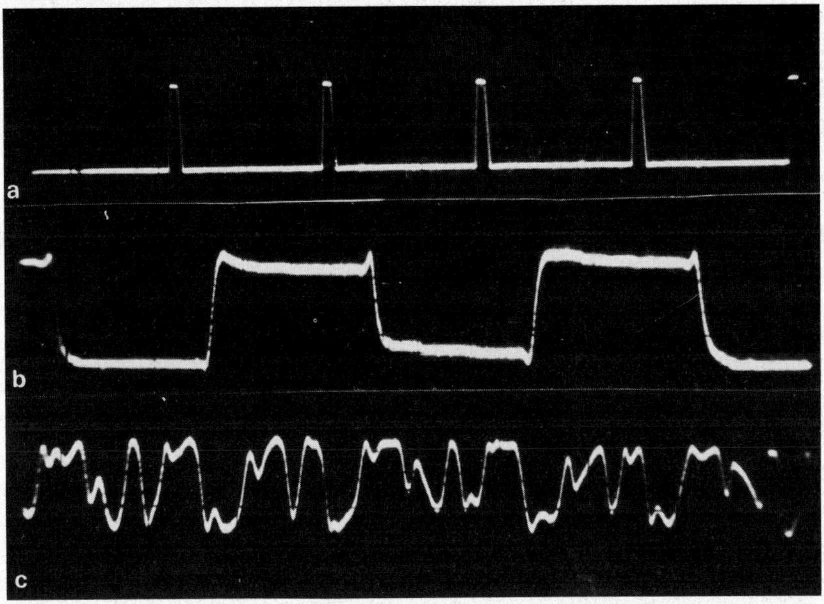

Fig. 18. Observation of chaotic behavior in a hybrid device (from GIBBS, HOPF, KAPLAN and SHOEMAKER [1981]). (a) Time calibration; one pulse every cavity transit time, equal to 40 ms. (b) Output intensity versus time in the periodic domain. (c) Intensity versus time in the chaotic regime.

IKEDA [1979] arises in a context of difference equations (nonlinear maps), for a long time the two instabilities have been considered as substantially different matters. The Bonifacio–Lugiato instability was usually considered as associated with absorptive optical bistability, and the Ikeda instability with dispersive optical bistability. However, this is wrong because, as we have shown in § 2.4.1, the Bonifacio–Lugiato instability also arises in the dispersive case. As was first proven in LUGIATO, ASQUINI and NARDUCCI [1982] and we have explicitly reported here, the Ikeda instability is a special case of the Bonifacio–Lugiato instability and it arises precisely in the limit (96′). The fundamental contribution of IKEDA [1979] was the prediction of chaos in optical bistability.

As we have seen in § 2.4.3, in the case when $\alpha L \ll 1$, $T \ll 1$, the fundamental pulsation frequency is α_1, which corresponds to a period equal to the cavity transit time $\mathscr{L}/c$. On the other hand, in the Ikeda instability the period is twice the cavity transit time. This feature can be easily understood from eq. (107). In fact,the instability occurs for $\mathscr{S} < -1$ and concerns the eigenvalues λ_{n+}.

Hence, we have

$$\operatorname{Im} \lambda_{n+} = \frac{c}{\mathscr{L}} \pi + \alpha_n = 2\pi \left(2\frac{\mathscr{L}}{c}\right)^{-1} (1 + 2n),$$

which gives a period $2\mathscr{L}/c$.

The expression (107) of the eigenvalues holds for general values of αL, T and δ_0, in the limit (96′). Now, let us consider the case (44) which is perfectly compatible with (96′). In this situation the eigenvalues reduce to

$$\lambda_{n,\pm} = -\mathrm{i}\alpha_n - k\left\{1 + \frac{2C(1+\Delta^2)}{(1+\Delta^2+X)^2} \pm \mathscr{H}^{1/2}\right\}, \tag{109}$$

with

$$\mathscr{H} = \frac{4C(1+\Delta^2)}{(1+\Delta^2+X)^2}\left\{\frac{X^2 - \Delta^2(1+\Delta^2)}{(1+\Delta^2+X)^2} + C\Delta\theta\right\} - \theta^2. \tag{110}$$

The expressions (109) coincide with those of λ_{n1}, λ_{n4} given by eqs. (78–81) if one introduces the limit (96′), that is if one sets $\tilde{\alpha}_n = \alpha_n/\gamma_\perp = 0$. Hence, the real part of λ_{n+}, λ_{n-}, as given by (109) coincides with Re λ_{04} and Re λ_{01} and therefore, as demonstrated in § 2.4.2, is never positive in the positive slope parts of the graph $X = X(Y)$. This means that the Ikeda instability disappears in the mean field limit (27) and requires αL and T to be large enough. However, as we have shown in § 2.4.1, one can still have a self-pulsing instability if one drops the condition (96).

On the other hand, Re $\lambda_{n,\pm}$, as given by eq. (109), can be positive when $C < 0$, as in the case of the laser with injected signal (see § 2.1.7). Hence, in this case the Ikeda instability can arise even in the mean field limit (27).

Recently, it has been shown that *chaotic behavior* in dispersive optical bistability can also arise *in the mean field limit* (27) with *single-mode operation*, both for Kerr medium (IKEDA and AKIMOTO [1982]) and in the mean field model (55) when $\gamma_\perp \gg \gamma_\parallel, k$, while k and $\gamma_\parallel$ have the same order of magnitude (LUGIATO, NARDUCCI, BANDY and PENNISE [1982]). Decreasing the incident field along the high transmission branch, one finds the appearance of regular pulsations that undergo a sequence of period doubling bifurcations, which finally terminates into chaos.

2.4.5. *Observability of self-pulsing in all-optical systems*

Let us now briefly describe the ideal conditions in which to observe self-pulsing in all-optical systems. First, one must use a ring cavity, because the instability region is strongly reduced in the case of a Fabry–Perot (CASAGRANDE, LUGIATO and ASQUINI [1980]). Second, one should fulfil as closely as possible the plane wave condition, since one cannot exclude the fact that the radial shape of the electric field decreases or destroys the self-pulsing instability region. Third, one must use a homogeneously broadened system, because inhomogeneous broadening drastically reduces the gain $\mathscr{G}_n$ (see eq. (92)) (MCCALL [1974]). Furthermore, the elastic collision broadening should not be large, because otherwise no positive slope instability arises.

The dispersive situation Δ, $\theta \neq 0$ is preferable because, by playing with parameters, one can obtain the self-pulsing instability without bistability, and therefore one can exclude a priori the possibility of precipitation to the lower branch, which is a competitive process with respect to self-pulsing.

In order to ensure that at least one off-resonance cavity frequency lies in the instability range $\alpha_{min} < \alpha_n < \alpha_{max}$, it is necessary that the frequency spacing $2\pi c/\mathscr{L}$ is not larger than the power broadened linewidth $\gamma_\perp(1+X)^{1/2}$, because $\alpha_{max} < \gamma_\perp(1+X)^{1/2}$. For $\gamma_\perp = \gamma_\parallel, X \gg 1$, we have $\gamma_\perp(1+X)^{1/2} \approx \gamma_\perp x = \Omega_T$. where Ω_T is the Rabi frequency of the internal field $\mu E_T/\hbar\sqrt{T}$ (compare eq. (61)). Hence, long cavities, and large power are necessary. We stress that the Ikeda instability discussed in the previous section requires the much more stringent condition $c/\mathscr{L} \ll \gamma_\perp$ (see eq. (96)).

Another disadvantage of the Ikeda instability is that it requires very large values of αL, on the order of 100. From all these considerations, it appears that the observation of self-pulsing is not easy, but not impossible. It remains an exciting challenge for the experimentalists.

2.5. DRESSED MODE THEORY OF OPTICAL BISTABILITY. ANALYTICAL TREATMENT OF SELF-PULSING

It is hardly reasonable to expect that a complete understanding of the self-pulsing behavior can emerge from the direct numerical solutions of the Maxwell–Bloch equations, which most often amount to a crude registration of data, without any predictive power. Furthermore, one meets with the following difficulty. In order to render the situation more transparent it is useful to consider the case (44). In this limit, however, one can identify two widely separated time scales: the characteristic duration time of each pulse, which is on the order of the cavity transit time $\mathscr{L}/c$, and the characteristic variation time of the envelope of the pulses, which is longer by at least a factor T^{-1}. Thus, the problem becomes numerically complicated. These considerations point to the need for a formulation of the problem that is analytical or quasi-analytical so as to reduce the amount of numerical work. This goal can be achieved by the so-called *dressed mode theory of optical bistability* (BENZA and LUGIATO [1979b]), which is a development of Haken's theory of generalized Ginzburg–Landau equations for phase transition-like phenomena in systems far from thermal equilibrium (HAKEN [1975a,b], HAKEN and OHNO [1976a,b]). This formalism allows one to select the few dominant variables out of the infinite number of degrees of freedom of the system, thereby reducing the problem from the level of Maxwell–Bloch equations to a set of equations *in time only* for these "order parameters". This selection of variables is obtained using *Haken's adiabatic elimination principle*. In this crucial point our procedure deviates substantially from that of HAKEN [1975b]. In fact, Haken's method is iterative and therefore it is strictly limited to a neighborhood of the critical point, where the steady state becomes unstable. This limitation is common to all the standard bifurcation theory techniques (NICOLIS and PRIGOGINE [1977]). On the contrary, our adiabatic elimination is exact in the limit (27). Hence, it allows us to consider arbitrary deviations from the critical point.

We devote our attention mainly to the simplest situation in which only the two modes that are symmetrically adjacent to the resonant one are unstable. In this case, we can reduce the problem to a set of only two differential equations, that nonetheless fully includes both nonlinearity and propagation. Thus, we obtain a simple description in terms of a two-dimensional phase space, that allows us to explore the full domain of existence of the self-pulsing solution, until it becomes itself unstable, thereby giving rise to a higher order bifurcation.

In this section, we shall consider only the resonant case $\Delta = \theta = 0$. In such a situation, we can safely consider F and $\tilde{P}$ in eqs. (49) as real quantities. In fact, for $\Delta = \theta = 0$, the instabilities arise only from the real part of the variables, which produce the eigenvalues λ_{n1}, λ_{n2}, and λ_{n3}. Consideration of the imaginary parts does not add any interesting new feature, because the eigenvalues λ_{n4} and λ_{n5} do not attain a positive real part.

2.5.1. *The dressed mode formalism*

Let us start from eqs. (49) with $F^* = F$, $\tilde{P}^* = P$. Similarly to eq. (74), let us introduce the three-component vector

$$\boldsymbol{q}(z, t') = \begin{pmatrix} \delta F(z, t') \\ \delta P(z, t') \\ \delta D(z, t') \end{pmatrix}, \tag{111}$$

where δF, δP and δD are defined in eq. (72). Equations (49) can be reformulated in terms of the deviations (compare eqs. (73) with $\Delta = \theta = 0$)

$$\frac{\partial \boldsymbol{q}}{\partial t'} = \hat{\mathscr{L}} \boldsymbol{q} + \boldsymbol{\psi}_{\mathrm{NL}}, \tag{112}$$

where $\hat{\mathscr{L}}$ is the operator that includes all the linear terms of the equations:

$$\hat{\mathscr{L}} = \begin{pmatrix} -c \dfrac{L}{\mathscr{L}} \dfrac{\partial}{\partial z} - k & -2Ck & 0 \\ \gamma_\perp \tilde{D}_{\mathrm{st}} & -\gamma_\perp & \gamma_\perp x \\ -\gamma_\parallel \tilde{P}_{\mathrm{st}} & -\gamma_\parallel x & -\gamma_\parallel \end{pmatrix}, \tag{113}$$

while $\boldsymbol{\psi}_{\mathrm{NL}}$ is a vector which contains all the nonlinear terms

$$\boldsymbol{\psi}_{\mathrm{NL}} = \begin{pmatrix} 0 \\ \gamma_\perp \delta F \, \delta D \\ -\gamma_\parallel \delta F \, \delta P \end{pmatrix}. \tag{114}$$

Note that $\mathscr{L}$ is not only a matrix, but also an operator on the variable z, because it contains $\partial/\partial z$.

Now, let us consider the Hilbert space, defined as follows: the generic element $\boldsymbol{\psi}$ of the space has the structure

$$\boldsymbol{\psi} = \begin{pmatrix} \psi_1(z) \\ \psi_2(z) \\ \psi_3(z) \end{pmatrix}, \tag{115}$$

where $\psi_i(z)$ $(i = 1, 2, 3)$ are square integrable functions in the interval $0 < z < L$. The scalar product of two elements $\boldsymbol{\varphi}$, $\boldsymbol{\psi}$ is

$$(\boldsymbol{\varphi}, \boldsymbol{\psi}) = \sum_{i=1}^{3} \int_0^L \mathrm{d}z\, \varphi_i^*(z)\, \psi_i(z). \tag{116}$$

The vectors $\boldsymbol{q}$ and $\boldsymbol{\psi}_{\mathrm{NL}}$ are elements of the Hilbert space. The eigenvalue equation for the linear operator $\hat{\mathscr{L}}$ is

$$\hat{\mathscr{L}}\, \boldsymbol{O}_{nj} = \lambda_{nj}\, \boldsymbol{O}_{nj}, \qquad j = 1, 2, 3, \tag{117}$$

where the eigenvalues λ_{n1} are given by eqs. (78) and (84.1), and the eigenvalues λ_{n2} and λ_{n3} are given by eqs. (78′) and (83). The eigenstates $\boldsymbol{O}_{nj}$ have the expression

$$\boldsymbol{O}_{nj} = \exp(\mathrm{i}k_n z) \begin{pmatrix} G(\lambda_{nj}, x) \\ \dfrac{(\lambda_{nj}/\gamma_\parallel) + 1 - x^2}{1 + x^2} \\ -\dfrac{x}{1 + x^2}\left(\dfrac{\lambda_{nj}}{\gamma_\perp} + 2\right) \end{pmatrix}, \tag{118}$$

where

$$G(\lambda, x) = \left(\frac{\lambda}{\gamma_\perp} + 1\right)\left(\frac{\lambda}{\gamma_\parallel} + 1\right) + x^2. \tag{119}$$

Of course, the eigenstates are defined up to a constant factor. This has been chosen in such a way that the three elements of $\boldsymbol{O}_{nj}$ remain finite in the limit (27). In fact, as one easily verifies, $G(\lambda_{n_3^2}^{(0)}, x) = 0$.

At this point, we say that the indices n and j label the *modes* of our system. We call them *dressed* because they incorporate an exact part of the atom-field interaction, namely the part that is contained in the linear operator $\hat{\mathscr{L}}$. With respect to the modes introduced in § 2.2.2, the dressed modes also include the index j, which diagonalizes the linearized part of the atom-field interaction.

We note that Re λ_{n1} is proportional to the field relaxation rate k, whereas Re λ_{n2} and Re λ_{n3} are proportional to the atomic relaxation rates $\gamma_\perp$ and $\gamma_\parallel$. Hence, the dressed modes with $j = 1$ have a dominant field character, and accordingly we shall call them *field modes*. Similarly, we shall call the dressed modes with $j = 2, 3$ *atomic modes*.

In order to introduce suitable dressed mode amplitudes, it is necessary to consider the adjoint operator $\hat{\mathscr{L}}^\dagger$. The spectrum of $\hat{\mathscr{L}}^\dagger$ is obtained from the spectrum of $\hat{\mathscr{L}}$ by the complex conjugation

$$\hat{\mathscr{L}}^\dagger \overline{\boldsymbol{O}}_{nj} = \lambda_{nj}^* \overline{\boldsymbol{O}}_{nj}. \tag{120}$$

Note the relations

$$\lambda_{n1}^* = \lambda_{-n1}, \qquad \lambda_{n2}^* = \lambda_{-n3}. \tag{121}$$

The second equation in (121) holds provided that $\gamma_\parallel$ has the same order of magnitude as $\gamma_\perp$ and x is not too small. In the following, we shall always assume that these conditions are satisfied.

The eigenstates $\overline{\boldsymbol{O}}_{nj}$ have the form

$$\overline{\boldsymbol{O}}_{nj} = \frac{\exp(\mathrm{i}k_n z)}{L\,\mathscr{N}_{nj}^*} \begin{pmatrix} 1 \\ -2C\dfrac{k}{\gamma_\perp}\dfrac{(\lambda_{nj}^*/\gamma_\parallel)+1}{G(\lambda_{nj}^*, x)} \\ -2C\dfrac{k}{\gamma_\parallel}\dfrac{x}{G(\lambda_{nj}^*, x)} \end{pmatrix}, \tag{122}$$

where the normalization constant $\mathscr{N}_{nj}$, whose explicit expression is given in Appendix 2, is chosen in such a way that

$$(\overline{\boldsymbol{O}}_{nj}, \overline{\boldsymbol{O}}_{n'j'}) = \delta_{nn'}\,\delta_{jj'}. \tag{123}$$

Hence, the vectors $\{\boldsymbol{O}_{nj}\}$, $\{\overline{\boldsymbol{O}}_{nj}\}$ constitute a complete bi-orthonormal set of vectors in the Hilbert state. We can expand the vector $\boldsymbol{q}$ on the basis $\{\boldsymbol{O}_{nj}\}$:

$$\boldsymbol{q}(t') = \sum_{nj} \xi_{nj}(t')\,\boldsymbol{O}_{nj}, \qquad \xi_{nj}(t') = (\overline{\boldsymbol{O}}_{nj}, \boldsymbol{q}(t')). \tag{124}$$

The dimensionless variables ξ_{nj}, which depend on time only, are the dressed mode amplitudes. In particular, using eqs. (72), (111) and (118) we have the following expression of the normalized electric field

$$F(z, t') = x + \sum_{nj} \xi_{nj}(t')\,G(\lambda_{nj}, x)\exp(\mathrm{i}k_n z). \tag{125}$$

The expression of the amplitude f_n, defined in eq. (52), is therefore

$$f_n(t') = x\delta_{n0} + \sum_{j} \xi_{nj}(t')\,G(\lambda_{nj}, x). \tag{126}$$

Next, we substitute the expansion (124) into (112) and multiply scalarly eq. (112) times $\overline{\boldsymbol{O}}_{nj}$ on the left. Using eqs. (123), (117), (116) and (114) we obtain the time evolution equations for the amplitudes ξ_{nj}:

$$\dot{\xi}_{nj} = \lambda_{nj}\xi_{nj} + \sum_{n'j'}\sum_{n''j''} \Gamma(nj, n'j', n''j'')\xi_{n'j'}\xi_{n''j''}, \tag{127}$$

where the coefficients Γ are given by

$$\Gamma(nj, n'j', n''j'') = \delta_{n,n'+n''} O_{n''j'',1}\{\gamma_\perp O^*_{nj,2} O_{n'j',3} - \gamma_\parallel O^*_{nj,3} O_{n'j',2}\}. \tag{128}$$

In eq. (128) we have called $O_{nj,i}$ ($i = 1, 2, 3$) the three components of the vector $\boldsymbol{O}_{nj}$, without the exponential factor (see eq. (118)), and correspondingly we have called $\overline{O}_{nj,i}$ ($i = 1, 2, 3$) the three components of the vector $\overline{\boldsymbol{O}}_{nj}$, without the factor $L^{-1} \exp(ik_n z)$.

Equation (127) governs the dressed mode dynamics. It has a linear term which arises from the part $\hat{\mathscr{L}}\boldsymbol{q}$ of eq. (112) and several nonlinear terms, which arise from $\boldsymbol{\psi}_{NL}$, and rule the mode–mode coupling. The time evolution, starting from a situation in which the system is slightly displaced from an unstable stationary state, can be described as follows. In the initial stage, due to the smallness of the deviation $\boldsymbol{q}$ and hence, of ξ_{nj}, the linear part of eq. (127) is dominant. Therefore the unstable modes (Re $\lambda_{nj} > 0$) grow exponentially while the stable modes are exponentially damped. In due time, the nonlinear terms become important and eventually lead the system to a periodic self-pulsing regime. The expressions for the relevant coefficients Γ are given in Appendix B. The only important feature, in view of the following treatment, is the order of magnitude of these coefficients with respect to T. One has for any j'

$$\begin{aligned}\Gamma(n1, n'j', n''1) &= O(T),\\ \Gamma(n1, n'j', n''{}^{2}_{3}) &= O(T^2),\end{aligned} \tag{129}$$

$$\begin{aligned}\Gamma(n{}^{2}_{3}, n'j', n''1) &= O(1),\\ \Gamma(n{}^{2}_{3}, n'j', n''2) &= O(T),\\ \Gamma(n{}^{2}_{3}, n'j', n''3) &= O(T).\end{aligned} \tag{130}$$

In the following we shall restrict ourselves to the case $\gamma_\perp = \gamma_\parallel = \gamma$ in which all the expressions attain a maximum of simplicity. In this case, one finds that the coefficients $\Gamma(nj, n'j', n''j'')$ get a further factor of T with respect to that indicated in eqs. (130) when $j = 2$ and $j' = 3$ or $j = 3$ and $j' = 2$. Hence, eq.

(130) must be corrected as follows

$$\begin{aligned}
&\Gamma(n2, n'1, n''1),\ \Gamma(n2, n'2, n''1) = O(1),\\
&\Gamma(n2, n'3, n''1) = O(T),\\
&\Gamma(n2, n'1, n''2),\ \Gamma(n2, n'2, n''2),\\
&\Gamma(n2, n'1, n''3),\ \Gamma(n2, n'2, n''3) = O(T),\\
&\Gamma(n2, n'3, n''2),\ \Gamma(n2, n'3, n''3) = O(T^2),
\end{aligned} \tag{131}$$

The order of magnitude of the coefficients $\Gamma(n3, n'j', n''j'')$ is immediately obtained by taking into account the relation

$$\Gamma^*(nj, n'j', n''j'') = \Gamma(-n\bar{j}, -n'\bar{j}', -n''\bar{j}''), \tag{132}$$

where $\bar{j} = 1$ for $j = 1$, $\bar{j} = 3$ for $j = 2$, $\bar{j} = 2$ for $j = 3$.

The dressed mode formalism outlined in this subsection is essentially a simplification of the procedure of HAKEN [1975a,b]. The distinction between field and atomic modes was introduced in BENZA and LUGIATO [1979b]. As we shall see in the next subsection, this distinction is important for the following developments.

2.5.2 *Adiabatic elimination of the atomic modes* (BENZA and LUGIATO [1982])

For the following steps, it is suitable to reformulate eq. (127) in terms of the variables S_{nj}, defined by

$$\xi_{nj}(t') = S_{nj}(t') \exp(-\mathrm{i}\alpha_n t'). \tag{133}$$

Using eq. (128), we obtain

$$\dot{S}_{nj} = (\lambda_{nj} + \mathrm{i}\alpha_n) S_{nj} + \sum_{n'j'} \sum_{n''j''} \Gamma(nj, n'j', n''j'') S_{n'j'} S_{n''j''}. \tag{134}$$

Accordingly, eq. (125) becomes

$$F(z, t') = x + \sum_{nj} S_{nj}(t') G(\lambda_{nj}, x) \exp\left[-\mathrm{i}\alpha_n \left(t' - \frac{z\mathscr{L}}{cL}\right)\right]. \tag{135}$$

Equation (134) gives us an infinite system of equations that is fully equivalent to the original Maxwell–Bloch equations. In order to reduce the problem to a tractable dimension, the first step is to select the frequencies to be considered. Of course this selection is guided by physical intuition, on the basis of the frequencies that are unstable. Essentially, one must guess which frequencies n have non-negligible dressed mode amplitudes S_{nj}. Of course, for each selected frequency n, one must also consider the corresponding frequency $-n$, because $S_{-n1} = S^*_{n1}$, $S_{-n2} = S^*_{n3}$.

After this selection the number of equations (134) for the amplitudes S_{nj} becomes finite, but the system is still too complicated even in the simplest setting of two unstable frequencies that we shall consider in the next section. The second crucial step to simplify the problem is the systematic use of the limit (27). More precisely, we shall perform the limit (27) simultaneously with the limit

$$t' \to \infty, \qquad \tau = kt' = \text{const.} \tag{136}$$

In fact, if one performs only the limit (27), one trivializes the time evolution of the field variables, because one eliminates the interaction of the field with the atoms ($\alpha L \to 0$). This is no longer the case if we also let $t' \to \infty$, with $\tau = cTt'/\mathscr{L}$ kept constant.

Let us now analyze in detail the consequences that the use of the limits (27) and (136) imposes on eqs. (134). The first consequence is that the atomic modes, which vary at a rate γ, in this limit attain a stationary situation (i.e. $\mathrm{d}S_{n2}/\mathrm{d}t' = \mathrm{d}S_{n3}/\mathrm{d}t' = 0$). On the contrary, the field modes, which vary at a rate k, still possess a time evolution as function of τ. Hence, in the limits (27) and (136) the atomic modes are *adiabatically eliminated in an exact way*. After setting $\mathrm{d}S_{n2}/\mathrm{d}t' = \mathrm{d}S_{n3}/\mathrm{d}t' = 0$ one obtains an algebraic system of equations for the atomic mode amplitudes

$$0 = (\lambda_{n^2_3} + \mathrm{i}\alpha_n)S_{n^2_3} + \sum_{n'j'} \sum_{n''j''} \Gamma(n^2_3, n'j', n''j'')S_{n'j'}S_{n''j''}. \tag{137}$$

By solving this system one finds the expressions of the atomic amplitudes S_{n2}, S_{n3} as functions of the field amplitudes S_{n1}. In general, the nonlinearity of the algebraic equations renders it impossible to solve these equations analytically, and therefore one must use approximate methods, such as the iterative procedure of HAKEN [1975a,b]. However, in our case the limits (27) and (136) again introduce a decisive simplification. In fact, let us consider the terms in eq. (137) such that both factors $S_{n'j'}$ and $S_{n''j''}$ are of atomic type, that is both j' and j'' are equal to 2 or 3. As one sees from eqs. (131), the corresponding factor $\Gamma(n^2_3, n'j', n''j'')$ always vanishes in the limit (27). Thus, in this limit the system (137) becomes *linear* with respect to the atomic variables S_{n2} and S_{n3}, and therefore can be solved in a standard way. A further simplification is that eqs. (137) decouple into separate subsystems, one for the amplitudes S_{n2} and one for S_{n3}. In fact, as one sees again from eqs. (131), all the terms that couple the two sets of variables contain coefficients that vanish for $T \to 0$.

Hence, if in eqs. (137) we introduce the limit (27), the adiabatic elimination is performed *exactly and analytically*. The expressions for the atomic modes S_{n2} and S_{n3} as functions of the field modes S_{n1} are given by rational functions that do not depend on αL and T. By substituting these expressions into eqs. (134) with $j = 1$, one obtains a closed system of equations for the field modes, that play the role of order parameters in our problem. In the limit $T \ll 1$, all the terms in the thus obtained equations for the variables S_{n1} are proportional to k. In fact, as one sees from eqs. (78), $\lambda_{n1} + \mathrm{i}\alpha_n$ is proportional to k, apart from corrections of order T^2 that vanish in the limits (27) and (136). Furthermore, as one sees from eqs. (129), the coefficients $\Gamma(n1, n'j', n''j'')$, which are the only ones that survive in the limit (27), are proportional to T and hence, to k.

Therefore, using (136) the equations for the field modes have the form

$$\frac{\mathrm{d}S_{n1}}{\mathrm{d}\tau} = h_n(S_{01}, S_{-11}, S_{11}, \ldots). \tag{138}$$

where

$$h_n = \lambda_n^{(+)} S_{n1} + k^{-1} \sum_{n'j'} \Gamma(n1, n'j', (n-n')1)\, S_{n'j'}\, S_{(n-n')1}. \tag{139}$$

For $j' = 2, 3$, $S_{n'j'}$ is the rational function of S_{n1}, previously calculated in the adiabatic elimination of the atomic variables. Hence, h_n depends on αL and T only via $C = \alpha L/2T$.

Even when the number of quantities S_{n1} in play is not small, the system (138) can be numerically handled more easily than the Maxwell–Bloch equations, because it involves only the time variable, and most important of all, it solves the problem of separating the two main time scales of our problem (see § 2.5). The first time scale, on the order of the cavity transit time $\mathscr{L}/c$, is that of the pulses, and appears in the factor $\exp(-\mathrm{i}\alpha_n t')$ in eq. (135). The second time scale, on the order of k^{-1}, is that of the pulse envelope. Equations (138) govern the time evolution of the quantities S_{n1}, that evolve on the time scale k^{-1}.

We stress that the adiabatic elimination performed in this subsection is not so restrictive as that performed in § 2.4.4. Namely, it does not require that $\gamma\mathscr{L}/c \gg 1$, but only that $\gamma/k = \gamma\mathscr{L}/cT \gg 1$. In fact, thanks to the transformation (133), the variation of the field dressed mode amplitudes S_{n1} in time is ruled by $\lambda_{n1} + \mathrm{i}\alpha_n$, which is proportional to k (see eqs. (78)).

2.5.3. *The case of two unstable modes*

From now on we shall consider the simplest situation, that is the case in which only the two modes immediately and symmetrically adjacent to the resonant one are unstable. In this case, it is reasonable to guess that the dressed modes that play a dominant role are the three modes corresponding to the resonant frequency $n = 0$ plus the six modes corresponding to the adjacent frequencies $n = \pm 1$. This suggestion has been supported both by comparison with the numerical solutions of the Maxwell–Bloch equations (see next subsection) and by showing that inclusion of the adjacent modes with $n = 2$ never changes the results qualitatively, and quantitatively the correction is on the order of 10% (LUGIATO, BENZA, NARDUCCI and FARINA [1983]).

Let us now consider the nine equations (134) for the amplitudes $S_{nj}(n = +1, 0, -1; j = 1, 2, 3)$, taking into account only the terms that are relevant in the limits (27) and (136). As discussed in the previous section, in this limit one automatically has $\dot{S}_{n2} = \dot{S}_{n3} = 0$. Thus one obtains two independent triplets of algebraic equations that are linear with respect to the atomic amplitudes S_{n2} and S_{n3}. The solution of the first system provides the expressions of S_{n2} as functions of the field amplitudes S_{n1}, and the solution of the second system the expressions of S_{n3}. Note that $S_{n3} = S^*_{-n2}$. By substituting these expressions into the equations for the field amplitudes one obtains a closed set of three equations for S_{11}, S_{01} and S_{-11} such as (138). These equations read

$$\frac{\mathrm{d}S_{11}}{\mathrm{d}\tau} = S_{11}\,\bar{f}\{(S_{11}\cdot S_{-11}), S_{01}\},$$

$$\frac{\mathrm{d}S_{01}}{\mathrm{d}\tau} = g\{(S_{11}\cdot S_{-11}), S_{01}\},$$

$$\frac{\mathrm{d}S_{-11}}{\mathrm{d}\tau} = S_{-11}\,\bar{f}^*\{(S_{11}\cdot S_{-11}), S_{01}\}, \tag{140}$$

where the functions $\bar{f}$ and g depend on S_{11} and S_{-11} only via the product $S_{11}\cdot S_{-11}$, and g is real. Furthermore, the only parameters that appear in $\bar{f}$ and g are C, x and $\tilde{\alpha}_1 = 2\pi c/\mathscr{L}\gamma$. By setting

$$S_{11} \equiv \rho_1 \mathrm{e}^{\mathrm{i}\varphi}, \qquad S_{01} \equiv \sigma \text{ real}, \tag{141}$$

we obtain from eqs. (140) (BENZA and LUGIATO [1982])

$$\frac{\mathrm{d}\rho_1}{\mathrm{d}\tau} = f(\rho_1, \sigma), \tag{142a}$$

$$\frac{\mathrm{d}\sigma}{\mathrm{d}\tau} = g(\rho_1, \sigma), \tag{142b}$$

$$\frac{\mathrm{d}\varphi}{\mathrm{d}\tau} = l(\rho_1, \sigma), \tag{143}$$

where

$$\begin{aligned} f(\rho_1, \sigma) &= \rho_1 \operatorname{Re} \bar{f}(\rho_1^2, \sigma), \\ l(\rho_1, \sigma) &= \operatorname{Im} \bar{f}(\rho_1^2, \sigma). \end{aligned} \tag{144}$$

The explicit expressions for the functions f and g are given in Appendix C. Clearly, eqs. (142a) and (142b) form a closed system of equations for ρ_1 and σ. On the other hand, eq. (143) has the solution

$$\varphi(\tau) = \varphi(\tau = 0) + \int_0^\tau \mathrm{d}\tau' \, l(\rho_1(\tau'), \sigma(\tau')), \tag{145}$$

so that the expression of $\varphi(\tau)$ is obtained once the system (142) has been solved.

Let us now come back to eq. (135). Since $G(\lambda_{n\frac{2}{3}}^{(0)}, x) = 0$ (because $\lambda_{n2}^{(0)}$ and $\lambda_{n3}^{(0)}$ are the roots of the equation $G(\lambda, x) = 0$)) in the limit (27) only the terms with $j = 1$ contribute. Using (141) and setting

$$\begin{aligned} \tilde{\rho}(\tau) &= 2\rho_1(\tau) |G(-\mathrm{i}\alpha_1, x)|, \\ \tilde{\sigma}(\tau) &= \sigma(\tau) \, G(0, x), \end{aligned} \tag{146}$$

eq. (135) reads

$$\begin{aligned} F(z, t') &= x + \tilde{\sigma}(\tau) \\ &\quad + \tilde{\rho}(\tau) \cos\left\{ -\frac{2\pi c}{\mathscr{L}} \left(t' - \frac{z\mathscr{L}}{cL} \right) + \varphi(\tau) + \tilde{\theta} \right\}, \\ \tau &= kt', \qquad \tilde{\theta} = \arg G(-\mathrm{i}\alpha_1, x). \end{aligned} \tag{147}$$

Hence, the upper and lower envelopes of the time evolution are given by

$$F_{\substack{\text{upper} \\ \text{lower}}}(\tau) = x + \tilde{\sigma}(\tau) \pm \tilde{\rho}(\tau). \tag{148}$$

Equation (148) shows that $\tilde{\sigma}(\tau)$ is the half-amplitude of the oscillations, while $\tilde{\sigma}(\tau)$ is the difference between the mean value of the oscillations and the stationary value x. For $\tau \to \infty$, when the system precipitates to the low transmission state, $\tilde{\rho}(\infty) = 0$ and $x + \tilde{\sigma}(\infty)$ is the value of the normalized transmitted light in the low transmission branch. When the system approaches a steady self-pulsing behavior $\tilde{\rho}(\infty) \neq 0$, and as one sees from eq. (145) $\varphi(\tau)$ has the form

$$\tau \to \infty, \qquad \varphi(\tau) = c_1 + c_2 \tau, \tag{149}$$

where c_1 and c_2 are constant. Now

$$-\frac{2\pi c}{\mathscr{L}} t' + c_2 \tau = \left(-\frac{2\pi c}{\mathscr{L}} + c_2 k \right) t' .$$

In the limit (27), $c_2 k$ gives a vanishing correction to the frequency $2\pi c/\mathscr{L}$ and hence, we can drop it. Furthermore, we can safely assume that $\varphi(\tau = 0)$ is such that $c_1 = 0$. Therefore for $\tau \to \infty$ (147) gives

$$F(z, t') = x + \tilde{\sigma}(\infty) + \tilde{\rho}(\infty) \cos \left\{ -\frac{2\pi c}{\mathscr{L}} \left(t' - \frac{z}{c'} \right) + \tilde{\theta} \right\} . \tag{150}$$

Note that $t' - (z/c') = t - (z/c)$ (see eq. (47)). Equation (150) describes the asymptotic pulses which have a period equal to the cavity transit time $\mathscr{L}/c$. The envelope for $\tau \to \infty$ becomes perfectly flat and is given by $x + \tilde{\sigma}(\infty) \pm \tilde{\rho}(\infty)$.

Thus, we can reason in terms of the two-dimensional phase space of the variables ρ_1 and σ. In particular, the stationary solutions $\rho_{1\text{st}}$ and σ_{st} of eqs. (142) (i.e. the solutions of the equations $f(\rho_{1\text{st}}, \sigma_{\text{st}}) = g(\rho_{1\text{st}}, \sigma_{\text{st}}) = 0$) correspond to asymptotic (i.e. $\rho_{1\text{st}} = \rho_1(\infty)$, $\sigma_{\text{st}} = \sigma(\infty)$) self-pulsing solutions if they lie out of the σ-axis ($\rho_{1\text{st}} > 0$), and correspond to cw solutions if $\rho_{1\text{st}} = 0$, because in the latter case the amplitude $\tilde{\rho}$ of the oscillations vanishes (see eqs. (146) and (150) with $\rho_1(\infty) = \rho_{\text{st}}$). The stability of a steady self-pulsing solution is checked by linearizing eqs. (142) around such a solution. At this point, we can appreciate what degree of simplification we have achieved by the dressed mode theory. In fact, we have reduced the problem from the level of Maxwell–Bloch equations to the following one: to draw the two lines $f(\rho_1, \sigma) = 0$ and $g(\rho_1, \sigma) = 0$ in order to find their intersections, that give the stationary solutions $\rho_{1\text{st}}, \sigma_{\text{st}}$. The position of these solutions depends on the values of the externally controllable parameters x and $\tilde{\alpha}_1$. Hence, the behavior of the system when we vary the external parameters is understood by simply looking at the displacements of the stationary solutions in the phase plane. This will be illustrated in the following section.

2.5.4. *Periodic self-pulsing behavior in the long time limit* $\tau \to \infty$ (BENZA and LUGIATO [1981, 1982])

Figure 14 (above) shows the S-shaped curve of transmitted versus incident field for $C = 20$, and indicates the points that are unstable when condition (88) is satisfied. In the following, we shall consider only the case in which at most the two modes $n = \pm 1$ adjacent to the resonant one are unstable, and we shall analyze the behavior of the system when we vary the two externally controllable parameters x and $\tilde{\alpha}_1$. We note that x is controlled by varying the incident field y; $\tilde{\alpha}_1 = 2\pi c/\mathscr{L}\gamma$ is controlled by varying the total length $\mathscr{L}$ of the ring cavity, while keeping constant the length L of the atomic sample (see Fig. 1). Hence, we shall explore the plane of the variables x and $\tilde{\alpha}_1$ shown in Fig. 15 for $C = 20$. The stationary state in the high transmission branch is unstable when the point x, $\tilde{\alpha}_1$ lies in the region bounded by the lines $\alpha_{max}(x)/\gamma$, $\alpha_{min}(x)/\gamma$, and $x = x_m$, where α_{max} and α_{min} are defined by eq. (88), and x_m corresponds to the lower bistability threshold (compare Fig. 8). We stress that we must not explore the whole plane x, $\tilde{\alpha}_1$, because in correspondence to a part of this plane other modes, different from the adjacent modes $n = \pm 1$, are unstable. For instance, in correspondence to the shaded region in Fig. 15, the modes $n = \pm 2$ are unstable. Hence, we must leave this region out of consideration. Similarly there is a region of the plane, not indicated in Fig. 15, in correspondence to which the modes $n = \pm 3$ are unstable, and so on. All these regions must be excluded from our analysis, because we use the equations of § 2.5.3 which consider only the modes with $n = 0, \pm 1$.

In the following, we shall show the phase plane of the variables ρ_1, σ in correspondence to several values of the variables x, $\tilde{\alpha}_1$. In the phase plane, we shall draw the two lines $f(\rho_1, \sigma) = 0$ and $g(\rho_1, \sigma) = 0$. Note from eqs. (144) that the line $f(\rho_1, \sigma) = 0$ is composed by the line $\rho_1 = 0$, and the line $\mathrm{Re}\, \bar{f}(\rho_1^2, \sigma) = 0$. Hence, the points $\{\rho_1 = 0, \sigma = \bar{\sigma}\}$, where $\bar{\sigma}$ are the solutions of the equation $g(0, \sigma) = 0$, are stationary solutions of the system (142) which correspond to the cw states of the system. Namely, for any given value of y we have three such solutions, which correspond to the three points A, B and C in Fig. 14. In particular, the point A in the high transmission branch corresponds to the origin of the phase plane $\rho_1 = \sigma = 0$. In all the graphs of the plane $\{\rho_1, \sigma\}$ which will follow, these three solutions will be always indicated by the same letters A, B and C.

First, let us analyze the behavior of the system when we move along the horizontal line (a) in Fig. 15. This corresponds to varying continuously the incident field along the high transmission branch, while the total length $\mathscr{L}$ of the cavity is kept fixed, so that $\tilde{\alpha}_1 = 8$. Figure 19 shows the lines Re $\bar{f}(\rho_1^2, \sigma) = 0$ (solid line) and $g(\rho_1, \sigma) = 0$ (dashed line) for four different values of y in increasing order (see Fig. 15). Each intersection of the two lines corresponds to a self-pulsing state provided that it is stable. We see that, as soon as we enter into the instability region from the left along the line (a) in Fig. 15, a stable self-pulsing state S bifurcates from the cw state A, which has become unstable (Fig. 19a). By increasing the incident field, the amplitude of the oscillations for the solution S increases. When we go out of the instability region a second solution U bifurcates from the cw state A which is now again stable (Fig. 19b). On the contrary, the solution U is unstable. By further increasing y the two

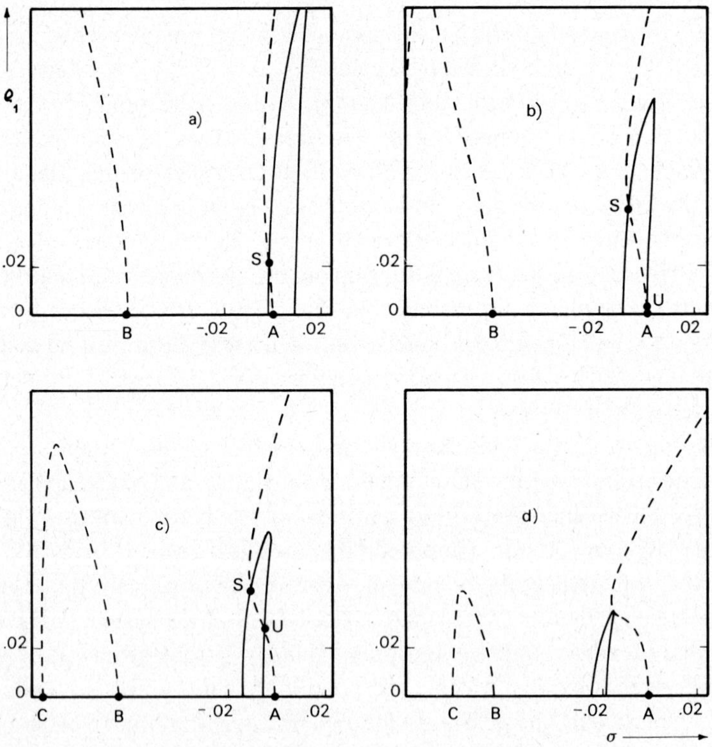

Fig. 19. Lines Re $\bar{f}(\rho_1^2, \sigma) = 0$ (solid line) and $g(\rho_1, \sigma)$ (dashed line) in the phase plane $\{\rho_1, \sigma\}$ for $C = 20$, $\tilde{\alpha}_1 = 8$ and: (a) $y = 13.262$ ($x = 8.75$); (b) $y = 13.664$ ($x = 9.75$); (c) $y = 13.96$ ($x = 10$); (d) $y = 15.31$ ($x = 12$).

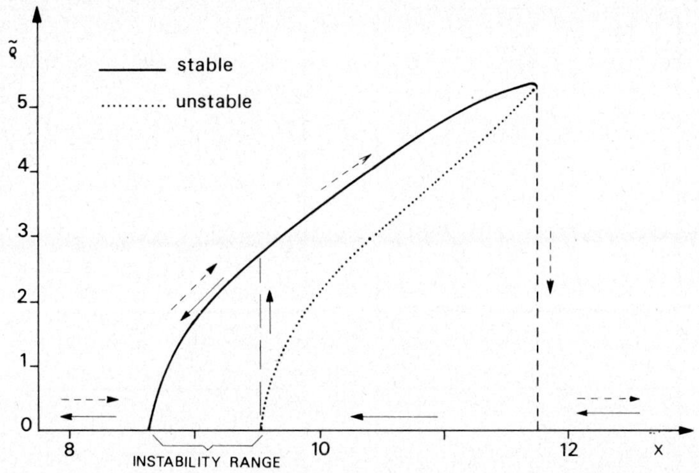

Fig. 20. Hysteresis cycle of the half-amplitude of the oscillations $\tilde{\rho}$ (in the long-time limit $\tau \to \infty$) as a function of the stationary transmitted field x in the high transmission branch for $C = 20$, $\tilde{\alpha}_1 = 8$. Solid (dashed) arrows indicate the behavior of the system for decreasing (increasing) incident field.

solutions S and U become nearer and nearer (Fig. 19c) until they coalesce and disappear together (Fig. 19d) so that the system can no longer show self-pulsing. The behavior of the system when we move along the line (a) in Fig. 15 is summarized in Fig. 20, where the half-amplitude of the oscillations $\tilde{\rho}$, for solution S (full line), is graphed versus the steady state value x of the transmitted light in the high transmission branch. As shown by the arrows, one finds hysteresis when one increases and decreases the incident field. Thus, we find a *hysteresis cycle of novel type* which involves both cw and pulsing states. Precisely, Fig. 20 exhibits a second-order and a first-order phase transition. The second-order one occurs at the left boundary of the instability range. The first-order transition is manifested by the hysteresis cycle that begins on the right boundary of the instability range, where the unstable self-pulsing solution U (dotted curve in Fig. (20)) bifurcates.

Let us now consider the behavior of the system when we move along the line (b) in Fig. 15, which corresponds to $\tilde{\alpha}_1 = 5$. Figure 21 shows the phase plane for the three values of y (that is x) shown in Fig. 15. We see that now, when we enter into the instability region, no self-pulsing solution is there (Fig. 21a). In this situation, the only stable solution of system (142) is the one corresponding to the low transmission cw state C, so that if the system is initially near to the unstable high transmission state A, it precipitates to the low transmission

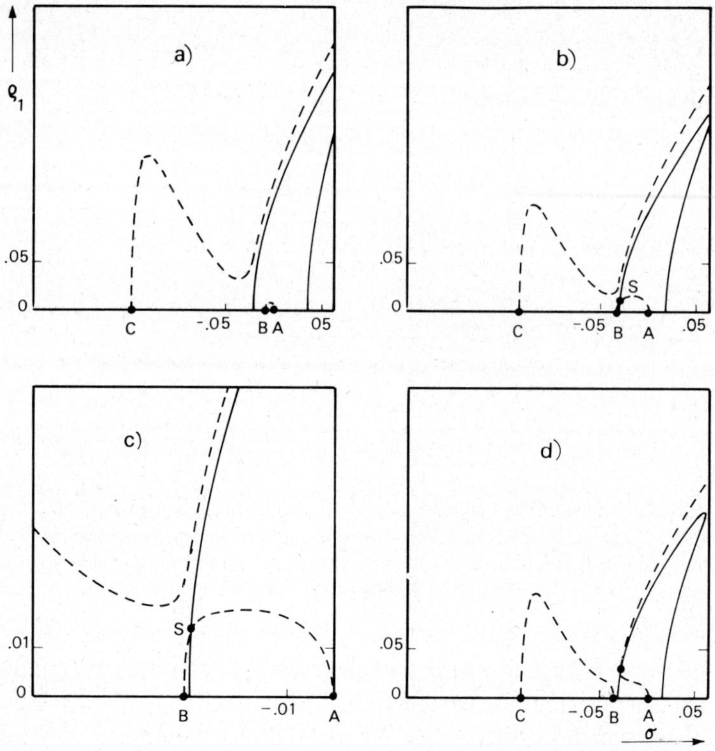

Fig. 21. Same as Fig. 19 but for $\tilde{\alpha}_1 = 5$. (a) $y = 12.49$ ($x = 6.25$); (b, c) $y = 12.573$ ($x = 6.875$); (d) $y = 12.6$ ($x = 7$). Figure 21c is an enlarged view of a part of Fig. 21b.

branch. For a suitable value of the incident field, a self-pulsing solution S appears (Figs. 21b,c). In this case, S bifcates from the unstable cw state B, and also S is *unstable*. Increasing y slightly, the state S *becomes stable*. Note also that the shape of the curve $g(\rho_1, \sigma) = 0$ undergoes a qualitative change in the connection of its two parts (compare Figs. 21b,c,d). For larger values of y, the behavior is quite similar to that found in the case of line (a) in Fig. 15. Namely, when we go out of the instability region an unstable state U bifurcates from the cw state A. The solutions S and U become nearer and nearer until they coalesce. Figure 22 shows the behavior of the half-amplitude of the oscillations when the steady state value x, of the transmitted light in the high transmission state, is varied along the line (b) in Fig. 15. Note that in this case there is no second-order phase transition at the left boundary of the instability range, because the self-pulsing solution S is unstable. When we increase the incident

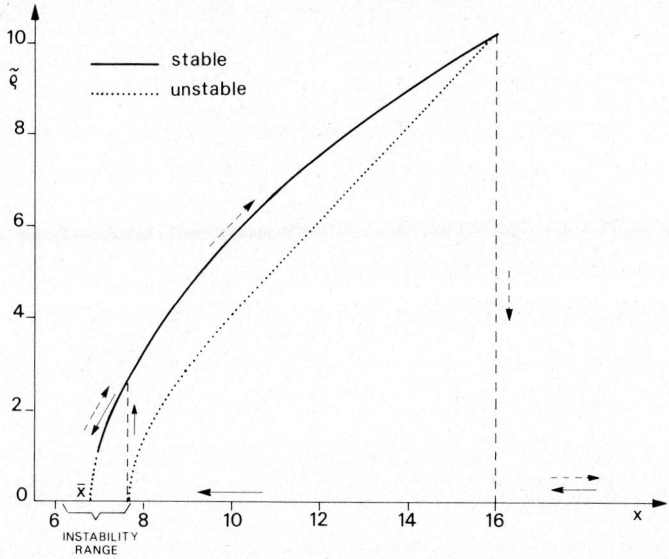

Fig. 22. Same as Fig. 20, but for $\tilde{\alpha}_1 = 5$.

field, the self-pulsing behavior appears discontinuously as soon as the solution S becomes stable.

The most striking feature that emerges from Figs. 20 and 22 is that the self-pulsing state can also exist outside the instability region. The full domain of existence of the stable self-pulsing state S is in fact the region bounded by the lines DEF and DGH in Fig. 23. We have already seen what happens when we cross the line DE from left to right (or from up to down): the self-pulsing state S bifurcates from the cw solution A in the high transmission branch. Thus, DE is a line of second-order phase transitions from cw to pulsed. On the other hand, GE is a line of first-order phase transitions, because when the system is in the cw state A and we cross it from right to left (or from down to up), we find a *discontinuous* transition from cw to pulsed. Note that line GE belongs to the boundary of the instability region but *does not belong* to the boundary of the domain of existence of the self-pulsing state.

When we cross the lines EF and GH from the inside to the outside of the domain of existence, the stable self-pulsing state coalesces with an unstable self-pulsing solution, and the two solutions disappear together while the system discontinuously jumps from the self-pulsing state to the cw regime. When we cross the line DG from right to left (or from up to down) the self-pulsing state becomes unstable. The stability analysis of eqs. (142), linearized around the

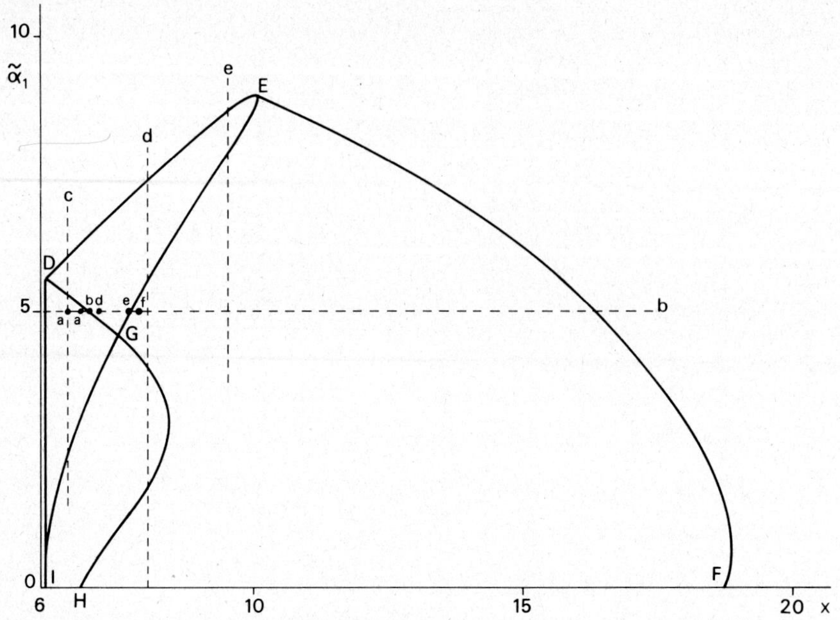

Fig. 23. The domain of existence of the stable self-pulsing state is the region DEFHGD. The lines **c**, **d**, and **e** refer to Figs. 24, 25, and 26 respectively. The points on line **b** (which coincides with line **b** in Fig. 15) correspond to Figs. 27a–f, respectively.

self-pulsing solution S, reveals that the two eigenvalues of the linearized equations are complex conjugate in correspondence to line DG. Hence, when crossing DG, we have *a Hopf bifurcation.* This will be discussed in detail in § 2.5.6.

In correspondence to the part DGI of the instability region there is always precipitation to the low transmission branch. Note that the domain of existence DEFHGD partially overlaps with the shaded region of Fig. 15, in correspondence to which the modes $n = \pm 2$ are unstable. As discussed before, this common part must be left out of consideration. The same must be done for the regions of the plane $(x, \tilde{\alpha}_1)$ in correspondence to which the modes $n = \pm 3$, $n = \pm 4$, etc. are unstable.

Figures 24 and 25 show the variation of the half-amplitude of the oscillations $\tilde{\rho}$ as a function of $\tilde{\alpha}_1$ when we move along the vertical lines (c) and (d) in Fig. 23, which correspond to $x = 6.5$ and $x = 8$ respectively. In order to move along these lines, one must keep the incident field fixed and vary the total length of the cavity. Note that the left-hand part of the graph in Fig. 25 cannot be reached by moving along the vertical line (d) because it is disconnected from the

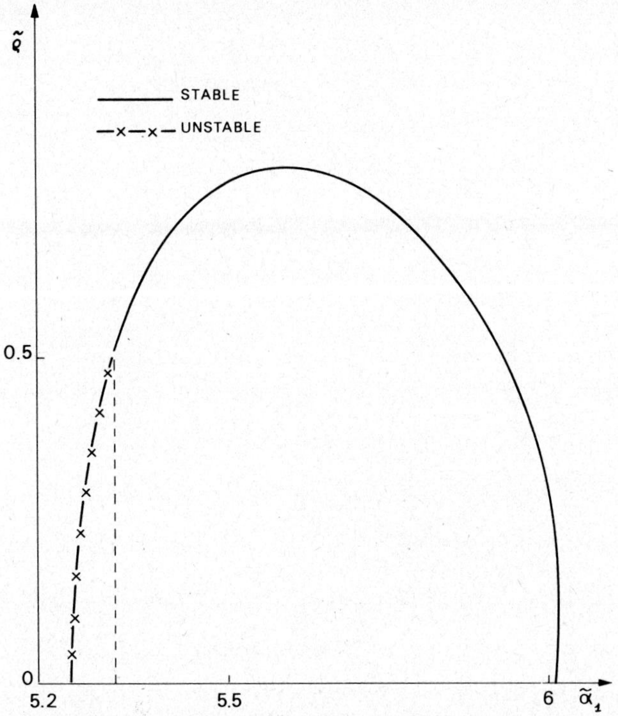

Fig. 24. The half-amplitude of the oscillations $\tilde{\rho}$ is graphed as a function of $\tilde{\alpha}_1$ for $C = 20$, $x = 6.5$.

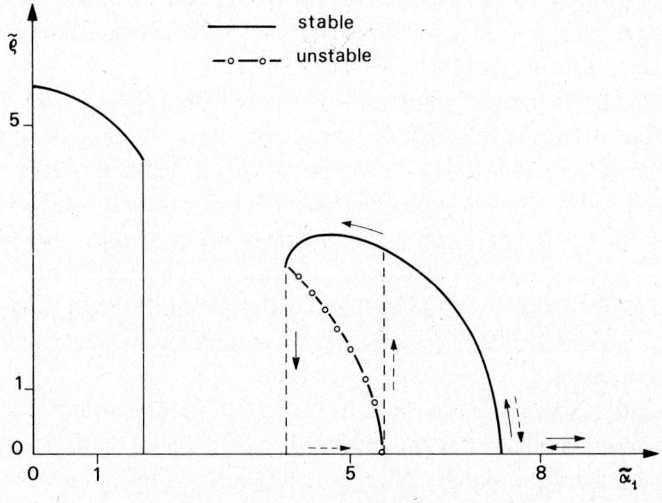

Fig. 25. Same as Fig. 24, but for $x = 8$. Solid (dashed) arrows indicate the behavior of the system when we decrease (increase) the incident field.

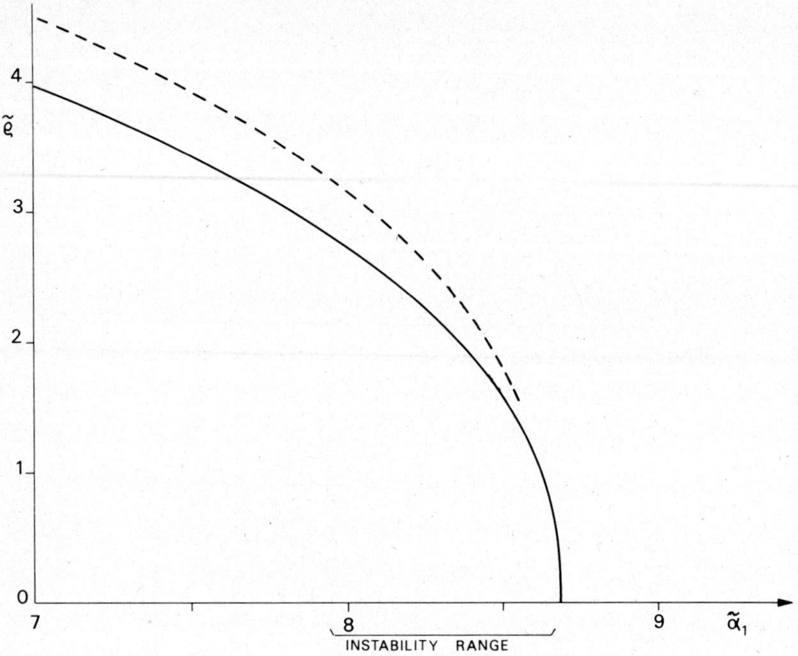

Fig. 26. Same as Fig. 24, but for $x = 9.5$. The solid line is obtained using eqs. (142), the dashed line by numerically solving the Maxwell–Bloch equations.

remaining part of the curve. The self-pulsing state corresponding to the left-hand branch in Fig. 25 can be reached only by changing the parameters x and α_1 simultaneously, in such a way that the point $(x, \tilde{\alpha}_1)$ always remains inside the domain of existence of the self-pulsing state.

Figure 26 shows the half-amplitude of the oscillations $\tilde{\rho}$ as a function of α_1 along the line (e) in Fig. 23, which corresponds to $x = 9.5$. The broken line is obtained by numerically solving the Maxwell–Bloch equations and the full line by using eqs. (142). The agreement between the two curves is satisfactory; the 10% discrepancy is due in part to numerical errors in the solution of the Maxwell–Bloch equations, and in part to the neglect of higher order frequencies in the derivation of eqs. (142). Note that the dressed mode theory also predicts the correct behavior when the amplitude of the oscillations is comparable with the stationary field.

The results of this subsection lead us to look at optical bistability with new eyes. Now our bistable system no longer appears simply as a device with two different stationary states, but rather as a *multistable* system, in which some of the stable steady states are cw, and some are pulsing. One can reach all the (cw and pulsing) branches of the system by suitably controlling the external parameters. In fact, in correspondence to each point in the region GEFH of the plane of the control parameters (see Fig. 23), we have three possible steady states, of which one is pulsing and two are cw (both the low and the high transmission steady states are stable in this situation).

Thus the phenomenology of optical bistability is deeply enriched. From a formal viewpoint, this is already evident from eqs. (142) that treat cw and pulsing solutions on the same footing.

2.5.5. *Transient approach to self-pulsing and precipitation* (LUGIATO, BENZA, NARDUCCI and FARINA [1981, 1983])

Up to this point, we have considered the self-pulsing behavior only in the long time limit $\tau \to \infty$. In this subsection we describe the main features of the time evolution of the envelope (148), obtained by numerically solving eqs. (142). Specifically, let us analyze how the transient changes when we move along the line (b) in Figs. 15 and 23, which corresponds to $\tilde{\alpha}_1 = 5$.

With reference to the numbered dots in Fig. 23, consider the sequence of figures 27a–f. The upper and lower solid lines in each computer solution represent the envelope of self-pulsing, while the horizontal axis is placed in correspondence with the unstable steady state amplitude x of the transmitted field. Figure 27a corresponds to operating conditions to the left of the threshold line DG in Fig. 23. The system undergoes a transient oscillation, but eventually it terminates to the stable low transmission branch. As we change the value of x in such a way that we approach the line DG from the left, the time that the system takes to precipitate becomes longer and longer, showing a pronounced critical slowing down effect. As one moves the operating point to the right of the threshold line DG, the character of the transmitted field changes drastically. Figure 27b shows the evolution of the self-pulsing envelope just to the right of the threshold. The output field, which eventually approaches a stable oscillation, undergoes an extensive transient modulation of its envelope. Sometimes this behavior is called "breathing" (MAYR, RISKEN and VOLLMER [1981]); here the effect is only a transient one. It is especially interesting to examine the envelope breathing in the phase space of the $\{\rho_1, \sigma\}$ variables. This is done in Fig. 27c. In this way the nature of the focal attractor, which is responsible for the stable self-pulsing, is made especially evident. As the operating point is moved farther away from the threshold line, the envelope modulation gradually disappears (Fig. 27d), while the overall time scale of the approach to steady pulsing state shortens considerably. The focus changes into a node. On the other hand, when we bring the system close to the right boundary of the instability region (line EG in Fig. 23), the time scale begins to grow again (Fig. 27e), and diverges. This new critical slowing down effect arises as usual from a stability changeover. In fact, on the boundary of the instability region the real part of the eigenvalue λ_{11} vanishes.

Next, let us move to the right of the instability region, but still inside the domain of existence of the stable self-pulsing solution (point f in Fig. 23). In this case, if the system is initially only slightly displaced from the steady state in the high transmission branch (*soft excitation*), it simply returns to it, because it is now stable. On the other hand, if the initial displacement is large enough (*hard excitation*), the system approaches the stable self-pulsing state (Fig. 27f). Hence, the region DEG in Fig. 23 is the soft excitation domain, and the region EFHG the hard excitation domain. When we sweep the incident field adiabatically back and forth, as indicated in Figs. 20 and 22, we are automatically in hard excitation conditions.

Finally, if we move to the right of the line EF, no self-pulsing state exists any longer, and hence, the system returns to the steady state in the high transmission state.

Fig. 27. (facing page). (a) Transient evolution of the self-pulsing envelope, followed by precipitation to the low transmission branch. The total run time is 50 units of τ. The picture shows two runs, one for $\tilde{\alpha}_1 = 5$ and $x = 6.5$, the other for $\tilde{\alpha}_1 = 5$ and $x = 6.86$. Both correspond to operating points to the left of the threshold line DG in Fig. 23. Note the critical slowing down exhibited by the solution with $x = 6.86$, whose operating parameters are very close to the line DG. (b) Self-pulsing envelope corresponding to operating values of the parameters which are slightly to the right of the line DG ($\tilde{\alpha}_1 = 5$, $x = 6.867$). The total run time is 600 units of τ, more than ten times the length of the characteristic time scale of Fig. 27a. (c) Phase-space portrait of the solution shown in Fig. 27b. The trajectory lies in the second quadrant of the $\{\rho_1, \sigma\}$ plane. Here we are actually plotting $\tilde{\rho}$ versus $\tilde{\sigma}$ in order to allow an easy comparison of the different numerical features. (d) Same as Fig. 27b, with $\tilde{\alpha}_1 = 5$ and $x = 6.87$. The total run time is 300 units of τ. Note the rapid decrease of both the breathing effect and the overall time scale. (e) Same as Fig. 27b, with $\tilde{\alpha}_1 = 5$ and $x = 7.65$. Here the operating point is slightly on the left of the line EG in Fig. 23. The total run time is 300 units of τ. Note the marked critical slowing down. (f) Same as Fig. 27b, with $\tilde{\alpha}_1 = 5$, $x = 7.68$. Here the operating point lies in the hard excitation domain. The total run time is 15 units of τ.

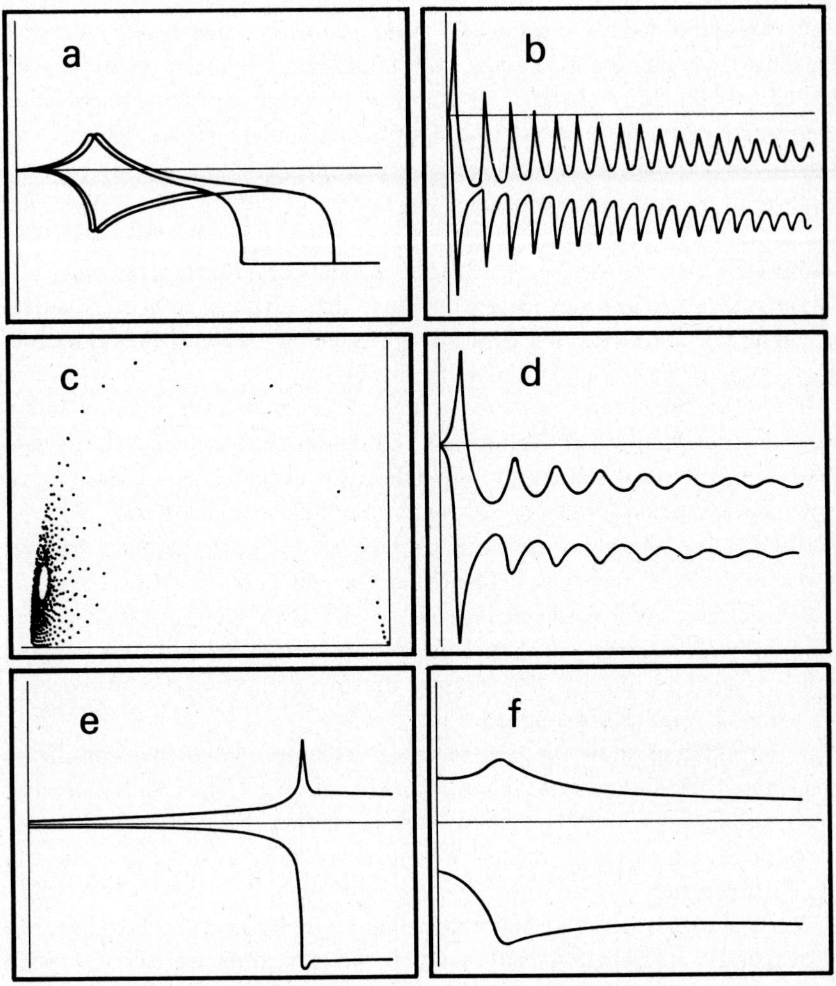
a
b
c
d
e
f

2.5.6. *Hopf bifurcations and unstable limit cycles*

As we said in § 2.5.4, the real part of the complex conjugate eigenvalues of the linearized counterparts of eqs. (142) undergoes a sign change as the scan line crosses the threshold line DG. Because the imaginary part remains finite, the conditions for the existence of a Hopf bifurcation are met all along the line DG. Upon close inspection of the region immediately above the line DG one finds that precipitation persisted even under conditions that insured the existence of a focal attractor. This effect is consistent with the existence of an *unstable limit cycle* that erects a barrier around the attractor and makes it invisible to all trajectories, except to those that originate in the immediate neighborhood of the focus itself. This unstable limit cycle exists in a narrow layer in the plane of Fig. 23 above the line DG.

In order to confirm the presence of the unstable limit cycle in a convincing way, one can produce (LUGIATO, BENZA, NARDUCCI and FARINA [1981, 1983]) time-reversed solutions of the differential equations (142). By moving again along the line (b) of Fig. 23, we find the following features:

i) The size of the limit cycle increases very rapidly over a very small interval above the value x_{thr} where the line (b) crosses the line DG (Figs. 28b,c).

ii) The limit cycle is identical in shape and size when approached in a time-reversed fashion, both from its interior and from the outside.

iii) At a sharply defined value slightly above x_{thr}, the limit cycle disappears abruptly, leaving behind a naked stable focus.

In Fig. 28d we show the time-reversed evolution of the envelope of the transmitted field for the same values of the parameters of Fig. 28c. In this case, the breathing behavior has an infinite duration in time. However, this does not correspond to a physically realizable behavior of the system because the limit cycle is unstable.

We stress that the Hopf bifurcation that we described in this subsection corresponds to the *third instability* that we found in the framework of absorptive optical bistability. The first instability is responsible for the switching between low and high transmission branches, and hence, for the emergence of the hysteresis loop for the transmitted field. The second instability is responsible for the self-pulsing, and can be induced by forcing the control parameters to fall within a well-defined domain of instability. As we have seen, stable self-pulsing can also be induced over a wider domain in the space of control parameters, provided that the system is initially perturbed by a sufficiently large excitation. The third instability occurs at the boundary between the stable self-pulsing and precipitation domains. This is connected with the bifurcation

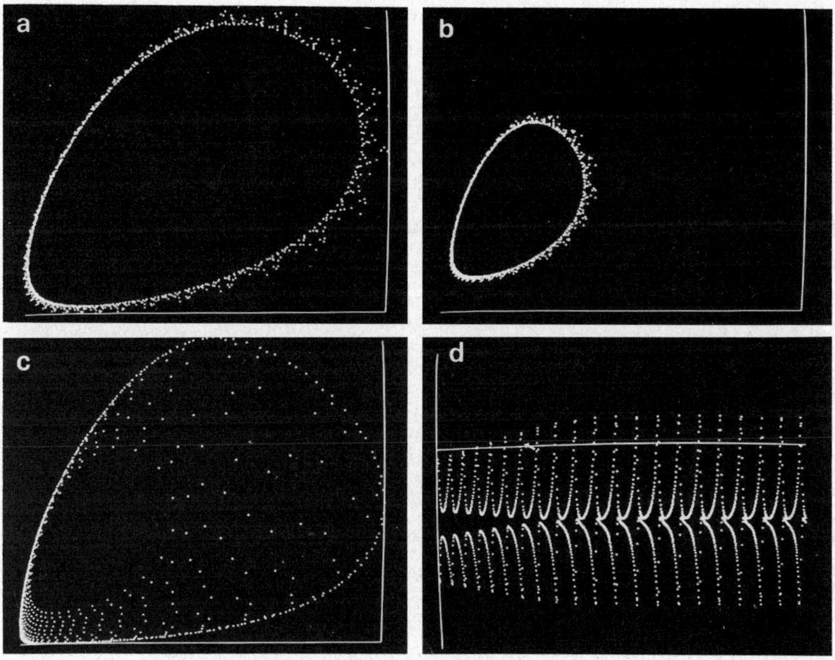

Fig. 28. (a, b) Time reversed phase space trajectory in the $\{\rho_1, \sigma\}$ phase plane reveals the presence of a limit cycle. The operating parameters are $\tilde{\alpha}_1 = 5$, $x = 6.8663$. The total run time is 1000 units of τ. Fig. 28a is an enlarged view of a part of Fig. 28b. Fig. 28b shows the window $0.0044326 \leqslant \rho_1 \leqslant 1.81622$, $-1.5024 \leqslant \sigma \leqslant -1.1139$. (c) Same as Fig. 28b, but with $x = 6.8669$. The window is the same as in Fig. 28b. Note the size increase of the limit cycle as one moves very slightly away from line DG in Fig. 23. (d) Time-reversed evolution of the envelope of the transmitted field for the same values of the parameters as in Fig. 28c. Here the envelope behaves as a perfect clock that ticks backwards in time.

of a line of stable foci and a manifold of unstable limit cycles. The stability changeover is characterized by pronounced critical slowing down and by strong self-pulsing envelope modulation.

It is resonable to expect that, by allowing a larger number of unstable modes, one finds a richer bifurcation pattern in absorptive optical bistability. On the other hand, in the case of dispersive optical bistability for proper choices of the parameters (see for instance (96)) one has a sequence of period doubling bifurcations which eventually leads to chaotic behavior, as we discussed in § 2.4.4.

§ 3. Quantum Statistical Treatment

So far, we have systematically neglected any fluctuations, treating the time evolution of the system as completely deterministic. In this section we drop this limitation. This is necessary in order to describe the spectra of transmitted light and of fluorescent light, and the photon statistics of the transmitted light. In particular, under suitable conditions the quantum mechanical fluctuations give rise to nonclassical effects, as for instance photon antibunching or the so-called "squeezing". Furthermore, the discussion of the stability properties of the stationary states of the system is necessarily incomplete when fluctuations are not included. Finally, fluctuations become particularly important in small systems, like the miniaturized bistable devices which are presently the object of very active research in the trend toward practical optical logic devices (GIBBS, MCCALL and VENKATESAN [1980], ABRAHAM and SMITH [1982b]).

There are three main sources of fluctuations in our system: (a) quantum mechanical noise, (b) thermal noise and (c) external noise, that is the fluctuations in the incident field.

3.1. THE MANY-MODE AND THE ONE-MODE MASTER EQUATIONS

The starting point of our quantum statistical theory of optical bistability is a suitable master equation that governs the time evolution of the statistical operator $W(t)$ of the system composed by the electric field and the atoms (LUGIATO [1981]). This equation, which includes all the longitudinal modes of the cavity, generalizes the well known one-mode model for the laser (WEIDLICH and HAAKE [1965a,b], HAKEN [1970]) and for optical bistability (BONIFACIO and LUGIATO [1978d]). It holds in the limit (44) and, as we shall show, in the semiclassical approximation it reproduces the Maxwell–Bloch equations (45) and (1b,c).

3.1.1. *The many-mode master equation*

We call $A_n, A_n^\dagger$ the annihilation and creation operators of photons of the longitudinal cavity mode with frequency $\omega_c + \alpha_n$ (see (51)). We have obviously

$$[A_n, A_{n'}^\dagger] = \delta_{nn'} \tag{151}$$

On the other hand, let us consider the atoms. The ith two-level atom ($i = 1 \ldots \text{N}$) is associated to the raising and lowering operators r_i^+, r_i^- and to the inversion operator

$$r_{3i} = \tfrac{1}{2}(r_i^+ r_i^- - r_i^- r_i^+).$$

One has the commutation rules

$$[r_{3i}, r_j^\pm] = \pm r_i^\pm \delta_{ij}, \qquad [r_i^+, r_j^-] = 2r_{3i}\delta_{ij}. \tag{152}$$

The atoms are placed inside a sample of length L (Fig. 1). Let z_i be the position of the ith atom. Following BONIFACIO and LUGIATO [1975] we introduce the collective dipole operators

$$R_n^\pm = \sum_{j=1}^{N} r_j^\pm \exp\left\{\pm \mathrm{i}\left(\frac{\omega_c}{c'} + k_n\right) z_j\right\}, \tag{153}$$

where k_n is defined by eq. (53) and $c' = cL/\mathscr{L}$. Simultaneously we consider the collective inversion operators

$$R_{3,n} = \sum_{j=1}^{N} r_{3j} \exp\{-\mathrm{i}k_n z_j\}. \tag{154}$$

In particular, $R_{3,0}$ is the one (one half) total inversion operator. The operators (153) and (154) obey the following commutation relations

$$[R_n^+, R_{n'}^-] = 2R_{3,n'-n}, \qquad [R_{3,n}, R_{n'}^\pm] = \pm R_{n'\mp n}^\pm. \tag{155}$$

In particular, for each n the triple $R_n^\pm$, $R_{3,0}$ obeys angular momentum commutation relations

$$(R_n^+, R_n^-] = 2R_{3,0}, \qquad [R_{3,0}, R_n^\pm] = \pm R_n^\pm. \tag{156}$$

Now let W_s be the statistical operator of the coupled system atom + radiation field inside the cavity (or equivalently atoms + cavity modes) in the Schrödinger picture. We use a picture in which the frequency of the incident field is eliminated. To this aim we define

$$W(t) = \exp\left[\frac{\mathrm{i}}{\hbar} H_0 t\right] W_s(t) \exp\left[-\frac{\mathrm{i}}{\hbar} H_0 t\right],$$

$$H_0 = \hbar\omega_0\left(\sum_n A_n^\dagger A_n + R_{3,0}\right). \tag{157}$$

We formulate the following master equation for $W(t)$

$$\frac{\mathrm{d}W}{\mathrm{d}t'} = -\mathrm{i}\mathscr{L}_\mathrm{F} W + \Lambda_\mathrm{F} W - \mathrm{i}\mathscr{L}_\mathrm{A} W$$

$$+ \Lambda_\mathrm{A} W - \mathrm{i}\mathscr{L}_\mathrm{AF} W - \mathrm{i}\mathscr{L}_\mathrm{ext} W + \Lambda_\mathrm{th,ext} W, \tag{158}$$

where t' is the time defined by eq. (47) and:

a) $\mathscr{L}_F$ describes the free time evolution of the modes of the cavity

$$\mathscr{L}_F W = \hbar^{-1}[H_F, W],$$
$$H_F = \sum_n \hbar((\omega_c - \omega_0) + \alpha_n) A_n^\dagger A_n, \tag{159}$$

where α_n is defined in eq. (51), and ω_0 is the frequency of the incident field.

b) Λ_F describes the damping of the modes due to the photon escape from the cavity

$$\Lambda_F W = k \sum_n \{[A_n W, A_n^\dagger] + [A_n, W A_n^\dagger]\}, \tag{160}$$

where the cavity damping constant k is defined by eq. (33).

c) $\mathscr{L}_A$ describes the free time evolution of the atoms:

$$\mathscr{L}_A W = \hbar^{-1}[H_A, W], \qquad H_A = \hbar(\omega_a - \omega_0) R_{3,0}, \tag{161}$$

where as usual ω_a is the atomic transition frequency.

d) Λ_A describes the radiative and collisional decay of the atoms:

$$\Lambda_A W = \sum_{i=1}^{N} \left\{ \frac{\gamma_\parallel}{2} ([r_i^-, W r_i^+] + [r_i^- W, r_i^+]) \right.$$
$$\left. + \left(\gamma_\perp - \frac{\gamma_\parallel}{2}\right) ([r_{3i}, W r_{3i}] + [r_{3i} W, r_{3i}]) \right\} \tag{162}$$

where the part proportional to $\gamma_\perp - \gamma_\parallel/2$ is a dephasing term.*

e) $\mathscr{L}_{AF}$ describes the interaction between the atomic system and the cavity modes in the dipole and rotating wave approximations

$$\mathscr{L}_{AF} = \hbar^{-1}[H_{AF}, W],$$
$$H_{AF} = i\hbar \bar{g} \sum_n (A_n^\dagger R_n^- - A_n R_n^+), \tag{163}$$

where the coupling constant $\bar{g}$ is given by

$$\bar{g} = \left(\frac{2\pi\omega_0 L}{\hbar V \mathscr{L}}\right)^{1/2} \mu. \tag{164}$$

* Strictly speaking, when the temperature is greater than zero there is also an upward transition term, which has the same form as the first term in eq. (162), with r_i^+ and r_i^- exchanged. We neglect this term, which produces only a negligible correction to the quantum fluctuation term in eq. (177), below.

f) $\mathscr{L}_{\text{ext}}$ takes into account the presence of the external field

$$\mathscr{L}_{\text{ext}} = \hbar^{-1}[H_{\text{ext}}, W], \qquad H_{\text{ext}} = \mathrm{i}\hbar k \alpha_0 (A_0^+ - A_0), \tag{165}$$

where α_0 is given by

$$\alpha_0 = \left(\frac{V\mathscr{L}}{8\pi\hbar\omega_0 L}\right)^{1/2} \frac{E_{\mathrm{I}}}{\sqrt{T}}. \tag{166}$$

This term injects a coherent field inside the cavity. In fact, if one neglects $\mathscr{L}_{\mathrm{AF}}$ and $\Lambda_{\text{th, ext}}$ one easily verifies that in the long time limit all the modes approach the vacuum state, except the resonant mode $n = 0$ which approaches the coherent state $|\alpha_0\rangle$.

g) $\Lambda_{\text{th, ext}}$ takes into account thermal fluctuations and simulates the fluctuations of the incident field (SCHENZLE and BRAND [1978], DRUMMOND, MCNEIL and WALLS [1980b]) in the limit in which they can be described as a white noise:

$$\Lambda_{\text{th, ext}} W = 2k \sum_n \bar{n}_n [A_n, [W, A_n^\dagger]]. \tag{167}$$

In the case of thermal fluctuations, $\bar{n}$ is given by

$$\bar{n}_n = \frac{1}{\exp[\beta\hbar(\omega_c + \alpha_n)] - 1}, \qquad \beta = \frac{1}{k_{\mathrm{B}} T}, \tag{168}$$

where k_{B} is Boltzmann's constant and T is the temperature (not to be confused with the transmissivity coefficient of the mirrors, which is indicated by the same symbol in this chapter). In the case of fluctuations of the incident field $\bar{n}_n = 0$ for $n \neq 0$, while $\bar{n}_0$ measures the strength of these fluctuations. In the general case, $\bar{n}_n$ is the sum of the contributions from thermal and external fluctuations.

3.1.2. *Connection with the semiclassical theory*

Let us now derive from the master equation (158) the time evolution equations for the mean values $\langle A_n\rangle(t') = \mathrm{Tr}(A_n W(t'))$, $\langle R_n^-\rangle(t')$, and $\langle R_{3n}\rangle(t')$. This is easily done using the commutation rules (151), (152), (155) and the definitions (153), (154). We obtain

$$\langle \dot{A}_n\rangle = -\mathrm{i}((\omega_c - \omega_0) + \alpha_n)\langle A_n\rangle - k[\langle A_n\rangle - \alpha_0 \delta_{n0}] + \bar{g}\langle R_n^-\rangle, \tag{169a}$$

$$\langle \dot{R}_n^- \rangle = 2\bar{g} \sum_{n'} \langle A_{n'} R_{3,n-n'} \rangle - \gamma_\perp (1 + \mathrm{i}\Delta) \langle R_n^- \rangle, \tag{169b}$$

$$\langle \dot{R}_{3,n} \rangle = -\bar{g} \sum_{n'} [\langle A_{n'}^\dagger R_{n'+n}^- \rangle + \langle A_{n'} R_{n'-n}^+ \rangle] - \gamma_\parallel [\langle R_{3,n} \rangle + \tfrac{1}{2} N \delta_{n0}], \tag{169c}$$

where we have assumed that $\sum_{i=1}^N \exp(\mathrm{i}k_n z_i) = N\delta_{n0}$, as one has for a lattice structure of the atomic system (BONIFACIO and LUGIATO [1975]). The equation for $\langle R_n^+ \rangle$ is immediately obtained from eq. (169b), by taking into account that

$$\langle R_n^- \rangle = \langle R_n^+ \rangle^*, \qquad \langle A_n \rangle = \langle A_n^\dagger \rangle^*, \qquad \langle R_{3,n} \rangle = \langle R_{3,-n} \rangle^*.$$

Next, we introduce the *semiclassical approximation*, that is, we factorize the mean values of the products of a field and an atomic operator into the corresponding products of mean values (for instance $\langle A_{n'} R_{3n''} \rangle \to \langle A_{n'} \rangle \langle R_{3n''} \rangle$). This approximation becomes exact in the thermodynamic limit $N \to \infty$, $V \to \infty$ with N/V = constant. Thus, equations (169) become identical to eq. (54), provided one makes the following correspondences

$$f_n = \mu \left(\frac{8\pi\omega_0 L}{\hbar \gamma_\perp \gamma_\parallel V \mathscr{L}} \right)^{1/2} \langle A_n \rangle,$$

$$p_n = -\left(\frac{N}{2} \sqrt{\frac{\gamma_\parallel}{\gamma_\perp}} \right)^{-1} \langle R_n^- \rangle,$$

$$d_n = -\left(\frac{N}{2} \right)^{-1} \langle R_{3,n} \rangle. \tag{170}$$

This is easily verified using eqs. (53), (166), (2), (164), (7a), and (33), and the definitions $k\theta = \omega_n - \omega_0$, and $C = \alpha L/2T$. This result proves that, in the semiclassical approximation, the master equation reproduces eqs. (54) and hence, the Maxwell–Bloch equations (45) and (1b,c) which are equivalent to eq. (54).

3.1.3. *The one-mode master equation*

With only one exception (BENZA and LUGIATO [1981]), all the papers that deal with the quantum statistical treatment of optical bistability use the *one-mode approximation*. Accordingly, in the following we shall use the master equation (158), but neglecting all the off-resonance modes $n \neq 0$. Furthermore, we shall use the notations A, $R^{\pm}$, R_3 and $\bar{n}$ instead of A_0, $R_0^{\pm}$, $R_{3,0}$ and $\bar{n}_0$. The model that we obtain in such a way is the quantum statistical analogue of the semiclassical mean field model (55) (BONIFACIO and LUGIATO [1978d]).

In introducing the one-mode approximation, we assume that the values of the parameters are such that no self-pulsing instability can arise (see § 2.4). Furthermore, since the phenomena that we are going to describe occur on a time scale much longer than the cavity transit time, we can replace the time t', defined by eq. (47), by the real time t.

For the following treatment it is useful to introduce an alternative expression of the interaction Hamiltonian H_{AF} restricted to the resonant mode $n = 0$:

$$H_{AF} = i\hbar \bar{g} \sum_{i=1}^{N} (A^{\dagger}\tilde{r}_i^{-} - A\tilde{r}_i^{+}), \tag{163'}$$

where

$$\tilde{r}_i^{\pm} = r_i^{\pm} \exp\left(\pm i \frac{\omega_c}{c'} z_i \right). \tag{171}$$

The three operators $\tilde{r}_i^{\pm}$, r_{3i} obey the same commutation rules (152). In the following, we shall always use $\tilde{r}_i^{\pm}$, but we shall systematically drop the superscript $\sim$.

Finally, in the following sections, we shall always *restrict ourselves to the resonant situation* $\Delta = \theta = 0$, because no particularly interesting novelty arises in the general case Δ, $\theta \neq 0$.

3.2. THE FOKKER–PLANCK EQUATION FOR THE GENERALIZED WIGNER FUNCTION

The master equation (158) is too complicated to be solved exactly, even in the one-mode approximation. Further approximations must be introduced. The first one exploits the presence of a smallness parameter in our system. This parameter is the inverse of the saturation photon number (which corresponds to the saturation intensity I_s defined in eq. (7.3)), given by

$$N_s = \frac{\gamma_{\perp}\gamma_{\parallel}}{4\bar{g}^2}. \tag{172}$$

In order to introduce this approximation, we begin translating eq. (158), restricted to $n = 0$, into a c-number partial differential equation. This can be performed via the characteristic function technique devised by Haken and coworkers (HAKEN [1970]). Precisely, one introduces a characteristic function by which five c-number quantities are associated with the operators $R^{\pm}$, R_3, A and $A^{\dagger}$. The Fourier transform of this function is a quasi-probability distribution in five variables. Its time evolution is governed by a classical looking partial differential equation, which is derived from the master equation by suitable techniques. By means of this distribution, one can calculate the expectation values of products of operators in suitable order as classical mean values of the corresponding c-number variables. This procedure, which creates a bridge between the quantum mechanical density operator and the classical density function, generalizes the notions of the Wigner function (WIGNER [1932]), and of the Glauber P-function (GLAUBER [1963a,b]). The time evolution equation for the distribution function contains derivatives of all orders. The condition $N_s \gg 1$ allows us to introduce the *Fokker–Planck approximation*, which consists of neglecting the terms with derivatives of order higher than second.

A crucial point concerns the ordering prescription for the operators in the characteristic function. Actually, different choices such as normal, antinormal, or symmetrical ordering, lead to different quasiprobability distributions, which in turn give normal, antinormal or symmetrical-ordered expectation values, respectively. The distributions themselves obey different equations according to the chosen ordering prescription.

In order to treat laser fluctuations, Haken and coworkers (HAKEN [1970]) adopted the normal ordering prescription, which is the most natural one in quantum optics, and therefore used a generalized Glauber distribution. However, as was first shown in GRONCHI and LUGIATO [1978], this procedure is not suitable in the case of optical bistability, because it leads to a Fokker–Planck equation whose diffusion matrix is not in general positive definite. In order to avoid this difficulty, the proper choice is that of symmetrical ordering. Hence, the characteristic function is

$$C(\xi, \xi^*, \eta, \zeta, \zeta^*, t) = \\ = \mathrm{Tr}[\exp(\mathrm{i}\xi R^- + \mathrm{i}\xi^* R^+ + \mathrm{i}\eta R_3 + \mathrm{i}\zeta A + \mathrm{i}\zeta^* A^\dagger)\, W(t)], \tag{173}$$

whose Fourier transform $P_w(\bar{v}, \bar{v}^*, \overline{m}, \beta, \beta^*, t)$ is a generalization of the Wigner function to include the atomic variables. P_w is normalized to unity and is always real but not necessarily positive, hence, it is a quasiprobability distribution. The moments of P_w give symmetrized expectation values, for instance

$$\int \mathrm{d}_2\bar{v}\, \mathrm{d}\overline{m}\, \mathrm{d}_2\beta\, P_w(\bar{v}, \bar{v}^*, \overline{m}, \beta, \beta^*, t)\cdot \beta^*\beta = \tfrac{1}{2}[\langle A^\dagger A\rangle(t) + \langle AA^\dagger\rangle(t)], \tag{174}$$

where $d_2\bar{v} = d(\mathrm{Re}\,\bar{v})\,d(\mathrm{Im}\,\bar{v})$, etc. The partial differential equation which rules the evolution of P_w can be derived either via the simple constructive procedure of GRONCHI and LUGIATO [1978], which is summarized in Appendix D, or via suitable techniques (GILMORE [1974]) which allow us to disentangle the exponential in eq. (173) into a product of exponentials, and hence, allow the operator approach of HAKEN [1970]. For $\Delta = \theta = 0$, and in terms of the normalized variables, corresponding to eqs. (10) and (48),

$$x = \frac{\beta}{\sqrt{N_s}}, \qquad y = \frac{\alpha_0}{\sqrt{N_s}},$$
$$v = -\left(\frac{N}{2}\sqrt{\frac{\gamma_\parallel}{\gamma_\perp}}\right)^{-1}\bar{v}, \qquad m = -\left(\frac{N}{2}\right)^{-1}\bar{m}, \qquad . \tag{175}$$

this equation reads (GRONCHI and LUGIATO [1978])

$$\begin{aligned}
&\frac{\partial}{\partial t}P_w(v, v^*, m, x, x^*, t) = \\
&\left\{-\frac{\partial}{\partial v}[-\gamma_\perp(v - mx)] - \frac{\partial}{\partial v^*}[-\gamma_\perp(v^* - mx^*)]\right. \\
&\left.-\frac{\partial}{\partial m}\{-\gamma_\parallel[m - 1 + \tfrac{1}{2}(v^*x + vx^*)]\}\right\} \\
&-\frac{\partial}{\partial x}[-k(x - y + 2Cv)] - \frac{\partial}{\partial x^*}[-k(x^* - y + 2Cv^*)] \\
&+\frac{\gamma_\perp^2}{k}\frac{1}{2CN_s}\left[\frac{\partial^2}{\partial v^*\partial v} - \frac{\bar{d}^2}{4}\frac{\partial}{\partial m}\left(\frac{\partial}{\partial v}v + \frac{\partial}{\partial v^*}v^*\right) + \frac{\bar{d}^2}{4}\frac{\partial^2}{\partial m^2}(1 - m)\right] \\
&\left.+\frac{k}{N_s}(1 + 2\bar{n})\frac{\partial^2}{\partial x^*\partial x}\right\}P_w(v, v^*, m, x, x^*, t) + O\left(\frac{1}{N_s^{3/2}}\right),
\end{aligned} \tag{176}$$

where $\bar{d} = \gamma_\parallel/\gamma_\perp$ and we have taken into account the relation

$$C = \frac{g^2 N}{2k\gamma_\perp} \tag{176'}$$

which follows immediately from the definition $C = \alpha L/2T$ and eqs. (7a), (2), (33) and (164). Since $N_s \gg 1$, we can introduce the Fokker–Planck approximation, neglecting the terms which are not explicitly written in eq. (176). Thus eq. (176) becomes a Fokker–Planck equation. Furthermore, the drift (i.e. first-order derivative) terms are larger than the diffusion (i.e. second-order derivative) terms by a factor of N_s. This fact ensures the smalness of fluctuations. If we also drop the diffusion terms, we recover the deterministic semiclassical description (55).

While the drift coefficients do not depend on the ordering prescription, the diffusion coefficients do. The matrix of the diffusion coefficients in eq. (176) is positive definite for all values of the parameters, at least in the physically relevant domain $|m| \leqslant 1$.

3.2.1. *Good cavity case*: *adiabatic elimination of the atomic variables*

The Fokker–Planck equation (176) is still too complicated to be solved, even at steady state. This situation is greatly simplified in the two limit situations, that in § 2.3.1 were called "good cavity case" and "bad cavity case". In fact, in the first case we can *adiabatically eliminate* the atomic variables from the Fokker–Planck equation, in the second the field variables. The adiabatic elimination can be performed following the procedure of GORDON [1967] and HAKEN [1977].

In the good cavity case $k \ll \gamma_\perp, \gamma_\parallel$ we obtain the following closed Fokker–Planck equation for the field variables (LUGIATO [1979], LUGIATO, CASAGRANDE and PIZZUTO [1982]:

$$k^{-1}\frac{\partial}{\partial t}P_w(x, x^*, t) = \left\{-\frac{\partial}{\partial x}\left[y - x - \frac{2Cx}{1+|x|^2}\right] + \text{c.c.}\right.$$
$$-\frac{C}{2N_s}\frac{\partial^2}{\partial x^2}x^2\frac{(1+|x|^2)^2 + 1 + \bar{d}}{(1+|x|^2)^3} + \text{c.c.}$$
$$\left.+\frac{1}{N_s}\frac{\partial^2}{\partial x^*\partial x}\left[1 + 2\bar{n} + C\frac{(1+|x|^2)^2(2+|x|^2) - (1+\bar{d})|x|^2}{(1+|x^2|)^3}\right]\right\}P_w(x, x^*, t), \tag{177}$$

where $P_w(x, x^*, t)$ is obtained from the full distribution by integration over Re v, Im v and m. It will be useful to rephrase eq. (177) in polar coordinates

$$x = r\exp(\mathrm{i}\varphi). \tag{178}$$

We obtain

$$k^{-1}\frac{\partial}{\partial t}P_{\mathrm{w}}(r,\varphi,t)=\frac{1}{r}\left\{\frac{\partial}{\partial r}r\left[r+\frac{2Cr}{1+r^2}-y\cos\varphi\right]+\frac{\partial}{\partial\varphi}y\sin\varphi\right.$$

$$+\frac{1}{4N_{\mathrm{s}}}\left[\frac{\partial}{\partial r}r\frac{\partial}{\partial r}\left(1+2\bar{n}+2C\frac{r^4+(1-\bar{d})r^2+1}{(1+r^2)^3}\right)\right.$$

$$\left.\left.+(1+2\bar{n}+2C)\frac{1}{r}\frac{\partial^2}{\partial\varphi^2}\right]\right\}P_{\mathrm{w}}(r,\varphi,t), \qquad (177')$$

from which one sees by simple inspection that the diffusion matrix is positive definite, since $\bar{d}\leqslant 2$. In eq. (177′) we have neglected a contribution proportional to $N_{\mathrm{s}}^{-1}\ll 1$ in the drift term. The same approximation is used in eq. (237), below.

3.2.2. *Bad cavity case: adiabatic elimination of the field variables*

In the bad cavity case $k\gg\gamma_{\perp},\gamma_{\parallel}$, we can adiabatically eliminate the field variables from eq. (176). Thus, we obtain the following Fokker–Planck equation for the atomic variables (LUGIATO [1979])

$$\frac{\partial}{\partial t}P_{\mathrm{w}}(v,v^*,m,t)=\left\{-\gamma_{\perp}\left[\frac{\partial}{\partial v}(-v+ym-2Cmv)+\text{c.c.}\right]\right.$$

$$-\gamma_{\parallel}\frac{\partial}{\partial m}[1-m-\tfrac{1}{2}y(v+v^*)+2Cv^*v]$$

$$+\frac{\gamma_{\perp}^2}{k}\frac{1}{2CN_{\mathrm{s}}}\left[\frac{\partial^2}{\partial v^*\partial v}(1+2Cm^2)-\left(\frac{\partial^2}{\partial v^*\partial m}\bar{d}(\tfrac{1}{4}\bar{d}v+Cmv)+\text{c.c.}\right)\right.$$

$$\left.\left.+\frac{\partial^2}{\partial m^2}\frac{\bar{d}^2}{2}\left(\frac{1-m}{2}+Cv^*v\right)\right]\right\}P_{\mathrm{w}}(v,v^*,m,t), \qquad (179)$$

where $P_{\mathrm{w}}(v,v^*,m,t)$ is obtained from the full distribution by integration over Re x and Im x.

Furthermore, we obtain the following "adiabatic formula"

$$\hat{x}=y-2C\hat{r}^{-}, \qquad (180)$$

where the operator $\hat{x}$ corresponds to the c-number variable x, while $\hat{r}$ corresponds to v:

$$\hat{x} = \frac{A}{\sqrt{N_s}}, \qquad \hat{r} = -\left(\frac{N}{2}\sqrt{\frac{\gamma_\parallel}{\gamma_\perp}}\right)^{-1} R^{-}. \tag{181}$$

Equation (180) reduces the calculation of expectation values and time correlation functions of field quantities to expectation values and correlation functions of atomic quantities. For instance, $\langle x \rangle(t) = y - 2C\langle \hat{r}^{-} \rangle$, etc. Of course eq. (180) holds only for $k \gg \gamma_\perp, \gamma_\parallel$.

3.3. SPECTRUM OF TRANSMITTED LIGHT

The spectrum $S(\omega)$ of the transmitted light is given by the Fourier transform of the time correlation function at steady state $\langle A^\dagger(t)A(0)\rangle_{st}$:

$$S(\omega) = \frac{1}{\pi}\,\mathrm{Re}\int_0^\infty \mathrm{d}t \exp[-\mathrm{i}(\omega - \omega_0)t]\langle A^\dagger(t)A(0)\rangle_{st}. \tag{182}$$

Hence, to obtain the spectrum one must calculate the fluctuations of the system around the steady state. More specifically, for any given incident field y let us choose one of the two stable steady states, and let us call x_{st} the normalized transmitted field in the chosen state. Subdividing $A(t)$ into the stationary mean value $\langle A\rangle_{st} = x_{st}\sqrt{N_s}$ (see eq. (181)), and the fluctuation $\delta A(t) = A(t) - \langle A\rangle_{st}$, we have that $S(\omega)$ is composed of a coherent and an incoherent part

$$S(\omega) = S_{coh}(\omega) + S_{inc}(\omega), \tag{183}$$

$$S_{coh}(\omega) = N_s\, x_{st}^2\, \delta(\omega - \omega_0), \tag{184}$$

$$S_{inc}(\omega) = \frac{1}{\pi}\,\mathrm{Re}\int_0^\infty \mathrm{d}t \exp[-\mathrm{i}(\omega - \omega_0)t]\langle \delta A^\dagger(t)\,\delta A(0)\rangle_{st}. \tag{185}$$

The coherent or classical part has the same frequency as the injected field and is proportional to the intensity x_{st}^2 of the transmitted field. The incoherent part is the quantum mechanical contribution and arises from the fluctuations around the steady state.

It is useful to also introduce the hermitian and antihermitian parts of the operator $\hat{x}$ defined by eq. (181):

$$\hat{x}_1 = \frac{\hat{x} + \hat{x}^\dagger}{2}, \qquad \hat{x}_2 = \frac{\hat{x} - \hat{x}^\dagger}{2i}. \tag{186}$$

In terms of these operators, the incoherent part of the spectrum has the expression

$$S_{\text{inc}}(\omega) = \frac{N_s}{\pi} \operatorname{Re} \int_0^\infty dt \exp[-i(\omega - \omega_0)t] \times [\langle \delta\hat{x}_1(t)\, \delta\hat{x}_1(0)\rangle_{\text{st}} + \langle \delta\hat{x}_2(t)\, \delta\hat{x}_2(0)\rangle_{\text{st}} + \mathrm{i}\langle \delta\hat{x}_1(t)\, \delta\hat{x}_2(0)\rangle_{\text{st}} - \mathrm{i}\langle \delta\hat{x}_2(t)\, \delta\hat{x}_1(0)\rangle_{\text{st}}]. \tag{185'}$$

In the bad cavity case, taking into account that from eq. (180) we have $\delta\hat{x} = -2C\delta\hat{r}^-$, the expression of $S_{\text{inc}}(\omega)$ to be used is

$$S_{\text{inc}}(\omega) = \frac{4C^2 N_s}{\pi} \operatorname{Re} \int_0^\infty dt \exp[-\mathrm{i}(\omega - \omega_0)t] \langle \delta\hat{r}^+(t)\, \delta\hat{r}^-(0)\rangle_{\text{st}}. \tag{187}$$

The integrated incoherent spectrum is given by

$$S_{\text{inc}}^{(\text{int})} = \int_0^\infty d\omega\, S_{\text{inc}}(\omega) = \langle \delta A^\dagger\, \delta A\rangle_{\text{st}}; \tag{188}$$

in the bad cavity case it becomes

$$S_{\text{inc}}^{(\text{int})} = 4C^2 N_s \langle \delta\hat{r}^+\, \delta\hat{r}^-\rangle_{\text{st}}. \tag{188'}$$

The reflected field (Fig. 1) is given by (compare eqs. (3a,b))

$$E_R(t) = -\sqrt{R}\, E_I + \sqrt{RT}\, E(L, t - \Delta t) = \sqrt{R}\,[E_T(t - \Delta t) - E_I). \tag{189}$$

Hence, the coherent part of the spectrum of reflected light is given by $RN_s(y - x_{\text{st}})^2\, \delta(\omega - \omega_0)$, and is in a sense complementary to the coherent part of the transmitted light. On the other hand, the incoherent part of the spectrum of reflected light is identical to that of transmitted light, apart from a factor of $R \approx 1$.

In this section we assume that thermal fluctuations and external field fluctuations are negligible, hence, we set $\bar{n} = 0$ in eqs. (176) and (177).

3.3.1. *Good cavity case*

In the case $k \ll \gamma_\perp, \gamma_\parallel$ we start from eq. (177). As for $N_s \gg 1$ the fluctuations are small, we can linearize the Fokker–Planck equation (177) around steady state. Hence, we introduce the deviations from steady state

$$x' = x - x_{st}, \qquad (x^*)' = x^* - x_{st}. \tag{190}$$

The linearized Fokker–Planck equation takes a particularly simple form in terms of the variables (compare eq. (186))

$$x_1' = \frac{x' + (x^*)'}{2}, \qquad x_2' = \frac{x_1' - (x^*)'}{2i} \tag{191}$$

It reads (LUGIATO [1979], CASAGRANDE and LUGIATO [1980])

$$k^{-1}\frac{\partial}{\partial t}P_w(x_1', x_2', t) = \left\{\frac{\partial}{\partial x_1'}\lambda x_1' + \frac{\partial}{\partial x_2'}\lambda_\varphi x_2' + \frac{1}{4N_s}\left[1 + 2C\frac{x_{st}^4 + (1-\bar{d})x_{st}^2 + 1}{(1+x_{st}^2)^3}\right]\frac{\partial^2}{\partial x_1'^2} + \frac{1+2C}{4N_s}\frac{\partial^2}{\partial x_2'^2}\right\}P_w(x_1', x_2', t), \tag{192}$$

where

$$\lambda = \left.\frac{\mathrm{d}y}{\mathrm{d}x}\right|_{x_{st}} = 1 + 2C\frac{1 - x_{st}^2}{(1+x_{st}^2)^2}, \qquad \lambda_\varphi = \frac{y}{x_{st}} = 1 + \frac{2C}{1+x_{st}^2} \tag{193}$$

and the function $y(x)$ is defined by eq. (32). Note that $\tilde{\lambda}$ and λ_φ coincide with $-\lambda_n^{(+)}$ and $-\lambda_n^{(-)}$ respectively, if one puts $n = 0$ (resonant mode) in eqs. (84). They also coincide with $-k^{-1}\lambda_1$ and $-k^{-1}\lambda_2$ given by eq. (58).

To obtain the incoherent part of the spectrum, according to eq. (185′) we must calculate the time correlation functions $\langle \delta\hat{x}_1(t)\,\delta\hat{x}_1(0)\rangle_{st}$, etc. To do that we use the regression theorem (LAX [1967]) which ensures that the time correlation functions $\langle \delta\hat{x}_1(t)\,\delta\hat{x}_1(0)\rangle_{st}$, and $\langle \delta\hat{x}_1(t)\,\delta\hat{x}_2(0)\rangle_{st}$, obey the same time evolution equation as $\langle \delta\hat{x}_1\rangle(t)$, and similarly, the time correlation functions $\langle \delta\hat{x}_2(t)\,\delta\hat{x}_2(0)\rangle_{st}$, and $\langle \delta\hat{x}_2(t)\,\delta\hat{x}_1(0)\rangle_{st}$, obey the same time evolution equation as $\langle \delta\hat{x}_2\rangle(t)$. Since $\langle \delta\hat{x}_i\rangle(t) = \langle x_i'\rangle(t)$ $(i = 1, 2)$, the equations for $\langle \delta\hat{x}_i\rangle(t)$ are immediately obtained from eq. (192), and read

$$\langle \delta\dot{\hat{x}}_1\rangle(t) = -k\tilde{\lambda}\langle \delta\hat{x}_1\rangle(t), \qquad \langle \delta\dot{\hat{x}}_2\rangle(t) = -k\lambda_\varphi\langle \delta\hat{x}_2\rangle(t). \tag{194}$$

Hence, from eq. (185′) we have

$$S_{\text{inc}}(\omega) = \frac{N_s}{\pi} \operatorname{Re} \int_0^{\infty} \mathrm{d}t \exp[-\mathrm{i}(\omega - \omega_0)t] \times [\mathrm{e}^{-k\bar{\lambda}t}(\langle \delta\hat{x}_1^2 \rangle_{\text{st}} + \mathrm{i}\langle \delta\hat{x}_1\, \delta\hat{x}_2 \rangle_{\text{st}}) + \mathrm{e}^{-k\lambda_\varphi t}(\langle \delta\hat{x}_2^2 \rangle_{\text{st}} - \mathrm{i}\langle \delta\hat{x}_2\, \delta\hat{x}_1 \rangle_{\text{st}})]. \quad (195)$$

The variances $\langle \delta\hat{x}_1^2 \rangle_{\text{st}}$, etc. are also easily obtained from eq. (192). In fact, taking into account that $\langle (x_i')^2 \rangle(t) = \langle (\delta\hat{x}_i)^2 \rangle(t)$ $(i = 1, 2)$ and $\langle x_1' x_2' \rangle(t) = (1/2)\langle \delta\hat{x}_1\, \delta\hat{x}_2 + \delta\hat{x}_2\, \delta\hat{x}_1 \rangle(t)$ (compare eq. (174)) we have

$$\langle \delta\hat{x}_1^2 \rangle_{\text{st}} = \frac{1}{4N_s\bar{\lambda}}\left(1 + 2C\frac{x_{\text{st}}^4 + (1-\bar{d})x_{\text{st}}^2 + 1}{(1+x_{\text{st}}^2)^3}\right), \quad (196a)$$

$$\langle \delta\hat{x}_2^2 \rangle_{\text{st}} = \frac{1}{4N_s\lambda_\varphi}(1 + 2C), \quad (196b)$$

$$\tfrac{1}{2}(\langle \delta\hat{x}_1\, \delta\hat{x}_2 \rangle_{\text{st}} + \langle \delta\hat{x}_2\, \delta\hat{x}_1 \rangle_{\text{st}}) = 0. \quad (196c)$$

By combining eq. (196c) with the commutation rule $[\delta\hat{x}_1, \delta\hat{x}_2] = \mathrm{i}/2N_s$, we have

$$\langle \delta\hat{x}_1\, \delta\hat{x}_2 \rangle_{\text{st}} = -\langle \delta\hat{x}_2\, \delta\hat{x}_1 \rangle_{\text{st}} = \frac{1}{4N_s} \quad (196c')$$

Finally, by inserting eq. (196a,b) and (196c′) into eq. (195) we obtain (LUGIATO [1979])

$$S_{\text{inc}}(\omega) = \frac{Ck}{2\pi}\frac{x_{\text{st}}^2}{1+x_{\text{st}}^2}\left\{\frac{2x_{\text{st}}^2 + 1 - \bar{d}}{(1+x_{\text{st}}^2)^2}\frac{1}{(\omega-\omega_0)^2 + k^2\bar{\lambda}^2} + \frac{1}{(\omega-\omega_0)^2 + k^2\lambda_\varphi^2}\right\}. \quad (197)$$

Hence, $S_{\text{inc}}(\omega)$ is given by the superposition of two Lorentzians, whose weights are inversely proportional to $\bar{\lambda}$ and λ_φ, respectively. Note, however, that the weight of the first Lorentzian is not always positive.

According to eq. (188), the integrated incoherent spectrum is given by

$$S_{\text{inc}}^{(\text{int})} = \frac{C}{2}\frac{x_{\text{st}}^2}{1+x_{\text{st}}^2}\left\{\frac{2x_{\text{st}}^2 + 1 - \bar{d}}{(1+x_{\text{st}}^2)^2}\frac{1}{\bar{\lambda}} + \frac{1}{\lambda_\varphi}\right\}. \quad (198)$$

Let us now analyze the spectrum for $C \gg 1$, first in the purely radiative case $\gamma_{\|} = 2\gamma_{\perp} \equiv \gamma$ (i.e. $\bar{d} = 2$), then for $\gamma_{\|} \ll \gamma_{\perp}$ (i.e. $\bar{d} \ll 1$):

a) $\bar{d} = 2$ ($\gamma_{\|} = 2\gamma_{\perp}$). For $y \ll C$ (so that $x_{st} \ll 1$, $y \approx 2Cx_{st}$) it follows from eqs. (198) and (193) that $S_{inc}^{(int)}$ is porportional to x_{st}^4, and hence, to y^4. On the other hand, for $y \gg C$ (i.e. $x_{st} \approx y$), $S_{inc}^{(int)}$ saturates to the value $C/2$. Furthermore $S_{inc}^{(int.)}$ diverges when the bistability thresholds y_M and y_m (see Fig. 8) are approached, because $\tilde{\lambda} \to 0$. (compare § 2.3.1). In the neighborhood of y_M and y_m the linearization of the Fokker–Planck equation fails.

Let us now consider the lineshape $S_{inc}(\omega)$. As shown by eq. (197), for $k \ll \gamma_{\perp}, \gamma_{\|}$ the linewidth is scaled by the empty cavity halfwidth k. The relevant eigenvalues $\tilde{\lambda}$ and λ_{φ} are always real. For $y \ll C$ it follows from eq. (193) that $\lambda_{\varphi} \approx \tilde{\lambda} \approx 2C$. Hence, the width of the spectrum is $4Ck$, much larger than the empty-cavity width $2k$. Since $C \propto N$, in this situation the linewidth is proportional to the number of atoms (*cooperative line broadening*). As one sees from eq. (197), in this situation the weight of the first Lorentzian is negative. Consequently the line gets a hole in the middle, so that it takes the shape of a doublet (Fig. 29a,b). This is a purely quantum mechanical effect, as we shall show in § 3.3.3.

Increasing y the doublet character disappears (Fig. 29c). Approaching the upper bistability threshold y_M the spectrum becomes a narrow line (Fig. 29d). In fact $\tilde{\lambda} \to 0$ so that the soft-mode contribution from the first term in eq. (197) dominates the cooperative background from the second term. This *line narrowing* is clearly a manifestation of the critical slowing down illustrated in § 2.3.1.

When we cross the threshold $y = y_M$, jumping to the one-atom branch, the spectrum changes discontinuously from a narrow line to a line whose width coincides with the empty cavity width $2k$ (Fig. 29e). In fact, for $y \gtrsim y_M$ one has $x \approx y$, and from eq. (193) we have $\lambda_{\varphi} \approx \tilde{\lambda} \approx 1$. For $y \gg y_M$, the incoherent part of the spectrum does not change increasing y (Fig. 29f). In fact, one finds

$$S_{inc}(\omega) \approx \frac{C}{2\pi} \frac{k}{(\omega - \omega_0)^2 + k^2}. \tag{197'}$$

Let us now decrease y along the one-atom branch. Approaching the lower bistability threshold $y = y_m$ the line becomes narrower and narrower, because again $\tilde{\lambda} \to 0$ (see Fig. 22g,h). Finally, crossing the lower threshold, the spectrum suddenly broadens because the system jumps to the cooperative branch.

b) $\bar{d} \ll 1$ ($\gamma_{\|} \ll \gamma_{\perp}$). In this case, the behavior of $S_{inc}(\omega)$ is qualitatively similar to that of the case $\bar{d} = 2$, except for $y \ll C$. In fact, for $\bar{d} \ll 1$, both weights in eq. (197) are always positive, so that for $y \ll C$ the spectrum is not a doublet but a single line of width $4Ck$.

These results have been generalized to the case $\Delta, \theta \neq 0$ in DRUMMOND and WALLS [1981].

The spectrum of transmitted light in the good cavity case, including all the longitudinal modes of the cavity, is given in BENZA and LUGIATO [1981], on the basis of the many-mode master equation (158). Their paper also describes the behavior of this spectrum when we approach the self-pulsing instability domain.

3.3.2. *Bad cavity case*

The details of the calculation of $S_{\text{inc}}(\omega)$ for $k \gg \gamma_\perp, \gamma_\parallel$ starting from the Fokker–Planck equation (179) are given in LUGIATO [1979]. Again, one finds that $S_{\text{inc}}(\omega)$ is given by the superposition of few Lorentzians which are peaked at $\omega = \omega_0 + \text{Im}\,\lambda_i$, and have width $-\text{Re}\,\lambda_i$, where λ_i are the eigenvalues of the linearized semiclassical mean field equations (see § 2.3.1). Namely, the relevant eigenvalues $-\lambda_i$ are

$$\lambda_{\substack{\alpha\\ \beta}} = \frac{\gamma_\perp}{2}\left\{\bar{d} + \frac{y}{x_{\text{st}}} \pm \left[\left(\bar{d} - \frac{y}{x_{\text{st}}}\right)^2 - 4\bar{d}x_{\text{st}}(2x_{\text{st}} - y)\right]^{1/2}\right\}, \tag{199}$$

$$\lambda_\gamma = \gamma_\perp \frac{y}{x_{\text{st}}} = \gamma_\perp \lambda_\varphi. \tag{200}$$

The real part of the root λ_β coincides with the damping constant $\bar{\lambda}$ discussed in § 2.3.1; when (x, y) lies on the cooperative branch λ_β is well approximated by expression (59). When $\lambda_{\substack{\alpha\\ \beta}}$ are real $S_{\text{inc}}(\omega)$ is given by the superposition of three Lorentzians:

$$S_{\text{inc}}(\omega) = \frac{C\gamma_\perp^2}{\pi k} \sum_{i=\alpha,\beta,\gamma} w_i \frac{\lambda_i}{(\omega - \omega_0)^2 + \lambda_i^2}, \tag{201}$$

where

$$w_\gamma = \frac{1}{2\lambda_\gamma}\frac{x_{\text{st}}^2}{1 + x_{\text{st}}^2},$$
$$w_{\substack{\alpha\\ \beta}} = \mp \frac{1}{2\lambda_{\substack{\alpha\\ \beta}}} \frac{x_{\text{st}}^2}{1 + x_{\text{st}}^2} \frac{1}{\lambda_\alpha^2 - \lambda_\beta^2}[(\bar{d} - 1)\lambda_{\substack{\alpha\\ \beta}}^2 - \gamma_\parallel(\bar{d} - 1 - 2x_{\text{st}}^2)]. \tag{202}$$

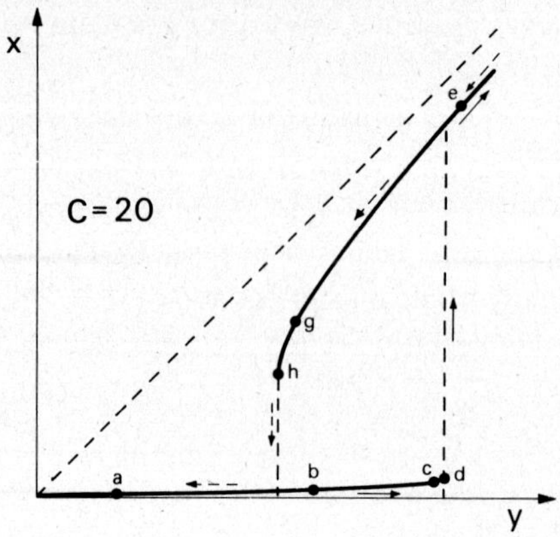
x
C = 20
e
g
h
a
b
c
d
y

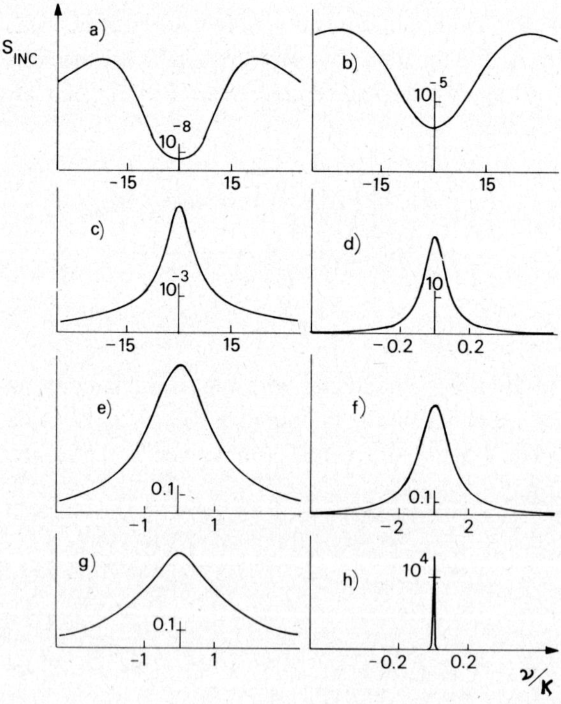
S_INC
a)
10^-8
-15
15
b)
10^-5
-15
15
c)
10^-3
-15
15
d)
10
-0.2
0.2
e)
0.1
-1
1
f)
0.1
-2
2
g)
0.1
-1
1
h)
10^4
-0.2
0.2
ν/κ

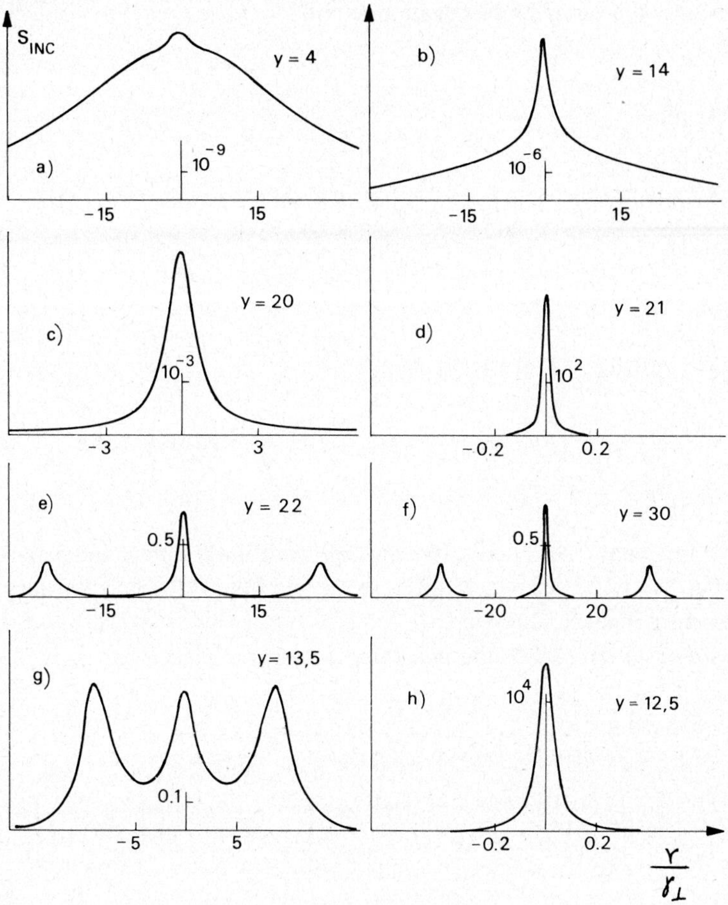

Fig. 30. Hysteresis cycle of the incoherent part of the spectrum $S_{\mathrm{inc}}(\omega)$ of the transmitted light in the *bad cavity case* $k \gg \gamma_\perp, \gamma_\parallel$ for $\gamma_\parallel = 2\gamma_\perp$, $C = 20$. The points of the (x, y) plane corresponding to (a)–(h) are indicated in Fig. 29. See also caption of Fig. 29.

Fig. 29. (facing page). Hysteresis cycle of the incoherent part of the spectrum $S_{\mathrm{inc}}(\omega)$ of the transmitted light in the *good cavity case* $k \ll \gamma_\perp, \gamma_\parallel$ for $\gamma_\parallel = 2\gamma_\perp$, $C = 20$. The point of the (x, y) plane corresponding to (f) does not appear because y is too large. Here and in figure 30 we have (i) $\nu = \omega - \omega_0$, where ω_0 is the frequency of the incident field, (ii) S_{inc} is given in units $C/2\pi k$, (iii) the scale varies from diagram to diagram as indicated, (iv) y has the values (a) $y = 4.06$, (b) $y = 14.193$, (c) $y = 20.589$, (d) $y = 21.026$, (e) $y = 21.995$, (f) $y = 31.33$, (g) $y = 13.39$, (h) $y = 12.48$.

When λ_α, λ_β are complex conjugate one puts

$$\lambda_{\substack{\alpha\\ \beta}} = \lambda_R \pm i\lambda_I \tag{203}$$

and obtains

$$S_{inc}(\omega) = \frac{C\gamma_\perp^2}{\pi k}\left\{w_\gamma \frac{\lambda_\gamma}{(\omega-\omega_0)^2+\lambda_\gamma^2} + g(\omega-\omega_0) + g(\omega_0-\omega)\right\}, \tag{204}$$

where

$$g(v) = \frac{1}{8}\frac{1}{\lambda_R^2-\lambda_I^2}\frac{\gamma_\perp^3}{(v-\lambda_I)^2+\lambda_R^2}\left\{2\bar{d}^2(2x_{st}^2+1-\bar{d})\right.$$
$$\left. - \frac{v}{\gamma_\perp\lambda_R}\left[(\bar{d}-1)(\lambda_R^2+\lambda_I^2) - \gamma_\parallel^2(\bar{d}-1-2x_{st}^2)\right]\right\}\frac{x_{st}^2}{1+x_{st}^2}. \tag{205}$$

For $\gamma_\parallel = 2\gamma_\perp$ eqs. (201) and (204) coincide with the formulas independently derived in AGARWAL, NARDUCCI, GILMORE and FENG [1978a,b] via quantum mechanical Langevin equations.

According to eq. (188′), the integrated spectrum is given by

$$S_{inc}^{(int)} = C\frac{\gamma_\perp}{k}\frac{x_{st}^4}{(1+x_{st}^2)y(y/x_{st}+\bar{d})[y/x_{st}+x_{st}(2x_{st}-y)]}$$
$$\times\left\{\frac{\bar{d}}{2}(2x_{st}+y) + \left(1-\frac{\bar{d}}{2}\right)\frac{y}{x_{st}^2}\left[x_{st}(2x_{st}-y)+\frac{y}{x_{st}}+\bar{d}\right]\right\}. \tag{206}$$

For $y \ll C$ we have $S_{inc}^{(int)} \propto x_{st}^4$ as in the good cavity case. For $y \gg C$ $S_{inc}^{(int)}$ saturates to the value $\gamma_\perp C/k$. Furthermore $S_{inc}^{(int)}$ diverges as usual when we approach the bistability thresholds y_M and y_m.

As we see from eqs. (199) and (200), in the bad cavity case the linewidth of the transmitted field is scaled by the atomic linewidth $\gamma_\perp$. The hysteresis cycle of the spectrum for $\gamma_\parallel = 2\gamma_\perp \equiv \gamma$ is shown in Fig. 30 for $C \gg 1$. When the system is on the cooperative branch, λ_α, λ_β are real, so that the spectrum $S_{inc}(\omega)$ is a single line. For $y \ll C$ one has $\lambda_\alpha \approx \lambda_\gamma \approx C\gamma$, $\lambda_\beta \approx \gamma$. In these conditions the contribution of the term proportional to w_β in eq. (201) is negligible, so that the spectrum is a broad line (Fig. 30a) whose halfwidth is γC, which coincides with the cooperative linewidth γ_R of pure superfluorescence (BONIFACIO and LUGIATO [1975]). Since $C \propto N$, in this situation the linewidth is proportional to the number of atoms (cooperative line broadening) exactly as in the good cavity case. Increasing y along the cooperative branch, the peak corresponding to the soft mode λ_β emerges from the cooperative background (Fig. 30b). Approaching the upper bistability threshold we have the usual line narrowing, in which the soft mode dominates and the cooperative background is completely negligible (Fig. 30c,d). This line narrowing is accompanied by strong atom–atom correlations (LUGIATO [1979]).

Let us now cross the threshold $y = y_M$, so that the system jumps to the one-atom branch. The roots λ_α, λ_β are complex conjugate, so that the spectrum suddenly becomes a triplet (Fig. 30e). This means a *discontinuous* appearance of a *Dynamical Stark Effect*. For $y \gg y_M$ one has $x_{st} \approx y$ and $\lambda_\gamma \approx \frac{1}{2}\gamma$, $\lambda_R \approx \frac{3}{4}\gamma$, $\lambda_I \approx \frac{1}{2}\gamma y\sqrt{2}$ (compare eqs. (60) and (61)). Hence, since $\lambda_I \gg \lambda_R$, $x_{st} \gg 1$, $S_{inc}(\omega)$ takes the simple form

$$S_{inc}(\omega) \propto \left\{ \frac{\frac{1}{2}\gamma}{(\omega - \omega_0)^2 + \frac{1}{4}\gamma^2} + \frac{1}{2}\left[\frac{\frac{3}{4}\gamma}{(\omega - \omega_0 - \Omega_I)^2 + \frac{9}{16}\gamma^2} + \frac{\frac{3}{4}\gamma}{(\omega - \omega_0 + \Omega_I)^2 + \frac{9}{16}\gamma^2} \right] \right\}, \tag{207}$$

where Ω_I is the Rabi frequency of the incident field, defined in eq. (61). Equation (207) coincides with the lineshape predicted for the spectrum of fluorescent light in the high intensity situation by the one-atom theory of resonance fluorescence (MOLLOW [1969]). For $C \gg 1$ one has $\Omega_I \gg \gamma$, so that the sidebands are well separated from the central line (Fig. 30f).

Let us now decrease y along the one-atom branch. The two sidebands get nearer and nearer to the central line (Fig. 30g) until in the vicinity of the lower threshold $y = y_m$ the roots λ_α, λ_β become real and one again has a line narrowing, because the linewidth $2\lambda_\beta$ tends to zero (Fig. 30h).

3.3.3. *Nonclassical effects in the transmitted field*

In the previous subsection we have seen that, in the good cavity case (eq. (197)), the weight of the first Lorentzian in the expression for $S_{inc}(\omega)$ is negative in the purely radiative case $\bar{d} = 2$ for $y \ll C$. This is a consequence of the fact that, if we consider the Fokker–Planck equation for the Glauber P-function equivalent to eq. (177′), its radial diffusion coefficient is negative in these conditions. In fact, this coefficient is proportional to the weight of the first

Lorentzian in eq. (197). Moreover, in general, the lack of positive definiteness in the Fokker–Planck equation for the Glauber *P*-function (or generalized Glauber P-function) is an indication of the rise of nonclassical effects in the transmitted field. These effects arise specifically from the noncommutative character of the creation and annihilation operators.

The most well known effect is *photon antibunching* which is defined as follows. Let us consider the second-order correlation function

$$g^{(2)}(t) = \frac{\langle A^\dagger(0)\, A^\dagger(t)\, A(t)\, A(0)\rangle_{\mathrm{st}}}{\langle A^\dagger A\rangle_{\mathrm{st}}^2}.$$

For purely coherent light $g^{(2)}(t) = 1$; one has *bunching* when $g^{(2)}(t) > 1$ (GLAUBER [1963]) and *antibunching* for $g^{(2)}(t) < 1$. Let us now analyze this problem in the case of optical bistability (DRUMMOND and WALLS [1980], CASAGRANDE and LUGIATO [1980]). By taking into account the first of eqs. (181), and that $x_{\mathrm{st}} = O(1)$, $\delta\hat{x} = O(N_{\mathrm{s}}^{-1/2})$, we obtain, after simple calculations,

$$g^{(2)}(0) = 1 + \frac{4}{x_{\mathrm{st}}^2}\left(\langle \delta\hat{x}_1^2\rangle_{\mathrm{st}} - \frac{1}{4N_{\mathrm{s}}}\right) + O(N_{\mathrm{s}}^{-3/2}). \tag{208}$$

Using eqs. (196a) and (193) we have

$$\langle \delta\hat{x}_1^2\rangle_{\mathrm{st}} - \frac{1}{4N_{\mathrm{s}}} = \frac{C}{2\tilde{\lambda}N_{\mathrm{s}}}\,\frac{x_{\mathrm{st}}^2(2x_{\mathrm{st}}^2 + 1 - \bar{d})}{(1 + x_{\mathrm{st}}^2)^3}. \tag{209}$$

For $\bar{d} = 2$ (purely radiative case) and $x_{\mathrm{st}} \ll 1$ (i.e. $y \ll C$), it follows from eq. (209) that $\langle \delta\hat{x}_1^2\rangle - (1/4N_{\mathrm{s}})$ is negative, hence, there is antibunching. As we see from eqs. (208) and (209), this effect is small, since $N_{\mathrm{s}} \gg 1$. Note that the expression (209) is proportional to the weight of the first Lorentzian in (197). As shown by CASAGRANDE and LUGIATO [1980] $g^{(2)}(t) = g^{(2)}(0)\exp(-k\tilde{\lambda}t)$. Furthermore, the antibunching effect in the cooperative branch also arises in the bad cavity case.

Another nonclassical effect that has become popular in recent years is the so-called "squeezing". The squeezed states are characterized by the fact that the variance in one of the two quadrature components is smaller than in the coherent Glauber states. Namely, let us consider the two operators

$$A_1 = \frac{A + A^\dagger}{2}, \qquad A_2 = \frac{A - A^\dagger}{2\mathrm{i}}. \tag{210}$$

In the coherent states, which are minimum uncertainty states, one has

$$\langle \delta A_1^2 \rangle_{st} = \langle \delta A_2^2 \rangle_{st} = 1/4. \tag{211}$$

The state is squeezed when $\langle \delta A_1^2 \rangle < 1/4$ or $\langle \delta A_2^2 \rangle < 1/4$ (one cannot have both simultaneously because of Heisenberg's indetermination rule). This type of state is important in the framework of optical communication, interferometry and gravitational wave detection (YUEN [1976], MEYSTRE and SCULLY [1983]).

Now, since $\delta A_1 = \sqrt{N_s}\, \delta \hat{x}_1$ we see immediately from eq. (209) that, in the case of optical bistability (for $\Delta = \theta = 0$) when there is photon antibunching there is also squeezing in the component A_1. This effect is small because $\langle \delta A_1^2 \rangle_{st}$ always remains larger than 0.21 (LUGIATO and STRINI [1982a]). However, it increases in the case of multiphoton bistability where one finds that $\langle \delta A_1^2 \rangle_{st}$ can be as small as 1/16 (LUGIATO and STRINI [1982b,c]). In optical bistability, antibunching and squeezing are cooperative effects which arise from atom–atom correlations.

We observe that thermal and external fluctuations tend to destroy the antibunching and squeezing effects. In fact, if $\bar{n} \neq 0$ in eq. (177) we obtain, instead of eq. (209),

$$\langle \delta \hat{x}_1^2 \rangle - \frac{1}{4N_s} = \frac{1}{2\lambda N_s} \left[\frac{x_{st}^2 (2x_{st}^2 + 1 - \bar{d})}{(1 + x_{st}^2)^3} + \bar{n} \right]. \tag{209'}$$

Hence, if $\bar{n}$ is large enough expression (209′) is positive for all values x_{st}. Also atomic number fluctuations tend to destroy nonclassical effects (DRUMMOND and WALLS [1981].

We stress that the nonclassical effects discussed in this section arise from atom–atom correlations. In fact, previous treatments which neglected atom–atom correlations (see for instance BONIFACIO, GRONCHI and LUGIATO [1978]) led to a Fokker–Planck equation for the Glauber P-function with a positive definite diffusion matrix, and hence, without antibunching and squeezing. In LUGIATO, CASAGRANDE and PIZZUTO [1982] it is shown that the procedure based on the generalized Wigner distribution is more complete and accurate than these previous treatments.

3.4. SPECTRUM OF FLUORESCENT LIGHT

In this section we consider the spectrum $I(\omega)$ of the fluorescent light emitted at 90° with respect to the incident light. It is proportional to the Fourier transform of the time correlation function $\sum_{i=1}^{N}\langle r_i^+(t)\, r_i^-(0)\rangle_{st}$, where r_i^+ are the raising and lowering operators of the ith two-level atom (see § 3.1.3)

$$I(\omega) \propto \frac{1}{\pi}\,\mathrm{Re}\int_0^{\infty} \mathrm{d}t \exp[-\mathrm{i}(\omega-\omega_0)t] \sum_{i=1}^{N}\langle r_i^+(t)\, r_i^-(0)\rangle_{st}. \tag{212}$$

By subdividing $r_i^\pm(t)$ into the stationary mean value $\langle r_i^\pm\rangle_{st} = \langle R^\pm\rangle_{st}/N$ and the fluctuation $\delta r_i^\pm(t) = r_i^\pm(t) - \langle r_i^\pm\rangle_{st}$, we show that $I(\omega)$ is composed of a coherent and an incoherent part:

$$I(\omega) = I_{coh}(\omega) + I_{inc}(\omega), \tag{213}$$

$$I_{coh}(\omega) \propto \sum_{i=1}^{N} |\langle r_i^-\rangle|^2\, \delta(\omega-\omega_0) = N^{-1}|\langle R^-\rangle_{st}|^2\, \delta(\omega-\omega_0), \tag{214a}$$

$$I_{inc}(\omega) \propto \frac{N}{\pi}\,\mathrm{Re}\int_0^{\infty} \mathrm{d}t \exp[-\mathrm{i}(\omega-\omega_0)t]\langle \delta r_1^+(t)\, \delta r_1^-(0)\rangle_{st}, \tag{214b}$$

where in eq. (214b) we have replaced the index i by 1 on account of the symmetry between the atoms.

Using eqs. (170) and (50) we show immediately that

$$I_{coh}(\omega) \propto \frac{N}{4}\bar{d}\frac{x_{st}^2}{(1+x_{st}^2)}\,\delta(\omega-\omega_0). \tag{215}$$

The integrated spectrum $I_F = \int \mathrm{d}\omega\, I(\omega)$ is given by

$$I_F \propto \sum_{i=1}^{N}\langle r_i^+ r_i^-\rangle_{st} = \frac{N}{2} + \langle R_3\rangle_{st} = \frac{N}{2}\,\frac{x_{st}^2}{1+x_{st}^2}, \tag{216}$$

where we have again used eqs. (170) and (50). On the other hand, from eq. (215) we obtain

$$I_{coh}^{(int)} = \int \mathrm{d}\omega\, I_{coh}(\omega) \propto \frac{N}{4}\bar{d}\frac{x_{st}^2}{(1+x_{st}^2)^2}. \tag{217}$$

Hence, the integrated incoherent spectrum is given by

$$I_{inc}^{(int)} = I_F - I_{coh}^{(int)} \propto \frac{N}{2}\,\frac{x_{st}^2}{(1+x_{st}^2)^2}\left(1-\frac{\bar{d}}{2}+x_{st}^2\right). \tag{218}$$

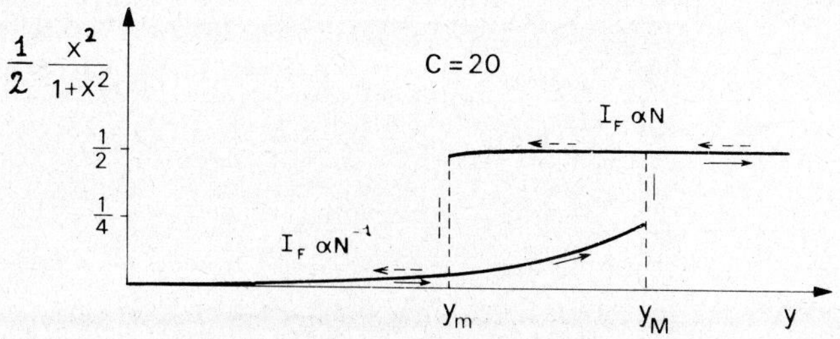

Fig. 31. Hysteresis cycle of the total fluorescent intensity I_F, which is proportional to the population of the upper level, $(N/2)\,(x^2/1 + x^2)$.

From eqs. (217) and (218) we get

$$I_{\text{coh}}^{(\text{int})}/I_{\text{inc}}^{(\text{int})} = \frac{\bar{d}/2}{1 - \dfrac{\bar{d}}{2} + x_{\text{st}}^2}. \tag{218'}$$

For $\bar{d} \ll 1$ one always has $I_{\text{coh}}^{(\text{int})} \ll I_{\text{inc}}^{(\text{int})}$. For $\bar{d} = 2$ one has that $I_{\text{coh}}^{(\text{int})} \gg I_{\text{inc}}^{(\text{int})}$ in the cooperative stationary state, and the contrary in the one-atom stationary state.

Let us now focus our attention on the fluorescent intensity emitted in a unit solid angle at 90°, which is proportional to (216). Its hysteresis cycle is shown in Fig. 31. In the one-atom branch, where $x_{\text{st}} = y$, this intensity is proportional to the number of atoms as usual. On the other hand, in the cooperative branch, where $x_{\text{st}} \approx y/2C \propto N^{-1}$, *the fluorescent intensity is inversely proportional to the number of atoms*. This is a remarkable cooperative effect (BONIFACIO and LUGIATO [1976]). The first observations of the hysteresis cycle of the fluorescent intensity have been made recently by GRANT and KIMBLE [1982] (Fig. 32) and ARECCHI, GIUSFREDI, PETRIELLA and SALIERI [1982].

3.4.1. *Calculation of the incoherent part of the spectrum*

The time evolution equation for the time correlation function $\langle r_1^+(t)\, r_1^-(0)\rangle_{\text{st}}$ in eq. (214b) is immediately obtained from the master eq. (158), restricted to the resonant mode $n = 0$, via the regression theorem (LAX [1967]). In fact, on using the commutation rules (151), (152) and the form (163′) of the interaction Hamiltonian we obtain the equation

$$\langle r_1^+\rangle(t) = 2\bar{g}\langle A^\dagger r_{31}\rangle(t) - \gamma_\perp \langle r_1^+\rangle(t).$$

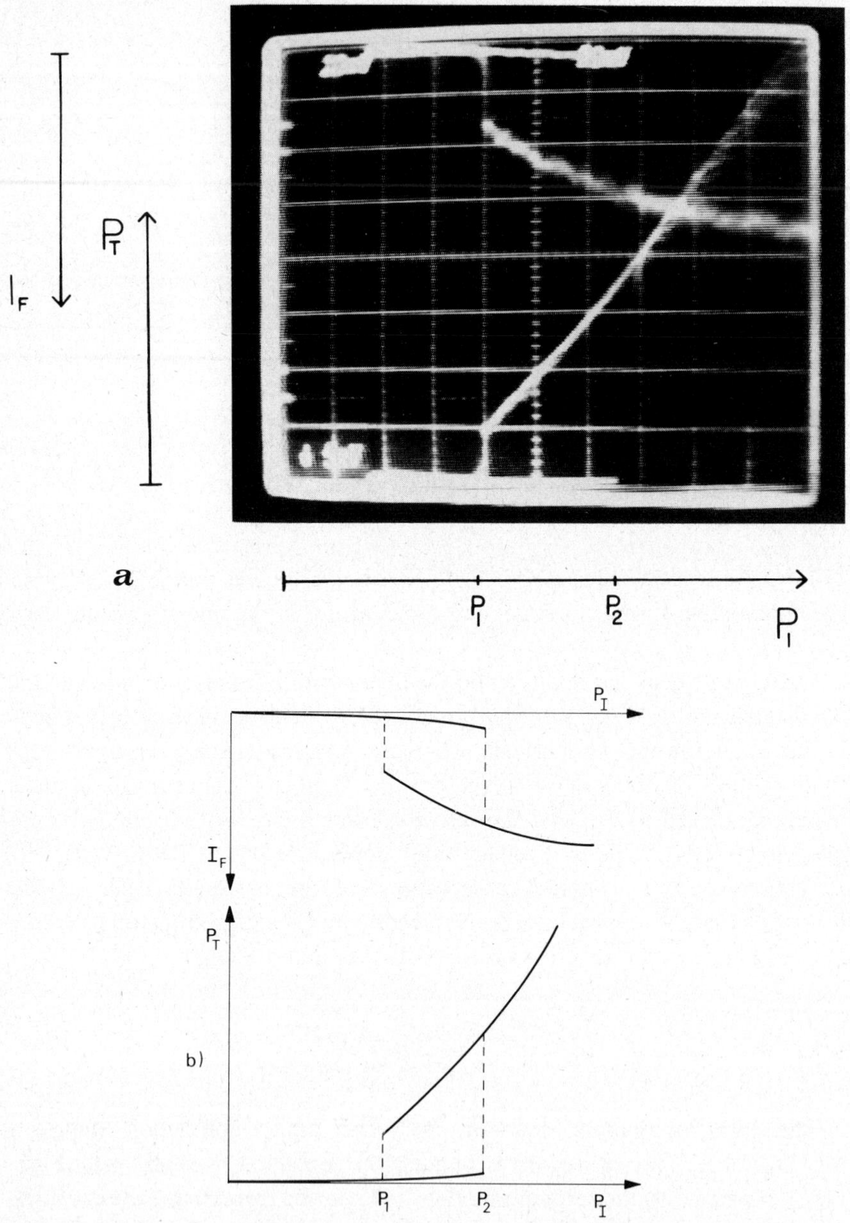

Fig. 32. Simultaneous observation of output power P_T and fluorescent intensity I_F in an atomic beam of sodium traveling through a confocal cavity (from KIMBLE [1982]). The bistability is purely absorptive. Figure 32b explains the picture in Fig. 32a. In this experiment $C = \alpha L \mathscr{F}/2\pi = 35$, where the cavity finesse $\mathscr{F}$ is $\mathscr{F} \approx 210$. The time taken to switch the incident power up and down is 10 ms. The switching points occur at $P_1 = 0.63$ mW and $P_2 = 1$ mW.

The regression theorem states that the time evolution equation of $\langle r_1^+(t)\, r_1^-(0)\rangle_{st}$ has the same form as that of $\langle r_1^+\rangle(t)$. Hence, it reads

$$\frac{d}{dt}\langle r_1^+(t)\, r_1^-(0)\rangle_{st} = 2\bar{g}\langle A^\dagger(t)\, r_{31}(t)\, r_1^-(0)\rangle_{st} - \gamma_\perp \langle r_1^+(t)\, r_1^-(0)\rangle_{st}. \tag{219a}$$

We shall consider eq. (219a) together with the time evolution equations for $\langle r_1^-(t)\, r_1^-(0)\rangle_{st}$ and $\langle r_{31}(t)\, r_1^-(0)\rangle_{st}$, which read

$$\frac{d}{dt}\langle r_1^-(t)\, r_1^-(0)\rangle_{st} = 2\bar{g}\langle A(t)\, r_{31}(t)\, r_1^-(0)\rangle_{st} - \gamma_\perp \langle r_1^-(t)\, r_1^-(0)\rangle_{st}, \tag{219b}$$

$$\frac{d}{dt}\langle r_{31}(t)\, r_1^-(0)\rangle_{st} = -\bar{g}(\langle A(t)\, r_1^+(t)\, r_1^-(0)\rangle_{st} - \langle A^\dagger(t)\, r_1^-(t)\, r_1^-(0)\rangle_{st}) - \gamma_\parallel(\langle r_{31}(t)\, r_1^-(0)\rangle_{st} + \tfrac{1}{2}\langle r_1^-\rangle_{st}). \tag{219c}$$

Next, we introduce the deviations of $r_1^\pm$, r_{31}, A and $A^\dagger$ from their respective stationary values,

$$\begin{aligned} r_1^\pm(t) &= \langle r_1^\pm\rangle_{st} + \delta r_1^\pm(t), \qquad r_{31}(t) = \langle r_{31}\rangle_{st} + \delta r_{31}(t), \\ A(t) &= \langle A\rangle_{st} + \delta A(t), \end{aligned} \tag{220}$$

and insert eq. (220) into eqs. (219). By taking into account the fact that, at steady state, one has (using eq. (151), (152) and (158)) and setting $d\langle r_1^\pm\rangle/dt = d\langle r_{31}\rangle/dt = 0$)

$$\begin{aligned} 0 &= 2\bar{g}\langle A^\dagger r_{31}\rangle_{st} - \gamma\langle r_1^+\rangle_{st} \\ &= 2\bar{g}(\langle A^\dagger\rangle_{st}\langle r_{31}\rangle_{st} + \langle \delta A^\dagger\, \delta r_{31}\rangle_{st}) - \gamma_\perp \langle r_1^+\rangle_{st}, \end{aligned} \tag{221a}$$

$$0 = 2\bar{g}(\langle A\rangle_{st}\langle r_{31}\rangle_{st} + \langle \delta A\, \delta r_{31}\rangle_{st}) - \gamma_\perp\langle r_1^-\rangle_{st}, \tag{221b}$$

$$\begin{aligned} 0 = &-\bar{g}(\langle A\rangle_{st}\langle r_1^+\rangle_{st} + \langle \delta A\, \delta r_1^+\rangle_{st} + \langle A^\dagger\rangle_{st}\langle r_1^-\rangle_{st} + \langle \delta A^\dagger\, \delta r_1^-\rangle_{st}) \\ &- \gamma_\parallel(\langle r_{31}\rangle_{st} + 1/2), \end{aligned} \tag{221c}$$

we obtain the equations

$$\begin{aligned} \frac{d}{dt}\langle \delta r_1^+(t)\, \delta r_1^-(0)\rangle_{st} = 2\bar{g}\{&\langle A^\dagger\rangle_{st}\langle \delta r_{31}(t)\, \delta r_1^-(0)\rangle_{st} \\ &+ \langle r_{31}\rangle_{st}\langle \delta A^\dagger(t)\, \delta r_1^-(0)\rangle_{st} + \langle \delta A^\dagger(t)\, \delta r_{31}(t)\, \delta r_1^-(0)\rangle_{st}\} \\ &- \gamma_\perp\langle \delta r_1^+(t)\, \delta r_1^-(0)\rangle_{st}, \end{aligned} \tag{222a}$$

$$\frac{\mathrm{d}}{\mathrm{d}t}\langle \delta r_1^-(t)\,\delta r_1^-(0)\rangle_{\mathrm{st}} = 2\bar{g}\{\langle A\rangle_{\mathrm{st}}\langle \delta r_{31}(t)\,\delta r_1^-(0)\rangle_{\mathrm{st}} + \langle r_{31}\rangle_{\mathrm{st}}\langle \delta A(t)\,\delta r_1^-(0)\rangle_{\mathrm{st}} + \langle \delta A(t)\,\delta r_{31}(t)\,\delta r_1^-(0)\rangle_{\mathrm{st}}\} - \gamma_\perp\langle \delta r_1^-(t)\,\delta r_1^-(0)\rangle_{\mathrm{st}}, \tag{222b}$$

$$\frac{\mathrm{d}}{\mathrm{d}t}\langle \delta r_{31}(t)\,\delta r_1^-(0)\rangle_{\mathrm{st}} = -\bar{g}\{\langle A\rangle_{\mathrm{st}}\langle \delta r_1^+(t)\,\delta r_1^-(0)\rangle_{\mathrm{st}} + \langle r_1^+\rangle_{\mathrm{st}}\langle \delta A(t)\,\delta r_1^-(0)\rangle_{\mathrm{st}} + \langle \delta A(t)\,\delta r_1^+(t)\,\delta r_1^-(0)\rangle_{\mathrm{st}} + \langle A^\dagger\rangle_{\mathrm{st}}\langle \delta r_1^-(t)\,\delta r_1^-(0)\rangle_{\mathrm{st}} + \langle r_1^-\rangle_{\mathrm{st}}\langle \delta A^\dagger(t)\,\delta r_1^-(0)\rangle_{\mathrm{st}} + \langle \delta A^\dagger(t)\,\delta r_1^-(t)\,\delta r_1^-(0)\rangle_{\mathrm{st}} - \gamma_\parallel\langle \delta r_{31}(t)\,\delta r_1^-(0)\rangle_{\mathrm{st}}. \tag{222c}$$

A crucial point in the quantum statistical treatment of a system with normal fluctuations is the consideration of the scaling properties of the quantities in play, with respect to the size of the system (this is precisely what we did in § 3.2 when we took into account the smallness of the quantity N_s^{-1}, which is proportional to N^{-1}). For instance, one has $\langle A\rangle_{\mathrm{st}} \propto N^{1/2}$, $\delta A \propto N^0$. In fact, $\langle \delta A^\dagger(t)\,\delta A(0)\rangle_{\mathrm{st}} \propto N^0$ (see § 3.3.1). Similarly, if we consider the macroscopic polarization operators $R^\pm = \sum_i r_i^\pm$, one has $\langle R^\pm\rangle_{\mathrm{st}} \propto N$, $\delta R^\pm \propto N^{1/2}$. In fact, $\langle \delta R^+(t)\,\delta R^-(0)\rangle_{\mathrm{st}} \propto N$, $\langle \delta A^\dagger(t)\,\delta R^-(0)\rangle_{\mathrm{st}} \propto N^{1/2}$, etc. Following this line, one would be tempted to guess that $\delta r_1^\pm \propto N^{-1/2}$. However this is not true in general. In fact, let us consider $\langle \delta r_1^+\,\delta r_1^-\rangle_{\mathrm{st}}$. We have

$$\langle \delta r_1^+\,\delta r_1^-\rangle_{\mathrm{st}} = \langle r_1^+ r_1^-\rangle_{\mathrm{st}} - \langle r_1^+\rangle_{\mathrm{st}}\langle r_1^-\rangle_{\mathrm{st}} = \langle r_{31}\rangle_{\mathrm{st}} + 1/2 - \langle r_1^+\rangle_{\mathrm{st}}\langle r_1^-\rangle_{\mathrm{st}} \propto N^0.$$

Hence, $\langle \delta r_1^+(t)\,\delta r_1^-(0)\rangle_{\mathrm{st}} \propto N^0$. On the other hand,

$$\langle \delta r_i^+(t)\,\delta r_j^-(0)\rangle_{\mathrm{st}} \propto N^{-1} \quad \text{for} \quad i \neq j, \text{ because}$$

$$\sum_{i\neq j}\langle \delta r_i^+(t)\,\delta r_j^-(0)\rangle_{\mathrm{st}} = \langle \delta R^+(t)\,\delta R^-(0)\rangle_{\mathrm{st}} - \sum_{i=1}^{N}\langle \delta r_i^+(t)\,\delta r_i^-(0)\rangle_{\mathrm{st}} \propto N.$$

On the basis of these scaling arguments, let us consider the order of magnitude of the various terms in the right hand side of eq. (222a). Taking into account that

$$\bar{g} \propto N^{-1/2}, \qquad \langle A^\dagger\rangle_{\mathrm{st}} \propto N^{1/2}, \qquad \langle \delta r_{31}(t)\,\delta r_1^-(0)\rangle_{\mathrm{st}} \propto N^0,$$
$$\langle r_{31}\rangle_{\mathrm{st}} \propto N^0, \qquad \langle \delta A^\dagger(t)\,\delta r_1^-(0)\rangle_{\mathrm{st}} \propto N^{-1/2},$$

$$\langle \delta A^\dagger(t)\, \delta r_{31}(t)\, \delta r_1^-(0)\rangle_{st} \propto N^{-1/2},$$

$$\langle \delta r_1^+(t)\, \delta r_1^-(0)\rangle_{st} \propto N^0.$$

one concludes that the first and last terms in the right-hand side of eq. (222a) are of order N^0, whereas the others are of order N^{-1}. Quite similar considerations can be repeated for eqs. (222b) and (222c). Hence, in the limit of large N eqs. (222) reduce to

$$\frac{d}{dt}\langle \delta r_1^+(t)\, \delta r_1^-(0)\rangle_{st} = 2\bar{g}\,\langle A^\dagger\rangle_{st}\,\langle \delta r_{31}(t)\, \delta r_1^-(0)\rangle_{st} - \gamma_\perp \langle \delta r_1^+(t)\, \delta r_1^-(0)\rangle_{st}, \tag{223a}$$

$$\frac{d}{dt}\langle \delta r_1^-(t)\, \delta r_1^-(0)\rangle_{st} = 2\bar{g}\,\langle A\rangle_{st}\,\langle \delta r_{31}(t)\, \delta r_1^-(0)\rangle_{st} - \gamma_\perp \langle \delta r_1^-(t)\, \delta r_1^-(0)\rangle_{st}, \tag{223b}$$

$$\frac{d}{dt}\langle \delta r_{31}(t)\, \delta r_1^-(0)\rangle_{st} = -\bar{g}\{\langle A\rangle_{st}\,\langle \delta r_1^+(t)\, \delta r_1^-(0)\rangle_{st} + \langle A^\dagger\rangle_{st}\,\langle \delta r_1^-(t)\, \delta r_1^-(0)\rangle_{st}\} - \gamma_\parallel \langle \delta r_{31}(t)\, \delta r_1^-(0)\rangle_{st}. \tag{223c}$$

Equations (223) give a closed system of equations which has the same form as the corresponding set of equations for the fluorescence of the single atom (MOLLOW [1969]) apart from the replacement of the incident field α_0 with the selfconsistent internal field $\langle A\rangle_{st}$ (LUGIATO [1980b], CARMICHAEL [1981]). The initial conditions for eqs. (223) are

$$\langle \delta r_1^+(0)\, \delta r_1^-(0)\rangle_{st} = \langle r_1^+ r_1^-\rangle_{st} - \langle r_1^+\rangle_{st}\langle r_1^-\rangle_{st} = \frac{1}{2}\,\frac{x_{st}^2}{(1+x_{st}^2)^2}\,(1 - \tfrac{1}{2}\bar{d} + x_{st}^2), \tag{224a}$$

$$\langle \delta r_1^-(0)\, \delta r_1^-(0)\rangle_{st} = -\frac{\bar{d}}{4}\,\frac{x_{st}^2}{(1+x_{st}^2)^2}, \tag{224b}$$

$$\langle \delta r_{3i}(0)\, \delta r_1^-(0)\rangle_{st} = \frac{1}{4}\,\bar{d}^{1/2}\,\frac{x_{st}^3}{(1+x_{st}^2)^2}. \tag{224c}$$

Since the initial conditions also coincide with those of the one-atom resonance fluorescence after replacement of α_0 by $\langle A \rangle_{st}$, the expression of the incoherent part of the spectrum is immediately obtained from that given in MOLLOW [1969].

These results are valid for general values of the parameters k, $\gamma_\perp$, $\gamma_\parallel$ and hence, hold both for a bad and for a good quality cavity. Usually, the regression theorem transfers to the steady state spectrum the behavior (relaxation times, etc.) of the transient approach to steady state. This occurs, for instance, for the one-atom resonance fluorescence and for the spectrum of *transmitted* light in optical bistability. However, as shown by the previous results, this is not the case for the spectrum of *fluorescent* light in optical bistability. This fact is due to the anomalous scaling of the one-atom deviations $\delta r_i^\pm$, δr_{3i} when they enter into a correlation function for a single atom only. This implies that atom–atom correlations do not play any role in determining the spectrum of fluorescent light, contrary to what occurs for the spectrum of transmitted light.

3.4.2. *Comparison of the spectra of fluorescent and transmitted light*

In the good cavity case $k \ll \gamma_\perp, \gamma_\parallel$, the spectrum of fluorescent light is quite different from that of transmitted light. For instance, in the high transmission branch, the spectrum of fluorescent light has a three-peaked structure. On the contrary, the incoherent part of the spectrum of transmitted light is one-peaked and for sufficiently large values of the incident field the linewidth is the empty cavity width (see eq. (197′)).

However, in the bad cavity case $k \ll \gamma_\perp, \gamma_\parallel$, the hysteresis cycle of the incoherent part of the spectrum of fluorescent light shows some similarities with that of the transmitted light, especially in the high transmission branch. In particular, one finds in both cases the discontinuous appearance of the resolved triplet when we increase the incident field, starting from zero (Discontinuous Dynamical Stark Effect). However, the spectrum of fluorescent light does not show line narrowing at the boundaries of the hysteresis cycle, contrary to what we found for the spectrum of the transmitted light. Furthermore, there is no cooperative broadening in the low transmission branch for $y \ll C$. In fact, in the case of fluorescent light, the cooperative behavior only arises via the internal field $\langle A \rangle_{st}$, which is in general quite different from the incident field. Additional cooperative effects such as line narrowing or broadening are absent, since, as we have seen, atom–atom correlations do not enter into play in the calculation of the spectrum of fluorescent light.

We end this subsection with a comparison between the energy emitted by the system as incoherent transmitted light and the energy emitted as incoherent fluorescent light. This comparison comes naturally because, while in the case of the fluorescent light the incoherent part has, roughly speaking, the same order of magnitude as the coherent one, in the case of the transmitted light the coherent part, which is proportional to the transmitted number of photons $N_s x_{st}^2$, always dominates the incoherent one.

From the master equation (158), as well as from the Fokker–Planck equation (176), one easily obtains the energy balance equation

$$\frac{d}{dt}(\langle A^\dagger A\rangle + \langle R_3\rangle) = \\ = k\alpha_0^2 - k\langle A^\dagger A\rangle - k\langle(A^\dagger - \alpha_0)(A - \alpha_0)\rangle - \gamma_\parallel(\tfrac{1}{2}N + \langle R_3\rangle). \tag{225}$$

The meaning of the various terms on the right-hand side of eq. (225) is:

$$\begin{aligned} &k\alpha_0^2 && \propto \text{ incident energy per unit time,} \\ &k\langle A^\dagger A\rangle && \propto \text{ transmitted energy per unit time,} \\ &k(\langle A^\dagger - \alpha_0)(A - \alpha_0)\rangle && \propto \text{ reflected energy per unit time,} \\ &\gamma(\tfrac{1}{2}N + \langle R_3\rangle) && \propto \text{ fluorescent light emitted per unit time in all directions,} \\ &(\gamma_\parallel - \gamma)(\tfrac{1}{2}N + \langle R_3\rangle) && \propto \text{ energy dissipated in the atomic sample per unit time;} \end{aligned} \tag{226}$$

γ is the natural lifetime, $(\gamma_\parallel - \gamma)$ is the inelastic collision rate.

In turn, the transmitted energy is subdivided into a coherent part $k|\langle A\rangle|^2$ and an incoherent part $k|\langle\delta A^\dagger\rangle|^2$, and the total fluorescent energy is composed of a coherent part $\gamma N\langle r_1^+\rangle\langle r_1^-\rangle$ and an incoherent part $\gamma N\langle\delta r_1^+\,\delta r_1^-\rangle$. Let us now compare at steady state the energy emitted as incoherent fluorescent light and the energy emitted as incoherent transmitted light, both per unit time and solid angle. In doing that, we must take into account that the fluorescent light is emitted in all directions, that is in a solid angle 4π, whereas the transmitted light is emitted in a diffraction solid angle λ_0^2/S, where λ_0 is the wavelength of the incident field and $S = V/L$ is the section of the region containing the atom.

We consider the case $k \gg \gamma_\perp, \gamma_\parallel, \gamma_\parallel = 2\gamma_\perp = \gamma$. Furthermore, we consider for definiteness the situation $y \gg y_M$ (one-atom stationary state). In this condition, the incoherent fluorescent light practically coincides with the total fluorescent light. From eqs. (216) and (226) we show that the fluorescent energy emitted per unit time and solid angle is proportional to

$$\mathscr{P}_{\text{inc}}^{(\text{fl})} = \gamma\frac{N}{2}\frac{1}{4\pi}, \tag{227}$$

where we have taken into account that, in the present conditions, $x_{st}^2/(1+x_{st}^2) \approx 1$. On the other hand, from eqs. (226), (188) and (206) we show that the incoherent transmitted energy per unit time and solid angle is proportional to

$$\mathscr{P}_{\text{inc}}^{(\text{tr})} = C\gamma_\perp \frac{S}{\lambda_0^2}. \tag{228}$$

Since for $\gamma_\perp = \gamma/2$ from eq. (33), (176′), (164) and the relation $\gamma = 4\omega_0^3\mu^2/3\hbar c^3$ (ALLEN and EBERLY [1975]) we have

$$C = \frac{3}{8\pi T}\frac{N\lambda_0^2}{S}, \tag{229}$$

and the ratio between eqs. (227) and (228) is

$$\frac{\mathscr{P}_{\text{inc}}^{(\text{fl})}}{\mathscr{P}_{\text{inc}}^{(\text{tr})}} = \frac{2}{3}T. \tag{230}$$

Hence, the two energies have roughly the same order of magnitude.

3.5. PHOTON STATISTICS OF THE TRANSMITTED LIGHT AT STEADY STATE

As we anticipated, optical bistability is an example of a first-order-like phase transition in an open system, far from thermal equilibrium. As is well known, this behavior is shown also by other systems in quantum optics, for instance the laser with saturable absorber (KASANTSEV, RAUTIAN and SURDUTOVICH [1970], SALOMAA and STENHOLM [1973], SCOTT, SARGENT and CANTRELL [1975], LUGIATO, MANDEL, DEMBINSKI and KOSSAKOWSKI [1978]), the dye laser (BACZYNSKI, KOSSAKOWSKI and MARSZALEK [1976], SCHAEFER and WILLIS [1976]), sub/second harmonic generation (WOO and LANDAUER [1971], DRUMMOND, MCNEIL and WALLS [1980a]), and the bidirectional ring laser (MANDEL, ROY and SINGH [1981]). The characteristic feature of optical bistability is that it occurs in a purely passive system and that it never exhibits a second-order transition. Hence, optical bistability plays the role of a prototype of first-order transition in optical systems, exactly as the usual laser, with active atoms only, is the prototype of second-order phase transition (DEGIORGIO and SCULLY [1970], GRAHAM and HAKEN [1970]). To work out this analogy, one must analyze in full detail the fluctuations of the system. In fact, in the bistable situation, only one of the two stationary solutions is absolutely stable, while the other is only metastable. The semiclassical treatment is unable to tell us which one of the two is absolutely stable. Actually, the linear stability analysis checks the stability of the stationary solutions only against the "small" fluctuations around each steady state. Also, in the linearized treatment of the previous sections, we have analyzed only the small fluctuations around the stationary solutions, thereby treating stable and metastable states on the same footing. However, the system can also develop "large" fluctuations which make the system "tunnel" from the metastable to the stable solution. The probability of such large fluctuations is extremely small, as we shall show in § 3.6.1; however, to analyze the thermodynamic stability of the steady states, one must develop a treatment which works out the full spectrum of fluctuations. This treatment has been given so far only for the good quality case $k \ll \gamma_\perp, \gamma_\parallel$ (BONIFACIO, GRONCHI and LUGIATO [1978], LUGIATO, CASAGRANDE and PIZZUTO [1982]), the only exception being DRUMMOND [1982]. In the following subsections, we shall discuss the steady state solution of the Fokker–Planck equation (177).

3.5.1. *The case of thermal and external fluctuations*

In this subsection, we shall assume that the intrinsic quantum mechanical fluctuations are negligible with respect to the thermal and the external field fluctuations. Hence, in the diffusion coefficients of eq. (177) we keep only the terms proportional to $\bar{n}$. It is suitable to express this Fokker–Planck equation in terms of the real part x_1 and the imaginary part x_2 of the variable x (compare eqs. (186) and (191)). We obtain

$$k^{-1}\frac{\partial P_{\mathrm{w}}(x_1, x_2, t)}{\partial t} = \left\{\frac{\partial}{\partial x_1}\frac{\partial \tilde{V}_y}{\partial x_1} + \frac{\partial}{\partial x_2}\frac{\partial \tilde{V}_y}{\partial x_2} + \bar{q}\left(\frac{\partial^2}{\partial x_1^2} + \frac{\partial^2}{\partial x_2^2}\right)\right\} P_{\mathrm{w}}(x_1, x_2, t), \tag{231}$$

where $\tilde{V}_y$ is the mechanical potential defined in eq. (63) and

$$\bar{q} = \frac{1}{2}\frac{\bar{n}}{N_{\mathrm{s}}}. \tag{232}$$

In this situation the diffusion coefficient is constant. This case is usually called "additive noise" (SCHENZLE and BRAND [1978]).

At steady state we put $\partial P_{\mathrm{w}}/\partial t = 0$. Since eq. (231) obeys the detailed balance conditions (HAKEN [1977]), the steady state equation can easily be solved. The solution is (BONIFACIO, GRONCHI and LUGIATO [1978], SCHENZLE and BRAND [1978])

$$P_{\mathrm{w}}^{(\mathrm{st})}(x_1, x_2) = \mathcal{N} \exp\left\{-\frac{1}{\bar{q}}\tilde{V}_y(x_1, x_2)\right\}, \tag{233}$$

where $\mathcal{N}$ is the normalization constant determined by the condition $\int \mathrm{d}x_1\,\mathrm{d}x_2\,P_{\mathrm{w}}^{(\mathrm{st})}(x_1, x_2) = 1$. The function $P_{\mathrm{w}}^{(\mathrm{st})}$ describes the fluctuations of the transmitted field at steady state, and in particular its photon statistics.

As we see from eq. (233), in the situation of thermal and external noise the mechanical potential $\tilde{V}_y$ plays the role of a generalized free energy, which determines the stability of the steady states. In fact, the extrema of the function $\tilde{V}_y$ (i.e. the points where $\partial\tilde{V}_y/\partial x_1 = \partial\tilde{V}_y/\partial x_2 = 0$) coincide with the semiclassical solutions. The unstable state (if any) corresponds to a saddle point of the free energy, the absolutely stable stationary state corresponds to the absolute minimum of the function $\tilde{V}_y(x_1, x_2)$, while the metastable state (if any) corresponds to the relative minimum.

The number of peaks of the distribution $P_{\mathrm{w}}^{(\mathrm{st})}$ is equal to the number of minima of the potential $\tilde{V}_y$. Hence, for $0 \leqslant y \leqslant y_{\mathrm{m}}$ (see Fig. 8), the probability distribution has one peak in correspondence with the cooperative stationary solution x_{a}. For $y_{\mathrm{m}} < y < y_{\mathrm{M}}$, $P_{\mathrm{w}}^{(\mathrm{st})}$ has two peaks at $x_1 = x_{\mathrm{a}}$, $x_2 = 0$ and $x_1 = x_{\mathrm{c}}$, $x_2 = 0$. For $y \geqslant y_{\mathrm{M}}$, the probability distribution again has one peak, corresponding to the one-atom stationary solution x_{c}. The parameter $\bar{q}$ controls the width of the peaks; the smaller is $\bar{q}$, the narrower are the peaks. The smallness of $\bar{q}$ also has another important consequence: the range of values of y in which the two peaks have comparable areas is very small. In other words, in the largest part of the bistable region $y_{\mathrm{m}} < y < y_{\mathrm{M}}$, one of the two peaks is absolutely dominant over the other. Only in a narrow transition region is the probability distribution really double peaked. In this region, it is more proper to speak of two metastable states instead of one stable and one metastable state. In fact, the system can jump randomly from one to the other minimum of the free energy, due to the action of fluctuations. These jumps occur with characteristic times τ_{L} and τ_{R}, discussed in § 3.6.1. The width of the transition region tends to zero in the thermodynamic limit $N \to \infty$, $V \to \infty$ with N/V constant. In this limit $\bar{q}$ tends to zero. If we consider the mean value $\langle x \rangle_{\mathrm{st}} = \langle x_1 \rangle_{\mathrm{st}}$ of the transmitted field, calculated from eq. (233), we find the discontinuous transition (Fig. 33)

$$\langle x \rangle_{\mathrm{st}} \xrightarrow{\bar{q} \to 0} \begin{cases} x_{\mathrm{a}} & \text{for } y < \tilde{y}, \\ x_{\mathrm{c}} & \text{for } y > \tilde{y}, \end{cases} \tag{234}$$

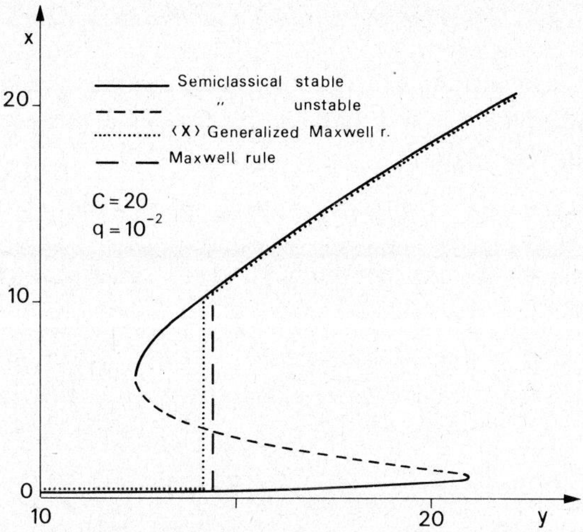

Fig. 33. Semiclassical stationary solutions, Maxwell rule and mean value of the normalized field amplitude x.

where the value $\tilde{y}$ is specified by the condition

$$V_{\tilde{y}}(x_{\mathrm{a}}) = V_{\tilde{y}}(x_{\mathrm{c}}). \tag{234'}$$

The definition of $V_y(x)$ is given in eq. (64), and we have taken into account that the minima of the generalized free energy $\tilde{V}_y$ are on the real axis.

Clearly this behavior strongly resembles first-order phase transitions in equilibrium systems. Equation (234) coincides with the well known Maxwell rule, because it cuts the semiclassical curve in the (x, y) plane in such a way that one obtains two regions of equal area. In fact, the latter condition amounts to fixing the value $\tilde{y}$ of y in such a way that

$$\int_{x_{\mathrm{b}}}^{x_{\mathrm{a}}} \mathrm{d}x \left(x + \frac{2Cx}{1+x^2} - \tilde{y} \right) = \int_{x_{\mathrm{b}}}^{x_{\mathrm{c}}} \mathrm{d}x \left(\tilde{y} - x - \frac{2Cx}{1+x^2} \right)$$

$$\Rightarrow \int_{x_{\mathrm{a}}}^{x_{\mathrm{c}}} \mathrm{d}x \left(x + \frac{2Cx}{1+x^2} - \tilde{y} \right) = 0. \tag{235}$$

From eq. (64), one sees that condition (235) coincides with (234′).

3.5.2. *General case. Effects of the intrinsic quantum fluctuations*

Let us now also include the intrinsic quantum fluctuations, so that we consider the full eq. (177′). It is suitable to reformulate this equation in terms of the probability distribution

$$\bar{P}_{\mathrm{w}}(r, \varphi) = r P_{\mathrm{w}}(r, \varphi), \tag{236}$$

which obeys the normalization condition $\int_0^\infty \mathrm{d}r \int_0^{2\pi} \mathrm{d}\varphi\, \bar{P}(r, \varphi, t) = 1$. We have (see last sentence in § 3.2.1)

$$k^{-1} \frac{\partial \bar{P}_{\mathrm{w}}(r, \varphi, t)}{\partial t} = \left\{ \frac{\partial}{\partial r} \left[r \left(1 + \frac{2C}{1 + r^2} \right) - y \cos \varphi + q \frac{\partial}{\partial r} D(r) \right] + \frac{\partial}{\partial \varphi} \left[\frac{y}{r} \sin \varphi + q \left(1 + \frac{1 + 2\bar{n}}{2C} \right) \frac{1}{r^2} \frac{\partial}{\partial \varphi} \right] \right\} \bar{P}_{\mathrm{w}}(r, \varphi, t), \tag{237}$$

where

$$q = C/2N_{\mathrm{s}} \tag{238a}$$

$$D(r) = \frac{1 + 2\bar{n}}{2C} + \frac{r^4 + (1 - \bar{d})r^2 + 1}{(1 + r^2)^3}. \tag{238b}$$

The parameter q rules the strength of the fluctuations. From eq. (238b) we see that now the amplitude diffusion coefficient is not constant, but is intensity dependent. This is because intrinsic fluctuations give rise to "multiplicative noise" (SCHENZLE and BRAND [1978]). In this situation saturation effects are important not only in the average motion, but also in the fluctuations.

Unfortunately, eq. (237) cannot be solved exactly at steady state because it involves two variables and does not obey the detailed balance conditions. However, a very well approximated expression for the amplitude stationary distribution can be easily obtained in the following way. At a semiclassical level, the phase has only one stationary value, $\varphi = 0$. At a quantum statistical level, the phase will fluctuate around $\varphi = 0$, but these fluctuations are small because the diffusion constant q is small. Hence, at steady state, one can linearize eq. (237) *with respect to the phase only*, so that $\cos \varphi$ is simply replaced by 1. At this point, one can integrate eq. (237) with respect to the phase, obtaining the following closed equation for the amplitude distribution $P(r, t) = \int_0^{2\pi} \mathrm{d}\varphi\, \bar{P}_{\mathrm{w}}(r, \varphi, t)$:

$$k^{-1} \frac{\partial P(r, t)}{\partial t} = -\frac{\partial}{\partial r} \left\{ K(r) - q \frac{\partial}{\partial r} D(r) \right\} P(r, t), \quad K(r) = y - r - \frac{2Cr}{1 + r^2}. \tag{239}$$

In steady state, from eq. (239) we obtain

$$\left\{K(r) + q\frac{\partial}{\partial r}D(r)\right\} P^{\mathrm{st}}(r) = 0. \tag{239'}$$

The solution of this equation is (BONIFACIO, GRONCHI and LUGIATO [1978], LUGIATO, CASAGRANDE and PIZZUTO [1982]):

$$P^{(\mathrm{st})}(r) = \mathcal{N}\, D^{-1}(r) \exp\left[-\frac{1}{q}U_y(r)\right], \tag{240}$$

where $\mathcal{N}$ is the normalization constant and

$$U_y(r) = \int \mathrm{d}r'\; K(r')/D(r'). \tag{241}$$

Hence, in this case, the role of generalized free energy is played by $U_y(r)$, which is different from the mechanical potential $V_y(r)$, because $D(r)$ is not constant. Clearly the equation $\mathrm{d}U_y/\mathrm{d}r = 0$, which determines the extrema of the potential, coincides with the semiclassical state equation (32). For $q \ll 1$, the factor $D^{-1}(r)$ in eq. (240) produces a negligible shift in the position of the extrema of distribution $P^{(\mathrm{st})}(r)$, which then coincide with the extrema of $U_y(r)$. Hence, the stable semiclassical solutions correspond to most probable values (i.e. peaks of the distribution function), while the unstable solutions correspond to least probable values. In particular, for $C > 4$ in the bistable situation $y_{\mathrm{m}} < y < y_{\mathrm{M}}$ (see Fig. 8), $P^{(\mathrm{st})}(r)$ has two peaks at $r = x_{\mathrm{a}}$ and $r = x_{\mathrm{c}}$. Exactly as described in the previous subsection, the parameter q controls the width of the peaks and of the transition region. For $q \ll 1$, distribution (240) can be very well approximated by the superposition of two Gaussians

$$P^{(\mathrm{st})}(r) = w_{\mathrm{a}}\, G_{\mathrm{a}}(r) + w_{\mathrm{c}}\, G_{\mathrm{c}}(r), \tag{242}$$

where (i = a, c)

$$G_i(r) = \frac{1}{\sigma_i\sqrt{2\pi}} \exp\{-(r - x_i)^2/2\sigma_i^2\},$$

$$\sigma_i^{-2} = \frac{1}{q}\left(\frac{\mathrm{d}^2 U_y}{\mathrm{d}r^2}\right)_{x_i}, \tag{243a}$$

while the weights w_i, which obey the normalization condition $w_a + w_c = 1$, are given by

$$w_i = \frac{\sigma_i D^{-1} x_i \exp\left[-\frac{1}{q} U_y(x_i)\right]}{\sum_{j=a,b} \sigma_j D^{-1}(x_j) \exp\left[-\frac{1}{q} U_y(x_j)\right]}. \tag{243b}$$

Of course, for $y < y_m$ and $y > y_M$ one of the two Gaussians in (242) is missing. When $\bar{n} = 0$, for $y \to 0$ the probability distribution approaches the Wigner function of the vacuum state, as it must be, while for $y \to \infty$ it approaches the Wigner function corresponding to the coherent state $|\alpha_0\rangle = |\sqrt{N_s}\, y\rangle$.

The approximation (242) fails only in the neighborhood of the two discontinuity points $y = y_m$ and $y = y_M$. From eq. (242) we obtain the following expressions for the first moments:

$$\langle r\rangle_{st} = \langle \hat{x}\rangle_{st} = w_a x_a + w_c x_c,$$

$$\sigma^2 \equiv \langle \hat{x}^\dagger \hat{x}\rangle_{st} - \langle \hat{x}\rangle_{st}^2 = \langle r\rangle^2{}_{st} - \langle r^2\rangle_{st} - \frac{1}{2N_s}$$

$$= w_a \sigma_a^2 + w_c \sigma_c^2 + w_a w_c (x_a - x_c)^2 - \frac{1}{2N_s}. \tag{244}$$

where we have used the first of eqs. (181), (178) and (174). Figure 34 shows the mean value $\langle x\rangle_{st}$ and the relative fluctuation $\sigma^2/\langle \hat{x}\rangle_{st}^2$ as a function of y for $C = 20$, $\bar{n} = 0$ (negligible thermal and external fluctuations), and $q = 10^{-2}$ (this value of q is chosen for pedagogical reasons, in order not to have a too narrow transition region). The mean value coincides with one of the two semiclassical solutions everywhere except in the transition region, which is centered on the value of y, such that

$$w_a = w_c \Rightarrow \langle x\rangle_{st} = \tfrac{1}{2}(x_a + x_c). \tag{244$'$}$$

This condition requires that the two peaks have equal areas. Out of the transition region the absolutely stable stationary solution is the one which practically coincides with $\langle x\rangle_{st}$ while the other is metastable.

As we see from Fig. 34, the fluctuation $\sigma^2/\langle \hat{x}\rangle_{st}^2$ is always very small except in the narrow transition region, where we find a remarkable peak. This behavior is easily understood from eq. (244). In fact, out of the transition region one has

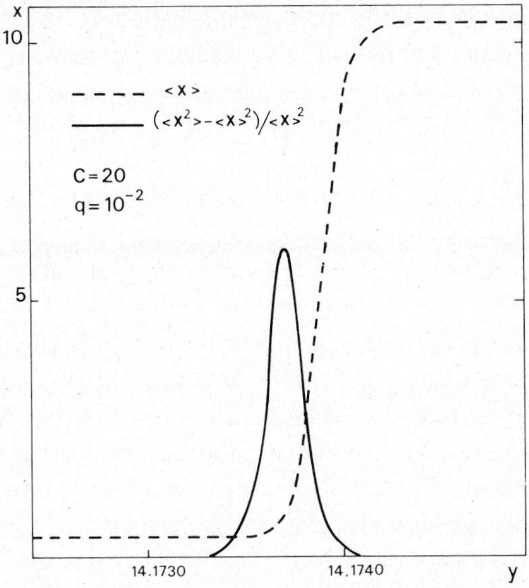

Fig. 34. Mean value and relative fluctuation of the transmitted field.

practically $w_a = 0$ or $w_c = 0$, so that $\sigma^2 = \sigma_c^2 - 1/2N_s$ or $\sigma^2 = \sigma_a^2 - 1/2N_s$, respectively. In the transition region, both w_a and w_c are of the order of unity, so that $\sigma^2 \approx w_a w_c (x_a - x_c)^2 \gg \sigma_a^2, \sigma_c^2$. In other words, the strong competition between the two peaks of the probability distribution gives the drastic increase of fluctuations in the transition region.

In the thermodynamic limit $q \to 0$, we again obtain (234), where now however, the value $\tilde{y}$ is determined by the condition

$$U_{\tilde{y}}(x_a) = U_{\tilde{y}}(x_b) \tag{245}$$

instead of (234′). Since $U_y \neq V_y$, condition (245) gives a *generalized Maxwell rule* different from the usual one, as is clearly indicated in Fig. 33 (Bonifacio, Gronchi and Lugiato [1978]). This clearly indicates the nonthermodynamic character of the transition. A similar phenomenon arises in chemical reactions (Nicolis and Lefever [1977]).

We end this subsection with a remark on the interpretation of $\langle \hat{x} \rangle_{st}$. In the usual situation (one-peaked distribution function), the mean value practically coincides with the most probable value. This is no longer the case in the transition region, in which the probability distribution has two peaks of comparable areas. In fact, in this region, there is a practically vanishing probability of finding $x = \langle \hat{x} \rangle_{st}$ as a result of a single measurement. Hence, the mean value

$\langle \hat{x} \rangle_{st}$ must be interpreted in a strictly statistical sense. That is, the results of the single experiments will be (with overwhelming probability for $q \ll 1$) very near to x_a and x_c. The statistical average of these results gives $\langle \hat{x} \rangle_{st}$.

3.6. TRANSIENT BEHAVIOR: QUANTUM STATISTICAL TREATMENT

In this subsection we illustrate the main features of the transient behavior that depend on fluctuations. In particular, the discussion of the metastable states necessarily requires the calculation of their lifetime. This general problem was first considered many years ago by KRAMERS [1940], and later in great detail by LANDAUER (see for instance LANDAUER [1962]). We shall restrict ourselves to the good cavity case, and for simplicity we shall base our discussion on the one-dimensional Fokker–Planck equation (239). For the sake of definiteness, we consider the same problem as in § 2.3.1, but in a quantum statistical context (LUGIATO, FARINA and NARDUCCI [1980]). Namely, we consider an ensemble of systems which are initially in the stationary state, corresponding to the value y_0 of the incident field, so that $P(r, 0) = P^{(st)}_{y_0}(r)$. At $t = 0$, the external field is suddenly switched to a new value y_{op}, which is slightly different from y_0. Hence, in the course of time, the distribution $P(r, t)$ will approach the new stationary solution of eq. (239), $P^{(st)}_{y_{op}}(r)$ corresponding to the new value y_{op} of the incident field. We wish to illustrate the main features of the time evolution of $P(r, t)$ for $q \ll 1$.

We assume that both y_0 and y_{op} are well inside the bistability region $y_m < y < y_M$ (Fig. 8). Hence, from eq. (242), $P(r, 0)$ is well represented by the superposition of two Guassians

$$P(r, 0) = w_a^{(0)} G_a^{(0)}(r) + w_c^{(0)} G_c^{(0)}(r), \tag{246}$$

where w_a, $w_c = 1 - w_a$, and G_a and G_c, are given by eqs. (243a) and (243b). The index (0) indicates that these quantities are calculated with x_a and x_c being the stable solutions of eq. (32), corresponding with the value y_0 of the incident field.

For $q \ll 1$, the time evolution of $P(r, t)$ occurs in two quite separate stages (LUGIATO, FARINA and NARDUCCI [1980]). In the first one, there is a *local relaxation process* in which the two peaks in eq. (246) evolve independently of each other. If the difference between y_0 and y_{op} is small enough, the evolution of each peak is described by the Fokker–Planck equation (239), linearized around the corresponding semiclassical steady state. That is, the evolution, say, of peak ***a*** is governed by eq. (239), linearized around the value $r = x_a^{(op)}$, where

$x_a^{(op)}$ is the semiclassical solution corresponding with the value y_{op} of the incident field. Hence, this stage occurs on a time scale of the order of $(k\tilde{\lambda}_a)^{-1}$, $(k\tilde{\lambda}_c)^{-1}$, where $\tilde{\lambda}_a$ and $\tilde{\lambda}_c$ are given by the first of eqs. (193) with $x_{st} = x_a$, $x_{st} = x_c$.

At the end of this stage the probability distribution $P(r)$ has the form

$$P(r) = w_a^{(0)} G_a^{(op)}(r) + w_c^{(0)} G_c^{(op)}(r), \tag{247}$$

where now the two Gaussians are given by eq. (243a), with x_a and x_c corresponding to the final value y_{op} of the external field. Hence, the shape of the two peaks in eq. (247) is identical to that of the corresponding peaks in the stationary distribution $P_{y_{op}}^{(st)}(r)$, but the weights w_a and w_c are different, because they still correspond to the initial value y_0 of the incident field. This is because, up to this point, the time evolution has been only a local relaxation without any flux of probability between the two peaks.

In the second stage of the time evolution, which occurs on a much longer time scale, one has a transfer of probability from one peak to the other, which leads eventually to the final distribution, that is given by

$$P(r, \infty) = w_a^{(op)} G_a^{(op)}(r) + w_c^{(op)} G_c^{(op)}(r). \tag{248}$$

This stage is usually called "tunneling", because it consists of a flux of probability across the "barrier" of the free energy $U_{y_{op}}(r)$ (see eq. (241)). In fact, as we know, this potential has two minima, at $r = x_a^{(op)}$ and $r = r_c^{(op)}$, separated by a barrier centered on the unstable point $r = x_b^{(op)}$. This long-time stage is discussed in the next subsection.

3.6.1. *The tunneling process*

The picture of the time evolution given above is consistent with the spectrum of eigenvalues of the Fokker–Planck operator. In fact, if we write eq. (239) in the form

$$\frac{\partial P(r, t)}{\partial t} = \Lambda P(r, t), \tag{249}$$

for $q \ll 1$ the linear operator Λ exhibits a nearly degenerate pair of lowest lying eigenvalues $\lambda_0 = 0$ and $\lambda_1 \neq 0$, and a large gap between λ_1 and the remaining part of the spectrum. As usual, the lowest eigenvalue $\lambda_0 = 0$ corresponds to the steady state solution, while λ_1, in this case, is related to the eigenfunction of the Fokker–Planck operator that describes the long-time approach to steady state. On the other hand, the remaining eigenvalues are related to the short-time, local relaxation process that we described above. In the long time limit, in which only the lowest two eigenvalues are important for the description of our bistable system, the function $P(r, t)$ is well approximated by

$$P(r, t) = P^{(\mathrm{st})}_{y_{\mathrm{op}}}(r) + \exp(-\lambda_1 t)\, u_1(r), \tag{250}$$

where u_1 is the eigenfunction of the Fokker–Planck operator Λ, corresponding to λ_1. In this case, the probability of occupation of the left peak

$$w_{\mathrm{a}}(t) \equiv \int_0^{x_{\mathrm{b}}^{(\mathrm{op})}} P(r, t)\, \mathrm{d}r, \tag{251}$$

where $x_{\mathrm{b}}^{(\mathrm{op})}$ is the unstable point of the potential $U_{y_{\mathrm{op}}}(r)$, obeys the very simple equation

$$\frac{\mathrm{d}w_{\mathrm{a}}(t)}{\mathrm{d}t} = -\lambda_1 (w_{\mathrm{a}}(t) - w_{\mathrm{a}}(\infty)), \tag{252}$$

where

$$w_{\mathrm{a}}(\infty) = \int_0^{x_{\mathrm{b}}^{(\mathrm{op})}} P^{(\mathrm{st})}_{y_{\mathrm{op}}}(r)\, \mathrm{d}r \approx w_{\mathrm{a}}^{(\mathrm{op})}. \tag{253}$$

Equation (252) can be reformulated as a rate equation of the form

$$\frac{\mathrm{d}w_{\mathrm{a}}(t)}{\mathrm{d}t} = -\frac{1}{\tau_{\mathrm{L}}} w_{\mathrm{a}} + \frac{1}{\tau_{\mathrm{R}}}(1 - w_{\mathrm{a}}), \tag{254}$$

where τ_{L} and τ_{R} are the escape times out of the left and right wells of the potential $U_{y_{\mathrm{op}}}(r)$, respectively. In fact, on comparing eq. (252) with eq. (254), one finds

$$\frac{1}{\tau_{\mathrm{L}}} + \frac{1}{\tau_{\mathrm{R}}} = \lambda_1, \tag{255a}$$

$$\frac{1}{\tau_{\mathrm{R}}} = \lambda_1 w_{\mathrm{a}}(\infty), \tag{255b}$$

from which follows

$$\frac{1}{\tau_{\mathrm{L}}} = \lambda_1 (1 - w_{\mathrm{a}}(\infty)). \tag{255c}$$

In addition, it is worth observing that

$$\frac{\tau_L}{\tau_R} = \frac{w_a(\infty)}{1 - w_a(\infty)}. \tag{256}$$

We now calculate the eigenvalue λ_1 by a procedure that takes its premises from KRAMERS [1940], after allowance is made for the r-dependence of the diffusion coefficient (BONIFACIO, LUGIATO, FARINA and NARDUCCI [1981]). An alternative approach is that based on the first passage time method (STRATONOVICH [1963], FARINA, NARDUCCI, YUAN and LUGIATO [1980], ENGLUND, SCHIEVE, ZUREK and GRAGG [1981]). The probability current density $J(r, t)$ is given by (see eq. (239))

$$J(r, t) = \left\{ K(r) - q \frac{\partial}{\partial r} D(r) \right\} P(r, t). \tag{257}$$

On using eq. (251), we at once obtain

$$\frac{\mathrm{d}w_a(t)}{\mathrm{d}t} = -J(x_b^{(op)}, t), \tag{258}$$

where we have taken into account that $J(0, t) = 0$, because $r = 0$ is a reflecting boundary of the potential. As shown in BONIFACIO, LUGIATO, FARINA and NARDUCCI [1981], in the long-time limit $J(r, t)$ is practically independent of r in the neighborhood of $x_b^{(op)}$; this property holds, in practice, from $x_b^{(op)}$ to almost $x_a^{(op)}$ and $x_c^{(op)}$.

In order to evaluate $J(x_b^{(op)}, t)$ we now consider the following identity of KRAMERS [1940] after a suitable modification has been applied to accommodate the r-dependence of the diffusion coefficient

$$\exp\left(\frac{U_{y_{op}}(r)}{q}\right) J(r, t) = -q \frac{\partial}{\partial r} \left[D(r) P(r, t) \exp\left(\frac{U_{y_{op}}(r)}{q}\right)\right]. \tag{259}$$

If we integrate eq. (259) between the two minima $x_a^{(op)}$ and $x_c^{(op)}$ of the potential, and observe that the function $\exp[U_{y_{op}}(r)/q]$ is very sharply peaked around $x_b^{(op)}$, where $J(r, t)$ is practically constant, we obtain

$$\int_{x_a^{(op)}}^{x_c^{(op)}} \mathrm{d}r \exp\left[\frac{U_{y_{op}}(r)}{q}\right] J(r, t) \approx \int_{x_b^{(op)} - \varepsilon}^{x_b^{(op)} + \varepsilon} \mathrm{d}r \exp\left[\frac{U_{y_{op}}(r)}{q}\right] J(r, t)$$

$$\approx J(x_b^{(op)}, t) \int_{x_b^{(op)} - \varepsilon}^{x_b^{(op)} + \varepsilon} \mathrm{d}r \exp\left[\frac{U_{y_{op}}(r)}{q}\right] \simeq J(x_b^{(op)}, t) \int_{x_a^{(op)}}^{x_c^{(op)}} \mathrm{d}r \exp\left[\frac{U_{y_{op}}(r)}{q}\right]. \tag{260}$$

The parameter ε, which is of the order of $\sqrt{q}$, is introduced only to guarantee the validity of the step by which $J(r,t)$ is factorized outside the integral sign. The last step is justified because $\exp[U_{y_{op}}(r)/q]$ is very small with respect to $\exp[U_{y_{op}}(x_b)/q]$ around both $x_a^{(op)}$ and $x_c^{(op)}$. Hence, after integrating both sides of eq. (259) from $x_a^{(op)}$ to $x_c^{(op)}$ and using eq. (260), we obtain

$$J(x_b^{(op)},t) = q\left\{\left[DP\exp\left(\frac{U_{y_{op}}(r)}{q}\right)\right]_{r=x_a^{(op)}} - \left[DP\exp\left(\frac{U_{y_{op}}(r)}{q}\right)\right]_{r=x_c^{(op)}}\right\}\Big/\int_{x_a^{(op)}}^{x_c^{(op)}} \mathrm{d}r\exp\left(\frac{U_{y_{op}}(r)}{q}\right). \tag{261}$$

Finally, from eqs. (258), (261) and (251), we conclude that

$$\frac{\mathrm{d}w_a}{\mathrm{d}t} = -\frac{q\left[DP\exp\left(\frac{U_{y_{op}}(r)}{q}\right)\right]_{r=x_a^{(op)}}}{\left[\int_{x_a^{(op)}}^{x_c^{(op)}}\mathrm{d}r\exp\left(\frac{U_{y_{op}}(r)}{q}\right)\right]\left[\int_0^{x_b^{(op)}}\mathrm{d}r\,P(r)\right]}w_a + \frac{q\left[DP\exp\left(\frac{U_{y_{op}}(r)}{q}\right)\right]_{r=x_c^{(op)}}}{\left[\int_{x_a^{(op)}}^{x_c^{(op)}}\mathrm{d}r\exp\left(\frac{U_{y_{op}}(r)}{q}\right)\right]\left[\int_{x_b^{(op)}}^{\infty}\mathrm{d}r\,P(r)\right]}(1-w_a). \tag{262}$$

As a consequence, upon identification of eq. (254) with eq. (262) we find

$$\frac{1}{\tau_L} = \frac{q\left[DP\exp\left(\frac{U_{y_{op}}(r)}{q}\right)\right]_{r=x_a^{(op)}}}{\left[\int_{x_a^{(op)}}^{x_c^{(op)}}\mathrm{d}r\exp\left(\frac{U_{y_{op}}(r)}{q}\right)\right]\left[\int_0^{x_b^{(op)}}\mathrm{d}r\,P(r)\right]}, \tag{263a}$$

$$\frac{1}{\tau_R} = \frac{q\left[DP\exp\left(\frac{U_{y_{op}}(r)}{q}\right)\right]_{r=x_c^{(op)}}}{\left[\int_{x_a^{(op)}}^{x_c^{(op)}}\mathrm{d}r\exp\left(\frac{U_{y_{op}}(r)}{q}\right)\right]\left[\int_{x_b^{(op)}}^{\infty}\mathrm{d}r\,P(r)\right]}. \tag{263b}$$

Note that even if P is time-dependent, the right-hand sides in eqs. (263a,b) are actually independent of time. In fact, for example,

$$P(x_a^{(op)}, t) \Big/ \int_0^{x_b^{(op)}} dr\, P(r, t)$$

is the ratio between the height and the area of the left peak of the distribution. As the peak remains undeformed in shape during the long time evolution, this ratio remains constant in time. This fact can also be seen immediately from the long-time expression for $P(r, t)$ (compare eqs. (247) and (248))

$$P(r, t) = w_a(t)\, G_a^{(op)}(r) + (1 - w_a(t))\, G_c^{(op)}(r). \tag{264}$$

In fact, for $q \ll 1$ we have $G_c^{(op)}(x_a^{(op)}) \approx 0$. Thus τ_L and τ_R can be evaluated by setting $P = P_{y(op)}^{(st)}(r)$ so that, from eq. (240) we obtain

$$\tau_L = \left[\int_{x_a^{(op)}}^{x_c^{(op)}} dr \exp\left(\frac{U_{y_{op}}(r)}{q}\right)\right]\left[\int_0^{x_b^{(op)}} dr\, D^{-1}(r) \exp\left(-\frac{U_{y_{op}}(r)}{q}\right)\right],$$

$$\tau_R = \left[\int_{x_a^{(op)}}^{x_c^{(op)}} dr \exp\left(\frac{U_{y_{op}}(r)}{q}\right)\right]\left[\int_{x_b^{(op)}}^{\infty} dr\, D^{-1}(r) \exp\left(-\frac{U_{y_{op}}(r)}{q}\right)\right]. \tag{265}$$

It is simple to produce an analytic approximation to eq. (265) in the limit of small fluctuations ($q \ll 1$). In the neighborhood of $x_b^{(op)}$ we can approximate $U_{y_{op}}(r)$ with

$$U_{y_{op}}(r) \simeq U_{y_{op}}(x_b^{(op)}) - \frac{1}{2}\left|\frac{d^2 U_{y_{op}}}{dr^2}\right|_{x_b^{(op)}} (r - x_b^{(op)})^2, \tag{266}$$

so that the first integral in eq. (265) becomes

$$\int_{x_a^{(op)}}^{x_c^{(op)}} dr \exp\left(\frac{U_{y_{op}}(r)}{q}\right) \simeq \frac{\sqrt{2\pi}}{\left\{\left|\dfrac{d^2 U_{y_{op}}}{dr^2}\right|_{x_b^{(op)}}\right\}^{1/2}} \exp\left(\frac{U_{y_{op}}(x_b^{(op)})}{q}\right). \tag{267}$$

The second integral can be handled in a similar way after expanding the sharply peaked integral around $r = x_a^{(op)}$ or $r = x_c^{(op)}$. The final result for τ_L takes the form

$$\tau_L \simeq \frac{2\pi}{D(x_a^{(op)})\left\{\left(\dfrac{d^2 U_{y_{op}}}{dr^2}\right)_{x_a^{(op)}} \left|\dfrac{d^2 U_{y_{op}}}{dr^2}\right|_{x_b^{(op)}}\right\}^{1/2}} \exp\left(\frac{U_{y_{op}}(x_b^{(op)}) - U_{y_{op}}(x_a^{(op)})}{q}\right). \tag{268a}$$

In a similar way, the escape time from the right to the left well can be shown to be approximately equal to

$$\tau_R \simeq \frac{2\pi}{D(x_c^{(op)})\left\{\left(\frac{d^2 U_{y_{op}}}{dr^2}\right)_{x_c^{(op)}} \left|\frac{d^2 U_{y_{op}}}{dr^2}\right|_{x_b^{(op)}}\right\}^{1/2}} \exp\left(\frac{U_{y_{op}}(x_b^{(op)}) - U_{y_{op}}(x_c^{(op)})}{q}\right). \tag{268b}$$

We note the appearance of the characteristic ratio $\Delta U/q$, which plays a similar role in this problem as the activation thermal-energy ratio in chemical reactions.

As we pointed out in eq. (255a), the sum of the reciprocals of the escape times τ_L and τ_R gives the first nonzero eigenvalue λ_1 of the Fokker–Planck operator in the limit when the fluctuations are small. The eigenvalue λ_1, in turn, governs the global evolution of the entire probability distribution for long times.

As we see from eq. (268), the escape times are not determined by the mechanical potential V_y, but by the generalized free energy U_y. The difference between the two of them lies in the fact that U_y incorporates the dependence of the diffusion coefficient on the amplitude r (eq. 241)). The importance of the variation of the noise along the path in determining the escape times has been emphasized by LANDAUER (see for instance LANDAUER [1978]).

If one plots λ_1 as a function of y in the bistable domain $y_m < y < y_M$ (see Fig. 8), one sees that it exhibits a sharp minimum in the region where $\tau_L \approx \tau_R$. In fact, as we see from eq. (268), λ_1 is mainly determined by the larger of the left and right "barriers" $U_y(x_b) - U_y(x_a)$ and $U_y(x_b) - U_y(x_c)$. When $\tau_L \approx \tau_R$, the two barriers are substantially equal and attain their minimum value. The region where $\tau_L \approx \tau_R$ coincides in turn with the transition region, because from eq. (256) it follows that $\tau_L = \tau_R$ means $w_a(\infty) = 1 - w_a(\infty) = w_c(\infty)$, which is precisely the condition that characterizes the transition region (compare eq. (244')). This can also be checked from the explicit expression (268) and (243a,b). Thus, λ_1 shows a kind of critical slowing down behavior (SCHENZLE and BRAND [1979]) of a purely statistical nature, in contrast with the semi-classical critical slowing down described in § 2.3. Furthermore, the value of y_{op} at which λ_1 is minimum is the one which insures the greatest stability of the bistable system against spontaneous switching due to noise, and appears to be optimum as a bias for logical operation. That is, it corresponds to the so-called holding intensity.

We finally note that a tunneling effect has been nicely observed and analyzed in bidirectional ring lasers (MANDEL, ROY and SINGH [1981]).

3.7. REMARKS ON THE OBSERVABILITY OF QUANTUM STATISTICAL EFFECTS

The relevance of fluctuations in our system is "measured" by the parameters $\bar{q}$ (see eq. (232)), which governs the thermal fluctuations and the external field fluctuations, and q (see eq. (238a)), which rules intrinsic quantum fluctuations. The quantity $\bar{q}$ is controlled by the temperature and the stability of the injected laser field. For a large system, the parameter q is ordinarily very small. On the other hand, the miniaturized optical bistable devices, presently considered as candidates for memory elements, may well present much more pronounced quantum effects. It must be said, however, that the description of fluctuations in these systems is necessarily much more complicated than the one given here, because it involves the dynamics of semiconductors instead of two-level systems (see GOLL and HAKEN [1980], and STEYN-ROSS and GARDINER [1982]).

The only quantum phenomenon which does not present basic difficulties in its observation is the spectrum of the fluorescent light. However, in order to observe the spectrum of the transmitted or reflected light one meets the difficulty that the classical coherent part is much more intense than the incoherent part, which is the interesting one and arises from fluctuations. One can conceive of depressing the coherent part by some interferometric method or by exploiting the fact that the coherent part is polarized, whereas the incoherent part is not, but an adequate elimination seems difficult. Perhaps heterodyne detection can help in this connection. In order to observe the nonclassical effects described in § 3.3.3 (i.e., photon antibunching and "squeezing"), one has the further difficulty of ensuring that the intrinsic quantum fluctuations are much larger than the thermal and external fluctuations.

The analysis in § 3.6.1 is basic to discuss the *reliability* of our system as an optical memory. In fact, τ_L and τ_R give the lifetime of the metastable states. From eq. (268) we see that, for $q \ll 1$, these lifetimes are extremely (not to say astronomically) long. Hence, the system will not switch spontaneously from one branch to the other, and this ensures the reliability of our optical bistable system. On the other hand, this very desirable feature turns out to be very bad with respect to the observation of the effects described in § 3.5 (bimodal distribution function, generalized Maxwell rule, and anomalous fluctuations in the transition region). In fact, it is necessary to wait for times that are longer than τ_L and τ_R in order to allow the system to reach steady state from a statistical viewpoint.

In conclusion, the fluctuation effects become easily accessible only when either the external fluctuations are large (but in this case the intrinsic quantum fluctuations remain hidden), or when the system is small enough.

Concluding Note

This article reflects the state of the art in the theory of optical bistability in early June 1982, when it was mailed to the editor. Therefore, with a few exceptions, the references do not cover the papers appeared after that time, in particular many contributions to the 12th International Quantum Electronics Conference in Munich, Germany, June 1982, and to the Topical Conference on optical bistability in Rochester, N.Y., June 1983. Here I can only mention the paper by GRANT and KIMBLE [1983] on the observation of critical slowing down at optical frequencies. Doctors Hyatt Gibbs and Sam McCall are preparing a book on optical bistability, which will contain an up-to-date list of references.

Acknowledgements

First of all I wish to thank Rodolfo Bonifacio, with whom I shared the satisfaction of working out many of the results described in this article, and Hermann Haken, because all these results have been directly inspired by his fundamental work in quantum optics and synergetics. I am grateful to Lorenzo Narducci for his present precious collaboration and for his help in the preparation of this manuscript. Many thanks are due also to all the other friends with whom I had the privilege of collaborating on some aspects of my research on optical bistability. At the University of Milan my thanks go to Gino Benza, Frederico Casagrande, Marziale Milani, Laura Asquini and Giuliano Strini; abroad to Stan Dembinski, Jim Farina, Andrzej Kossakowski, Paul Mandel, Pierre Meystre, Murray Sargent and Jimmy Yuan. Furthermore, I thank friends with whom I had particularly stimulating discussions on this subject: Eitan Abraham, Tito Arecchi, Ennio Arimondo, Charles Bowden, Ernst Brun, Claude Cohen-Tannoudji, Vittorio Degiorgio, Joe Eberly, Fritz Haake, John Hermann, Gerd Leuchs, Sam McCall, Ben Mollow, Marlan Scully, Dan Walls and Herbert Walther. I am grateful to Hyatt Gibbs, Adriano Gozzini, Fred Hopf, Jeff Kimble and Richard Shoemaker for giving me permission to include in this paper pictures of some of their experiments. Last but not least, I wish to thank Piero Caldirola, because his wise leadership of our theoretical group created the ideal conditions to work on Physics.

Appendix A. Derivation of eqs. (45) and (46) in the limit of small absorption, transmission and detuning

Let us consider a stationary solution $E_{st}(z)$, $P_{st}(z)$ and $D_{st}(z)$ of the Maxwell–Bloch equations (1) with the boundary condition (3.2). We introduce the deviations from the stationary values

$$\begin{aligned} \delta E(z, t) &= E(z, t) - E_{st}(z), \\ \delta P(z, t) &= P(z, t) - P_{st}(z), \\ \delta D(z, t) &= D(z, t) - D_{st}(z). \end{aligned} \tag{A.1}$$

In terms of these deviations the Maxwell–Bloch equations can be rephrased as

$$\frac{\partial \delta E}{\partial t} + c\frac{\partial \delta E}{\partial z} = -g\delta P, \tag{A.2a}$$

$$\frac{\partial \delta P}{\partial t} = \frac{\mu}{\hbar}[E_{st}(z)\delta D + D_{st}(z)\delta E + \delta E\,\delta D] - [\gamma_\perp + \mathrm{i}(\omega_a - \omega_0)]\delta P, \tag{A.2b}$$

$$\frac{\partial \delta D}{\partial t} = -\frac{\mu}{2\hbar}[(E_{st}(z)\delta P^* + P^*_{st}(z)\delta E + \delta E\,\delta P^*) + \text{c.c.}] - \gamma_\parallel \delta D. \tag{A.2c}$$

On the other hand, from eq. (3.2) we obtain the following boundary condition for δE:

$$\delta E(0, t) = R\exp(-\mathrm{i}\delta_0)\,\delta E(L, t - \Delta t), \tag{A.3}$$

where we have taken into account that

$$E_{st}(0) = E_I\sqrt{T} + R\exp(-\mathrm{i}\delta_0)\,E_{st}(L).$$

Condition (A.3) differs from a pure periodicity condition in space due to the presence of the factor $R\exp(-\mathrm{i}\delta_0)$ and of the retardation Δt. Since it is convenient to work with the periodicity condition, we introduce the following transformation of variables

$$\tilde{\delta E} = w(z, T)\delta E, \qquad \tilde{\delta P} = w(z, T)\delta P, \qquad \tilde{\delta D} = \delta D, \tag{A.4}$$

$$w(z, T) = \exp\left[\frac{z}{L}\ln(R\exp(-\mathrm{i}\delta_0))\right], \qquad t' = t + \Delta t\frac{z}{L} = t + \frac{\mathscr{L} - L}{c}\frac{z}{L}. \tag{A.5}$$

Using (A.4) and (A.5) one verifies that (A.3) reduces to

$$\delta\tilde{E}(0, t') = \delta\tilde{E}(L, t'). \tag{A.6}$$

while the time evolution equations (A.2) become

$$\frac{\partial\delta\tilde{E}}{\partial t'} + c\frac{L}{\mathscr{L}}\frac{\partial\delta\tilde{E}}{\partial z} = \frac{c}{\mathscr{L}}[\ln R - \mathrm{i}\delta_0]\delta\tilde{E} - g\frac{L}{\mathscr{L}}\delta P, \tag{A.7a}$$

$$\frac{\partial\delta\tilde{P}}{\partial t'} = \frac{\mu}{\hbar}\{w(z, T)E_{\mathrm{st}}(z)\,\delta\tilde{D} + D_{\mathrm{st}}(z)\delta\tilde{E} + \delta\tilde{E}\,\delta\tilde{D}\}$$
$$-[\gamma_\perp + \mathrm{i}(\omega_{\mathrm{a}} - \omega_0)]\delta P, \tag{A.7b}$$

$$\frac{\partial\delta\tilde{D}}{\partial t'} = -\frac{\mu}{2\hbar}\{[(w^*(z, T))^{-1}E_{\mathrm{st}}(z)\,\delta P^* + (w(z, T))^{-1}P^*(z)\delta E$$
$$+|w(z, T)|^{-2}\,\delta\tilde{E}\,\delta\tilde{P}^*] + \mathrm{c.c.}\} - \gamma_\parallel\,\delta\tilde{D}. \tag{A.7c}$$

For $T \ll 1$, $\ln R \simeq -T$, so that using eq. (33) and the relation $\delta_0 = \theta T$, eq. (A.7a) becomes

$$\frac{\partial\delta\tilde{E}}{\partial t'} + c\frac{L}{\mathscr{L}}\frac{\partial\delta\tilde{E}}{\partial z} = -k(1 + \mathrm{i}\theta)\delta\tilde{E} - g\frac{L}{\mathscr{L}}\delta\tilde{P}. \tag{A.8}$$

Furthermore, for $T \ll 1$ one has that $w(z, T)$ can be replaced by unity in eqs. (A.7b) and (A.7c).

Let us now consider the stationary fields in the limit (44), in which they are practically uniform in space. By setting the derivatives with respect to time equal to zero in the Maxwell–Bloch equations (1), and integrating eq. (1a) with respect to z to first order in αL (i.e. gL), we obtain

$$E_{\mathrm{st}}(L) - E_{\mathrm{st}}(0) = -g\frac{L}{c}P_{\mathrm{st}}, \tag{A.9}$$

$$0 = \frac{\mu}{\hbar}E_{\mathrm{st}}D_{\mathrm{st}} - [\gamma_\perp + \mathrm{i}(\omega_{\mathrm{a}} - \omega_0)]P_{\mathrm{st}}, \tag{A.10a}$$

$$0 = -\frac{\mu}{2\hbar}(E_{\mathrm{st}}P^*_{\mathrm{st}} + E^*_{\mathrm{st}}P_{\mathrm{st}}) - \gamma_\parallel(D_{\mathrm{st}} - \tfrac{1}{2}N). \tag{A.10b}$$

Next, we combine eq. (A.9) with the boundary condition (3b). By the relation $\delta_0 = \theta T$, neglecting the terms proportional to powers of T higher than first and using eq. (33), we obtain the equation

$$0 = -ik\theta E_{\text{st}} - k\left(E_{\text{st}} - \frac{E_{\text{I}}}{\sqrt{T}}\right) - g\frac{L}{\mathscr{L}}P_{\text{st}}, \tag{A.11}$$

where we have replaced $E_{\text{st}}(\text{L})$ by E_{st}.

Finally, we take into account that, in the limit (44), $\delta\tilde{E} = \delta E$, etc. since $w(z, T) = 1$. Hence, using eq. (A.1) and summing eqs. (A.8) and (A.11), we obtain eq. (45). On the other hand, by summing eqs. (A.7.2) (with $w(z, T) = 1$) and (A.10.1), eqs. (A.7.3) and (A.10.2), we recover eqs. (1.2) and (1.3) respectively. Furthermore, since E_{st} is independent of z we obtain, from eqs. (A.6) and (A.1), the standard periodicity boundary condition (46).

Appendix B. The coefficients Γ in eq. (127)

Let us enlist the explicit expressions of the coefficients $\Gamma(nj, n'j', n''j'')$ of eq. (127) in the limit (44). First, let us define the following two functions

$$\tilde{H}(\lambda, x) = x^2\left(\frac{\lambda}{\gamma_\perp} + 2\right) - \frac{\lambda + \gamma_\parallel}{\gamma_\perp}\left(\frac{\lambda}{\gamma_\parallel} + 1 - x^2\right), \tag{B.1}$$

$$\tilde{K}(\lambda, \lambda', x) = \left(\frac{\lambda}{\gamma_\parallel} + 1\right)\left(\frac{\lambda'}{\gamma_\perp} + 2\right) + \left(\frac{\lambda'}{\gamma_\parallel} + 1 - x^2\right). \tag{B.2}$$

The explicit expression of the normalization constant $\mathscr{N}_{nj}$ in eq. (122) is

$$\mathscr{N}_{nj} = G(\lambda_{nj}, x) + \frac{2C}{1 + x^2}\frac{k}{\gamma_\parallel}\frac{\tilde{H}(\lambda_{nj}, x)}{G(\lambda_{nj}, x)}. \tag{B.3}$$

Furthermore one has:

(i) For $j = j'' = 1$, any j', $\Gamma(n1, n'j', n''1)$

$$= \delta_{n,n'+n''}2Ck\frac{x}{1 + x^2}\frac{\tilde{G}(-\mathrm{i}\alpha''_n, x)}{\tilde{G}^2(-\mathrm{i}\alpha_n, x)}\tilde{K}(-\mathrm{i}\alpha_n, \lambda^{(0)}_{n'j'}, x) + O(T^2). \tag{B.4}$$

(ii) For $j = 1$, $j'' = 2, 3$, any j',

$$\Gamma(n1, n'j', n''{}^2_3) = O(T^2)\delta_{n,n'+n''}. \tag{B.5}$$

(iii) For $j = 2, 3$, $j'' = 1$, any j',

$$\Gamma(n^2_3, n'j', n''1) = \delta_{n,n'+n''}\gamma_\parallel x\frac{\tilde{G}(-\mathrm{i}\alpha_{n''}, x)}{\tilde{H}(\lambda^{(0)}_{n^2_3}, x)}\tilde{K}(\lambda^{(0)}_{n^2_3}, \lambda^{(0)}_{n'j'}, x) + O(T). \tag{B.6}$$

(iv) For $j = 2, 3, \quad j'' = 2, 3, \quad$ any j',

$$\Gamma(n\tfrac{2}{3}, n'j', n''j'') = O(T)\delta_{n,n'+n''}. \tag{B.7}$$

In the case $\gamma_\perp = \gamma_\parallel \equiv \gamma$ that we consider in this chapter the first term in eq. (B.6) vanishes when $j = 3$ and $j' = 2$, or $j = 2$ and $j' = 3$. In fact, in this case, it follows from eq. (83) that

$$\lambda^{(0)}_{n\frac{2}{3}} = -\gamma \pm \mathrm{i}\gamma x. \tag{B.8}$$

By substituting eq. (B.8) into eq. (B.3) one easily verifies that

$$\tilde{K}(\lambda^{(0)}_{n2}, \lambda^{(0)}_{n'3}, x) = \tilde{K}(\lambda^{(0)}_{n3}, \lambda^{(0)}_{n'2}, x) = 0. \tag{B.9}$$

Appendix C. Explicit expression for eqs. (142)

Let us consider the nine equations for the amplitudes $S_{nj}(n = 1, 0, -1;$ $j = 1, 2, 3)$ with only the terms that survive in the limits (27) and (136). In this limit, one has exactly $\dot{S}_{n2} = \dot{S}_{n3} = 0$. Thus the equations for $S_{n2}(n = 1, 0, -1)$ form a closed set of three algebraic equations, whose solution gives the expressions of $\{S_{n2}\}$ as functions of $\{S_{n1}\}$. The same holds for the equations for $S_{n3}(n = 1, 0, -1)$.

Next, we define

$$\delta(n1, n'j', n''1) = k^{-1}\Gamma(n1, n'j', n''1),$$
$$\delta(n2, n'j', n''1) = \gamma^{-1}\Gamma(n2, n'j', n''1), \tag{C.1a}$$

and

$$\tilde{\lambda}_{n\frac{2}{3}} = \gamma^{-1}\lambda^{(0)}_{n\frac{2}{3}} = -1 \pm \mathrm{i}x, \qquad P = S_{12}\exp(-\mathrm{i}\varphi),$$
$$R = S_{-12}\exp(\mathrm{i}\varphi), \qquad Q = S_{02}, \tag{C.1b}$$

and use eq. (141) and the relation $S_{n3} = S^*_{-n2}$. By inserting the expressions for $\{S_{n2}\}$ and $\{S_{n3}\}$ into the time evolution equations for $S_{n1}(n = 1, 2, 3)$, we obtain the equations

$$\begin{aligned}\frac{\partial \rho_1}{\partial \tau} = {} & \mathrm{Re}\,\lambda^{(1)}_{11}\rho_1 + \mathrm{Re}\,[\delta(11, 01, 11) + \delta(11, 11, 01)]\rho_1\sigma \\ & + \mathrm{Re}\,[\delta(11, 02, 11)Q + \delta(11, 03, 11)Q^*]\rho_1 \\ & + \mathrm{Re}\,[\delta(11, 12, 01)P + \delta(11, 13, 01)R^*]\sigma,\end{aligned} \tag{C.2a}$$

$$\frac{\partial \sigma}{\partial \tau} = \lambda_{01}^{(1)}\sigma + 2\,\mathrm{Re}\,\delta(01, 11, -11)\rho_1^2 + \delta(01, 01, 01)\sigma^2 + 2\,\mathrm{Re}\,[\delta(01, 12, -11)P]\rho_1 + 2\,\mathrm{Re}\,[\delta(01, -12, 11)R]\rho_1 + 2\,\mathrm{Re}\,[\delta(01, 02, 01)Q]\sigma, \quad \text{(C.2b)}$$

where P, Q and R are solutions of the system of linear equations

$$\left\{\begin{array}{l} A_{11}P + A_{12}Q = -z_1, \\ A_{21}P + A_{22}Q + A_{23}R = -z_2, \\ A_{32}Q + A_{33}R = -z_3, \end{array}\right\} \quad \text{(C.3)}$$

where

$$\begin{aligned} A_{11} &= \tilde{\lambda}_{12} + \mathrm{i}\tilde{\alpha}_1 + \delta(12, 12, 01)\sigma, \\ A_{12} &= \delta(12, 02, 11)\rho_1, \qquad A_{21} = \delta(02, 12, -11)\rho_1, \\ A_{22} &= \tilde{\lambda}_{02} + \delta(02, 02, 01)\sigma, \qquad A_{23} = \delta(02, -12, 11)\rho_1, \\ A_{32} &= \delta(-12, 02, -11)\rho_1, \qquad A_{33} = \tilde{\lambda}_{-12} - \mathrm{i}\tilde{\alpha}_1 + \delta(-12, -12, 01)\sigma, \\ z_1 &= [\delta(12, 11, 01) + \delta(12, 01, 11)]\rho_1\sigma, \\ z_2 &= [\delta(02, 11, -11) + \delta(02, -11, 11)]\rho_1^2 + \delta(02, 01, 01)\sigma^2, \\ z_3 &= [\delta(-12, -11, 01) + \delta(-12, 01, -11)]\rho_1\sigma. \end{aligned} \quad \text{(C.4)}$$

Hence, eqs. (C.2) can be written in the form (142) where the functions f and g depend only on the parameters C, x and $\tilde{\alpha}_1$.

Appendix D. Derivation of the Fokker–Planck equation (176)

Let us consider the moments of the generalized Wigner distribution $P_{\mathrm{w}}(\bar{v}, \bar{v}^*, \bar{m}, \beta, \beta^*, t)$:

$$\langle (R^+)^p (R_3)^l (R^-)^q (A^+)^r A^s \rangle_{\mathrm{s}}(t) \equiv \int \mathrm{d}_2\bar{v}\, \mathrm{d}\bar{m}\, \mathrm{d}_2\beta (\bar{v}^*)^p \bar{m}^l \bar{v}^q (\beta^*)^r \beta^s\, P_{\mathrm{w}}(\bar{v}, \bar{v}^*, \bar{m}, \beta, \beta^*, t), \quad \text{(D.1)}$$

where the suffix "s" means "symmetrized" (see eq. (174)). Using the commutation rules (151) and (156) one derives from the master equation (158), restricted to the single mode $n = 0$, the hierarchy of time evolution equations for the moments (D.1), which has the structure

$$\frac{\mathrm{d}}{\mathrm{d}t}\langle (R^+)^p (R_3)^l (R^-)^q (A^\dagger)^r A^s \rangle_\mathrm{s}$$

$$= \sum_{p'l'q'r's'} C^{plqrs}_{p'l'q'r's'} \langle (R^+)^{p'} (R_3)^{l'} (R^-)^{q'} (A^\dagger)^{r'} A^{s'} \rangle_\mathrm{s}, \qquad \text{(D.2)}$$

where the coefficients C are suitable constants.

The set of equations (D.2) is equivalent to the partial differential equation for P_w

$$\frac{\partial}{\partial t} P_\mathrm{w}(\bar{v}, \bar{v}^*, \bar{m}, \beta, \beta^*, t) = \sum_{\bar{p}\bar{l}\bar{q}\bar{r}\bar{s}} \frac{\partial^n}{\partial(\bar{v}^*)^{\bar{p}}\, \partial \bar{v}^{\bar{q}}\, \partial \bar{m}^{\bar{l}}\, \partial(\beta^*)^{\bar{r}}\, \partial \beta^{\bar{s}}}$$

$$\left(\sum_{p_0 l_0 q_0 r_0 s_0} d^{\bar{p}\bar{l}\bar{q}\bar{r}\bar{s}}_{p_0 l_0 q_0 r_0 s_0} (\bar{v}^*)^{p_0}\, \bar{m}^{l_0}\, \bar{v}^{q_0}\, (\beta^*)^{r_0}\, \beta^{s_0} \right)$$

$$\times P_\mathrm{w}(\bar{v}, \bar{v}^*, \bar{m}, \beta, \beta^*, t), \qquad n = \bar{p} + \bar{q} + \bar{l} + \bar{r} + \bar{s}, \qquad \text{(D.3)}$$

provided the coefficients d are chosen in such a way that eqs. (D.3), with definition (D.1), exactly reproduce eqs. (D.2). The determination of these coefficients can be done in steps, by calculating first the terms with $n = 1$ (drift coefficients), secondly the terms with $n = 2$ (diffusion coefficients), etc. Since we are interested in obtaining a Fokker–Planck equation it is enough to calculate the coefficients d only for $n = 1$ and $n = 2$. On comparing eqs. (D.2) and (D.3) via eq. (D.1) we easily obtain, for $n = 1$

$$d^{\bar{p}\bar{l}\bar{q}\bar{r}\bar{s}}_{p_0 l_0 q_0 r_0 s_0} = -C^{\bar{p}\bar{l}\bar{q}\bar{r}\bar{s}}_{p_0 l_0 q_0 r_0 s_0}, \qquad \text{(D.4a)}$$

and for $n = 2$

$$d^{\bar{p}\bar{l}\bar{q}\bar{r}\bar{s}}_{p_0 l_0 q_0 r_0 s_0} = \frac{1}{\bar{p}!\bar{l}!\bar{q}!\bar{r}!\bar{s}!} C^{\bar{p}\bar{l}\bar{q}\bar{r}\bar{s}}_{p_0 l_0 q_0 r_0 s_0}$$

$$- C^{\bar{p}-1\,\bar{l}\bar{q}\bar{r}\bar{s}}_{p_0-1\,l_0 q_0 r_0 s_0} - C^{\bar{p}\bar{l}-1\,\bar{q}\bar{r}\bar{s}}_{p_0 l_0-1\,q_0 r_0 s_0} - C^{\bar{p}\bar{l}\bar{q}-1\,\bar{r}\bar{s}}_{p_0 l_0 q_0-1\,r_0 s_0}$$

$$- C^{\bar{p}\bar{l}\bar{q}\bar{r}-1\,\bar{s}}_{p_0 l_0 q_0 r_0-1\,s_0} - C^{\bar{p}\bar{l}\bar{q}\bar{r}\bar{s}-1}_{p_0 l_0 q_0 r_0 s_0-1}, \qquad \text{(D.4b)}$$

where one must take $C = 0$ when one of its indices is negative. By substituting into eqs. (D.4) the explicit expressions of the coefficients C, and neglecting the terms with derivatives of order higher than the second, eq. (D.3) reduces to a

Fokker–Planck equation for $P_w(\bar{v}, \bar{v}^*, \bar{m}, \beta, \beta^*, t)$. Finally, by transforming to the variables v, v^*, m, x, x^* defined by eq. (175), one obtains eq. (176).

References

Abraham, E. and S. S. Hassan, 1980, Opt. Commun. **35,** 291.

Abraham, E. and S. D. Smith, 1982a, J. Phys. **E15,** 33.

Abraham, E. and S. D. Smith, 1982b, Optical bistability and related devices, Rep. Prog. Phys. **45,** 815.

Abraham, E., R. K. Bullough and S. S. Hassan, 1979, Opt. Commun. **29,** 109.

Abraham, E., S. S. Hassan and R. K. Bullough, 1980, Opt. Commun. **33,** 93.

Abraham, E., W. J. Firth and E. M. Wright, 1982, Proc. Vth Nat. Quantum Electronics Conf., Hull, England, ed. P. L. Knight (Wiley, New York).

Agarwal, G. S., L. M. Narducci, D. H. Feng and R. Gilmore, 1977, Dynamical Approach to Steady state and Fluctuations in Optical Bistable Systems, in: Coherence and Quantum Optics IV, Proc. Fourth Rochester Conf., 1977, eds. L. Mandel and E. Wolf (Plenum, New York, 1978).

Agarwal, G. S., L. M. Narducci, R. Gilmore and D. H. Feng, 1978a, Opt. Lett. **2,** 88.

Agarwal, G. S., L. M. Narducci, R. Gilmore and D. H. Feng, 1978b, Phys. Rev. **A18,** 620.

Agarwal, G. S., L. M. Narducci, R. Gilmore and D. H. Feng, 1979, Phys. Rev. **A20,** 545.

Agrawal, G. P., 1981, Appl. Phys. Lett. **38,** 505.

Agrawal, G. P. and H. J. Carmichael, 1979, Phys. Rev. **A19,** 2074.

Agrawal, G. P. and C. Flytzanis, 1980, Phys. Rev. Lett. **44,** 1058.

Agrawal, G. P. and C. Flytzanis, 1981, IEEE J. Quantum Electron. **QE-17,** 374.

Allen, L. and J. H. Eberly, 1975, Optical Resonance and Two-Level Atoms (Wiley, New York).

Arecchi, F. T. and R. Bonifacio, 1965, IEEE J. Quantum Electron. **QE-1,** 169.

Arecchi, F. T. and A. Politi, 1978, Lett. Nuovo Cim. **23,** 65.

Arecchi, F. T. and A. Politi, 1979, Opt. Commun. **29,** 361.

Arecchi, F. T., G. Giusfredi, E. Petriella and P. Salieri, 1982, J. Appl. Phys. **29,** 79.

Arimondo, E., A. Gozzini, L. Lovitch and E. Pistelli, 1981, Microwave Dispersive Bistability in a Confocal Fabry–Perot Microwave Cavity, in: Bowden *et al.* [1981].

Asquini, M. L. and F. Casagrande, 1981, Z. Phys. **B44,** 233.

Austin, J. W. and L. G. Deshazer, 1971, J. Opt. Soc. Am. **61,** 650.

Baczynski, A., A. Kossakowski and T. Marszalek, 1976, Z. Phys. **B23,** 205.

Ballagh, R. I., J. Cooper, M. W. Hamilton, W. J. Sandle and D. M. Warrington, 1981, Opt. Commun. **37,** 143.

Barbarino, S., A. Gozzini, I. Longo, F. Maccarrone and R. Stampacchia, 1982, Nuovo Cim. **B71,** 183.

Benza, V. and L. A. Lugiato, 1979a, Lett. Nuovo Cim. **26,** 405.

Benza, V. and L. A. Lugiato, 1979b, Z. Phys. **B35,** 383.

Benza, V., L. A. Lugiato and P. Meystre, 1980, Opt. Commun. **33,** 113.

Benza, V. and L. A. Lugiato, 1981, Semiclassical and Quantum Statistical Dressed Mode Description of Optical Bistability, in: Bowden *et al.* [1981].

Benza, V. and L. A. Lugiato, 1982, Z. Phys. **47,** 79.

Bjorkholm, J. E., P. W. Smith and W. J. Tomlinson, 1981, Opt. Lett. **6,** 345.

Bischofberger, T. and Y. R. Shen, 1978, Appl. Phys. Lett. **32,** 156.

Bischofberger, T. and Y. R. Shen, 1979, Phys. Rev. **A19,** 1169.

Bösiger, P., E. Brun and D. Meier, 1978, Phys. Rev. **A18,** 671.

BONIFACIO, R., ed., 1982, Dissipative Systems in Quantum Optics – Resonance Fluorescence, Optical Bistability, Superfluorescence (Springer, Berlin).
BONIFACIO, R. and L. A. LUGIATO, 1975, Phys. Rev. **A11,** 1507.
BONIFACIO, R. and L. A. LUGIATO, 1976, Opt. Commun. **19,** 172.
BONIFACIO, R. and L. A. LUGIATO, 1977, Cooperative Effects in Optical Bistability and Resonance Fluorescence, in: Coherence and Quantum Optics IV, Proc. Fourth Rochester Conf., 1977, eds. L. Mandel and E. Wolf (Plenum, New York).
BONIFACIO, R. and L. A. LUGIATO, 1978a, Lett. Nuovo Cim. **21,** 505.
BONIFACIO, R. and L. A. LUGIATO, 1978b, Lett. Nuovo Cim. **21,** 510.
BONIFACIO, R. and L. A. LUGIATO, 1978c, Lett. Nuovo Cim. **21,** 517.
BONIFACIO, R. and L. A. LUGIATO, 1978d, Phys. Rev. **A18,** 1129.
BONIFACIO, R. and L. A. LUGIATO, 1978e, Phys. Rev. Lett. **40,** 1023, 1538.
BONIFACIO, R. and P. MEYSTRE, 1978, Opt. Commun. **27,** 147.
BONIFACIO, R. and P. MEYSTRE, 1979, Opt. Commun. **29,** 131.
BONIFACIO, R., M. GRONCHI and L. A. LUGIATO, 1978, Phys. Rev. **A18,** 2266.
BONIFACIO, R., L. A. LUGIATO and M. GRONCHI, 1979, Theory of Optical Bistability, in: Laser Spectroscopy IV, Proc. Fourth Int. Conf., Rottach-Egern 1979, eds. H. Walther and K. W. Rothe (Springer, Berlin).
BONIFACIO, R., M. GRONCHI and L. A. LUGIATO, 1979a, Nuovo Cim. **B53,** 311.
BONIFACIO, R., M. GRONCHI and L. A. LUGIATO, 1979b, Opt. Commun. **30,** 129.
BONIFACIO, R., L. A. LUGIATO, J. D. FARINA and L. M. NARDUCCI, 1981, IEEE J. Quantum Electron. **QE-17,** 357.
BONIFACIO, R., F. CASAGRANDE and L. A. LUGIATO, 1981, Opt. Commun. **36,** 159.
BORN, M. and E. WOLF, 1970, Principles of Optics, 4th Ed. (Pergamon Press, Elmsford, NY).
BOWDEN, C. M., 1981, Optical Bistability Based upon Atomic Correlation in a Small Volume, in: BOWDEN *et al.* [1981].
BOWDEN, C. M. and C. C. SUNG, 1979, Phys. Rev. **A19,** 2392.
BOWDEN, C. M., M. CIFTAN and H. R. ROBL, eds., 1981, Optical Bistability, Proc. Int. Conf. on Optical Bistability, Asheville, 1980 (Plenum, New York).
CARMICHAEL, H. J., 1980, Opt. Acta **27,** 147.
CARMICHAEL, H. J., 1981, Z. Phys. **B42,** 183.
CARMICHAEL, H. J. and G. P. AGRAWAL, 1980, Opt. Commun. **34,** 293.
CARMICHAEL, H. J. and J. A. HERMANN, 1980, Z. Phys. **B38,** 365.
CARMICHAEL, H. J. and D. F. WALLS, 1977, J. Phys. **B10,** L685.
CARMICHAEL, H. J., R. R. SNAPP and W. C. SCHIEVE, 1982, Phys. Rev. **A26,** 3408.
CASAGRANDE, F. and L. A. LUGIATO, 1980, Nuovo Cim. **B55,** 173.
CASAGRANDE, F., L. A. LUGIATO and M. L. ASQUINI, 1980, Opt. Commun. **32,** 492.
CHOW, W. W., M. O. SCULLY and E. W. VAN STRYLAND, 1975, Opt. Commun. **15,** 6.
COLLINS, Jr., S. A. and K. C. WASMUNDT, 1980, Opt. Eng. **19,** 478.
DEGIORGIO, V. and M. O. SCULLY, 1970, Phys. Rev. **A2,** 1170.
DRUMMOND, P. D., 1981, IEEE J. Quantum Electron. **QE-17,** 301.
DRUMMOND, P. D., 1982, Opt. Commun. **40,** 224.
DRUMMOND, P. D. and D. F. WALLS, 1980, J. Phys. **A13,** 725.
DRUMMOND, P. D. and D. F. WALLS, 1981, Phys. Rev. **A23,** 2563.
DRUMMOND, P. D., K. J. MCNEIL and D. F. WALLS, 1980a, Opt. Acta **27,** 321.
DRUMMOND, P. D., K. J. MCNEIL and D. F. WALLS, 1980b, Phys. Rev. **A22,** 1672.
ENGLUND, J. C., W. C. SCHIEVE, W. ZUREK and R. F. GRAGG, 1981, Fluctuations and Transitions in the Absorptive Optical Bistability, in: BOWDEN *et al.* [1981].
FARINA, J. D., L. M. NARDUCCI, J. M. YUAN and L. A. LUGIATO, 1980, Opt. Eng. **19,** 469.
FEIGENBAUM, M. J., 1978, J. Stat. Phys. **19,** 25.

FEIGENBAUM, M. J., 1979, J. Stat. Phys. **21,** 669.
FELBER, F. S. and J. H. MARBURGER, 1976, Appl. Phys. Lett. **28,** 731.
FIRTH, W. J., 1981, Opt. Commun. **39,** 343.
FIRTH, W. J. and E. M. WRIGHT, 1982, Opt. Commun. **40,** 233.
FLECK, J. A., 1968, Appl. Phys. Lett. **13,** 365.
GARMIRE, E., J. H. MARBURGER, S. D. ALLEN and H. G. WINFUL, 1979, Appl. Phys. Lett. **34,** 374.
GIBBS, H. M., S. L. MCCALL and T. N. C. VENKATESAN, 1976, Phys. Rev. Lett. **36,** 113.
GIBBS, H. M., S. L. MCCALL and T. N. C. VENKATESAN, 1979, Opt. News **5,** 6.
GIBBS, H. M., S. L. MCCALL, T. N. C. VENKATESAN, A. C. GOSSARD, A. PASSNER and W. WIEGMANN, 1979, Appl. Phys. Lett. **35,** 451.
GIBBS, H. M., S. L. MCCALL and T. N. C. VENKATESAN, 1980, Opt. Eng. **19,** 463.
GIBBS, H. M., F. A. HOPF, D. L. KAPLAN and R. L. SHOEMAKER, 1981, Phys. Rev. Lett. **46,** 474.
GILMORE, R., 1974, Lie Groups, Lie Algebras and Some of their Applications (Wiley, New York).
GLAUBER, R. J., 1963a, Phys. Rev. **130,** 2529.
GLAUBER, R. J., 1963b, Phys. Rev. **131,** 2766.
GOLDSTONE, J. A., P. T. HO and E. GARMIRE, 1981, Transient Phenomena in Bistable Devices, in: BOWDEN *et al.* [1981].
GOLL, J. and H. HAKEN, 1980, Phys. Stat. Sol. (b) **101,** 489.
GORDON, J. P., 1967, Phys. Rev. **161,** 367.
GOZZINI, A., 1982, preprint, Ist Fisica dell'Univ., Pisa.
GOZZINI, A., I. LONGO and F. MACCARRONE, 1982, Nuovo Cim. **D1,** 489.
GRAGG, R. F., W. C. SCHIEVE and A. R. BULSARA, 1978, Phys. Lett. **68A,** 294.
GRAGG, R. F., W. C. SCHIEVE and A. R. BULSARA, 1979, Phys. Rev. **A19,** 2052.
GRAHAM, R. and H. HAKEN, 1968, Z. Phys. **213,** 420.
GRAHAM, R. and H. HAKEN, 1970, Z. Phys. **237,** 31.
GRAHAM, R. and A. SCHENZLE, 1981, Phys. Rev. **A23,** 1302.
GRANT, D. E. and H. J. KIMBLE, 1982, Opt. Lett. **7,** 353.
GRANT, D. E. and H. J. KIMBLE, 1983, Opt. Commun. **44,** 415.
GRISCHKOWSKI, D., 1978, J. Opt. Soc. Am. **68,** 641.
GRONCHI, M. and L. A. LUGIATO, 1978, Lett. Nuovo Cim. **23,** 593.
GRONCHI, M. and L. A. LUGIATO, 1980, Opt. Lett. **5,** 108.
GRONCHI, M., V. BENZA, L. A. LUGIATO, P. MEYSTRE and M. SARGENT III, 1981, Phys. Rev. **A24,** 1419.
GRYNBERG, G., E. GIACOBINO, M. DEVAUD and F. BIRABEN, 1980, Phys. Rev. Lett. **45,** 434.
GRYNBERG, G., F. BIRABEN and E. GIACOBINO, 1981, Appl. Phys. **B26,** 155.
HAAG, G., M. MUNZ and G. MAROWSKY, 1981, IEEE J. Quantum Electron. **QE-17,** 349.
HAKEN, H., 1970, Laser Theory, in: Light and Matter Ic, ed. L. Genzel, Handbuch der Physik, vol. XXV/2c (Springer, Berlin).
HAKEN, H., 1975a, Z. Phys. **B21,** 105.
HAKEN, H., 1975b, Z. Phys. **B22,** 69.
HAKEN, H., 1977, Synergetics – An Introduction (Springer, Berlin).
HAKEN, H. and H. OHNO, 1976a, Opt. Commun. **16,** 205.
HAKEN, H. and H. OHNO, 1976b, Phys. Lett. **59A,** 261.
HAMILTON, M. W., R. J. BALLAGH and W. J. SANDLE, 1982, Z. Phys. **B49,** 263.
HANGGI, P., A. R. BULSARA and R. JANDA, 1980, Phys. Rev. **A22,** 671.
HASSAN, S. S., P. D. DRUMMOND and D. F. WALLS, 1978, Opt. Commun. **27,** 480.
HEER, C. V. and R. D. GRAFT, 1965, Phys. Rev. **A140,** 1088.
HERMANN, J. A., 1980, Opt. Acta **27,** 159.

HERMANN, J. A. and B. V. THOMPSON, 1980, Phys. Lett. **79A,** 153.
HERMANN, J. A. and B. V. THOMPSON, 1981, Analytic description of multiphoton optical bistability in a ring cavity, in: BOWDEN *et al.* [1981].
HOPF, F., P. MEYSTRE, P. D. DRUMMOND and D. F. WALLS, 1979, Opt. Commun. **31**, 245.
IKEDA, K., 1979, Opt. Commun. **30,** 257.
IKEDA, K. and O. AKIMOTO, 1982, Phys. Rev. Lett. **48,** 617.
IKEDA, K., H. DAIDO and O. AKIMOTO, 1980, Phys. Rev. Lett. **45,** 709.
JEWELL, J. L., H. M. GIBBS, S. S. TARNG, A. C. GOSSARD and W. WIEGMANN, 1982, Appl. Phys. Lett. **40,** 291.
KAPLAN, A. E., 1977, Sov. Phys. JETP **45,** 896.
KAPLAN, A. E. and P. MEYSTRE, 1982, Opt. Commun. **40,** 229.
KASANTSEV, A. P., S. G. RAUTIAN and G. I. SURDUTOVICH, 1968, Sov. Phys. JETP **27,** 756.
KITANO, M., T. YABUZAKI and T. OGAWA, 1981a, Phys. Rev. Lett. **46,** 926.
KITANO, M., T. YABUZAKI and T. OGAWA, 1981b, Phys. Rev. **A24,** 3156.
KRAMERS, H. A., 1940, Physica **7,** 284.
LAMB, W. E., Jr., 1965, Theory of Optical Maser Oscillators, in: Proc. Int. School of Physics "E. Fermi", Course XXI, Varenna 1963, ed. P. A. Miles (Academic, New York).
LANDAUER, R., 1962, J. Appl. Phys. **33,** 2209.
LANDAUER, R., 1978, Phys. Today **31,** 23.
LANDAUER, R. and J. W. F. WOO, 1973, in: Synergetics, ed. H. Haken (Teubner, Stuttgart).
LAX, M., 1967, Phys. Rev. **157,** 213.
LUGIATO, L. A., 1978, Lett. Nuovo Cim. **23,** 609.
LUGIATO, L. A., 1979, Nuovo Cim. **B50,** 89.
LUGIATO, L. A., 1980a, Opt. Commun. **33,** 108.
LUGIATO, L. A., 1980b, Lett. Nuovo Cim. **29,** 375.
LUGIATO, L. A., 1981, Z. Phys. **B41,** 85.
LUGIATO, L. A. and G. STRINI, 1982a, Opt. Commun. **41,** 67.
LUGIATO, L. A. and G. STRINI, 1982b, Opt. Commun. **41,** 374.
LUGIATO, L. A. and G. STRINI, 1982c, Opt. Commun. **41,** 447.
LUGIATO, L. A., P. MANDEL, S. DEMBINSKI and A. KOSSAKOWSKI, 1978, Phys. Rev. **A18,** 238.
LUGIATO, L. A., J. D. FARINA and L. M. NARDUCCI, 1980, Phys. Rev. **A22,** 253.
LUGIATO, L. A., V. BENZA, L. M. NARDUCCI and J. D. FARINA, 1981, Opt. Commun. **39,** 405.
LUGIATO, L. A., M. MILANI and P. MEYSTRE, 1982, Opt. Commun. **40,** 307.
LUGIATO, L. A., F. CASAGRANDE and L. PIZZUTO, 1982, Phys. Rev. **A26,** 3438.
LUGIATO, L. A., M. L. ASQUINI and L. M. NARDUCCI, 1982, Opt. Commun. **41,** 450.
LUGIATO, L. A., L. M. NARDUCCI, D. K. BANDY and C. A. PENNISE, 1982, Opt. Commun. **43,** 281.
LUGIATO, L. A., V. BENZA, L. M. NARDUCCI and J. FARINA, 1983, Z. Phys. **B49,** 351.
LUGOVOI, V. N., 1979, Phys. Stat. Sol. (b) **94,** 79.
LUGOVOI, V. N., 1981, IEEE J. Quantum Electron. **QE-17,** 384.
MANDEL, L., R. ROY and S. SINGH, 1981, Optical Bistability Effects in a Dye Ring Laser, in: BOWDEN *et al.* [1981].
MANDEL, P. and T. ERNEUX, 1982, Opt. Commun. **42,** 362; **44,** 55.
MARBURGER, J. H. and F. S. FELBER, 1978, Phys. Rev. **A17,** 335.
MAYR, M., H. RISKEN and H. D. VOLLMER, 1981, Opt. Commun. **36,** 480.
MCCALL, S. L., 1974, Phys. Rev. **A9,** 1515.
MCCALL, S. L., 1978, Appl. Phys. Lett. **32,** 284.
MCCALL, S. L. and H. M. GIBBS, 1978, J. Opt. Soc. Am. **68,** 1378.
MCCALL, S. L. and H. M. GIBBS, 1980, Opt. Commun. **33,** 335.
MEIER, D., R. HOLZNER, B. DERIGHETTI and E. BRUN, 1982, Bistability and chaos in NMR

Systems, in: Proc. Int. Symposium on Synergetics, Schloss Elmau 1982, ed. H. Haken (Springer, Berlin).

MEYSTRE, P., 1978, Opt. Commun. **26,** 277.

MEYSTRE, P. and H. HOPF, 1979, Opt. Commun. **29,** 235.

MEYSTRE, P. and M. O. SCULLY, eds., 1983, Quantum Optics, Experimental Gravitation and Measurement Theory, Proc. NATO ASI, Bad Windsheim, Germany, 1981 (Plenum, New York).

MILLER, D. A. B., 1981, IEEE J. Quantum Electron. **QE-17,** 306.

MILLER, D. A. B. and S. D. SMITH, 1979, Opt. Commun. **31,** 101.

MILLER, D. A. B., S. D. SMITH and A. JOHNSTON, 1979, Appl. Phys. Lett. **35,** 658.

MILLER, D. A. B., S. D. SMITH and C. T. SEATON, 1981, IEEE J. Quantum Electron. **QE-17,** 312.

MOLLOW, B. R., 1969, Phys. Rev. **188,** 1969.

MOLONEY, J. V., M. R. BELIC and H. M. GIBBS, 1982, Opt. Commun. **41,** 379.

NICOLIS, G. and R. LEFEVER, 1977, Phys. Lett. **62A,** 469.

NICOLIS, G. and I. PRIGOGINE, 1977, Self-organization in Non-Equilibrium Systems – From Dissipative Structures to Order Through Fluctuations (Wiley, New York).

OKADA, M. and K. TAKIZAWA, 1981, IEEE J. Quantum Electron. **QE-17,** 2135.

POWELL, H. T. and G. J. WOLGA, 1971, IEEE J. Quantum Electron. **QE-7,** 213.

RISKEN, H. and K. NUMMEDAL, 1968, J. Appl. Phys. **49,** 4662.

ROHART, F. and B. MACKE, 1980, J. Physique **41,** 837.

ROSANOV, N. N. and V. E. SEMENOV, 1981, Opt. Commun. **38,** 435.

ROY, R. and M. S. ZUBAIRY, 1980a, Phys. Rev. **A21,** 274.

ROY, R. and M. S. ZUBAIRY, 1980b, Opt. Commun. **32,** 163.

RUSHIN, S. and S. H. BAUER, 1979, Chem. Phys. Lett. **66,** 100.

SALOMAA, R. and S. STENHOLM, 1973, Phys. Rev. **A8,** 2695.

SANDLE, W. J. and A. GALLAGHER, 1981, Phys. Rev. **A24,** 2017.

SANDLE, W. J., R. J. BALLAGH and A. GALLAGHER, 1981, Optical Bistability Experiments and Mean-Field Theories, in: BOWDEN *et al.* [1981].

SARGENT III, M., 1978, Phys. Rep. **43,** 223.

SARGENT III, M., 1980, Sov. J. Quantum Electron. **10,** 1247.

SARGENT III, M., W. E. LAMB, Jr. and R. L. FORK, 1967, Phys. Rev. **164,** 450.

SARGENT III, M., M. O. SCULLY and W. E. LAMB, Jr., 1974, Laser Physics (Addison–Wesley, Reading, MA).

SARID, D., 1981, Opt. Lett. **6,** 552.

SCHAEFER, R. B. and C. R. WILLIS, 1976, Phys. Rev. **A13,** 1874.

SCHENZLE, A. and H. BRAND, 1978, Opt. Commun. **27,** 85.

SCHENZLE, A. and H. BRAND, 1979, Opt. Commun. **31,** 401.

SCHWENDIMANN, P., 1979, J. Phys. **A12,** L39.

SCOTT, J. F., M. SARGENT III and C. CANTRELL, 1975, Opt. Commun. **15,** 13.

SEIDEL, H., 1971, U.S. Patent 3, 610, 731.

SMITH, P. W., ed., 1981, Special Issue on Optical Bistability, IEEE J. Quantum Electron. **QE-17,** N3.

SMITH, P. W. and E. H. TURNER, 1977, Appl. Phys. Lett. **30,** 280.

SMITH, P. W., W. J. TOMLINSON, P. J. MALONEY and J. P. HERMANN, 1981, IEEE J. Quantum Electron. **QE-17,** 340.

SNAPP, R. R., H. J. CARMICHAEL and W. C. SCHIEVE, 1981, Opt. Commun. **40,** 68.

SPENCER, M. B. and W. E. LAMB, Jr., 1972, Phys. Rev. **A5,** 884.

SPILLER, E., 1972, J. Appl. Phys. **43,** 1673.

STAUPENDAHL, G. and K. A. SCHINDLER, 1982, Opt. Quantum Electron. **14,** 157.

STEYN-ROSS, M. L. and C. W. GARDINER, 1983, Phys. Rev. **A27,** 310.

STRATONOVICH, R. L., 1963, Topics in the Theory of Random Noise (Gordon and Breach, New York).
SZÖKE, A., V. DANEU, J. GOLDHAR and N. A. KURNIT, 1969, Appl. Phys. Lett. **15,** 376.
TEWARI, S. P., 1979, Opt. Acta **26,** 145.
VENKATESAN, T. N. C. and S. L. MCCALL, 1977, Appl. Phys. Lett. **30,** 282.
WALLS, D. F., P. ZOLLER and M. L. STEYN-ROSS, 1981, IEEE J. Quantum Electron. **QE-17,** 380.
WEIDLICH, W. and F. HAAKE, 1965a, Z. Phys. **185,** 30.
WEIDLICH, W. and F. HAAKE, 1965b, Z. Phys. **186,** 203.
WEYER, K. G., H. WIEDENMANN, M. RATEIKE, W. R. MCGILLIVRAY, P. MEYSTRE and H. WALTHER, 1981, Opt. Commun. **37,** 426.
WIGNER, E. P., 1932, Phys. Rev. **40,** 749.
WILLIS, C. R., 1978, Opt. Commun. **26,** 62.
WILLIS, C. R., 1981, Complex Order Parameters in Quantum Optics. First-Order Phase Transitions Analogies, in: BOWDEN *et al.* [1981].
WILLIS, C. R. and J. DAY, 1979, Opt. Commun. **28,** 137.
WINFUL, H. G. and G. D. COOPERMAN, 1982, Appl. Phys. Lett. **40,** 298.
WOO, J. W. F. and R. LANDAUER, 1971, IEEE J. Quantum Electron. **QE-7,** 435.
YUEN, H. P., 1976, Phys. Rev. **A13,** 2226.
ZARDECKI, A., 1980, Phys. Rev. **A22,** 1664.

E. WOLF, PROGRESS IN OPTICS XXI
© ELSEVIER SCIENCE PUBLISHERS B.V. 1984

III

THE RADON TRANSFORM AND ITS APPLICATIONS

BY

HARRISON H. BARRETT

Department of Radiology, Arizona Health Sciences Center, University of Arizona, Tucson, AZ 85724, U.S.A.

and

Optical Sciences Center, University of Arizona, Tucson, AZ 85721, U.S.A.

CONTENTS

§ 1. Introduction

The classic paper by RADON [1917] has had a far-reaching influence in many branches of science. What we now call the Radon transform arises in such diverse fields as X-ray crystallography, radio-astronomy, electron microscopy, diagnostic radiology, nuclear magnetic resonance, microwave scattering, and studies of the Fermi surface in metals. However, it is undoubtedly medical computed tomography (CT) that has attracted more attention than any other application of the Radon transform. The basic data obtained in a CT procedure are X-ray transmission measurements through a two-dimensional (2D) slice of the patient's body. Each measured transmission value is simply related to a line integral of the X-ray attenuation coefficient in this slice, and the set of all line integrals for a particular direction of the X-ray beam is a one-dimensional (1D) "projection" of the 2D object section. The set of all projections for all beam directions is the 2D Radon transform of the object, and reconstruction of the object from these projections is an implementation of the 2D inverse Radon transform.

Because of the overwhelming clinical and commercial success of CT, the 2D Radon transform has been studied in exhaustive detail. Its theoretical basis and practical limitations are very well understood. Much less attention has been given to the 3D Radon transform and its inverse, although many of the applications mentioned above are inherently three-dimensional.

The 3D Radon transform is a set of 1D functions obtained by integrating a 3D object over *planes*, as opposed to the line integrals that make up the 2D Radon transform. However, the structure of the inverse transform is different in the 2D and 3D cases, or, more generally, in spaces of odd and even dimensionality.

It is the purpose of this chapter to review our present state of knowledge about the Radon transform and to discuss some of its lesser-known applications. Although X-ray projections and tomographic reconstruction are used in § 2 as a starting point, this chapter is by no means a review of tomography. No consideration is given to such things as fan-beam geometries, iterative reconstruction algorithms, polyenergetic X-rays, detector noise, and clinical applications. These topics, which would be essential to any balanced treatment of tomography, are discussed in a number of books and review articles. See, for example, HERMAN [1980], BARRETT and SWINDELL [1981], KAK [1979], BROOKS and DICHIRO [1976], GORDON, HERMAN and JOHNSON [1975], or SWINDELL and BARRETT [1977]. On the other hand, the chapter is also not a mathematical treatise in the rigorous theorem-proof sense. Rather, an attempt has been made to present the essential mathematical points in a way that is understandable to a scientist or engineer versed in Fourier theory, and to show that the Radon transform is a very useful tool in many practical applications. Those who require more rigor should consult the work of JOHN [1955], HELGASON [1965, 1980], LUDWIG [1966], SHEPP and KRUSKAL [1978], SMITH, SOLOMON and WAGNER [1977], GEL'FAND, GRAEV and VILENKIN [1966a,b] and DEANS [1978, 1983]. The conference proceedings edited by HERMAN and NATTERER [1981] and the recent treatise by DEANS [1983] contain a wealth of further mathematical references.

§ 2. The two-dimensional Radon transform

2.1. LINE INTEGRALS

It is usually a good approximation to assume that a thin pencil beam of monoenergetic X-rays is attenuated in accordance with Beer's law. For a homogeneous medium of thickness L, this means that

$$I = I_0 \, e^{-\mu L}, \tag{2.1}$$

where I is the X-ray flux transmitted through the medium, I_0 is the flux that would be transmitted in the absence of the medium, and μ is the linear attenuation coefficient. For an inhomogeneous medium such as the human body, $\mu = \mu(\boldsymbol{r})$ is a function of position, and (2.1) must be replaced with

$$I = I_0 \exp\left[- \int_{\mathrm{S}}^{\mathrm{D}} \mu(\boldsymbol{r}) \, \mathrm{d}l \right], \tag{2.2}$$

where the line integral runs from the source position S to the detector position D. In computed tomography, intensity measurements are made for many line-integral paths confined to a plane, and $\boldsymbol{r}$ is therefore a 2D position vector with Cartesian components (x, y).

The logarithm of I/I_0 is linearly related to the line integral:

$$-\ln(I/I_0) = \int_{\mathrm{S}}^{\mathrm{D}} \mu(\boldsymbol{r}) \, \mathrm{d}l. \tag{2.3}$$

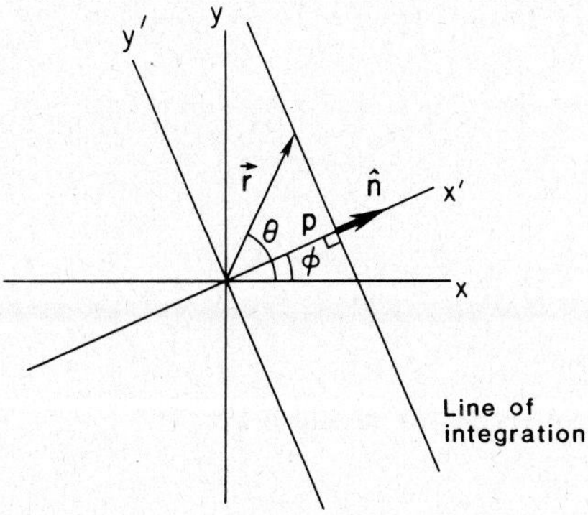

Fig. 1. Geometry for the 2D Radon transform.

Therefore the measured transmissions directly yield values for the line integrals, and it is the goal of CT to reconstruct or estimate $\mu(\boldsymbol{r})$ from a finite set of its line integrals.

Two parameters are required to specify a line of integration (Fig. 1). For example, we can use the angle ϕ between the line and the y-axis and its perpendicular distance from the origin p, or we can use p and the unit normal to the line $\hat{\boldsymbol{n}}$. We can then define the *projection* λ of a general 2D function $f(\boldsymbol{r})$ or $f(x, y)$ in several equivalent ways. If we choose a rotated coordinate system (x', y') such that the line of integration is parallel to the y'-axis, we can write

$$\lambda_\phi(p) \equiv \int_{-\infty}^{\infty} f_{\mathrm{r}}(p, y')\,\mathrm{d}y', \tag{2.4}$$

where $f_{\mathrm{r}}(x', y')$ denotes the same spatial distribution as $f(x, y)$, but expressed in the rotated system, i.e. $f_{\mathrm{r}}(x', y') = f[x(x', y'), y(x', y')]$. Equation (2.4) defines the integral of f along the line $x' = p$.

An equivalent definition of $\lambda_\phi(p)$ is

$$\lambda_\phi(p) = \int_\infty \mathrm{d}^2 r\, f(\boldsymbol{r})\delta(p - \boldsymbol{r}\cdot\hat{\boldsymbol{n}}), \tag{2.5}$$

where the subscript ∞ on the integral indicates that it runs over the infinite plane. The delta function in this equation is one-dimensional and has the effect of reducing the area integral to a line integral along the line $p = \boldsymbol{r}\cdot\hat{\boldsymbol{n}}$. The equivalence of (2.4) and (2.5) is easily demonstrated, since $f(\boldsymbol{r}) = f_{\mathrm{r}}(x', y')$,

$d^2r = dx'\,dy'$, and $x' = \boldsymbol{r}\cdot\hat{\boldsymbol{n}}$. Hence, (2.5) becomes

$$\lambda_\phi(p) = \int_{-\infty}^{\infty} dx' \int_{-\infty}^{\infty} dy'\, f_r(x', y')\delta(p - x')$$

$$= \int_{-\infty}^{\infty} dy'\, f_r(p, y'), \tag{2.6}$$

in agreement with (2.4).

It is sometimes convenient to define a vector

$$\boldsymbol{p} \equiv p\hat{\boldsymbol{n}} \tag{2.7}$$

and to write

$$\lambda_\phi(p) \equiv \lambda(\boldsymbol{p}). \tag{2.8}$$

We shall write $\lambda_\phi(p)$ when we wish to think of the projection as a 1D function with ϕ fixed, and $\lambda(\boldsymbol{p})$ when we wish to think of it as a 2D function. In the latter case, (2.5) may be regarded as an integral transform, the 2D Radon transform, connecting one 2D function, $f(\boldsymbol{r})$, to another, $\lambda(\boldsymbol{p})$. The space in which the vector $\boldsymbol{p}$ is defined is called *Radon space*, and a point in this space has polar coordinates (p, ϕ).

Note, however, that $\lambda(\boldsymbol{p})$ is not well behaved at the origin since $\lambda_{\phi_1}(0) \neq \lambda_{\phi_2}(0)$ (DEANS [1983]; LINDGREN and RATTEY [1981]).

Note also that $\lambda(\boldsymbol{p})$ and $f(\boldsymbol{r})$ share the same circular region of support (LUDWIG [1966]). More precisely, if $f(\boldsymbol{r}) = 0$ for $|\boldsymbol{r}| > R$, then $\lambda(\boldsymbol{p}) = 0$ for $|\boldsymbol{p}| > R$. This statement follows from (2.5) because the argument of the delta function does not vanish for any real $\hat{\boldsymbol{n}}$ if $|\boldsymbol{r}| < R$ and $|\boldsymbol{p}| > R$. More physically, if the line of integration does not pass through the region of support, the integral is zero.

Comparison of (2.6) with (2.3) shows that a single X-ray transmission measurement generates one point in the Radon transform of $\mu(\boldsymbol{r})$. (The limits of integration in (2.3) may be extended to $\pm\infty$ if $\mu(\boldsymbol{r})$ is zero outside a finite region.) In practice, only a finite set of transmission measurements is made, and each measurement suffers some inaccuracy, at the very least that due to shot noise in the X-ray beam. It is therefore impossible to determine $\mu(\boldsymbol{r})$ uniquely, and practical CT reconstruction algorithms seek the "best" estimate of $\mu(\boldsymbol{r})$. These algorithms are discussed fully by HERMAN [1980] and BARRETT and SWINDELL [1981], but they are not the concern of this chapter. Instead, we shall assume that $\lambda(\boldsymbol{p})$ is known exactly for all points in Radon space (i.e., all values of p and ϕ). We seek, therefore, the mathematically exact form of the

2D inverse Radon transform that will enable us to find $f(\boldsymbol{r})$ given $\lambda(\boldsymbol{p})$. For a detailed discussion of the *discrete* Radon transform, when only a finite set of line integrals is known, see LINDGREN and RATTEY [1981].

2.2. THE CENTRAL-SLICE THEOREM

What information about $f(\boldsymbol{r})$ is contained in each 1D projection $\lambda_\phi(p)$? To answer this question, we take the 1D Fourier transform of $\lambda_\phi(\mathrm{p})$:

$$\Lambda_\phi(\nu) = \mathscr{F}_1\{\lambda_\phi(p)\} = \int_{-\infty}^{\infty} \mathrm{d}p\, \lambda_\phi(p) \exp(-2\pi \mathrm{i} \nu p), \tag{2.9}$$

where $\mathscr{F}_m$ is the m-dimensional Fourier operator and ν is the frequency variable conjugate to p. Using (2.4) and changing the variable of integration from p to x', we find

$$\Lambda_\phi(\nu) = \int_{-\infty}^{\infty} \mathrm{d}x' \int_{-\infty}^{\infty} \mathrm{d}y'\, f_\mathrm{r}(x', y') \exp(-2\pi \mathrm{i} \nu x'). \tag{2.10}$$

This result is to be compared to the general definition of the 2D Fourier transform:

$$F(\boldsymbol{\rho}) = \mathscr{F}_2\{f(\boldsymbol{r})\} = \int_\infty \mathrm{d}^2 r\, f(\boldsymbol{r}) \exp(-2\pi \mathrm{i} \boldsymbol{\rho} \cdot \boldsymbol{r})$$

$$= \int_{-\infty}^{\infty} \mathrm{d}x \int_{-\infty}^{\infty} \mathrm{d}y\, f(x, y) \exp[-2\pi \mathrm{i}(\xi x + \eta y)], \tag{2.11}$$

where $\boldsymbol{\rho}$ is a 2D spatial-frequency vector with Cartesian components ξ and η. We shall consistently use capital letters to denote the Fourier transform of the corresponding lower-case function. Comparing (2.11) and (2.10), we see that

$$\Lambda_\phi(\nu) = F_\mathrm{r}(\xi', \eta')\Big|_{\substack{\xi' = \nu \\ \eta' = 0}} = F_\mathrm{r}(\nu, 0), \tag{2.12}$$

where ξ' and η' are the frequency components in a rotated system (Fig. 2) and are therefore conjugate to x' and y', respectively. This important result, known as the central-slice or projection-slice theorem, shows that *the* 1*D Fourier transform of a projection of a* 2*D function is directly one line through the* 2*D Fourier transform of the function itself.* The line is the ξ' axis in the rotated frequency-space coordinates. The entire 2D frequency space can thus be sampled on a set of lines passing through the origin (hence the word "central") by transforming projections at different angles ϕ.

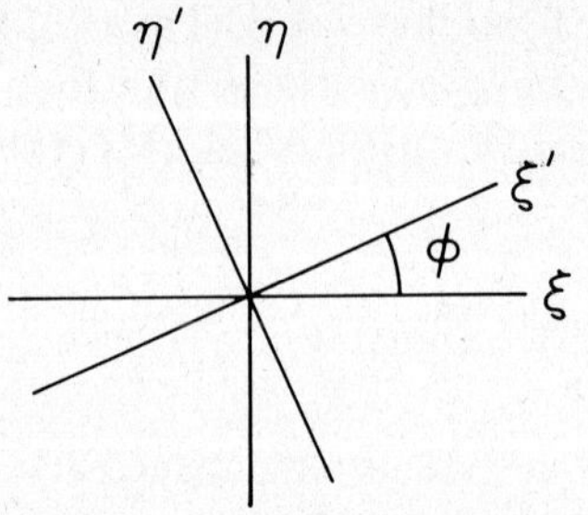

Fig. 2. Geometry of 2D Fourier space, showing rotated coordinates.

An alternative derivation of the central-slice theorem without rotated coordinates proceeds from (2.5). A 1D Fourier transform of this equation yields

$$\begin{aligned}\Lambda_\phi(\nu) &= \int_{-\infty}^{\infty} \mathrm{d}p \exp(-2\pi \mathrm{i} \nu p) \int_\infty \mathrm{d}^2 r\, f(\boldsymbol{r}) \delta(p - \boldsymbol{r}\cdot\hat{\boldsymbol{n}}) \\ &= \int_\infty \mathrm{d}^2 r\, f(\boldsymbol{r}) \exp(-2\pi \mathrm{i}\boldsymbol{r}\cdot\hat{\boldsymbol{n}}\nu) \\ &= F(\boldsymbol{\rho})|_{\boldsymbol{\rho}=\hat{\boldsymbol{n}}\nu} = F(\hat{\boldsymbol{n}}\nu). \end{aligned} \tag{2.13}$$

The condition $\boldsymbol{\rho} = \hat{\boldsymbol{n}}\nu$ is the same as $\xi' = \nu$ since the ξ' axis is parallel to $\hat{\boldsymbol{n}}$, and thus (2.13) is equivalent to (2.12).

With this theorem, we can now more precisely define a "complete" projection data set. The transform $F(\boldsymbol{\rho})$ can be determined for all $\boldsymbol{\rho}$, and hence $f(\boldsymbol{r})$ is uniquely determined, if $\lambda_\phi(p)$ is known for $-\infty < p < \infty$ and $0 < \phi \leq \pi$. It is not necessary for the range in ϕ to exceed π because of the symmetry relation,

$$\lambda_\phi(p) = \lambda_{\phi+\pi}(-p), \tag{2.14}$$

which is easily derived from (2.5) by noting that $\delta(x) = \delta(-x)$ and that the change $\phi \to \phi + \pi$ implies $\hat{\boldsymbol{n}} \to -\hat{\boldsymbol{n}}$. Equation (2.14) also follows from the truism $\boldsymbol{p} = p\hat{\boldsymbol{n}} = (-p)(-\hat{\boldsymbol{n}})$. Hence, polar coordinates (p, ϕ) and $(-p, \phi + \pi)$ denote the same point $\boldsymbol{p}$ in Radon space. Because of (2.14), the projection data set is also complete if $\lambda_\phi(p)$ is known for $0 \leq p < \infty$ and $0 < \phi \leq 2\pi$. In practice, the projection data are usually sampled in both p and ϕ, but we shall treat both as continuous variables.

2.3. THE INVERSE 2D RADON TRANSFORM

The quickest route to the inverse Radon transform is via the central-slice theorem. We need only take the inverse Fourier transform of (2.13), expressing $\boldsymbol{\rho}$ in polar coordinates (ρ, θ_ρ). However, it will simplify the algebra somewhat if we allow ρ to take on positive and negative values and restrict θ_ρ to $(0, \pi)$. Then the form of the 2D inverse Fourier transform is

$$f(\boldsymbol{r}) = \int_{-\infty}^{\infty} |\rho| \, \mathrm{d}\rho \int_0^{\pi} \mathrm{d}\theta_\rho \, F(\boldsymbol{\rho}) \exp(2\pi \mathrm{i} \boldsymbol{\rho} \cdot \boldsymbol{r}). \tag{2.15}$$

We now let $\boldsymbol{\rho} = \hat{\boldsymbol{n}} v$, which implies that $\rho = v$ and $\theta_\rho = \phi$, and use (2.13), yielding

$$f(\boldsymbol{r}) = \int_{-\infty}^{\infty} |v| \, \mathrm{d}v \int_0^{\pi} \mathrm{d}\phi \, \Lambda_\phi(v) \exp(2\pi \mathrm{i} v \boldsymbol{r} \cdot \hat{\boldsymbol{n}}). \tag{2.16}$$

This is one form of the 2D inverse Radon transform since (2.16), along with (2.9), allows us to determine $f(\boldsymbol{r})$ from $\lambda_\phi(p)$.

A more familiar form of the inverse Radon transform is obtained by rewriting (2.16) as

$$f(\boldsymbol{r}) = \int_0^{\pi} \mathrm{d}\phi \left[\int_{-\infty}^{\infty} K_\phi(v) \mathrm{e}^{2\pi \mathrm{i} v p} \, \mathrm{d}v \right]_{p = \boldsymbol{r} \cdot \hat{\boldsymbol{n}}}, \tag{2.17}$$

where

$$K_\phi(v) \equiv |v| \Lambda_\phi(v). \tag{2.18}$$

The v-integral in (2.17) is an inverse 1D Fourier transform, and, by the convolution theorem (GASKILL [1978]), we have

$$f(\boldsymbol{r}) = \int_0^{\pi} \mathrm{d}\phi \left[\lambda_\phi(p) * \mathscr{F}_1^{-1}\{|v|\} \right]_{p = \boldsymbol{r} \cdot \hat{\boldsymbol{n}}}, \tag{2.19}$$

with the asterisk denoting 1D convolution and $\mathscr{F}_m^{-1}$ being the inverse m-dimensional Fourier operator. From LIGHTHILL [1962],

$$\mathscr{F}_1^{-1}\{|v|\} = \frac{-1}{2\pi^2 p^2}, \tag{2.20}$$

and (2.19) therefore becomes

$$f(\boldsymbol{r}) = -\frac{1}{2\pi^2} \int_0^{\pi} \mathrm{d}\phi \left[\lambda_\phi(p) * \frac{1}{p^2} \right]_{p = \boldsymbol{r} \cdot \hat{\boldsymbol{n}}}, \tag{2.21}$$

which is an often-quoted form of the 2D inverse Radon transform.

The function $1/p^2$ in (2.20) and (2.21) must be interpreted as a generalized function. To see that it cannot be taken at face value, suppose that $f(\boldsymbol{r})$ is everywhere nonnegative. Then the projection $\lambda_\phi(p)$ must also be nonnegative and, if we interpret $1/p^2$ blindly as a positive number, the integral in (2.21) is nonnegative. However, there is an overall minus sign in front of the integral, and (2.21) predicts that $f(\boldsymbol{r})$ is negative, in contradiction to our original assumption.

One approach to the interpretation of $1/p^2$ is by use of the central-ordinate theorem of Fourier theory (GASKILL [1978]). We let $H(\nu) = |\nu|$ and denote the inverse Fourier transform of $H(\nu)$ by $h(p)$. Then

$$\int_{-\infty}^{\infty} h(p)\,\mathrm{d}p = H(0) = 0. \tag{2.22}$$

Hence, $h(p)$ cannot be everywhere negative, and (2.20) cannot be correct for all p. If we accept (2.20) for $p \neq 0$, we might be tempted to add a delta function $C\delta(p)$ to (2.20) and adjust C to satisfy (2.22). Then we could write

$$\int_{-\infty}^{\infty} h(p)\,\mathrm{d}p = \lim_{\varepsilon\to 0}\left[\int_{-\infty}^{-\varepsilon}\mathrm{d}p + \int_{\varepsilon}^{\infty}\mathrm{d}p\right]\left(\frac{-1}{2\pi^2p^2}\right) + \lim_{\varepsilon\to 0}\int_{-\varepsilon}^{\varepsilon} C\delta(p)$$

$$= \lim_{\varepsilon\to 0}\left[-\frac{1}{\pi^2\varepsilon} + C\right] = 0. \tag{2.23}$$

For this patch to work, we must let $C \to \infty$ as $\varepsilon \to 0$, which says that the generalized function $1/p^2$ behaves as a negative delta function of infinite weight at $p = 0$.

Another approach to the interpretation of this function is to write

$$|\nu| = \nu \operatorname{sgn} \nu = 2\pi\mathrm{i}\nu \frac{\operatorname{sgn} \nu}{2\pi\mathrm{i}}, \tag{2.24}$$

where

$$\operatorname{sgn} \nu \equiv \begin{cases} +1 & \nu > 0 \\ 0 & \nu = 0 \\ -1 & \nu < 0. \end{cases} \tag{2.25}$$

It is known (BRACEWELL [1965], GASKILL [1978]) that

$$\mathscr{F}_1^{-1}\{2\pi\mathrm{i}\nu\} = \delta'(p) \tag{2.26}$$

and

$$\mathscr{F}_1^{-1}\left\{\frac{\operatorname{sgn} \nu}{2\pi\mathrm{i}}\right\} = \frac{1}{2\pi^2}\mathscr{P}\left(\frac{1}{p}\right), \tag{2.27}$$

where $\delta'(p)$ is the first derivative of a 1D delta function and $\mathscr{P}$ denotes a Cauchy principal value. From these equations and the convolution theorem, we have

$$h(p) = \delta'(p) * \left(\frac{1}{2\pi^2}\right) \mathscr{P}\left(\frac{1}{p}\right), \tag{2.28}$$

Then we can rewrite (2.19) as

$$\begin{aligned} f(\boldsymbol{r}) &= \frac{1}{2\pi^2} \mathscr{P} \int_0^\pi \mathrm{d}\phi \left[\lambda_\phi(p) * \delta'(p) * \frac{1}{p} \right]_{p = \boldsymbol{r}\cdot\hat{\boldsymbol{n}}} \\ &= \frac{1}{2\pi^2} \mathscr{P} \int_0^\pi \mathrm{d}\phi \left[\lambda'_\phi(p) * \frac{1}{p} \right]_{p = \boldsymbol{r}\cdot\hat{\boldsymbol{n}}}, \end{aligned} \tag{2.29}$$

where $\lambda'_\phi(p) \equiv \partial\lambda_\phi(p)/\partial p$, and the last step in (2.29) follows from the definition of $\delta'(p)$. Note that $\mathscr{P}$ refers to the convolution, not the integral over ϕ.

Since convolution with $(-1/\pi)\mathscr{P}(1/p)$ is equivalent to Hilbert transformation (BRACEWELL [1965]), the quantity in brackets in (2.29) is proportional to the Hilbert transform of the derivative of the projection data.

An integration by parts leads from (2.29) back to (2.21) if we interpret $1/p^2$ by

$$\frac{1}{p^2} = -\frac{\mathrm{d}}{\mathrm{d}p} \mathscr{P}\left(\frac{1}{p}\right). \tag{2.30}$$

The origin of the infinite-weight delta now becomes clearer; it represents the derivative of the infinite discontinuity of $\mathscr{P}(1/p)$ at $p = 0$.

Another form of the inverse Radon transform is obtained by writing

$$\frac{\mathrm{d}}{\mathrm{d}p} \ln|p| = \mathscr{P}\left(\frac{1}{p}\right), \tag{2.31}$$

which is clearly correct for $p \neq 0$, and may be regarded as a definition for $p = 0$ (LIGHTHILL [1962]). An integration by parts in (2.29) then yields

$$f(\boldsymbol{r}) = \frac{1}{2\pi^2} \int_0^\pi \mathrm{d}\phi [\lambda''_\phi(p) * \ln|p|]_{p = \boldsymbol{r}\cdot\hat{\boldsymbol{n}}}, \tag{2.32}$$

where $\lambda''_\phi(p) = \partial^2\lambda_\phi(p)/\partial p^2$. It can also be shown that

$$\frac{\mathrm{d}^2 f(x)}{\mathrm{d}x^2} * g(x) = f(x) * \frac{\mathrm{d}^2 g(x)}{\mathrm{d}x^2} = \frac{\mathrm{d}^2}{\mathrm{d}x^2}[f(x) * g(x)], \tag{2.33}$$

where $f(x)$ and $g(x)$ are arbitrary differentiable functions. The proof of this equation follows by Fourier-transforming both sides. Using (2.33), we can rewrite (2.32) as

$$f(\boldsymbol{r}) = \frac{1}{2\pi^2} \int_0^{\pi} \mathrm{d}\phi \left\{ \frac{\partial^2}{\partial p^2} [\lambda_\phi(p) * \ln|p|] \right\}_{p = \boldsymbol{r}\cdot\hat{\boldsymbol{n}}} . \tag{2.34}$$

Still another version of the inverse Radon transform is

$$f(\boldsymbol{r}) = \frac{1}{2\pi^2} \nabla^2 \int_0^{\pi} \mathrm{d}\phi [\lambda_\phi(p) * \ln|p|]_{p = \boldsymbol{r}\cdot\hat{\boldsymbol{n}}} . \tag{2.35}$$

Equation (2.35) follows from (2.34) and the general result,

$$\nabla^2 g(\boldsymbol{r}\cdot\hat{\boldsymbol{n}}) = \left.\frac{\mathrm{d}^2 g(p)}{\mathrm{d}p^2}\right|_{p = \boldsymbol{r}\cdot\hat{\boldsymbol{n}}} = g''(\boldsymbol{r}\cdot\hat{\boldsymbol{n}}), \tag{2.36}$$

which is easily proven by using a Cartesian coordinate system with $\hat{\boldsymbol{n}}$ parallel to one of the axes. Equation (2.36) is valid in any number of dimensions and will prove useful in discussing the 3D Radon transform. (COURANT and HILBERT [1962], p. 681.)

2.4. BACK-PROJECTION AND SUMMATION

In all of the forms of the inverse Radon transform given in the last section, the substitution $p = \boldsymbol{r}\cdot\hat{\boldsymbol{n}}$ appeared. This operation has a simple physical interpretation; it converts a 1D function $g(p)$ into a 2D function $g(\boldsymbol{r}\cdot\hat{\boldsymbol{n}})$ that shows no variation in the direction perpendicular to $\hat{\boldsymbol{n}}$. That is, $g(p)$ is smeared or *back-projected* along the original projection direction, which is the y'-axis in Fig. 1. All points in the $\boldsymbol{r}$-plane with the same x'-coordinate are assigned the same value $g(x')$, since $x' = \boldsymbol{r}\cdot\hat{\boldsymbol{n}}$.

The *summation image* $b(\boldsymbol{r})$ results when all projections $\lambda_\phi(p)$ are back-projected and the resulting 2D image is summed over all projection angles ϕ. Since we are treating the data set as continuous, the sum becomes an integral and we have

$$b(\boldsymbol{r}) \equiv \int_0^{\pi} \mathrm{d}\phi\, \lambda_\phi(\boldsymbol{r}\cdot\hat{\boldsymbol{n}}). \tag{2.37}$$

An illustration of a summation image is given in Fig. 3. It is almost always a poor representation of $f(\boldsymbol{r})$.

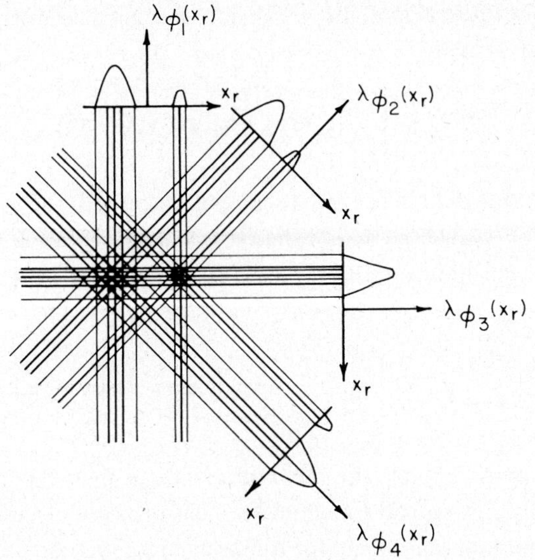

Fig. 3. Illustration of a summation image, from BARRETT and SWINDELL [1981]. The original object was two discs, and only four projections are shown. The rotated coordinate x_r corresponds to p in the text.

To see the relationship between $f(\boldsymbol{r})$ and $b(\boldsymbol{r})$, it is convenient to consider a point object,

$$f^{\delta}(\boldsymbol{r}) \equiv \delta(\boldsymbol{r} - \boldsymbol{r}_0), \tag{2.38}$$

where $\delta(\boldsymbol{r} - \boldsymbol{r}_0)$ is a 2D delta function. By (2.5), the projections of this object are

$$\lambda_{\phi}(p) = \int_{\infty} \mathrm{d}^2 r\, \delta(\boldsymbol{r} - \boldsymbol{r}_0)\delta(p - \boldsymbol{r} \cdot \hat{\boldsymbol{n}}) = \delta(p - \boldsymbol{r}_0 \cdot \hat{\boldsymbol{n}}). \tag{2.39}$$

Thus each projection of a 2D delta function is a 1D delta function. (Note that we distinguish 1D and 2D delta functions only by their arguments; if the argument is a scalar, the delta function is 1D, and if the argument is an mD vector, the delta function is mD.) Using (2.39) in (2.37), we see that the summation image is given by

$$b^{\delta}(\boldsymbol{r}) = \int_0^{\pi} \mathrm{d}\phi\, \delta[(\boldsymbol{r} - \boldsymbol{r}_0) \cdot \hat{\boldsymbol{n}}], \tag{2.40}$$

where the superscript δ is a reminder that we are dealing with a particular point object.

To perform the integral in (2.40), we denote the angle between $\boldsymbol{r} - \boldsymbol{r}_0$ and the x-axis by ψ and write

$$b^{\delta}(\boldsymbol{r}) = \int_0^{\pi} \mathrm{d}\phi\, \delta[\,|\boldsymbol{r} - \boldsymbol{r}_0| \cos(\psi - \phi)]. \tag{2.41}$$

The argument of the delta function is zero exactly once in the range $0 < \phi \leq \pi$, namely when $\phi = \phi_0$, where ϕ_0 is either $\psi \pm \pi/2$ or $\psi \pm 3\pi/2$, whichever falls in $(0, \pi)$. In the neighborhood of ϕ_0, $\cos(\psi - \phi) = \pm(\phi - \phi_0)$, and we have

$$b^{\delta}(\boldsymbol{r}) = \int_0^{\pi} \mathrm{d}\phi\, \delta[\pm|\boldsymbol{r} - \boldsymbol{r}_0|(\phi - \phi_0)] = \frac{1}{|\boldsymbol{r} - \boldsymbol{r}_0|}, \tag{2.42}$$

since $\delta(\alpha x) = \delta(x)/|\alpha|$ for α constant.

Thus, for a point object, the summation image has the cusp-like form $|\boldsymbol{r} - \boldsymbol{r}_0|^{-1}$. The point spread function for back-projection and summation is $|\boldsymbol{r}|^{-1}$, and the summation image for a more general object is

$$b(\boldsymbol{r}) = \int_{\infty} \mathrm{d}^2 r_0\, f(\boldsymbol{r}_0) \frac{1}{|\boldsymbol{r} - \boldsymbol{r}_0|} = f(\boldsymbol{r}) ** \frac{1}{|\boldsymbol{r}|}, \tag{2.43}$$

where the double asterisk denotes 2D convolution.

The blur function $1/|\boldsymbol{r}|$ can be removed by suitably filtering $b(\boldsymbol{r})$. Since $1/|\boldsymbol{r}|$ is rotationally symmetric, its 2D Fourier transform becomes a Hankel transform (GASKILL [1978]) given by

$$\mathscr{F}_2\left\{\frac{1}{|\boldsymbol{r}|}\right\} = 2\pi \int_0^{\infty} r\,\mathrm{d}r \cdot \frac{1}{r} J_0(2\pi\rho r) = \frac{1}{|\rho|} \int_0^{\infty} J_0(u)\,\mathrm{d}u = \frac{1}{|\rho|}, \tag{2.44}$$

where $u = 2\pi\rho r$ and $J_0(u)$ is the zero-order Bessel function of the first kind. Thus the frequency-domain counterpart of (2.43) is

$$B(\boldsymbol{\rho}) = F(\boldsymbol{\rho}) \cdot \frac{1}{|\rho|}. \tag{2.45}$$

Multiplication of (2.45) by $|\rho|$ and inverse Fourier transformation yields

$$f(\boldsymbol{r}) = \mathscr{F}_2^{-1}\{|\rho| B(\boldsymbol{\rho})\}. \tag{2.46}$$

This result, along with the definition (2.37) of $b(\boldsymbol{r})$, constitutes yet another form of the 2D inverse Radon transform.

It is interesting to compare (2.46) to (2.16). The steps involved in going from $\lambda_\phi(p)$ to $f(\boldsymbol{r})$ by use of (2.16) are:

1. Fourier transform (1D);
2. filter (1D) with transfer function $|\nu|$;
3. inverse Fourier transform (1D);
4. back-project (1D → 2D);
5. integrate over ϕ.

The corresponding steps with (2.46) and (2.37) are:

1. back-project (1D → 2D);
2. integrate over ϕ;
3. Fourier transform (2D);
4. filter (2D) with transfer function $|\rho|$;
5. inverse Fourier transform (2D).

Note that all of the same steps are involved, but in a different order. In the first case, the filtering occurs before back-projection and is therefore 1D, while in the second case it is after back-projection and therefore 2D. It is the central-slice theorem that guarantees the equivalence of 1D and 2D filtering. In both cases, of course, the filtering could also be carried out in the space domain.

In practice, neither the 1D nor the 2D filter can have the ideal form, $|\nu|$ or $|\rho|$, since that would imply an infinite boost as the frequency goes to infinity. Instead, the filter must be rolled off at high frequencies or *apodized* by some function $A(\nu)$ or $A(\rho)$. The resulting counterparts of eqs. (2.16) and (2.46) are no longer mathematically exact inverse transforms, but they perform better with real, noisy input data since they do not attempt an infinite boost. The literature on apodizing functions is, however, rapidly approaching infinity, and we shall not add to it here.

2.5. SUMMARY IN OPERATOR FORM

The 2D Radon transform may be written in shorthand as

$$\lambda = \mathscr{R}_2\{f\}, \tag{2.47}$$

where λ represents the set of functions $\lambda_\phi(p)$ for all ϕ and p, f represents $f(\boldsymbol{r})$ for all $\boldsymbol{r}$, and $\mathscr{R}_2$ is the integral operator defined by (2.4) or (2.5). All of the equivalent forms of the inverse Radon transform are then implementations of the operator $\mathscr{R}_2^{-1}$, where

$$f = \mathscr{R}_2^{-1}\{\lambda\}. \tag{2.48}$$

We have already introduced the 1D and 2D Fourier-transform operators, in terms of which (cf. (2.10) and (2.11)),

$$\Lambda = \mathscr{F}_1\{\lambda\}; \qquad F = \mathscr{F}_2\{f\}. \tag{2.49}$$

With these operators, the central slice theorem (2.12) or (2.13) reads

$$\mathscr{F}_2\{f\} = \mathscr{F}_1\,\mathscr{R}_2\{f\}, \tag{2.50a}$$

or, as a pure operator relation,

$$\mathscr{F}_2 = \mathscr{F}_1\,\mathscr{R}_2. \tag{2.50b}$$

By the usual rules for operator manipulation, (2.50b) becomes

$$\mathscr{R}_2^{-1} = \mathscr{F}_2^{-1}\,\mathscr{F}_1, \tag{2.51}$$

which is just (2.16).

To express the other equivalent forms of $\mathscr{R}_2^{-1}$, we must define two more operators. The 2D back-projection and summation operator $\mathscr{B}_2$ involves the substitution $p = \boldsymbol{r}\cdot\hat{\boldsymbol{n}}$, which is back-projection itself, and integration over ϕ ("summation"). Thus (2.37) may be written

$$b = \mathscr{B}_2\{\lambda\}, \tag{2.52}$$

where $b = b(\boldsymbol{r})$. Mathematically, $\mathscr{B}_2$ is the adjoint of $\mathscr{R}_2$ (LUDWIG [1966]).

The other operator we need is the Hilbert-transform operator $\mathscr{H}$, given by (BRACEWELL [1965])

$$\mathscr{H} = \frac{-1}{\pi p} *, \tag{2.53}$$

where the Cauchy principal value of $1/p$ is understood, and the asterisk still denotes 1D convolution.

Equations (2.17) and (2.21) may now be written

$$\mathscr{R}_2^{-1} = \mathscr{B}_2\,\mathscr{F}_1^{-1}|\nu|\,\mathscr{F}_1 = -\frac{1}{2\pi^2}\,\mathscr{B}_2\left[\frac{1}{p^2} *\right], \tag{2.54}$$

depending on whether the 1D convolution is to be carried out in the p-domain or the ν-domain. When applied to λ, (2.54) shows that the 1D projections must each be convolved with the generalized function $1/p^2$, then back-projected and summed in order to reconstruct f.

Equation (2.29) becomes

$$\mathscr{R}_2^{-1} = \frac{1}{2\pi^2}\mathscr{B}_2\left[\frac{1}{p}*\right]\frac{\partial}{\partial p} = -\frac{1}{2\pi}\mathscr{B}_2\mathscr{H}\frac{\partial}{\partial p}, \tag{2.55}$$

while (2.32) and (2.35) become

$$\mathscr{R}_2^{-1} = \frac{1}{2\pi^2}\mathscr{B}_2[\ln|p|*]\frac{\partial^2}{\partial p^2} = \frac{1}{2\pi^2}\nabla^2\mathscr{B}_2[\ln|p|*]. \tag{2.56}$$

To show the consistency of these relations, let us retrace the route from (2.56) to (2.54) with operators. The only additional information needed is the 1D Fourier transform of $\ln|p|$ (LIGHTHILL [1962]):

$$\mathscr{F}_1\{\ln|p|\} = \frac{-1}{2|\nu|}, \tag{2.57}$$

or, in operator form

$$\mathscr{F}_1[\ln|p|*] = \frac{-1}{2|\nu|}\mathscr{F}_1. \tag{2.58}$$

Note that the square brackets around $[\ln|p|*]$ indicate that it is an *operator*, while the curly brackets in (2.57) indicate that $\ln|p|$ is the *operand* of $\mathscr{F}_1$. The equivalence of (2.57) and (2.58) follows from the usual Fourier convolution theorem (GASKILL [1978]), and may easily be seen by operating on $\delta(p)$ with (2.58).

The Fourier derivative theorem (GASKILL, 1978) in operator form is

$$\mathscr{F}_1\frac{\partial^n}{\partial p^n} = (2\pi i\nu)^n\mathscr{F}_1. \tag{2.59}$$

Using (2.58) and (2.59) in the first form of (2.56), we find

$$\begin{aligned}\mathscr{R}_2^{-1} &= \frac{1}{2\pi^2}\mathscr{B}_2[\ln|p|*]\frac{\partial^2}{\partial p^2}\\ &= \frac{1}{2\pi^2}\mathscr{B}_2\mathscr{F}_1^{-1}\left[\frac{-1}{2|\nu|}\right]\mathscr{F}_1\mathscr{F}_1^{-1}[-4\pi^2\nu^2]\mathscr{F}_1\\ &= \frac{1}{2\pi^2}\mathscr{B}_2\mathscr{F}_1^{-1}[2\pi^2|\nu|]\mathscr{F}_1,\end{aligned} \tag{2.60}$$

which is just (2.54).

§ 3. The Three-Dimensional Radon Transform

3.1. PLANAR INTEGRALS

Three independent parameters are required to specify a plane in a 3D space. For example, we can use the unit normal $\hat{\mathbf{n}}$ and the distance from the origin p as shown in Fig. 4. The three parameters in this case are p and the polar angles θ_n and ϕ_n that specify $\hat{\mathbf{n}}$ (Fig. 4b), and the equation of the plane is $p = \mathbf{r}\cdot\hat{\mathbf{n}}$, where sans-serif type indicates 3D quantities, and $\mathbf{r} = (x, y, z)$ is the general 3D position vector. In a rotated coordinate system (x', y', z') such that x' is normal to the plane, the equation of the plane is $p = x'$.

An integral of a 3D function $f(\mathbf{r})$ or $f(x, y, z)$ over this plane is given by

$$\lambda_{\hat{n}}(p) = \int_{-\infty}^{\infty} \mathrm{d}y' \int_{-\infty}^{\infty} \mathrm{d}z'\; f_{\mathrm{r}}(p, y', z') = \int_{\infty} \mathrm{d}^3 r\, f(\mathbf{r})\delta(p - \mathbf{r}\cdot\hat{\mathbf{n}}), \tag{3.1}$$

where $f_{\mathrm{r}}(x', y', z')$ is the same physical distribution as $f(x, y, z)$, but expressed in the rotated coordinates.

As in the 2D case, it is sometimes convenient to define a 3D Radon space, with

$$\mathbf{p} = p\hat{\mathbf{n}} \tag{3.2}$$

specifying a point in that space. The spherical polar components of the vector $\mathbf{p}$ are (p, θ_n, ϕ_n). We shall write $\lambda(\mathbf{p})$ when we wish to regard λ as a 3D function and $\lambda_{\hat{n}}(p)$ when we wish to regard it as a set of 1D functions. In either case,

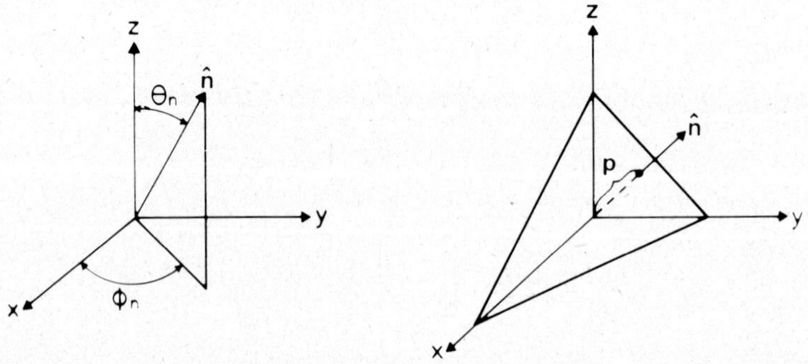

Fig. 4. (a) Geometry for the 3D Radon transform, showing a plane of integration with unit normal $\hat{\mathbf{n}}$ and perpendicular distance from the origin p. (b) The vector $\hat{\mathbf{n}}$ is specified by polar coordinates θ_n and ϕ_n.

(3.1) defines the 3D Radon transform, and we can write it symbolically as

$$\lambda = \mathcal{R}_3\{f\}, \tag{3.3}$$

where $\mathcal{R}_3$ is the 3D Radon operator.

Note that, as in the 2D case, $\lambda(\boldsymbol{p})$ and $f(\boldsymbol{r})$ share the same spherical region of support. If $f(\boldsymbol{r}) = 0$ for $|\boldsymbol{r}| > R$, then $\lambda(\boldsymbol{p}) = 0$ for $|\boldsymbol{p}| > R$.

3.2. THE CENTRAL-SLICE THEOREM

The derivation of the 3D central-slice theorem exactly follows the 2D derivation in § 2.2. A 1D Fourier transform of the first form of (3.1) is

$$\Lambda_{\hat{n}}(\nu) = \int_{-\infty}^{\infty} \mathrm{d}p\, \lambda_{\hat{n}}(p) \exp(-2\pi \mathrm{i} \nu p). \tag{3.4}$$

Letting $p = x'$ and using (3.1), we get

$$\begin{aligned} \Lambda_{\hat{n}}(\nu) &= \int_{-\infty}^{\infty} \mathrm{d}x' \int_{-\infty}^{\infty} \mathrm{d}y' \int_{-\infty}^{\infty} \mathrm{d}z'\, f_{\mathrm{r}}(x', y', z') \exp(-2\pi \mathrm{i} \nu x') \\ &= F_{\mathrm{r}}(\nu, 0, 0), \end{aligned} \tag{3.5}$$

where $F_{\mathrm{r}}(\xi', \eta', \zeta')$ is the 3D Fourier transform of $f_{\mathrm{r}}(x', y', z')$. In general, the 3D Fourier transform is defined by

$$F(\boldsymbol{\sigma}) = F(\xi, \eta, \zeta) = \int_{\infty} \mathrm{d}^3 r\, f(\boldsymbol{r}) \exp(-2\pi \mathrm{i} \boldsymbol{\sigma} \cdot \boldsymbol{r}). \tag{3.6}$$

Equation (3.5) is the 3D central-slice theorem in rotated coordinates, the 3D counterpart of (2.12).

An alternative expression of this theorem without rotated coordinates is found from a 1D Fourier transformation of the second form of (3.1):

$$\begin{aligned} \Lambda_{\hat{n}}(\nu) &= \int_{-\infty}^{\infty} \mathrm{d}p \exp(-2\pi \mathrm{i} p \nu) \int_{\infty} \mathrm{d}^3 r\, f(\boldsymbol{r}) \delta(p - \boldsymbol{r} \cdot \hat{\boldsymbol{n}}) \\ &= \int_{\infty} \mathrm{d}^3 r\, f(\boldsymbol{r}) \exp(-2\pi \mathrm{i} \boldsymbol{r} \cdot \hat{\boldsymbol{n}} \nu) \\ &= F(\hat{\boldsymbol{n}} \nu). \end{aligned} \tag{3.7}$$

Both (3.6) and (3.7) show that *the 1D Fourier transform of a planar projection of a 3D function is one line through the 3D Fourier transform of the function itself.*

The orientation of this line in 3D Fourier space is determined by $\hat{\boldsymbol{n}}$, the normal to the original plane of integration.

If $\hat{\boldsymbol{n}}$ explores all directions in a hemisphere and p ranges from $-\infty$ to ∞, enough information is obtained to determine $F(\boldsymbol{\sigma})$ and hence $f(\boldsymbol{r})$. Of course, physically meaningful functions have bounded support, so only a finite range in p is needed in practice. If measurements are made only for positive p, then $\hat{\boldsymbol{n}}$ must explore a full sphere of directions to completely determine $F(\boldsymbol{\sigma})$.

3.3. THE INVERSE 3D RADON TRANSFORM

As in the 2D case, the shortest route to the inverse 3D Radon transform is via the central-slice theorem. We must take a 3D inverse Fourier transform of (3.7) in polar coordinates, where $\boldsymbol{\sigma}$ has components $(\sigma, \theta_\sigma, \phi_\sigma)$. We allow σ to take on positive and negative values and restrict the direction of $\boldsymbol{\sigma}$ to a hemisphere. Then the form of the 3D inverse Fourier transform is

$$f(\boldsymbol{r}) = \int_{-\infty}^{\infty} \sigma^2 \, \mathrm{d}\sigma \int_{2\pi} \mathrm{d}\Omega_\sigma \, F(\boldsymbol{\sigma}) \exp(2\pi \mathrm{i} \boldsymbol{\sigma} \cdot \boldsymbol{r}), \tag{3.8}$$

where $\mathrm{d}\Omega_\sigma = \sin\theta_\sigma \, \mathrm{d}\theta_\sigma \, \mathrm{d}\phi_\sigma$, and the angular integral extends over 2π ster. In order to use (3.7), we let $\boldsymbol{\sigma} = \hat{\boldsymbol{n}}\nu$, which implies that $\sigma = \nu$, $\theta_n = \theta_\sigma$, $\phi_n = \phi_\sigma$. Then (3.8) becomes [cf. (2.16)]:

$$f(\boldsymbol{r}) = \int_{-\infty}^{\infty} \nu^2 \, \mathrm{d}\nu \int_{2\pi} \mathrm{d}\Omega_n \, \Lambda_{\hat{n}}(\nu) \exp(2\pi \mathrm{i} \nu \boldsymbol{r} \cdot \hat{\boldsymbol{n}}), \tag{3.9}$$

which is one form of the 3D inverse Radon transform.

An equivalent form is obtained by rewriting (3.9) as

$$\begin{aligned} f(\boldsymbol{r}) &= \int_{2\pi} \mathrm{d}\Omega_n \, \mathscr{F}_1^{-1}\{\nu^2 \Lambda_{\hat{n}}(\nu)\}_{p=\boldsymbol{r}\cdot\hat{\boldsymbol{n}}} \\ &= -\frac{1}{4\pi^2} \int_{2\pi} \mathrm{d}\Omega_n [\lambda''_{\hat{n}}(p)]_{p=\boldsymbol{r}\cdot\hat{\boldsymbol{n}}} \\ &= -\frac{1}{4\pi^2} \int_{2\pi} \mathrm{d}\Omega_n \, \lambda''_{\hat{n}}(\boldsymbol{r}\cdot\hat{\boldsymbol{n}}), \end{aligned} \tag{3.10}$$

where $\lambda''_{\hat{n}}(p) = \partial^2 \lambda_{\hat{n}}(p)/\partial p^2$, and the 1D Fourier derivative theorem (2.59) has been used.

Another form of the inverse transform follows at once from (3.10) and (2.36):

$$f(\boldsymbol{r}) = -\frac{1}{4\pi^2}\nabla^2 \int_{2\pi} \mathrm{d}\Omega_n\, \lambda_{\hat{n}}(\boldsymbol{r}\cdot\hat{\boldsymbol{n}}). \tag{3.11}$$

Note that (3.10) and (3.11) are somewhat simpler than their 2D counterparts, (2.32) and (2.35), since the convolution with $\ln|p|$ is not required in 3D.

3.4. BACK-PROJECTION AND SUMMATION

The substitution $p = \boldsymbol{r}\cdot\hat{\boldsymbol{n}}$ in (3.10) and (3.11) is 3D-back-projection. It assigns the same value to all points on the plane $p = \boldsymbol{r}\cdot\hat{\boldsymbol{n}}$, which is the original plane of integration. When each 1D projection $\lambda_{\hat{n}}(p)$ is back-projected and the resulting 3D function is summed (integrated) over all directions of $\hat{\boldsymbol{n}}$, the result is the 3D summation image $b(\boldsymbol{r})$, defined by [cf. (2.37)]

$$b(\boldsymbol{r}) \equiv \int_{2\pi} \mathrm{d}\Omega_n\, \lambda_{\hat{n}}(\boldsymbol{r}\cdot\hat{\boldsymbol{n}}). \tag{3.12}$$

Of course, $b(\boldsymbol{r})$ is not an accurate image of $f(\boldsymbol{r})$. To find the relationship between $b(\boldsymbol{r})$ and $f(\boldsymbol{r})$, we again consider a point object as in § 2.4:

$$f^{\delta}(\boldsymbol{r}) = \delta(\boldsymbol{r} - \boldsymbol{r}_0). \tag{3.13}$$

By arguments similar to those of (2.38)–(2.42), we can show (CHIU [1980]) that the projection of this 3D delta function is the 1D delta function,

$$\lambda_{\hat{n}}(p) = \delta(p - \boldsymbol{r}_0\cdot\hat{\boldsymbol{n}}), \tag{3.14}$$

and that the corresponding summation image is

$$b^{\delta}(\boldsymbol{r}) = \frac{\pi}{|\boldsymbol{r} - \boldsymbol{r}_0|}. \tag{3.15}$$

Thus, the point spread function in the summation image has the same functional form as in the 2D case [cf. (2.42)], and the summation image for a more general object is [cf. (2.43)]

$$b(\boldsymbol{r}) = \pi \int_{\infty} \mathrm{d}^3 r_0\, f(\boldsymbol{r}_0)\frac{1}{|\boldsymbol{r} - \boldsymbol{r}_0|}$$

$$= \pi f(\boldsymbol{r}) *** \frac{1}{|\boldsymbol{r}|}, \tag{3.16}$$

where the triple asterisk denotes 3D convolution.

One way of recovering $f(\boldsymbol{r})$ from $b(\boldsymbol{r})$ is by 3D frequency-domain filtering. The 3D Fourier transform of (3.16) is

$$\mathscr{B}(\boldsymbol{\sigma}) = \pi F(\boldsymbol{\sigma}) \mathscr{F}_3 \left\{ \frac{1}{|\boldsymbol{r}|} \right\}. \tag{3.17}$$

From Chiu [1980],

$$\mathscr{F}_3 \left\{ \frac{1}{|\boldsymbol{r}|} \right\} = \frac{1}{\pi \sigma^2}, \tag{3.18}$$

so that the required filter transfer function is proportional to σ^2, and

$$f(\boldsymbol{r}) = \mathscr{F}_3^{-1}\{F(\boldsymbol{\sigma})\} = \mathscr{F}_3^{-1}\{\sigma^2 \mathscr{B}(\boldsymbol{\sigma})\}. \tag{3.19}$$

Note, however, that multiplication by $-4\pi^2\sigma^2$ in the 3D frequency domain is the same as taking the Laplacian in the space domain. Therefore,

$$f(\boldsymbol{r}) = -\frac{1}{4\pi^2} \nabla^2 b(\boldsymbol{r}), \tag{3.20}$$

which, with (3.12), is the result previously derived in (3.11). Still another route to the same result is by applying the operator ∇^2 to (3.16), taking it under the integral sign, and recognizing that

$$\nabla^2 \frac{1}{|\boldsymbol{r} - \boldsymbol{r}_0|} = -4\pi\delta(\boldsymbol{r} - \boldsymbol{r}_0), \tag{3.21}$$

which is simply Poisson's equation for a point charge. The delta function in (3.21) is then used to perform the integral in (3.16), again yielding (3.20). The Laplacian therefore undoes the $1/|\boldsymbol{r}|$ blurring brought about by back-projection and summation.

It is interesting to note that the filtering operations required in the 3D Radon transform are *local*; they are simple second derivatives; ∇^2 or $\partial^2/\partial p^2$, that can be computed using only values of the function in the immediate neighborhood of the point at which the derivative is evaluated. By contrast, the 2D inverse transform requires convolution with either $1/p^2$, $1/p$, or $\ln|p|$, all of which have tails extending to $\pm\infty$. Basically, the local character of the 3D space-domain filter comes from the smoothness of the frequency-domain filter σ^2 or ν^2, while the slope discontinuity in the 2D filter, $|\rho|$ or $|\nu|$, causes the tails in the 2D space domain.

In practice, an exact Laplacian or second derivative would not be attempted because it would infinitely amplify the noise. Instead, some numerical approximation to the derivative or an apodized frequency-domain filter (CHIU, BARRETT and SIMPSON [1980]) would be used, but the resulting space-domain filter would still be much more compact than in the 2D case.

3.5. SUMMARY IN OPERATOR FORM

The 3D Radon transform is described by the operator $\mathscr{R}_3$,

$$\lambda = \mathscr{R}_3\{f\}. \tag{3.22}$$

The 3D central-slice theorem (3.5) or (3.7) in operator notation is

$$\mathscr{F}_3 = \mathscr{F}_1\,\mathscr{R}_3, \tag{3.23}$$

from which it follows that [cf. (3.9)]

$$\mathscr{R}_3^{-1} = \mathscr{F}_3^{-1}\,\mathscr{F}_1. \tag{3.24}$$

The summation image $b(\boldsymbol{r})$ is found by applying the operator $\mathscr{B}_3$ to the projection set as in (3.12), which becomes

$$b = \mathscr{B}_3\{\lambda\}. \tag{3.25}$$

With this operator, (3.10) and (3.11) may be written as

$$\begin{aligned} \mathscr{R}_3^{-1} &= -\frac{1}{4\pi^2}\,\mathscr{B}_3\,\frac{\partial^2}{\partial p^2} \\ &= -\frac{1}{4\pi^2}\,\nabla^2\,\mathscr{B}_3. \end{aligned} \tag{3.26}$$

If it is desired to carry out the filtering in the frequency domain, as in (3.10) or (3.19), we write [cf. (2.54)]

$$\mathscr{R}_3^{-1} = \mathscr{B}_3\,\mathscr{F}_1^{-1}\,\nu^2\,\mathscr{F}_1 = \mathscr{F}_3^{-1}\,\sigma^2\,\mathscr{F}_3\,\mathscr{B}_3, \tag{3.27}$$

where ν^2 and σ^2 are multiplicative operators in Fourier space.

§ 4. Related Transforms

4.1. THE GENERAL m-DIMENSIONAL RADON TRANSFORM

In the mathematical literature, considerable attention has been given to the general Radon transform in an m-dimensional Euclidean space. We shall briefly summarize this work here; for more detail, see DEANS [1983], LUDWIG [1966], JOHN [1955], HELGASON [1980], and COURANT and HILBERT [1962].

By analogy with (2.5) or (3.1), the m-dimensional (mD) Radon transform is defined by

$$\lambda_{\hat{\boldsymbol{n}}}(p) = \int_{\infty} \mathrm{d}^m r\, f(\boldsymbol{r})\delta(p - \boldsymbol{r}\cdot\hat{\boldsymbol{n}}), \tag{4.1}$$

where $\boldsymbol{r}$ is the general mD position vector, $\hat{\boldsymbol{n}}$ is an mD unit vector, and p is a scalar. Thus, $\delta(p - \boldsymbol{r}\cdot\hat{\boldsymbol{n}})$ is a 1D delta function that reduces the mD volume integral in (4.1) to an integral over the $(m-1)$-dimensional hyperplane having the equation $p = \boldsymbol{r}\cdot\hat{\boldsymbol{n}}$. In operator form, (4.1) becomes

$$\lambda = \mathscr{R}_m\{f\}. \tag{4.2}$$

The mD central-slice theorem is readily obtained by taking the 1D Fourier transform of (4.1), yielding

$$\begin{aligned}\Lambda_{\hat{\boldsymbol{n}}}(\nu) &= \mathscr{F}_1\{\lambda_{\hat{\boldsymbol{n}}}(p)\}\\ &= \int_{-\infty}^{\infty} \mathrm{d}p\, \mathrm{e}^{-2\pi i\nu p} \int_{\infty} \mathrm{d}^m r\, f(\boldsymbol{r})\delta(p - \boldsymbol{r}\cdot\hat{\boldsymbol{n}})\\ &= \int_{\infty} \mathrm{d}^m r\, f(\boldsymbol{r})\mathrm{e}^{-2\pi i\boldsymbol{r}\cdot\hat{\boldsymbol{n}}\nu}\\ &= F(\hat{\boldsymbol{n}}\nu),\end{aligned} \tag{4.3}$$

where $F(\boldsymbol{\rho})$ is the mD Fourier transform of $f(\boldsymbol{r})$. (Note that we are using the same notation in mD as in 2D.) Thus, once again, the 1D Fourier transform of a projection of a function yields one line through the mD transform of the function itself. The operator equivalent of (4.3) is [cf. (2.50) and (3.23)]:

$$\mathscr{F}_m = \mathscr{F}_1\,\mathscr{R}_m. \tag{4.4}$$

The mD inverse Radon transform follows from an mD inverse Fourier transform of (4.3). As in §§ 2.3 and 3.3, we allow the frequency variable ρ to take on both positive and negative values, so that the general mD inverse

Fourier transform is

$$f(\boldsymbol{r}) = \int_{-\infty}^{\infty} |\rho|^{m-1}\, \mathrm{d}\rho \int_{\frac{1}{2}\omega_m} \mathrm{d}\Omega_\rho\, F(\boldsymbol{\rho}) \exp(2\pi \mathrm{i} \boldsymbol{\rho}\cdot\boldsymbol{r}), \tag{4.5}$$

where $\mathrm{d}\Omega_\rho$ is the element of solid angle associated with the vector $\boldsymbol{\rho}$, and ω_m, the solid angle subtended by a full sphere in the mD hyperspace, is given by (JOHN [1955])

$$\omega_m = \frac{2\sqrt{\pi^m}}{\Gamma(m/2)}. \tag{4.6}$$

Note that $\omega_2 = 2\pi$ and $\omega_3 = 4\pi$ since $\Gamma(1) = 1$, and $\Gamma(3/2) = \sqrt{\pi/2}$.

We next let $\boldsymbol{\rho} = \hat{\boldsymbol{n}}\nu$, implying that $|\rho| = |\nu|$ and $\mathrm{d}\Omega_\rho = \mathrm{d}\Omega_n$. Combining (4.3) and (4.5) yields [cf. (3.9)]

$$f(\boldsymbol{r}) = \mathscr{R}_m^{-1}\{\lambda\} = \int_{-\infty}^{\infty} |\nu|^{m-1}\, \mathrm{d}\nu \int_{\frac{1}{2}\omega_m} \mathrm{d}\Omega_n \Lambda_{\hat{\boldsymbol{n}}}(\nu) \exp(2\pi\mathrm{i}\nu \boldsymbol{r}\cdot\hat{\boldsymbol{n}}). \tag{4.7}$$

At this point, a distinction must be made between odd and even m. For odd m, $|\nu|^{m-1} = \nu^{m-1}$, and the 1D Fourier derivative theorem (2.59) allows us to write [cf. (3.10)]

$$f(\boldsymbol{r}) = \left(\frac{1}{2\pi\mathrm{i}}\right)^{m-1} \int_{\frac{1}{2}\omega_m} \mathrm{d}\Omega_n \left[\frac{\partial^{m-1}}{\partial p^{m-1}} \lambda_{\hat{\boldsymbol{n}}}(p)\right]_{p=\boldsymbol{r}\cdot\hat{\boldsymbol{n}}}, \qquad (m \text{ odd}), \tag{4.8}$$

Thus, the mD inverse Radon transform in this case is implemented by differentiating each 1D projection $m-1$ times, back-projecting (substituting $\boldsymbol{r}\cdot\hat{\boldsymbol{n}}$ for p), and integrating over all projection directions $\hat{\boldsymbol{n}}$. In operator form [cf. (3.26)],

$$\mathscr{R}_m^{-1} = \left(\frac{1}{2\pi\mathrm{i}}\right)^{m-1} \mathscr{B}_m \frac{\partial^{m-1}}{\partial p^{m-1}}, \qquad (m \text{ odd}), \tag{4.9}$$

where $\mathscr{B}_m$ is the operator for mD back-projection and summation.

The case of even m is slightly more complicated because then $|\nu|^{m-1} = \nu^{m-1} \operatorname{sgn} \nu$. Multiplication of a 1D function by sgn ν in the frequency domain is equivalent to $-$i times the Hilbert transform operator $\mathscr{H}$ [see (2.27) and (2.53)] in the space domain. Hence,

$$\mathscr{R}_m^{-1} = -\mathrm{i}\left(\frac{1}{2\pi\mathrm{i}}\right)^{m-1} \mathscr{B}_m \mathscr{H} \frac{\partial^{m-1}}{\partial p^{m-1}}, \qquad (m \text{ even}). \tag{4.10}$$

Explicitly,

$$f(\boldsymbol{r}) = \frac{\mathrm{i}}{\pi}\left(\frac{1}{2\pi\mathrm{i}}\right)^{m-1} \int_{\frac{1}{2}\omega_m} \mathrm{d}\Omega_n \left[\frac{1}{p} * \frac{\partial^{m-1}\lambda_{\hat{\boldsymbol{n}}}(p)}{\partial p^{m-1}}\right]_{p=\boldsymbol{r}\cdot\hat{\boldsymbol{n}}}, \tag{4.11}$$

(m even),

where the Cauchy principal value of $1/p$ is understood. Many equivalent forms of (4.11) may be generated by integrating by parts.

The distinction between odd and even m is not necessary if the filtering is carried out in the frequency domain, and in both cases (4.7) becomes

$$\mathscr{R}_m^{-1} = \mathscr{B}_m \mathscr{F}_1^{-1} |\nu|^{m-1} \mathscr{F}_1. \tag{4.12}$$

4.2. THE ABEL TRANSFORM AND ITS 3D GENERALIZATION

The Abel transform, designated by the operator $\mathscr{A}$, is a special case of the 2D Radon transform, valid when the function being transformed is rotationally symmetric. If a function $f(r)$ is independent of the polar angle θ in polar coordinates, $\boldsymbol{r} = (r, \theta)$, then all projection directions are equivalent and we may as well take the vector $\hat{\boldsymbol{n}}$ parallel to the x-axis. Then (2.4) becomes

$$\lambda(x) = \mathscr{R}_2\{f(r)\} = \mathscr{A}\{f(r)\} = \int_{-\infty}^{\infty} f(r)\,\mathrm{d}y. \tag{4.13}$$

Since $x^2 + y^2 = r^2$, we also have

$$\lambda(x) = 2\int_{|x|}^{\infty} f(r)\frac{\mathrm{d}y}{\mathrm{d}r}\,\mathrm{d}r = 2\int_{|x|}^{\infty} \frac{f(r)r\,\mathrm{d}r}{[r^2 - x^2]^{1/2}}, \tag{4.14}$$

which is the explicit form of the Abel transform.

There are many ways to derive the inverse Abel transform. BRACEWELL [1965] makes a change of variables that reduces the integral in (4.14) to a convolution, which is then inverted by Fourier methods. It is also possible to carry out the ϕ integration in the general 2D inverse Radon transform (2.29) to get the inverse Abel transform. However, it is instructive to derive the inverse Abel transform directly from the central-slice theorem (BARRETT and SWINDELL [1981]). The operator form of the central-slice theorem (2.50) is

$$\mathscr{F}_2 = \mathscr{F}_1 \mathscr{R}_2. \tag{4.15}$$

For rotationally symmetric functions, $\mathscr{F}_2$ is the zero-order Hankel operator $\mathscr{H}_0$ (not to be confused with the Hilbert operator $\mathscr{H}$), defined by

$$\mathscr{F}_2\{f(r)\} = \mathscr{H}_0\{f(r)\} = 2\pi \int_0^\infty r\,\mathrm{d}r\, J_0(2\pi\rho r) f(r), \tag{4.16}$$

and $\mathscr{R}_2 = \mathscr{A}$. Note that $\mathscr{H}_0^{-1} = \mathscr{H}_0$ (GASKILL [1978]). Therefore, (4.15) becomes

$$\mathscr{H}_0 = \mathscr{F}_1 \mathscr{A}, \tag{4.17}$$

from which it follows that

$$\mathscr{A}^{-1} = \mathscr{H}_0 \mathscr{F}_1. \tag{4.18}$$

Explicitly, since $\lambda(x) = \lambda(-x)$,

$$\begin{aligned} f(r) = \mathscr{A}^{-1}\{\lambda(x)\} &= 2\pi \int_0^\infty \rho\,\mathrm{d}\rho\, J_0(2\pi\rho r) \int_{-\infty}^\infty \lambda(x) \exp(-2\pi \mathrm{i}\rho x)\,\mathrm{d}x \\ &= \int_0^\infty \mathrm{d}\rho\, J_0(2\pi\rho r) \int_{-\infty}^\infty \lambda(x)\, 2\pi\rho \cos(2\pi\rho x)\,\mathrm{d}x \\ &= \int_0^\infty \mathrm{d}\rho\, J_0(2\pi\rho r) \int_{-\infty}^\infty \lambda(x) \frac{\mathrm{d}}{\mathrm{d}x}[\sin(2\pi\rho x)]\,\mathrm{d}x \\ &= -\frac{1}{\pi} \int_r^\infty \frac{\mathrm{d}\lambda/\mathrm{d}x}{(x^2 - r^2)^{1/2}}\,\mathrm{d}x, \qquad (r > 0), \end{aligned} \tag{4.19}$$

where, in the last line, we have performed an integration by parts and made use of the integral,

$$\int_0^\infty J_0(at) \sin bt\,\mathrm{d}t = \begin{cases} 0, & b < a \\ (b^2 - a^2)^{-1/2}, & b > a. \end{cases} \tag{4.20}$$

Note that the lower limit in the last form of (4.19) is r. Therefore, to find $f(r)$, one needs only the values of $\lambda(x)$ for $x > r$. EIN-GAL [1975] refers to this result as the *hole theorem*; a hole in Radon space of radius r does not preclude an exact reconstruction of $f(r)$. We shall see in § 4.3 that this theorem holds even if the object is not rotationally symmetric.

A very interesting situation arises when we consider the 3D Radon transform of a spherically symmetric function $f(r)$. We shall denote the special form of $\mathscr{R}_3$ in this case as $\mathscr{V}$, given by

$$\lambda(x) = \mathscr{R}_3\{f(r)\} = \mathscr{V}\{f(r)\} = \int_{-\infty}^\infty \mathrm{d}y \int_{-\infty}^\infty \mathrm{d}z\, f(r). \tag{4.21}$$

Since $r^2 = x^2 + y^2 + z^2$, we can write

$$\int_{-\infty}^{\infty} \mathrm{d}y \int_{-\infty}^{\infty} \mathrm{d}z = 2\pi \int_0^{\infty} u\mathrm{d}u = 2\pi \int_{|x|}^{\infty} r\mathrm{d}r, \tag{4.22}$$

where $u^2 = y^2 + z^2$. Thus

$$\lambda(x) = 2\pi \int_{|x|}^{\infty} r\mathrm{d}r\, f(r). \tag{4.23}$$

Straightforward differentiation then shows that

$$\frac{\mathrm{d}\lambda(x)}{\mathrm{d}x} = -2\pi\, x\, f(x), \tag{4.24}$$

or

$$f(r) = \mathscr{V}^{-1}\{\lambda(x)\} = -\frac{1}{2\pi r}\lambda'(r) \tag{4.25}$$

where $\lambda'(r) = \mathrm{d}\lambda(r)/\mathrm{d}r$. In this case, *the inverse transform is not an integral transform at all, but just a simple derivative*. Back-projection is not required, and only projection values in the neighborhood of a single point in Radon space are required to determine $f(r)$ for a particular r.

This remarkable result is related to the local nature of the filtering operation in the 3D inverse Radon transform. Indeed, if $\lambda_{\hat{n}}(p)$ is independent of $\hat{\boldsymbol{n}}$, (3.10) becomes (CHIU, BARRETT and SIMPSON [1980])

$$f(r) = -\frac{1}{4\pi^2}\cdot 2\pi \int_0^1 \mathrm{d}(\cos\theta_n)\, \lambda''(r\cos\theta_n)$$

$$= -\frac{1}{2\pi r}\int_0^r \mathrm{d}t\, \lambda''(t) = -\frac{1}{2\pi r}\lambda'(r) \tag{4.26}$$

where $t = r\cos\theta_n$ and $\lambda'(0) = 0$, since λ is an even, differentiable function. Equation (4.26) agrees with (4.25), showing directly that $\mathscr{R}_3^{-1}$ reduces to $\mathscr{V}^{-1}$ for spherical symmetry.

Equation (4.25) was first given in the optics literature by VEST and STEEL [1978], but it was known much earlier in the literature on Compton scattering [DUMOND, 1929] and positron annihilation (STEWART [1957]; MIJNARENDS [1967]).

4.3. THE CORMACK TRANSFORM

Although RADON [1917] had laid the mathematical groundwork for reconstruction from projections, his work was not widely known and its importance to radiological imaging was not recognized until the late 1960s. An independent and mathematically rigorous formulation was obtained by CORMACK [1963, 1964], and this work may properly be regarded as the beginning of modern computed tomography.

Cormack's approach was to recognize that, in polar coordinates, any physically meaningful function $f(r, \theta)$ is periodic in θ with period 2π. Therefore, it can be expanded in a Fourier series or *circular harmonic decomposition* of the form

$$f(r, \theta) = \sum_{\ell=-\infty}^{\infty} f_\ell(r)\, \mathrm{e}^{\mathrm{i}\ell\theta}, \tag{4.27}$$

where

$$f_\ell(r) = \frac{1}{2\pi} \int_0^{2\pi} f(r, \theta)\, \mathrm{e}^{-\mathrm{i}\ell\theta}\, \mathrm{d}\theta. \tag{4.28}$$

Similarly, the projection $\lambda(p, \phi)$ can be expanded as

$$\lambda(p, \phi) = \sum_{\ell=-\infty}^{\infty} \lambda_\ell(p)\, \mathrm{e}^{\mathrm{i}\ell\phi}, \tag{4.29}$$

with

$$\lambda_\ell(p) = \frac{1}{2\pi} \int_0^{2\pi} \lambda(p, \phi)\, \mathrm{e}^{-\mathrm{i}\ell\phi}\, \mathrm{d}\phi. \tag{4.30}$$

(Note the slight change in notation from § 2; $\lambda_\ell(p)$ should not be confused with $\lambda_\phi(p)$, which is now denoted $\lambda(p, \phi)$.) By the same token, the 2D Fourier transform $F(\rho, \theta_\rho)$ of $f(r, \theta)$ may be written

$$F(\rho, \theta_\rho) = \sum_{\ell=-\infty}^{\infty} F_\ell(\rho)\, \mathrm{e}^{\,i\ell\theta_\rho}, \tag{4.31}$$

where

$$F_\ell(\rho) = \frac{1}{2\pi} \int_0^{2\pi} F(\rho, \theta_\rho)\, \mathrm{e}^{-\mathrm{i}\ell\theta_\rho}\, \mathrm{d}\theta_\rho. \tag{4.32}$$

We have already explored in some detail the relationships among $f(r, \theta)$, $F(\rho, \theta_\rho)$, and $\lambda(p, \phi)$, which are summarized in the operator equations (2.47),

(2.50), and (2.51). In this section, we derive the corresponding relationships connecting $f_\ell(\theta)$, $F_\ell(\theta_\rho)$, and $\lambda_\ell(\phi)$. The development here follows CHIU [1980].

Let us first investigate the relationship between $f_\ell(r)$ and $F_\ell(\rho)$. The 2D Fourier transform of $f(r, \theta)$ is given by

$$F(\rho, \theta_\rho) = \int_0^\infty r\,\mathrm{d}r \int_0^{2\pi} \mathrm{d}\theta\, f(r, \theta) \exp[-2\pi \mathrm{i} \rho r \cos(\theta - \theta_\rho)], \quad (4.33)$$

which, with (4.27) and (4.31), becomes

$$\sum_{\ell=-\infty}^{\infty} F_\ell(\rho)\, \mathrm{e}^{\mathrm{i}\ell\theta_\rho} = \sum_{\ell=-\infty}^{\infty} \mathrm{e}^{\mathrm{i}\ell\theta_\rho} \int_0^\infty r\,\mathrm{d}r \int_0^{2\pi} \mathrm{d}\theta\, \mathrm{e}^{\mathrm{i}\ell(\theta-\theta_\rho)} \times \exp[-2\pi \mathrm{i} \rho r \cos(\theta - \theta_\rho)]\, f_\ell(r). \quad (4.34)$$

Since the functions $\exp(\mathrm{i}\ell\theta_\rho)$ form a complete, orthogonal set on $0 < \theta_\rho \leq 2\pi$, (4.34) must hold term by term, i.e.

$$F_\ell(\rho) = \int_0^\infty r\,\mathrm{d}r \int_0^{2\pi} \mathrm{d}\psi\, \mathrm{e}^{\mathrm{i}\ell\psi} \exp(-2\pi \mathrm{i} \rho r \cos\psi) f_\ell(r) \quad (4.35)$$

where $\psi = \theta - \theta_\rho$. We recognize the integral over ψ as a standard representation of the Bessel function of order ℓ:

$$J_\ell(z) = \frac{(-\mathrm{i})^\ell}{2\pi} \int_0^{2\pi} \mathrm{e}^{\mathrm{i}\ell\psi}\, \mathrm{e}^{-\mathrm{i}z\cos\psi}\, \mathrm{d}\psi. \quad (4.36)$$

Hence,

$$F_\ell(\rho) = 2\pi \mathrm{i}^\ell \int_0^\infty r\,\mathrm{d}r\, f_\ell(r) J_\ell(2\pi\rho r) = \mathrm{i}^\ell \mathscr{H}_\ell\{f_\ell(r)\}, \quad (4.37)$$

where $\mathscr{H}_\ell$ is the operator for the ℓth-order Hankel transform. Thus $f_\ell(r)$ and $F_\ell(\rho)$ form an ℓth-order Hankel-transform pair. Since the Hankel transform is self-inverse, i.e. $\mathscr{H}_\ell^{-1} = \mathscr{H}_\ell$, we also have

$$f_\ell(r) = (-\mathrm{i})^\ell \mathscr{H}_\ell\{F_\ell(\rho)\}. \quad (4.38)$$

We already know from the central-slice theorem that $F(\rho, \theta_\rho)$ is the 1D Fourier transform of $\lambda(p, \phi)$ if we set $\nu = \rho$ and $\phi = \theta_\rho$. In terms of circular harmonics,

$$\mathscr{F}_1\{\lambda(p, \phi)\} = \int_{-\infty}^\infty \mathrm{d}p\, \mathrm{e}^{-2\pi \mathrm{i}\nu p} \sum_\ell \lambda_\ell(p)\, \mathrm{e}^{\mathrm{i}\ell\phi} = \sum_\ell F_\ell(\nu)\, \mathrm{e}^{\mathrm{i}\ell\phi}. \quad (4.39)$$

Again, this equation must hold term by term, and we see that $\lambda_\ell(p)$ and $F_\ell(\nu)$ are a 1D Fourier-transform pair, written symbolically as

$$F_\ell(\nu) = \mathscr{F}_1\{\lambda_\ell(p)\}. \tag{4.40}$$

We now know the operators connecting f_ℓ with F_ℓ and λ_ℓ with F_ℓ. All that remains is to find the connection between λ_ℓ and f_ℓ, which we refer to as the Cormack transform of order ℓ and denote by the operator $\mathscr{C}_\ell$:

$$\lambda_\ell(p) = \mathscr{C}_\ell\{f_\ell(r)\}. \tag{4.41}$$

From the definition of the Radon transform, (2.5), we have

$$\lambda(p, \phi) = \int_0^\infty r\,\mathrm{d}r \int_0^{2\pi} \mathrm{d}\theta\, f(r, \theta)\delta[p - r\cos(\theta - \phi)]. \tag{4.42}$$

Using the circular harmonic expansions (4.27) and (4.28) and again invoking the orthogonality of the harmonics, we get

$$\lambda_\ell(p) = \int_0^\infty r\,\mathrm{d}r\, f_\ell(r) \int_0^{2\pi} d\psi\, \mathrm{e}^{\mathrm{i}\ell\psi}\delta(p - r\cos\psi), \tag{4.43}$$

where $\psi = \theta - \phi$. The ψ integral may be performed by steps similar to those leading to (2.42) (CHIU [1980]), and the result is

$$\begin{aligned}\lambda_\ell(p) &= 2\int_{|p|}^\infty r\,\mathrm{d}r\, f_\ell(r)\frac{\cos[\ell\cos^{-1}(p/r)]}{(r^2 - p^2)^{1/2}} \\ &= 2\int_{|p|}^\infty r\,\mathrm{d}r \frac{f_\ell(r)T_\ell(p/r)}{(r^2 - p^2)^{1/2}} = \mathscr{C}_\ell\{f(r)\},\end{aligned} \tag{4.44}$$

where $T_\ell(p/r)$ is the Tschebycheff polynomial of the first kind of order ℓ, defined by

$$T_\ell(x) = \begin{cases}\cos(\ell\cos^{-1}x), & |x| < 1 \\ \cosh(\ell\cosh^{-1}x), & |x| > 1.\end{cases} \tag{4.45}$$

For a thorough but very readable introduction to these polynomials, see VAN DER POL and WEIJERS [1934].

Note that if $f(r, \theta)$ is rotationally symmetric, $f_\ell(r)$ is non-zero only for $\ell = 0$; in this case (4.44) reduces to the Abel transform (4.14) since $T_0(u) = 1$ for all u.

To gain more insight into the Cormack transform, let us consider a specific object, namely a cylindrical shell of radius R with an azimuthal dependence given by $\cos k\theta$ (k integer). That is, $f(r, \theta) = \delta(r - R)\cos k\theta$. By (4.28), the circular-harmonic decomposition of this object is

$$f_\ell(r) = \tfrac{1}{2}\delta(r - R)[\delta_{\ell,k} + \delta_{\ell,-k}], \tag{4.46}$$

where $\delta_{\ell,k}$ is the usual Kronecker symbol. From (4.44),

$$\lambda_\ell(p) = \frac{R}{(R^2 - p^2)^{1/2}}\left[T_k\left(\frac{p}{R}\right) + T_{-k}\left(\frac{p}{R}\right)\right],$$

$$= \frac{2R}{(R^2 - p^2)^{1/2}} T_k\left(\frac{p}{R}\right), \qquad |p| < R, \tag{4.47}$$

where the last step follows because $T_k(u) = T_{-k}(u)$ for all k and u. The factor $(R^2 - p^2)^{-1/2}$ in (4.47) represents the projection of the cylindrical shell $\delta(r - R)$, as illustrated in Fig. 5a, while $T_k(p/R)$ is the projection of $\cos k\theta$ (Fig. 5b). A general object can then be constructed as a superposition of cylindrical shells with various radii and azimuthal frequencies, and its projection is a superposition of terms like (4.47).

The Tschebycheff polynomial may also be visualized as a Lissajous pattern

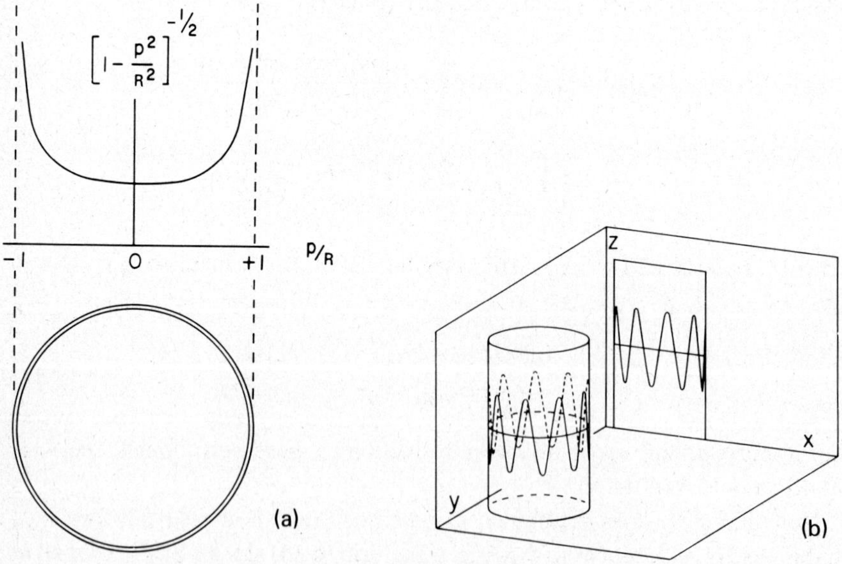

Fig. 5. (a) The projection of a thin circular shell $\delta(r - R)$ is $(R^2 - p^2)^{-1/2}$. (b) The projection of $\cos k\theta$ is $T_k(p/R)$. (After VAN DER POL and WEIJERS [1934.)

between two cosine waves having frequencies in the ratio $k:1$ (VAN DER POL and WEIJERS [1934]). If the y-axis of an oscilloscope is driven with a voltage $\cos k\omega t$ and the x-axis is driven with $\cos \omega t$, then the display is

$$y = \cos(k \cos^{-1} x) = T_k(x). \tag{4.48}$$

Exactly the same cosine distortion of the x-axis occurs when the function $\cos k\theta$ is projected from a cylinder to a plane.

There are many ways to derive the inverse Cormack transform. Cormack himself used the orthogonality properties of the Tschebycheff polynomials. EIN-GAL [1975], HANSEN [1981], and VERLY [1982] used the Mellin transform, and CHIU [1980] and HAWKINS [1982] performed a tedious contour integration. We shall approach the problem by using operator algebra. From (4.41),

$$f_\ell(r) = \mathscr{C}_\ell^{-1}\{\lambda_\ell(p)\}, \tag{4.49}$$

while, from (4.38) and (4.40),

$$\begin{aligned} f_\ell(r) &= (-\mathrm{i})^\ell \mathscr{H}_\ell\{F_\ell(\rho)\} \\ &= (-\mathrm{i})^\ell \mathscr{H}_\ell \mathscr{F}_1\{\lambda_\ell(p)\}. \end{aligned} \tag{4.50}$$

Therefore (EIN-GAL [1975]),

$$\mathscr{C}_\ell^{-1} = (-\mathrm{i})^\ell \mathscr{H}_\ell \mathscr{F}_1. \tag{4.51}$$

To convert this result to an explicit integral transform, we use the definitions of $\mathscr{H}_\ell$ and $\mathscr{F}_1$, (4.37) and (2.9) respectively, to obtain

$$f_\ell(r) = (-\mathrm{i})^\ell 2\pi \int_0^\infty \nu \,\mathrm{d}\nu\, J_\ell(2\pi\nu r) \int_{-\infty}^{\infty} \mathrm{d}p\, \mathrm{e}^{-2\pi \mathrm{i} p\nu} \lambda_\ell(p). \tag{4.52}$$

From (2.14) and (4.30), it follows that $\lambda_\ell(p)$ is an even (odd) function of p if ℓ is even (odd), i.e.

$$\lambda_\ell(-p) = (-1)^\ell \lambda_\ell(p). \tag{4.53}$$

Therefore, only the term proportional to $\cos(2\pi p\nu)$ contributes to the p integral in (4.52) for ℓ even, while only the term in $\sin(2\pi p\nu)$ contributes for ℓ odd. We can rewrite (4.52) as

$$f_\ell(r) = 2(-\mathrm{i})^\ell \int_0^\infty \mathrm{d}p\, \lambda_\ell(p) \frac{\partial I_\ell(p, r)}{\partial p}, \tag{4.54}$$

where

$$I_\ell(p,r) = \begin{cases} \displaystyle\int_0^\infty \mathrm{d}\nu\, J_\ell(2\pi\nu r)\sin(2\pi p\nu), & \ell \text{ even}, \\ \mathrm{i}\displaystyle\int_0^\infty \mathrm{d}\nu\, J_\ell(2\pi\nu r)\cos(2\pi p\nu), & \ell \text{ odd}. \end{cases} \tag{4.55}$$

From tabulated integrals and properties of the Tschebycheff polynomials (GRADSHTEYN and RYZHIK [1980], pp. 730 and 1032; VAN DER POL and WEIJERS [1934]), we find

$$I_\ell(p,r) = \begin{cases} \dfrac{-\mathrm{i}^\ell}{2\pi r} U_{\ell-1}\left(\dfrac{p}{r}\right) & r > p > 0, \\ \dfrac{\mathrm{i}^\ell}{2\pi r}\left[\dfrac{T_\ell(p/r)}{[(p^2/r^2)-1]^{1/2}} - U_{\ell-1}\left(\dfrac{p}{r}\right)\right], & p > r > 0 \end{cases} \tag{4.56}$$

where $U_\ell(x)$ is the Tschebycheff polynomial of the second kind, defined by

$$U_\ell(x) = \begin{cases} \sin[(\ell+1)\cos^{-1}x]/\sin(\cos^{-1}x), & x < 1 \\ \sinh[(\ell+1)\cosh^{-1}x]/\sinh(\cosh^{-1}x), & x > 1. \end{cases} \tag{4.57}$$

With (4.56) and an integration by parts, (4.54) becomes

$$f_\ell(r) = -\frac{1}{\pi}\int_r^\infty \mathrm{d}p\, \frac{\lambda_\ell'(p)T_\ell(p/r)}{(p^2-r^2)^{1/2}} + \frac{1}{\pi r}\int_0^\infty \mathrm{d}p\, \lambda_\ell'(p)U_{\ell-1}(p/r). \tag{4.58}$$

The second integral in (4.58) is actually zero. The proof of this contention (CHIU [1980]) rests on the fact that $U_\ell(x)$ is a polynomial of degree ℓ, and on the consistency condition, to be discussed below, which states that (CORMACK [1963]; EIN-GAL [1975]; VERLY [1981])

$$\int_0^\infty \lambda_\ell(p)p^k\,\mathrm{d}p = 0, \tag{4.59}$$

for $0 < k < |\ell|$ and $k + |\ell|$ even. Therefore, the final form of the inverse Cormack transform is

$$f_\ell(r) = \mathscr{C}_\ell^{-1}\{\lambda_\ell\} = -\frac{1}{\pi}\int_r^\infty \mathrm{d}p\, \frac{\lambda_\ell'(p)T_\ell(p/r)}{(p^2-r^2)^{1/2}}, \qquad (r > 0). \tag{4.60}$$

An interesting feature of this result is that only values of $\lambda(\boldsymbol{p})$ for $p > |r|$ are required to reconstruct $f(\boldsymbol{r})$ (LEWITT and BATES [1978]). This is again the hole theorem (EIN-GAL [1975]), discussed for rotationally symmetric functions below (4.20). Note also that $\mathscr{C}_0^{-1} = \mathscr{A}^{-1}$ [cf. (4.19) and (4.60)] since $T_0(x) = 1$.

We now derive the consistency condition (4.59). The derivation given here follows CHIU [1980].

From (2.39), a point object has as its projection the 1D delta function $\delta(p - \boldsymbol{r} \cdot \boldsymbol{n})$. Since a more general object can be considered as a superposition of points, it will suffice to show that the consistency relation holds when $\lambda(p, \phi) = \delta(p - \boldsymbol{r} \cdot \hat{\boldsymbol{n}})$. In other words, if we can show that

$$\int_0^\infty p^k \, \mathrm{d}p \int_0^{2\pi} \mathrm{d}\phi \delta(p - \boldsymbol{r} \cdot \hat{\boldsymbol{n}}) \mathrm{e}^{-\mathrm{i}\ell\phi} = 0 \tag{4.61}$$

for certain values of k and ℓ, then we can be guaranteed, by the principle of linear superposition, that (4.59) holds for the same values of k and ℓ.

Interchanging the order of integration in (4.61), we get

$$\begin{aligned}
&\int_0^{2\pi} \mathrm{d}\phi \int_0^\infty p^k \, \mathrm{d}p \, \mathrm{e}^{-\mathrm{i}\ell\phi} \delta[p - r\cos(\theta - \phi)] \\
&\quad = \mathrm{e}^{-\mathrm{i}\ell\theta} \int_0^{2\pi} \mathrm{d}\psi \, \mathrm{e}^{\mathrm{i}\ell\psi} \int_0^\infty p^k \, \mathrm{d}p \, \delta(p - r\cos\psi) \\
&\quad = \mathrm{e}^{-\mathrm{i}\ell\theta} \int_{-\pi/2}^{\pi/2} \mathrm{d}\psi \, \mathrm{e}^{\mathrm{i}\ell\psi} (r\cos\psi)^k \\
&\quad = r^k \mathrm{e}^{-\mathrm{i}\ell\theta} \int_{-\pi/2}^{\pi/2} \mathrm{d}\psi \cos\ell\psi \, (\cos\psi)^k \\
&\quad = \left(\frac{r}{2}\right)^k \mathrm{e}^{-\mathrm{i}\ell\theta} \int_{-\pi/2}^{\pi/2} \mathrm{d}\psi \, [\mathrm{e}^{\mathrm{i}(k-\ell)\psi} + k\,\mathrm{e}^{\mathrm{i}(k-\ell-2)\psi} + \dots \\
&\qquad + \mathrm{e}^{\mathrm{i}(-k-\ell)\psi}],
\end{aligned} \tag{4.62}$$

where $\psi = \theta - \phi$. Every term will vanish if $k - \ell, k - \ell - 2, \dots, -k - \ell$ are all nonzero even numbers. This occurs if $k + |\ell|$ is even and $0 \le k < |\ell|$, and, under these conditions, (4.61) and (4.59) must hold.

The consistency condition is of considerable theoretical importance because several different forms of the inverse Cormack transform appear in the literature. An example of two apparently different forms is (4.58) and (4.60). However, in this case as well as many others, application of the consistency condition removes the discrepancy (VERLY [1981]).

There are several other mathematical subtleties related to the stability and uniqueness of the inverse Cormack transform. For a thorough review, see HAWKINS [1982].

4.4. THE GEGENBAUER TRANSFORM

Just as the 2D functions $f(r, \theta)$, $\lambda(p, \phi)$, and $F(\rho, \theta_\rho)$ can be expanded in circular harmonics, so too can the corresponding 3D functions be expanded in spherical harmonics. The 3D counterpart of the Cormack transform is called the Gegenbauer transform. A brief survey of this theory is given here. For more details, see LUDWIG [1966], CHIU [1980], HAWKINS [1982], and DEANS [1983].

A 3D object in spherical coordinates, $f(r, \theta, \phi)$, can be expanded as

$$f(r, \theta, \phi) = \sum_{\ell=0}^{\infty} \sum_{m=-\ell}^{\ell} f_{\ell m}(r) Y_{\ell m}(\theta, \phi), \tag{4.63}$$

where $Y_{\ell m}(\theta, \phi)$ is a spherical harmonic (JACKSON [1975]). The expansion coefficient $f_{\ell m}(r)$ is given by

$$f_{\ell m}(r) = \int_0^{2\pi} \mathrm{d}\phi \int_0^{\pi} \mathrm{d}\theta \sin\theta f(r, \theta, \phi) Y^*_{\ell m}(\theta, \phi). \tag{4.64}$$

The projection data, $\lambda_{\hat{n}}(p) = \lambda(p, \theta_n, \phi_n)$, and the 3D Fourier transform of the object, $F(\sigma, \theta_\sigma, \phi_\sigma)$, have similar expansions:

$$\lambda(p, \theta_n, \phi_n) = \sum_{\ell,m} \lambda_{\ell m}(p) Y_{\ell m}(\theta_n, \phi_n), \tag{4.65}$$

$$F(\sigma, \theta_\sigma, \phi_\sigma) = \sum_{\ell,m} F_{\ell m}(\sigma) Y_{\ell m}(\theta_\sigma, \phi_\sigma), \tag{4.66}$$

with inverses analogous to (4.64).

The relation between $f_{\ell m}(r)$ and $F_{\ell m}(\sigma)$ is (CHIU [1980])

$$F_{m}(\sigma) = \mathrm{i}^\ell \mathscr{J}_\ell\{f_{\ell m}(r)\}, \tag{4.67}$$

where $\mathscr{J}_\ell$ is the operator for the ℓth-order spherical Bessel transform, given by [cf. (4.37)]

$$\mathscr{J}_\ell\{f_{\ell m}(r)\} = 4\pi \int_0^{\infty} f_{\ell m}(r) j_\ell(2\pi\sigma r) r^2 \, \mathrm{d}r, \tag{4.68}$$

where $j_\ell(u)$ is the spherical Bessel function of the first kind, which is related to the ordinary Bessel function by (MORSE and FESHBACH [1953] p. 1573)

$$j_\ell(u) = (\pi/2u)^{1/2} J_{\ell+1/2}(u). \tag{4.69}$$

Since $\mathscr{J}_\ell^{-1} = \mathscr{J}_\ell$ (MORSE and FESHBACH [1953] p. 781), we also have

$$f_{\ell m}(r) = (-\mathrm{i})^\ell \mathscr{J}_\ell\{F_{\ell m}(\sigma)\}. \tag{4.70}$$

If the object is spherically symmetric, the only nonzero harmonic is $\ell = m = 0$, and

$$F(\sigma) = 4\pi \int_0^\infty f(r)\,\mathrm{sinc}\,(2\sigma r) r^2\,\mathrm{d}r, \tag{4.71}$$

$$f(r) = 4\pi \int_0^\infty F(\sigma)\,\mathrm{sinc}\,(2\sigma r)\sigma^2\,\mathrm{d}\sigma, \tag{4.72}$$

since

$$j_0(\pi u) = \frac{\sin \pi u}{\pi u} \equiv \mathrm{sinc}\,(u). \tag{4.73}$$

From a spherical-harmonic decomposition of the 3D-central-slice theorem, (3.7) or (3.23), it follows readily that $\lambda_{\ell m}(p)$ and $F_{\ell m}(\sigma)$ are a 1D Fourier-transform pair, i.e. [cf. (4.40)]

$$F_{\ell m}(\sigma) = \mathscr{F}_1\{\lambda_{\ell m}(p)\}. \tag{4.74}$$

The remaining relation that is needed is the one connecting $f_{\ell m}(r)$ and $\lambda_{\ell m}(p)$. This relation is called the 3D Gegenbauer transform of order ℓ, and is denoted by the operator $\mathscr{G}_{3,\ell}$, so that

$$\lambda_{\ell m}(p) = \mathscr{G}_{3,\ell}\{f_{\ell m}(r)\}. \tag{4.75}$$

Explicitly (LUDWIG [1966]; CHIU [1980]):

$$\lambda_{\ell m}(p) = 2\pi \int_{|p|}^\infty f_{\ell m}(r) P_\ell(p/r) r\,\mathrm{d}r, \tag{4.76}$$

where P_ℓ is the Legendre polynomial.

The inverse Gegenbauer transform is given by

$$f_{\ell m}(r) = \mathscr{G}_{3,\ell}^{-1}\{\lambda_{\ell m}(p)\} = -\frac{1}{2\pi r}\int_0^r \lambda''_{\ell m}(p) P_\ell(p/r)\,\mathrm{d}p. \tag{4.77}$$

Note that $\mathscr{G}_{3,\ell}$ is independent of m.

If the object is spherically symmetric, (4.77) reduces to

$$f(r) = -\frac{1}{2\pi r}\int_0^r \lambda''(p)\,\mathrm{d}p = -\frac{1}{2\pi r}\lambda'(r), \tag{4.78}$$

which is just the VEST–STEEL relation (4.25). Thus (CHIU [1980])

$$\mathscr{G}_{3,0}^{-1} = \mathscr{V}^{-1}. \tag{4.79}$$

The m-dimensional generalization of these results has been given by LUDWIG [1966].

4.5. THE DIPOLE-SHEET TRANSFORM

The dipole-sheet transform (BARRETT [1982a]) is a symmetrized version of the 3D Radon transform. To see the asymmetry between the Radon transform and its inverse, let us rewrite (3.1) and (3.10) as

$$f_R(\boldsymbol{p}) = \mathcal{R}_3\{f(\boldsymbol{r})\} = \int_\infty \mathrm{d}^3 r\, f(\boldsymbol{r})\delta(p - \boldsymbol{r}\cdot\hat{\boldsymbol{n}}), \tag{4.80}$$

$$f(\boldsymbol{r}) = \mathcal{R}_3^{-1}\{f_R(\boldsymbol{p})\} = -\frac{1}{4\pi^2}\int_\infty \mathrm{d}^3 p\, f_R(\boldsymbol{p})\delta''(p - \boldsymbol{r}\cdot\hat{\boldsymbol{n}})/p^2, \tag{4.81}$$

where $f_R(\boldsymbol{p})$ is what we have previously called $\lambda(\boldsymbol{p})$ or $\lambda_{\hat{n}}(p)$. The equivalence of (4.81) and (3.10) follows since $\mathrm{d}^3 p = p^2\,\mathrm{d}p\,\mathrm{d}\Omega_n$ and since the delta function has the effect of removing the p integration in (4.81), taking the second derivative of f_R with respect to p, and making the substitution $p = \boldsymbol{r}\cdot\hat{\boldsymbol{n}}$.

Comparing the forward and inverse transforms in this form, we see that they involve the same kernel $\delta(p - \boldsymbol{r}\cdot\hat{\boldsymbol{n}})$, but that the inverse transform includes the second-derivative operation, the factor $1/p^2$, and the constant $-1/(4\pi^2)$. A symmetrical transform pair results if we move one derivative and a factor of $\mathrm{i}/(2\pi p)$ to the forward transform. Thus, we define a new transform pair by

$$f_D(\boldsymbol{p}) = \mathcal{D}\{f(\boldsymbol{r})\} = \int_\infty \mathrm{d}^3 r\, \psi(\boldsymbol{p}, \boldsymbol{r}) f(\boldsymbol{r}), \tag{4.82}$$

$$f(\boldsymbol{r}) = \mathcal{D}^{-1}\{f_D(\boldsymbol{p})\} = \int_\infty \mathrm{d}^3 p\, \psi^*(\boldsymbol{p}, \boldsymbol{r}) f_D(\boldsymbol{p}), \tag{4.83}$$

where the asterisk denotes complex conjugate, and

$$\psi(\boldsymbol{p}, \boldsymbol{r}) \equiv \frac{\mathrm{i}}{2\pi p}\delta'(p - \boldsymbol{r}\cdot\hat{\boldsymbol{n}}). \tag{4.84}$$

Regarded as a function of $\boldsymbol{r}$, the basis function $\psi(\boldsymbol{p}, \boldsymbol{r})$ vanishes except in an infinitesimal neighborhood of the plane $p = \boldsymbol{r}\cdot\hat{\boldsymbol{n}}$. It takes on the value $+\mathrm{i}\infty$ just to one side of this plane and $-\mathrm{i}\infty$ just to the other side of it. Physically, $\psi(\boldsymbol{p}, \boldsymbol{r})$ may be interpreted as a double layer or dipole sheet, and we refer to (4.82) as the dipole-sheet transform.

To derive the inverse dipole-sheet transform (4.83) from the inverse Radon transform, we note that

$$f_{\mathrm{D}}(\boldsymbol{p}) = \frac{\mathrm{i}}{2\pi p}\frac{\partial}{\partial p} f_{\mathrm{R}}(\boldsymbol{p}), \tag{4.85}$$

which follows by differentiating (4.80) under the integral sign. Next, we insert (4.85) into (4.81), yielding

$$\begin{aligned} f(\boldsymbol{r}) &= \frac{1}{4\pi^2}\int_{-\infty}^{\infty} \mathrm{d}p \int_{2\pi} \mathrm{d}\Omega_n \left[\frac{\partial f_{\mathrm{R}}(\boldsymbol{p})}{\partial p}\right] \delta'(p - \boldsymbol{r}\cdot\hat{\boldsymbol{n}}) \\ &= -\frac{\mathrm{i}}{2\pi}\int_{-\infty}^{\infty} \mathrm{d}p \int_{2\pi} \mathrm{d}\Omega_n\, p f_{\mathrm{D}}(\boldsymbol{p})\delta'(p - \boldsymbol{r}\cdot\hat{\boldsymbol{n}}) \\ &= -\frac{\mathrm{i}}{2\pi}\int_{\infty} \mathrm{d}^3 p f_{\mathrm{D}}(\boldsymbol{p})\delta'(p - \boldsymbol{r}\cdot\hat{\boldsymbol{n}})/p, \end{aligned} \tag{4.86}$$

in agreement with (4.83).

The functions $\psi(\boldsymbol{p}, \boldsymbol{r})$ are orthonormal and complete (BARRETT [1982a]). That is,

$$\int_{\infty} \psi^*(\boldsymbol{p}, \boldsymbol{r})\psi(\boldsymbol{p}', \boldsymbol{r})\,\mathrm{d}^3 r = \delta(\boldsymbol{p} - \boldsymbol{p}'), \tag{4.87}$$

$$\int_{\infty} \psi^*(\boldsymbol{p}, \boldsymbol{r})\psi(\boldsymbol{p}, \boldsymbol{r}')\,\mathrm{d}^3 p = \delta(\boldsymbol{r} - \boldsymbol{r}'). \tag{4.88}$$

The inverse transform (4.83) can also be derived by multiplying (4.82) by $\psi^*(\boldsymbol{p}, \boldsymbol{r}')$, integrating over $\boldsymbol{p}$, and using (4.88).

The following identities, given here without proof, are useful in manipulating the dipole-sheet transform:

$$\delta^{(n)}(x) = (-1)^n n!\, x^{-n}\delta(x), \tag{4.89}$$

$$\delta'(ax) = \frac{\delta'(x)}{a|a|}, \tag{4.90}$$

$$-\frac{1}{8\pi^2}\int_{4\pi} \mathrm{d}\Omega_n\, \delta''(\boldsymbol{r}\cdot\hat{\boldsymbol{n}}) = \delta(\boldsymbol{r}), \tag{4.91}$$

$$\int_{4\pi} \mathrm{d}\Omega_n\, \delta(\boldsymbol{r}\cdot\boldsymbol{p}) = \int_{4\pi} \mathrm{d}\Omega\, \delta(\boldsymbol{r}\cdot\boldsymbol{p}) = \frac{2\pi}{|rp|}, \tag{4.92}$$

$$\int_{4\pi} \mathrm{d}\Omega_n\, \delta(p - \boldsymbol{r}\cdot\hat{\boldsymbol{n}}) = \int_{4\pi} \mathrm{d}\Omega\, \delta(p - \boldsymbol{r}\cdot\hat{\boldsymbol{n}}) = \frac{2\pi}{|r|} \operatorname{rect}\left(\frac{p}{2r}\right), \tag{4.93}$$

$$\int_{\infty} \mathrm{d}^3 p\, \delta(p - \boldsymbol{r}\cdot\hat{\boldsymbol{n}}) = \frac{2\pi r^2}{3}, \tag{4.94}$$

$$\int_{4\pi} \mathrm{d}\Omega_n\, \delta'(p - \boldsymbol{r}\cdot\hat{\boldsymbol{n}})$$

$$= \int_{4\pi} \mathrm{d}\Omega\, \delta'(p - \boldsymbol{r}\cdot\hat{\boldsymbol{n}}) = \frac{-2\pi}{|r|} [\delta(p - |r|) - \delta(p + |r|)], \tag{4.95}$$

$$\int_{4\pi} \mathrm{d}\Omega_n (\boldsymbol{r}\cdot\hat{\boldsymbol{n}})^2\, \delta''(\boldsymbol{r}\cdot\hat{\boldsymbol{n}}) = \frac{4\pi}{|r|}, \tag{4.96}$$

$$\nabla^2 \frac{1}{|r|} = -4\pi\delta(\boldsymbol{r}), \tag{4.97}$$

$$\nabla^2 f(\boldsymbol{r}\cdot\hat{\boldsymbol{n}}) = f''(\boldsymbol{r}\cdot\hat{\boldsymbol{n}}). \tag{4.98}$$

Here, $\delta^{(n)}(x)$ denotes the nth derivative of a 1D delta function, and the vectors $\boldsymbol{p}$ and $\boldsymbol{r}$ have magnitudes $|p|$ and $|r|$, respectively, but p and r can be positive or negative. Recall also that $\delta(\boldsymbol{r})$ is a 3D delta function, while $\delta(r)$ is 1D.

Identity (4.89) can be used to derive an interesting theorem for derivatives of the Radon transform. Using (4.89) with $n = 1$ in (4.92) yields

$$f_{\mathrm{D}}(\boldsymbol{p}) = \frac{-\mathrm{i}}{2\pi p} \int_{\infty} \mathrm{d}^3 r f(\boldsymbol{r}) \frac{\delta(p - \boldsymbol{r}\cdot\hat{\boldsymbol{n}})}{p - \boldsymbol{r}\cdot\hat{\boldsymbol{n}}} = \frac{-\mathrm{i}}{2\pi p} \mathscr{R}_3 \left\{ \frac{f(\boldsymbol{r})}{p - \boldsymbol{r}\cdot\hat{\boldsymbol{n}}} \right\}, \tag{4.99}$$

or, with (4.85)

$$-\frac{\partial}{\partial p} \mathscr{R}_3\{\mathrm{f}(\boldsymbol{r})\} = \mathscr{R}_3 \left\{ \frac{f(\boldsymbol{r})}{p - \boldsymbol{r}\cdot\hat{\boldsymbol{n}}} \right\}. \tag{4.100}$$

This theorem is easily extended to the kth derivative of the mD Radon transform. Differentiating (4.1) and using (4.89) yields

$$\frac{\partial^k}{\partial p^k} \mathscr{R}_m\{f(\boldsymbol{r})\} = (-1)^k k!\, \mathscr{R}_m \left\{ \frac{f(\boldsymbol{r})}{(p - \boldsymbol{r}\cdot\hat{\boldsymbol{n}})^{\mathrm{k}}} \right\}, \tag{4.101}$$

where $(p - \boldsymbol{r}\cdot\hat{\boldsymbol{n}})^{-k}$ must be interpreted as a generalized function (Cauchy principal value for $k = 1$). Other formulas involving derivatives of the Radon transform are given by DEANS [1983].

The dipole sheet transform is very simple for spherically symmetric functions. Using identity (4.95) in (4.82) gives

$$f_{\mathrm{D}}(p) = \frac{\mathrm{i}}{2\pi p}\int_0^\infty r^2\,\mathrm{d}r f(r)\int_{4\pi}\mathrm{d}\Omega\,\delta'(p - \boldsymbol{r}\cdot\hat{\boldsymbol{n}})$$

$$= \frac{\mathrm{i}}{2\pi p}\int_0^\infty r^2\,\mathrm{d}r f(r)\left(\frac{-2\pi}{r}\right)\delta(p-r)$$

$$= -\mathrm{i}f(p). \tag{4.102}$$

Thus, except for the factor $-\mathrm{i}$, a spherically symmetric function is unchanged by the dipole-sheet transform. Any spherically symmetric function is an eigenfunction of $\mathscr{D}$.

Equation (4.102) is equivalent to the VEST–STEEL relation (4.25). From (4.85) and (4.102), we have

$$f_{\mathrm{D}}(p) = \frac{\mathrm{i}}{2\pi p}\frac{\mathrm{d}}{\mathrm{d}p}f_{\mathrm{R}}(p) = -\mathrm{i}f(p), \tag{4.103}$$

which is just (4.25).

From a mathematical point of view, the dipole-sheet transform is interesting because it is unitary, which means that its adjoint is the same as its inverse. For a general Hilbert-space operator $\mathscr{O}$, the adjoint $\mathscr{O}^\dagger$ is defined by

$$\langle u, \mathscr{O}v\rangle = \langle \mathscr{O}^\dagger u, v\rangle, \tag{4.104}$$

where u and v are vectors in the space and $\langle u, v\rangle$ denotes a scalar product. If u and v are represented as functions of $\boldsymbol{r}$,

$$\langle u, v\rangle = \int_\infty \mathrm{d}^3 r\, u^*(\boldsymbol{r})v(\boldsymbol{r}). \tag{4.105}$$

If $\mathscr{D}$ is unitary, then

$$\langle \mathscr{D}u, \mathscr{D}v\rangle = \langle \mathscr{D}^\dagger\mathscr{D}u, v\rangle = \langle \mathscr{D}^{-1}\mathscr{D}u, v\rangle = \langle u, v\rangle, \tag{4.106}$$

or

$$\int_\infty \mathrm{d}^3 p\, u_{\mathrm{D}}^*(\boldsymbol{p})v_{\mathrm{D}}(\boldsymbol{p}) = \int_\infty \mathrm{d}^3 r\, u^*(\boldsymbol{r})v(\boldsymbol{r}), \tag{4.107}$$

which is analogous to Parseval's theorem in Fourier theory. To prove (4.107), we simply express $u^*_D(\boldsymbol{p})$ and $v_D(\boldsymbol{p})$ by (4.82) and use (4.88), yielding

$$\begin{aligned}\langle \mathscr{D}u, \mathscr{D}v\rangle &= \int_\infty \mathrm{d}^3p \int_\infty \mathrm{d}^3r \int_\infty \mathrm{d}^3r'\ \psi^*(\boldsymbol{p}, \boldsymbol{r})u^*(\boldsymbol{r})\psi(\boldsymbol{p}, \boldsymbol{r}')v(\boldsymbol{r}') \\ &= \int_\infty \mathrm{d}^3r \int_\infty \mathrm{d}^3r'\ \delta(\boldsymbol{r} - \boldsymbol{r}')u^*(\boldsymbol{r})v(\boldsymbol{r}') \\ &= \int_\infty \mathrm{d}^3r\, u^*(\boldsymbol{r})v(\boldsymbol{r}) = \langle u, v\rangle, \end{aligned} \tag{4.108}$$

which shows that $\mathscr{D}$ is unitary.

The dipole-sheet operator is closely related to a family of unitary operators in mD space discussed by LUDWIG [1966].

§ 5. Applications

5.1. EMISSION COMPUTED TOMOGRAPHY

Computed tomography, as introduced in § 2.1, uses a source of X-rays outside the patient's body to measure the transmission of X-rays along many straight-line paths through the body. This transmission, as shown by (2.3), is simply related to the line integral of the X-ray attenuation coefficient $\mu(\boldsymbol{r})$ along the line from the source to the detector. The goal of the procedure is to reconstruct a 2D map of the attenuation coefficient.

Emission computed tomography (ECT), on the other hand, is a nuclear-medicine procedure in which the source of radiation is *inside* the patient's body. A radioactive pharmaceutical which concentrates preferentially in certain organs is administered, and the goal of the procedure is to map the distribution of radioactivity in the body.

ECT comes in two flavors, SPECT and PET. In SPECT (single-photon emission computed tomography), the radioisotope emits a single gamma ray in each nuclear decay event. The most common isotope is ^{99m}Tc, for which the gamma-ray energy is 140 kV. PET (positron emission tomography) uses isotopes like ^{11}C, ^{13}N, ^{15}O, and ^{19}F that emit positrons (antielectrons) when they decay. The positrons lose their kinetic energy within a few millimeters of where they originate and are thus not detectable outside the body. However, the positrons are annihilated by interaction with electrons in the tissue, producing two high-energy (511 keV) gamma rays. These photons are detected in coincidence by external detectors, producing line-integral data from which the isotope distribution can be reconstructed.

Although the hardware for PET is varied and complicated, the reconstruction methods are rather straightforward implementations of the 2D inverse Radon transform. (CORMACK [1973] suggested the use of the 3D transform for PET, but to date no system using this principle has been built.) Therefore, PET offers no essential new insights into the mathematics and will not be discussed further here. For a review, see BROWNELL, CORREIA and ZAMENHOF [1978].

SPECT, on the other hand, is mathematically novel in two respects: (1) it can involve either the 2D or the 3D Radon transform; (2) the attenuation of the radiation in the body leads to a new transform, the *attenuated Radon transform*.

A simple system for acquiring projection data in nuclear medicine is shown in Fig. 6. A position-sensitive detector such as an Anger scintillation camera (ANGER [1958, 1964]) is placed behind a parallel-hole collimator – essentially a thick block of lead with many holes bored in it. A particular element of the detector receives radiation that passes through a single bore of the collimator. If scattered radiation is negligible, the radiation reaching this detector element must have originated in a thin pencil-like region of the object. Therefore, if the collimator bores are long and slender and if the attenuation of the radiation by the patient's body can be neglected (a point to which we shall return below),

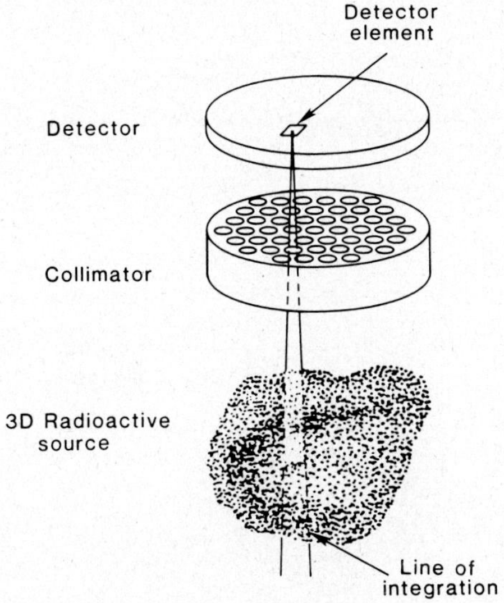

Fig. 6. A simple system for acquiring line-integral projection data in nuclear medicine. The detector is shown separated from the collimator for clarity, but in practice it would be in close proximity. Radiation reaching one detector element originates along a line in the object.

then the average flux at each detector element is, to a good approximation, proportional to a line integral of the isotope distribution. The entire 2D image for one position of the detector and collimator is thus a 2D line-integral projection of the 3D object; however, any one line of the 2D image is a 1D projection of a 2D slice of the object, which is just the usual CT projection, $\lambda_\phi(p)$ for fixed ϕ. To acquire the complete projection data set, the collimator and detector must be rotated around the object.

If a planar integral of the isotope distribution is desired, it can be obtained by integrating the 2D projection along lines. However, a more efficient way to get the same information is shown in Fig. 7. Here, an assembly of lead slats collimates the radiation in one direction only (KEYES [1975]). The 2D flux pattern on the detector varies in the direction perpendicular to the slats, but is constant parallel to them, at least on the average. The flux reaching one strip on the detector is, under the same assumptions as above, proportional to a planar integral of the isotope distribution. For one orientation of the collimator, the system measures $\lambda_{\hat{n}}(p)$ for fixed $\hat{\boldsymbol{n}}$, where $\hat{\boldsymbol{n}}$ is normal to the slats.

We saw in § 3.2 that $\hat{\boldsymbol{n}}$ must explore a full hemisphere of directions to produce a complete 3D Radon transform. Thus the collimator and detector must be systematically scanned. For example, the slats could be rotated 180° about a normal to the detector face, then this normal could be incremented by some small angle, and the process repeated many times.

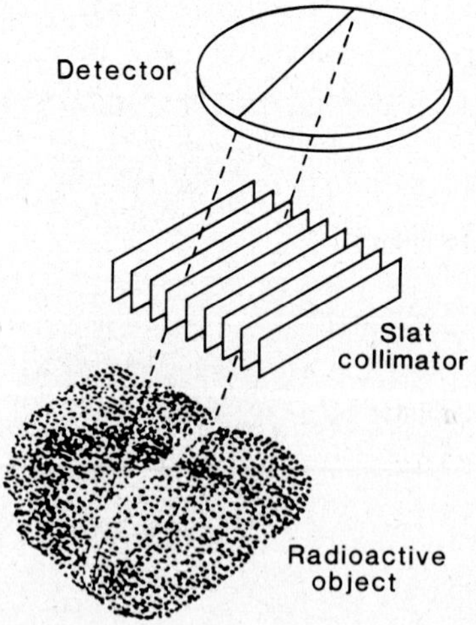

Fig. 7. System for acquiring planar-integral projection data in nuclear medicine. In practice, the detector would be in close proximity to the slat collimator. Radiation reaching a strip on the detector originates from a plane of the object.

A disadvantage of the slat collimator is that two scanning motions are required. The advantages are that the system efficiently collects photons and that the detector needs to have spatial resolution in only the direction perpendicular to the slats (GINDI, ARENDT, BARRETT, CHIU, ERVIN, GILES, KUJOORY, MILLER and SIMPSON [1982]). Furthermore planar integrals may be superior to line integrals when the data are noisy or incomplete. For example, RADULOVIC and VEST [1975] point out that good numerical reconstructions are still possible when $\hat{\boldsymbol{n}}$ covers significantly less than 2π steradians.

In discussing the parallel-hole and slat collimators above, we neglected the attenuation of the radiation in the patient's body. In fact, this is a terrible approximation. The attenuation coefficient of soft tissue at 140 keV, for example, is about $0.15\ \mathrm{cm}^{-1}$, so just 6 cm of tissue causes $1/e$ attenuation. What is measured with the parallel-hole collimator is not line integrals of the activity distribution, but exponentially weighted line integrals of the form

$$\lambda_{\phi}^{(e)}(p) = \int_{\infty} \mathrm{d}^2 r\, f(\boldsymbol{r}) a(\boldsymbol{r}; p, \phi) \delta(p - \boldsymbol{r} \cdot \hat{\boldsymbol{n}}), \tag{5.1}$$

where $f(\boldsymbol{r})$ is the radioisotope distribution in a 2D object section, and the attenuation factor $a(\boldsymbol{r}; p, \phi)$ is given by

$$a(\boldsymbol{r}; p, \phi) = \exp\left[\int_{\boldsymbol{r}}^{\mathrm{D}} \mu(\boldsymbol{r}')\, \mathrm{d}l'\right], \tag{5.2}$$

where the integral runs along a straight line from the source point $\boldsymbol{r}$ to the detector location D. Equation (5.1) is the attenuated Radon transform (GULLBERG [1979]). In general, $\lambda_{\phi}^{(e)}(p)$ depends on two arbitrary and independent 2D functions $f(\boldsymbol{r})$ and $\mu(\boldsymbol{r})$. There is a considerable literature on approximate and *ad hoc* inversions of (5.1). Good reviews are given by BUDINGER and GULLBERG [1977], GULLBERG [1979], and JASZCZAK, COLEMAN and LIM [1980].

A very important contribution to the theory of the attenuated Radon transform was recently made by TRETIAK and METZ [1980]. Their solution is exact when (1) the attenuation coefficient is independent of $\boldsymbol{r}$, and (2) the body contours are convex and accurately known. Since these conditions are often satisfied to a good approximation, we shall give a detailed account of the TRETIAK–METZ algorithm.

The geometry is shown in Fig. 8. If μ is constant and the body is convex, (5.1) becomes

$$\lambda_{\phi}^{(e)}(p) = \int_{\infty} \mathrm{d}^2 r\, f(\boldsymbol{r})\, \mathrm{e}^{-\mu(L_1 + L_2)} \delta(p - \boldsymbol{r} \cdot \hat{\boldsymbol{n}}), \tag{5.3}$$

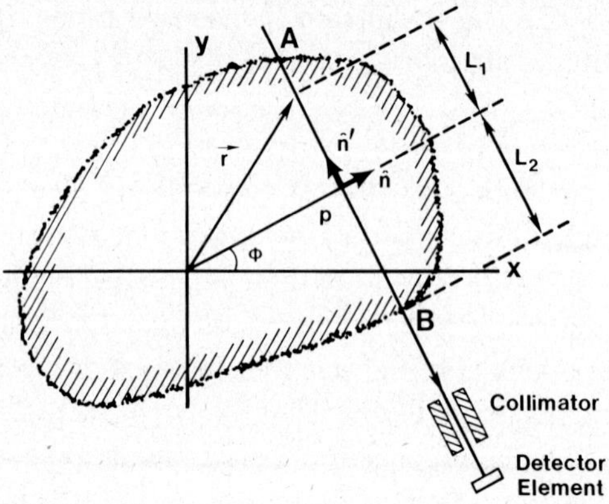

Fig. 8. Geometry for the attenuated Radon transform.

where L_2 is a presumably known function of p, ϕ, and the body contours, and $L_1 = \boldsymbol{r} \cdot \hat{\boldsymbol{n}}'$, with $\hat{\boldsymbol{n}}' \perp \hat{\boldsymbol{n}}$.

In the Tretiak–Metz algorithm, a modified projection $\tilde{\lambda}_\phi(p)$ is formed by multiplying the original projection $\lambda_\phi^{(e)}(p)$ by the exponential factor $\exp(\mu L_2)$. The modified projection must be convolved with a 1D filter function $h(p)$ and then back-projected with another exponential weighting factor $\exp(\mu L_1)$. This algorithm differs from the standard inverse Radon transform (2.21) by the presence of the weighting factors and in the functional form of $h(p)$. The estimate $f(\boldsymbol{r})$ of the activity distribution is thus

$$f(\boldsymbol{r}) = \int_0^{2\pi} \mathrm{d}\phi \exp(\mu L_1)[\tilde{\lambda}_\phi(p) * h(p)]_{p = \boldsymbol{r}\cdot\hat{\boldsymbol{n}}}, \tag{5.4}$$

where

$$\tilde{\lambda}_\phi(p) = \lambda_\phi^{(e)}(p) \exp(\mu L_2). \tag{5.5}$$

Note that the integral runs over 2π radians; $\lambda_\phi^{(e)}$ and $\lambda_{\phi+\pi}^{(e)}$ are not redundant because of the attenuation factor.

The filter $h(p)$ is most easily described in the frequency domain; its 1D Fourier transform is

$$H(\nu) = \tfrac{1}{2}|\nu| \left[1 - \mathrm{rect}\left(\frac{\pi\nu}{\mu} \right) \right]. \tag{5.6}$$

This function is illustrated in Fig. 9. It is the usual $|\nu|$ filter but with a gap in the interval $(-\mu/2\pi, \mu/2\pi)$.

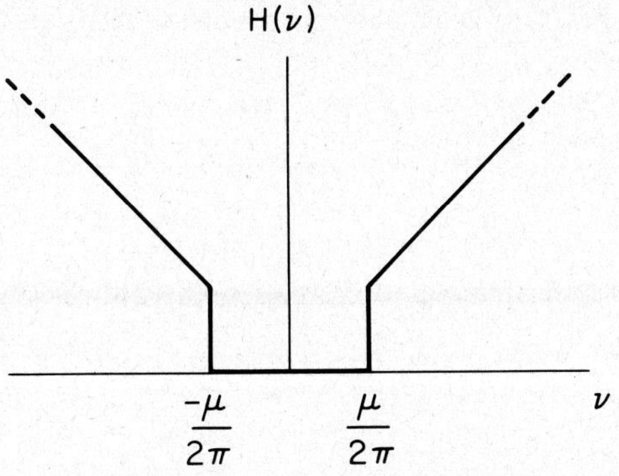

Fig. 9. Filter used in the TRETIAK–METZ algorithm.

To show that (5.4) and (5.5) constitute an exact inverse transform, let us consider a single point object at a general point $\boldsymbol{r}_0$:

$$f^{\delta}(\boldsymbol{r}) = \delta(\boldsymbol{r} - \boldsymbol{r}_0). \tag{5.7}$$

The attenuated Radon transform of this object is

$$\lambda_{\phi}^{(\mathrm{e},\delta)}(p) = \mathrm{e}^{-\mu(L_{10}+L_2)}\,\delta(p - \boldsymbol{r}_0 \cdot \hat{\boldsymbol{n}}), \tag{5.8}$$

where

$$L_{10} = \boldsymbol{r}_0 \cdot \hat{\boldsymbol{n}}'. \tag{5.9}$$

Inserting (5.8) into (5.4) yields

$$\hat{f}^{\delta}(\boldsymbol{r}) = \int_0^{2\pi} \mathrm{d}\phi\, \mathrm{e}^{\mu(\boldsymbol{r}-\boldsymbol{r}_0)\cdot\hat{\boldsymbol{n}}'}\, h[(\boldsymbol{r} - \boldsymbol{r}_0)\cdot\hat{\boldsymbol{n}}]. \tag{5.10}$$

This equation gives the point spread function (PSF) of the system. The PSF depends only on the difference $\boldsymbol{r} - \boldsymbol{r}_0$, and the system is therefore shift-invariant. Since the integral runs over 2π, we can choose the axes so that $\boldsymbol{r} - \boldsymbol{r}_0$ is along the x-axis. Then

$$\hat{f}^{\delta}(\boldsymbol{r}) = \int_0^{2\pi} \mathrm{d}\phi\, \mathrm{e}^{\mu R \sin\phi}\, h(R\cos\phi), \tag{5.11}$$

where $R = |\boldsymbol{r} - \boldsymbol{r}_0|$. If we represent $h(R\cos\phi)$ in terms of its Fourier transform, (5.11) becomes

$$\hat{f}^{\delta}(\boldsymbol{r}) = \int_0^{2\pi} \mathrm{d}\phi\, \mathrm{e}^{\mu R \sin\phi} \int_{-\infty}^{\infty} H(\nu)\, \mathrm{e}^{2\pi \mathrm{i} \nu R \cos\phi}\, \mathrm{d}\nu. \tag{5.12}$$

The integral over ϕ is (GRADSHTEYN and RYZHIK [1980] p. 488; TRETIAK and METZ [1980]),

$$\int_0^{2\pi} e^{\mu R \sin\phi}\, e^{2\pi i \nu R \cos\phi}\, d\phi = 2\pi J_0\{R[(2\pi\nu)^2 - \mu^2]^{1/2}\}, \qquad (2\pi\nu)^2 > \mu^2. \tag{5.13}$$

With (5.6), (5.13) and a change of variables, (5.12) becomes

$$\hat{f}^{\delta}(\boldsymbol{r}) = 2\pi \int_{\mu/2\pi}^{\infty} \nu\, d\nu\, J_0\{R[(2\pi\nu)^2 - \mu^2]^{1/2}\} = 2\pi \int_0^{\infty} \rho\, d\rho\, J_0(2\pi\rho R), \tag{5.14}$$

where

$$\rho^2 = \nu^2 - (\mu/2\pi)^2. \tag{5.15}$$

We recognize (5.14) as the Hankel transform of unity. Since the Hankel transform is the 2D Fourier transform for rotationally symmetric functions, we have

$$\hat{f}^{\delta}(\boldsymbol{r}) = \delta(\boldsymbol{r} - \boldsymbol{r}_0), \tag{5.16}$$

which is an exact reconstruction of the original $f^{\delta}(\boldsymbol{r})$. Since this result was obtained for an arbitrary $\boldsymbol{r}_0$, it holds for any object by linear superposition, and the TRETIAK–METZ formula is an exact inverse to the attenuated Radon transform with constant attenuation (CLOUGH and BARRETT. [1983]). In practice, some apodized approximation to $H(\nu)$ would be used, just as in the unattenuated case, but with the form (5.6), the inverse is mathematically exact.

It may seem surprising that an exact reconstruction can be obtained with a finite gap in the frequency space. This is possible because the central-slice theorem does not hold in the attenuated case. A frequency component ν in the projection does not correspond uniquely to a frequency $\boldsymbol{\rho} = \hat{\boldsymbol{n}}\nu$ in the object, and no object frequencies are irretrievably lost by the gap in $H(\nu)$.

Although many important properties of the attenuated Radon transform have been given by GULLBERG [1979], BELLINI, PIACENTINI, CAFFORIO and ROCCA [1979], and TRETIAK and DELANEY [1977], further work is still needed to generalize the TRETIAK–METZ theory to non-convex objects and inhomogeneous μ, and also to the 3D case.

A related transform, the attenuated Abel transform, was recently derived by CLOUGH and BARRETT [1983].

5.2. NMR IMAGING

Nuclear magnetic resonance (NMR) has long been an important research tool in physics and chemistry. It occurs because any nucleus with an unpaired proton or neutron possesses an intrinsic magnetic moment $\boldsymbol{\mu}$ and a spin angular momentum $\hbar \boldsymbol{I}$. These two quantities are related by

$$\boldsymbol{\mu} = \gamma \hbar \boldsymbol{I}, \tag{5.17}$$

where the constant γ is called the magnetogyric ratio. In a static magnetic field $\boldsymbol{B}_0$, the interaction energy of the nuclear magnetic dipole is

$$U = -\boldsymbol{\mu} \cdot \boldsymbol{B}_0. \tag{5.18}$$

For a nucleus with nuclear spin $I = \frac{1}{2}$ such as hydrogen, the component of $\boldsymbol{\mu}$ parallel to $\boldsymbol{B}_0$ can be only $\pm \frac{1}{2}\gamma\hbar$, and the energy difference between these two quantum states is

$$\Delta U = \gamma \hbar B_0. \tag{5.19}$$

If a radio-frequency (RF) magnetic field of frequency ω is applied, it can induce transitions between the energy levels provided

$$\hbar\omega = \Delta U \tag{5.20a}$$

or

$$\omega = \gamma B_0. \tag{5.20b}$$

This condition for resonant absorption is called the Larmor condition, and γB_0 is the Larmor frequency ω_L. For hydrogen, $\gamma = 2.675 \times 10^8\ s^{-1}\ \text{tesla}^{-1}$, and

$$\nu_L(\text{MHz}) = 42.58 B_0\ (\text{tesla}), \tag{5.21}$$

where $\nu_L = \omega_L / 2\pi$ and 1 tesla $= 10^4$ gauss.

NMR experiments involve a large number N of nuclei, N_+ of which are in the upper state and N_- in the lower state. In thermal equilibrium,

$$\frac{N_+}{N_-} = \exp[-\gamma\hbar B_0/k_B T] \approx 1 - \gamma\hbar B_0/k_B T, \tag{5.22}$$

where k_B is Boltzmann's constant and T is the absolute temperature. For example, if $T = 300\ ^\circ\text{K}$ and $B_0 = 0.1$ tesla, then $N_+/N_- = 1 - (6.7 \times 10^{-7})$. Thus, the populations are very nearly equal, and absorption and stimulated emission are almost equally probable, but the entire resonant effect comes from the small difference in population.

It is often convenient to consider the net macroscopic magnetization $\boldsymbol{M}$ rather than the magnetic moment of individual nuclei. Since magnetization is magnetic moment per unit volume, we have

$$\boldsymbol{M} = \frac{1}{V} \sum_{i=1}^{N} \boldsymbol{\mu}_i \qquad (5.23)$$

where $\boldsymbol{\mu}_i$ is the moment of the ith nucleus and V is the volume in which the N nuclei are contained. Because of the near balance between the two possible orientations of each $\boldsymbol{\mu}_i$, $|\boldsymbol{M}|$ is much less than $N\mu/V$. For most purposes, $\boldsymbol{M}$ may be treated as a classical vector.

A magnetic dipole $\boldsymbol{\mu}$ in a field $\boldsymbol{B}_0$ experiences a torque $\boldsymbol{\mu} \times \boldsymbol{B}_0$. The torque is the rate of change of the angular momentum, which in turn is connected to $\boldsymbol{\mu}$ by (5.17). Combining this result with (5.23) shows that

$$\frac{\mathrm{d}\boldsymbol{M}}{\mathrm{d}t} = \gamma \boldsymbol{M} \times \boldsymbol{B}_0. \qquad (5.24)$$

Thus, if $\boldsymbol{M}$ is initially not parallel to $\boldsymbol{B}_0$, its time derivative will be perpendicular to both $\boldsymbol{M}$ and $\boldsymbol{B}_0$; in other words, $\boldsymbol{M}$ precesses about $\boldsymbol{B}_0$ like a gyroscope. (See Fig. 10.) Equation (5.24) would be correct if no other forces besides that due to $\boldsymbol{B}_0$ acted on $\boldsymbol{M}$. In fact, we must consider three other effects: (1) the RF field; (2) longitudinal relaxation; (3) transverse relaxation.

The RF field may be accounted for by adding $\boldsymbol{B}_{\mathrm{RF}}$ to $\boldsymbol{B}_0$ in (5.24). However, $\boldsymbol{B}_{\mathrm{RF}}$ is usually applied in short pulses, and the total field is just $\boldsymbol{B}_0$ between pulses.

Longitudinal relaxation occurs because the spin system can exchange energy with its environment or "lattice", for example by emitting or absorbing phonons. This relaxation, which tends to reduce the angle between $\boldsymbol{M}$ and $\boldsymbol{B}_0$, may be

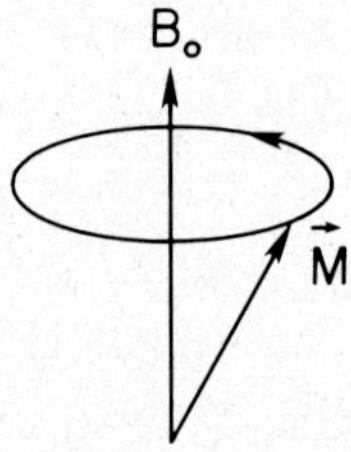

Fig. 10. The magnetization vector $\boldsymbol{M}$ precesses with angular velocity ω_{L} in a static magnetic field $\boldsymbol{B}_0$.

described phenomenologically by

$$\frac{\mathrm{d}M_z}{\mathrm{d}t} = \frac{M_0 - M_z}{T_1}, \tag{5.25}$$

where it is assumed that $\boldsymbol{B}_0$ is parallel to the z-axis, M_0 is the thermal-equilibrium value of M_z, and T_1 is called the spin-lattice or longitudinal relaxation time.

Relaxation of the transverse components of $\boldsymbol{M}$ occurs by spin-lattice interaction and also because the local magnetic field is not identical at each nucleus. In addition to $\boldsymbol{B}_0$ and $\boldsymbol{B}_{\mathrm{RF}}$, there is a field on a particular nucleus due to the magnetic moments of all other nuclei. This fluctuating field alters the precession frequency and broadens the absorption line. Because of the difference in precession frequency, the spins tend to "dephase" and the transverse components μ_x and μ_y are randomized after a characteristic time T_2. That means that the net transverse components of the magnetization, M_x and M_y, relax to zero with this time constant T_2, usually called the spin-spin or transverse relaxation time.

Accounting for all of these effects, we obtain the Bloch equations,

$$\frac{\mathrm{d}M_z}{\mathrm{d}t} = \gamma(\boldsymbol{M} \times \boldsymbol{B})_z + \frac{M_0 - M_z}{T_1}, \tag{5.26}$$

$$\frac{\mathrm{d}M_x}{\mathrm{d}t} = \gamma(\boldsymbol{M} \times \boldsymbol{B})_x - \frac{M_x}{T_2}, \tag{5.27}$$

$$\frac{\mathrm{d}M_y}{\mathrm{d}t} = \gamma(\boldsymbol{M} \times \boldsymbol{B})_y - \frac{M_y}{T_2}, \tag{5.28}$$

where $\boldsymbol{B} = \boldsymbol{B}_0 + \boldsymbol{B}_{\mathrm{RF}}$.

When the spin system absorbs energy from the RF pulse, the magnetization is tipped away from the z-axis and begins to precess. This creates transverse components of $\boldsymbol{M}$ oscillating at frequency ω_{L} and decaying with time constant T_2. A set of receiver coils can then pick up a signal, called the free-induction decay signal, from which one can deduce the spin density and T_2.

Many other ingenious schemes have been developed to measure the spin density N/V, magnetogyric ratio γ, and relaxation times T_1 and T_2. For a review, see ABRAGAM [1961], FARRAR and BECKER [1971], or SLICHTER [1963]. However, in all of these methods, it is assumed that the quantity being measured is homogeneously distributed throughout the specimen, and great care is exercised to make the applied field homogeneous as well. Thus the measurement involves a volume integral of some quantity over the volume of the specimen. More recently, great progress has been made in extracting information about the *spatial* distribution of spin density and relaxation time. These efforts have given rise to the new and exciting field of *NMR imaging* or *zeugmatography* (from the Greek ζευγμα, "that which joins together"). Recent reviews of this field, emphasizing clinical applications, have been given by PYKETT [1982], PYKETT, NEWHOUSE, BUONANNO, BRADY, GOLDMAN, KISTLER and POHOST [1982], MANSFIELD and MORRIS [1982], KAUFMANN, CROOKS and MARGULIS [1981], and KARSTAEDT, WITCOFSKI and PARTAIN [1981].

Since the human body is mainly water, almost all medical imaging applications of NMR have used the resonance of the proton or hydrogen nucleus. (^{16}O does not have a nuclear magnetic moment since it contains an even number of protons and neutrons.) The imaging instruments that are being developed thus have as their goal the mapping of proton density $n(\boldsymbol{r})$ (essentially, water density) or proton relaxation times $T_1(\boldsymbol{r})$ or $T_2(\boldsymbol{r})$. Either true 3D maps or a series of 2D slices can be produced, and all of the instruments have the capability of producing images that emphasize $n(\boldsymbol{r})$, $T_1(\boldsymbol{r})$, or $T_2(\boldsymbol{r})$ as desired, simply by using different RF pulse sequences.

NMR imaging systems are often classified as point, line, plane, or volume systems. Point systems collect data from a small, isolated region of the object at a time; 2D or 3D information is built up by scanning the sensitive point over the object. This procedure is straightforward but inefficient, and all modern systems collect data from a larger region simultaneously. They all, in one guise or another, measure either a 2D or a 3D Radon transform.

Let us consider first a few of the many ways to isolate a single plane. For this purpose, the magnetic field must be inhomogenous, since otherwise all points in the volume would be equivalent and no spatial resolution would be possible. The usual approach is to add a weak "gradient field" that varies linearly with position. More precisely, if the large static field $\boldsymbol{B}_0$ is in the z direction, the z component of the total field is

$$B_z = B_0 + \alpha(\boldsymbol{r}\cdot\hat{\boldsymbol{n}}), \tag{5.29}$$

where $\nabla B_z = \alpha\hat{\boldsymbol{n}}$ with α constant. The x and y components of the gradient field do not influence the resonant frequency in first order since

$$B = [B_x^2 + B_y^2 + B_z^2]^{1/2} \approx B_z + \frac{B_x^2 + B_y^2}{2B_z^2}, \tag{5.30}$$

which is negligibly different from B_z if $B_x, B_y \ll B_0$.

Now suppose that the nuclear spin system is excited with a relatively long RF pulse having a well-defined frequency ω_0. Only the spins with $\omega_L = \omega_0$ will resonantly absorb energy, and only these will contribute to the NMR signal. From (5.20) and (5.29), the resonant spins lie on the plane defined by

$$\omega_0 - \gamma B_0 = \alpha\gamma(\boldsymbol{r}\cdot\hat{\boldsymbol{n}}), \tag{5.31}$$

and the signal yields one point in a 3D Radon transform. The quantity being transformed depends in a complicated way on the RF pulse sequence and subsequent signal processing; in general, it is some mixture of $n(\boldsymbol{r})$, $T_1(\boldsymbol{r})$, and $T_2(\boldsymbol{r})$.

The method just described yields one planar integral in a single measurement, and a time of about $3T_1$ must be allowed for the system to return to equilibrium before another measurement can be made. Since several hundred measurements must be made to adequately sample the 3D Radon space, and T_1 is typically 0.1–1 s, the imaging time with this method is rather long. However, a slight modification of the method allows the simultaneous measurement of integrals on a set of parallel planes. If the RF pulse is very short so that it contains a broad spectrum of frequencies, all spins in the volume are simultaneously excited. Alternatively, if the gradient field is applied *after* excitation, all spins will have the same resonant frequency when they are excited. In either case, the precession frequency depends on the local field during precession, not during excitation. Therefore, each plane of constant field will contain spins precessing at a characteristic frequency and with a transverse component of $\boldsymbol{M}$ decaying with time constant T_2. The overall NMR signal will be a superposition of these free-induction decay signals (Fig. 11). A Fourier analysis serves to separate the composite signal into its individual frequencies, each frequency corresponding to one plane in the object. This procedure yields points along a line in 3D Radon space, and the direction of the field gradient $\hat{\boldsymbol{n}}$ can be varied on subsequent excitations to sample the full Radon volume. Reconstruction of the full 3D image of the object can then be carried out by any of the methods discussed in § 3 (SHEPP [1980]; MARR, CHEN and LAUTERBUR [1981]).

If only one or a few planes of the object are to be imaged, it is simpler to use line integrals and the 2D Radon transform. One way to accomplish this is to use two orthogonal field gradients, say one in the z direction and one in the x–y plane. Originally, only the z gradient is applied and a narrow-band RF pulse is used to selectively excite spins in a plane normal to the z-axis. Then the z gradient is switched off, the x–y gradient is turned on, and the free induction decay is observed. If this gradient is in direction $\hat{\boldsymbol{n}}$ in the x–y plane, all spins in a plane perpendicular to $\hat{\boldsymbol{n}}$ would have the same precession

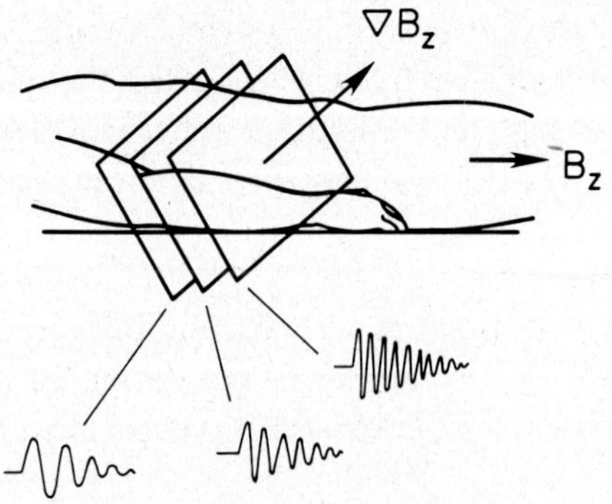

Fig. 11. Illustration of one way to acquire planar-integral data in NMR imaging. All spins in a volume are simultaneously excited. When they decay in a field gradient, each plane has a characteristic frequency, and the planes can be separated by a Fourier transform of the received signal.

frequency, but not all of these spins were excited. The signal at a particular frequency thus comes from spins along the line defined by the intersection of the plane of excitation and a plane normal to $\hat{\boldsymbol{n}}$. The complete Fourier transform of the free-induction decay signal is thus equivalent to a 1D projection of the 2D slice selected during excitation. In subsequent excitations, the direction $\hat{\boldsymbol{n}}$ is varied, and a full 2D Radon transform is sampled. Any variations of the 2D inverse Radon transform can be used for reconstruction.

5.3. ELECTROSTATICS

There is a close connection between Radon theory and electrostatics, as might be expected by the occurrence of the Laplacian operator ∇^2 in the inverse Radon transform. Indeed, Radon himself was motivated by this connection to consider the problem in the first place, and he originally found the inverse transform by using known results of potential theory.

Electrostatics is based on Poisson's equation,

$$\nabla^2 \Phi(\boldsymbol{r}) = -4\pi\rho(\boldsymbol{r}), \tag{5.32}$$

where $\Phi(\boldsymbol{r})$ is the electrostatic potential and $\rho(\boldsymbol{r})$ is the charge density. If there are no boundaries and $\rho(\boldsymbol{r})$ is known for all $\boldsymbol{r}$, the solution of Poisson's equation in 3D is well known to be

$$\Phi(\boldsymbol{r}) = -4\pi\nabla^{-2}\rho(\boldsymbol{r}) = \int_\infty \mathrm{d}^3r' \, \frac{\rho(\boldsymbol{r}')}{|\boldsymbol{r}-\boldsymbol{r}'|}, \tag{5.33}$$

where ∇^{-2} is the inverse of the Laplacian operator ∇^2. Comparing this result to (3.16), we see that $\Phi(\boldsymbol{r})$ is the summation image of $\rho(\boldsymbol{r})$, obtained by taking the 3D Radon transform, back-projecting, and summing over all projection directions in 2π steradians. In operator form (RADON [1917]),

$$\Phi = \frac{1}{\pi} \mathscr{B}_3 \mathscr{R}_3 \rho. \tag{5.34}$$

This is consistent with Poisson's equation since, by (3.26),

$$\mathscr{R}_3^{-1} = -\frac{1}{4\pi^2} \nabla^2 \mathscr{B}_3. \tag{5.35}$$

Therefore, we may formally write the inverse Laplacian as

$$\nabla^{-2} = -\frac{1}{4\pi^2} \mathscr{B}_3 \mathscr{R}_3. \tag{5.36}$$

We may verify that this result is consistent with (5.33) by writing the operators out in detail. This gives

$$\begin{aligned}
\Phi(\boldsymbol{r}) &= \frac{1}{\pi} \int_{2\pi} \mathrm{d}\Omega_n \left[\int_{\infty} \mathrm{d}^3 r' \, \rho(\boldsymbol{r}') \delta(p - \boldsymbol{r}' \cdot \hat{\boldsymbol{n}}) \right]_{p = \boldsymbol{r} \cdot \hat{\boldsymbol{n}}} \\
&= \frac{1}{\pi} \int_{\infty} \mathrm{d}^3 r' \, \rho(\boldsymbol{r}') \int_{2\pi} \mathrm{d}\Omega_n \, \delta[(\boldsymbol{r} - \boldsymbol{r}') \cdot \hat{\boldsymbol{n}})] \\
&= \int_{\infty} \mathrm{d}^3 r' \, \frac{\rho(\boldsymbol{r}')}{|\boldsymbol{r} - \boldsymbol{r}'|},
\end{aligned} \tag{5.37}$$

where the last step follows from (3.15). Thus, the operator form (5.36) is indeed equivalent to (5.33). The function $|\boldsymbol{r} - \boldsymbol{r}'|^{-1}$ plays the dual role of point spread function in the summation image and Green's function for Poisson's equation.

In 2D, the situation is a little more complicated. Poisson's equation has the same form,

$$\nabla^2 \Phi(\boldsymbol{r}) = -4\pi\rho(\boldsymbol{r}); \qquad \Phi(\boldsymbol{r}) = -4\pi\nabla^{-2}\rho(\boldsymbol{r}), \tag{5.38}$$

but we know from (2.56) that

$$\nabla^{-2} = \frac{1}{2\pi^2} \mathscr{B}_2 [\ln|p| *] \mathscr{R}_2. \tag{5.39}$$

Writing the operators out in detail, we find

$$\Phi(\boldsymbol{r}) = -\frac{2}{\pi}\int_0^{\pi} \mathrm{d}\phi \int_{-\infty}^{\infty} \mathrm{d}p \ln|p - \boldsymbol{r}\cdot\hat{\boldsymbol{n}}| \int_{\infty} \mathrm{d}^2r' \, \rho(\boldsymbol{r}')\delta(p - \boldsymbol{r}'\cdot\hat{\boldsymbol{n}})$$

$$= -\frac{2}{\pi}\int_0^{\pi} \mathrm{d}\phi \int_{\infty} \mathrm{d}^2r' \, \rho(\boldsymbol{r}') \ln|(\boldsymbol{r}' - \boldsymbol{r})\cdot\hat{\boldsymbol{n}}|. \tag{5.40}$$

The ϕ integral is given by

$$\int_0^{\pi} \mathrm{d}\phi \ln[\,|\boldsymbol{r} - \boldsymbol{r}'| \cdot \cos\psi|\,] = \pi \ln[\,|\boldsymbol{r} - \boldsymbol{r}'|\,] + \tfrac{1}{2}\int_0^{2\pi} \mathrm{d}\psi \ln[\,|\cos\psi|\,], \tag{5.41}$$

where ψ is the angle between $\boldsymbol{r} - \boldsymbol{r}'$ and $\hat{\boldsymbol{n}}$. The integral over ψ is just some constant C, independent of $\boldsymbol{r}$ and $\boldsymbol{r}'$, and (5.40) becomes

$$\Phi(\boldsymbol{r}) = -2\int_{\infty} \mathrm{d}^2r' \, \rho(\boldsymbol{r}')\{\ln|\boldsymbol{r} - \boldsymbol{r}'| + C\}. \tag{5.42}$$

The value of C is irrelevant since an additive constant in the potential does not influence the electric field.

To show that (5.42) is indeed a solution of (5.38), we can operate on it with ∇^2 and recognize that (MORSE and FESHBACH [1953] p. 891)

$$\nabla^2 \ln|\boldsymbol{r} - \boldsymbol{r}'| = 2\pi\delta(\boldsymbol{r} - \boldsymbol{r}'). \tag{5.43}$$

and, of course, $\nabla^2 C = 0$.

In 2D, the Green's function for Poisson's equation is *not* the same as the PSF in the summation image; the former is $-2\ln|\boldsymbol{r} - \boldsymbol{r}'|$ and the latter is $|\boldsymbol{r} - \boldsymbol{r}'|^{-1}$.

5.4. WAVE PROPAGATION

Before the advent of CT, the prime motivation for study of the Radon transform was its usefulness in radiation and scattering problems (JOHN [1955]; COURANT and HILBERT [1962]; LAX and PHILLIPS [1967]). These problems usually involve solution of the time-dependent homogeneous scalar wave equation,

$$\left(\nabla^2 - \frac{1}{c^2}\frac{\partial^2}{\partial t^2}\right)\psi(\boldsymbol{r}, t) = 0. \tag{5.44}$$

There is an intimate connection between the wave equation and the Radon transform because the general solution of the wave equation can be written as a superposition of plane waves. This statement is a commonplace if we think in terms of *harmonic* plane waves of the form $\exp(\mathrm{i}\boldsymbol{k}\cdot\boldsymbol{r} - \mathrm{i}\omega t)$, but the term plane waves has a much broader meaning than that. It refers to any function that is constant on a plane. Since $\boldsymbol{r}\cdot\hat{\boldsymbol{n}}$ is constant on any plane normal to $\hat{\boldsymbol{n}}$, we can construct plane waves at will by making functions of $\boldsymbol{r}\cdot\hat{\boldsymbol{n}}$. Examples of mathematical interest (JOHN [1955]; COURANT and HILBERT [1962]) include power-law plane waves $|p - \boldsymbol{r}\cdot\hat{\boldsymbol{n}}|^k$, logarithmic plane waves $\ln|p - \boldsymbol{r}\cdot\hat{\boldsymbol{n}}|$, and impulsive plane waves $\delta(p - \boldsymbol{r}\cdot\hat{\boldsymbol{n}})$ or $\delta'(p - \boldsymbol{r}\cdot\hat{\boldsymbol{n}})$. The latter, of course, form the basis functions for the Radon and dipole-sheet transforms, respectively. To make any of these functions into solutions of the homogeneous wave equation, we need only choose p = constant + ct. For example, from (4.98), it is readily seen that

$$\left(\nabla^2 - \frac{1}{c^2}\frac{\partial^2}{\partial t^2}\right)\delta(p_0 + ct - \boldsymbol{r}\cdot\hat{\boldsymbol{n}}) = 0. \tag{5.45}$$

This function represents a planar sheet moving with velocity c in the direction $\hat{\boldsymbol{n}}$ and at a perpendicular distance p_0 from the origin at $t = 0$.

To see the usefulness of plane-wave decompositions, let us use the tools assembled in §§ 3 and 4 to solve *Cauchy's problem*, in which we wish to find the solution to (5.44) subject to the initial conditions

$$\psi(\boldsymbol{r}, 0) = 0, \qquad \left.\frac{\partial\psi(\boldsymbol{r}, t)}{\partial t}\right|_{t=0} = v(\boldsymbol{r}). \tag{5.46}$$

There are no sources present and no spatial boundaries except at infinity. Once we have solved this problem, the more general solution where $\psi(\boldsymbol{r}, 0) \neq 0$ can be readily deduced (COURANT and HILBERT [1962] p. 682).

The general procedure is quite analogous to a Fourier decomposition. We write the solution as a superposition of plane waves with initially unknown expansion coefficients, then choose these coefficients to satisfy the initial conditions. Using impulsive plane waves as the basis, we can write

$$\psi(\boldsymbol{r}, t) = \int_{2\pi} \mathrm{d}\Omega_n \int_{-\infty}^{\infty} \mathrm{d}p\, g(p, \hat{\boldsymbol{n}})\delta(p + ct - \boldsymbol{r}\cdot\hat{\boldsymbol{n}}). \tag{5.47}$$

By differentiating under the integral sign, it is readily seen that this $\psi(\boldsymbol{r}, t)$ indeed satisfies the wave equation. To match the initial conditions, we must have

$$v(\boldsymbol{r}) = c\int_{2\pi} \mathrm{d}\Omega_n \int_{-\infty}^{\infty} \mathrm{d}p\, g(p, \hat{\boldsymbol{n}})\delta'(p - \boldsymbol{r}\cdot\hat{\boldsymbol{n}}). \tag{5.48}$$

From (4.83), we recognize the integral as an inverse dipole-sheet transform, so that

$$v(\boldsymbol{r}) = 2\pi i c\, \mathscr{D}^{-1}\{g(p, \hat{\boldsymbol{n}})/p\}. \tag{5.49}$$

Solving for $g(p, \hat{\boldsymbol{n}})$, we find

$$g(p, \hat{\boldsymbol{n}}) = \frac{p}{2\pi i c}\, \mathscr{D}\{v(\boldsymbol{r})\} = \frac{1}{4\pi^2 c} \int_\infty \mathrm{d}^3 r'\, v(\boldsymbol{r}')\, \delta'(p - \boldsymbol{r}' \cdot \hat{\boldsymbol{n}}). \tag{5.50}$$

Inserting (5.50) into (5.47) yields

$$\begin{aligned} \psi(\boldsymbol{r}, t) &= \frac{1}{4\pi^2 c} \int_{2\pi} \mathrm{d}\Omega_n \int_{-\infty}^{\infty} \mathrm{d}p\, \delta(p + ct - \boldsymbol{r} \cdot \hat{\boldsymbol{n}}) \int_\infty \mathrm{d}^3 r'\, v(\boldsymbol{r}') \delta'(p - \boldsymbol{r}' \cdot \hat{\boldsymbol{n}}) \\ &= \frac{-1}{4\pi^2 c} \int_{2\pi} \mathrm{d}\Omega_n \int_\infty \mathrm{d}^3 r'\, v(\boldsymbol{r}') \delta'[ct - (\boldsymbol{r} - \boldsymbol{r}') \cdot \hat{\boldsymbol{n}}]. \end{aligned} \tag{5.51}$$

To put this result in a more familiar form, we can use (4.95) to write

$$\int_{2\pi} \mathrm{d}\Omega_n\, \delta'[ct - (\boldsymbol{r} - \boldsymbol{r}') \cdot \hat{\boldsymbol{n}}] = \frac{-\pi}{R}\, \delta(ct - R), \tag{5.52}$$

for $t > 0$ with $R = |\boldsymbol{r} - \boldsymbol{r}'|$. Then (5.51) reduces to the well-known solution (MORSE and FESHBACH [1953] p. 837)

$$\psi(\boldsymbol{r}, t) = \frac{1}{4\pi c^2} \int_\infty \mathrm{d}^3 r'\, v(\boldsymbol{r}') \frac{1}{R}\, \delta\left(t - \frac{R}{c}\right). \tag{5.53}$$

An alternative solution can be obtained through the change of variables $\boldsymbol{R} = \boldsymbol{r}' - \boldsymbol{r}$, so that

$$\begin{aligned} \psi(\boldsymbol{r}, t) &= \frac{1}{4\pi c^2} \int_{4\pi} \mathrm{d}\Omega_R \int_0^\infty R\, \mathrm{d}R\, v(\boldsymbol{r} + \boldsymbol{R})\, \delta\left(t - \frac{R}{c}\right) \\ &= \frac{t}{4\pi} \int_{4\pi} \mathrm{d}\Omega_R\, v(\boldsymbol{r} + ct\hat{\boldsymbol{R}}), \end{aligned} \tag{5.54}$$

where $t > 0$ and $\hat{\boldsymbol{R}} = \boldsymbol{R}/R$. Equation (5.54), which is known as *Poisson's solution* (MORSE and FESHBACH [1953] p. 847), makes it clear that $\psi(\boldsymbol{r}, 0) = 0$, even though we did not explicitly use this condition.

The physical content of both (5.53) and (5.54) is the same: only points a distance ct away from the point $\boldsymbol{r}'$ are influenced by the initial conditions at that point at $t = 0$. Equation (5.54) shows clearly that $\psi(\boldsymbol{r}, t)$ is just the average of $v(\boldsymbol{r})$ over a sphere of radius ct about $\boldsymbol{r}$. It is a *spherical mean* (JOHN [1955]).

It is worth dwelling for a moment on (5.52), where we converted a plane wave to a spherical wave by averaging over directions. The average of all dipole sheets a distance ct from a point is a spherical shell at that distance. The shell is, of course, the Green's function for the wave equation, which satisfies

$$\left(\nabla^2 - \frac{1}{c^2}\frac{\partial^2}{\partial t^2}\right)\frac{1}{R}\,\delta\left(t - \frac{R}{c}\right) = -4\pi\delta(t)\delta(\boldsymbol{R}). \tag{5.55}$$

The statement that an average plane wave is a spherical wave holds also for harmonic waves, since

$$\begin{aligned}\int_{4\pi} \mathrm{d}\Omega_k\, \mathrm{e}^{\mathrm{i}(\boldsymbol{k}\cdot\boldsymbol{r} - \omega t)} &= \frac{2\pi \mathrm{e}^{-\mathrm{i}\omega t}}{\mathrm{i}k}\left[\frac{\mathrm{e}^{\mathrm{i}kr}}{r} - \frac{\mathrm{e}^{-\mathrm{i}kr}}{r}\right] \\ &= 4\pi \mathrm{e}^{-\mathrm{i}\omega t}\frac{\sin kr}{kr}. \end{aligned} \tag{5.56}$$

Once again we recognize a Green's function. this time for the Helmholtz equation:

$$(\nabla^2 + k^2)\frac{\mathrm{e}^{\pm \mathrm{i}kr}}{r} = -4\pi\delta(\boldsymbol{r}). \tag{5.57}$$

Equation (5.56) contains both incoming and outgoing spherical waves, so it is not a causal Green's function.

5.5. MICROWAVE SCATTERING

Following the lead of LAX and PHILLIPS, whose extensive work is detailed in their monograph (LAX and PHILLIPS [1967]), many theorists have applied the Radon transform to scattering studies, especially in the time domain. This topic is of great current interest because of its use in analyzing the "signature" of radar scatterers (KENNAUGH and MOFFATT [1965]; LEWIS [1969]; YOUNG [1976]; DAS and BOERNER [1978]; MAGER and BLEISTEIN [1978]; BOJARSKI [1979]; BOERNER [1979]; DEVANEY [1980]), and in microwave imaging (CHAN and FARHAT [1981]; ROCKMORE, DENTON and FRIEDLANDER [1979]). We shall indicate the general nature of the problem by calculating the temporal impulse response of a weak volume scatterer in the

Born approximation, but very similar results are obtained for metallic reflectors in the above references.

Consider a medium such as a tenuous vapor where the refractive index has the form

$$n(\boldsymbol{r}) = 1 + \Delta n(\boldsymbol{r}), \tag{5.58}$$

with $\Delta n(\boldsymbol{r}) \ll 1$. We assume also that the medium is nondispersive so that Δn does not depend on wavelength. The velocity of an electromagnetic wave is then a function of position, denoted $c_{\mathrm{m}}(\boldsymbol{r})$, and given by

$$c_{\mathrm{m}}(\boldsymbol{r}) = \frac{c}{1 + \Delta n(\boldsymbol{r})}, \tag{5.59}$$

and the scalar wave equation is

$$\left(\nabla^2 - \frac{1}{c_{\mathrm{m}}^2} \frac{\partial^2}{\partial t^2}\right) \psi(\boldsymbol{r}, t) = 0 \tag{5.60}$$

or, approximately,

$$\left(\nabla^2 - \frac{1}{c^2} \frac{\partial^2}{\partial t^2}\right) \psi(\boldsymbol{r}, t) \approx \frac{2\Delta n(\boldsymbol{r})}{c^2} \frac{\partial^2 \psi(\boldsymbol{r}, t)}{\partial t^2}. \tag{5.61}$$

The first Born approximation consists of replacing $\psi(\boldsymbol{r}, t)$ on the right-hand side of this equation with the unperturbed incident wave $\psi_{\mathrm{i}}(\boldsymbol{r}, t)$. The wave equation is then inhomogeneous with a known source term, and the Green's-function solution is

$$\psi_{\mathrm{s}}(\boldsymbol{r}, t) = -\frac{1}{2\pi c^2} \int_{-\infty}^{t} \mathrm{d}t' \int_{\infty} \mathrm{d}^3 r' \, \Delta n(\boldsymbol{r}') \frac{\partial^2 \psi_{\mathrm{i}}(\boldsymbol{r}', t')}{\partial t'^2} \frac{\delta(\tau - R/c)}{R}, \tag{5.62}$$

where $\tau = t - t'$, $R = |\boldsymbol{r} - \boldsymbol{r}'|$, and the total wave $\psi(\boldsymbol{r}, t) = \psi_{\mathrm{i}}(\boldsymbol{r}, t) + \psi_{\mathrm{s}}(\boldsymbol{r}, t)$.

Now we consider a point source located at $\boldsymbol{r} = \boldsymbol{r}_{\mathrm{s}}$ and a point detector at $\boldsymbol{r} = \boldsymbol{r}_{\mathrm{D}}$ as shown in Fig. 12. We shall regard the system of source, scatterer, and detector as a temporal filter and calculate its temporal impulse response $h(t)$, which depends on $\boldsymbol{r}_{\mathrm{s}}$, $\boldsymbol{r}_{\mathrm{D}}$, and $\Delta n(\boldsymbol{r})$. Thus, we assume that the source emits a very short pulse of radiation at $t = t_0$, so that the incident wave is

$$\psi_{\mathrm{i}}(\boldsymbol{r}', t') = \frac{1}{R_{\mathrm{s}}} \delta\left(t' - t_0 - \frac{R_{\mathrm{s}}}{c}\right), \tag{5.63}$$

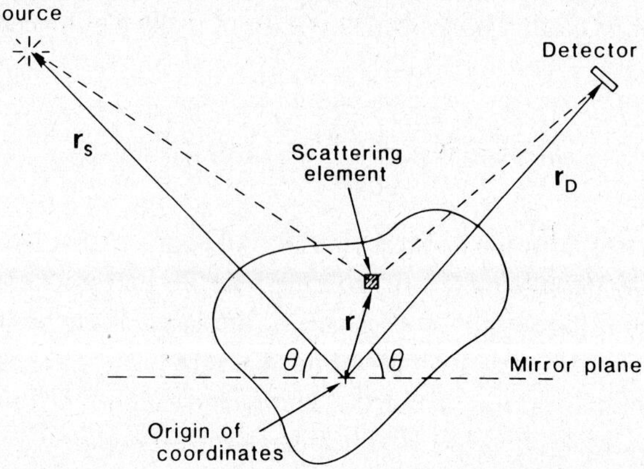

Fig. 12. Geometry for calculation of the temporal impulse response of a weak volume scatterer. If r_S and r_D are large, all points on the "mirror plane" or any plane parallel to it give approximately the same time delay for waves propagating from source to scattering point to detector, and the received signal at one time τ is proportional to the integral of $\Delta n(r)$ over this plane.

where $R_s = |\boldsymbol{r}' - \boldsymbol{r}_s|$. The impulse response $h(t)$ is the wave at the detector at time t, given from (5.62) and (5.63) by

$$h(t) = -\frac{1}{2\pi c^2}\int_{-\infty}^{t} \mathrm{d}t' \int_{\infty} \mathrm{d}^3 r' \frac{\Delta n(\boldsymbol{r}')}{R_s R_D}\,\delta''\left(t' - t_0 - \frac{R_s}{c}\right)\delta\left(t - t' - \frac{R_D}{c}\right), \tag{5.64}$$

where $R_D = |\boldsymbol{r}' - \boldsymbol{r}_D|$.

We next make the Fraunhofer approximation, valid for r_s and r_D large compared to the size of $\Delta n(r')$, so that

$$R_s = r_s - \hat{\boldsymbol{r}}_s \cdot \boldsymbol{r}' + \ldots, \tag{5.65}$$

$$R_D = r_D - \hat{\boldsymbol{r}}_D \cdot \boldsymbol{r}' + \ldots, \tag{5.66}$$

where $r_s = |\boldsymbol{r}_s|$, $\hat{\boldsymbol{r}}_s = \boldsymbol{r}_s/r_s$, and similarly for r_D. We retain only the first term in each expansion in the denominator of (5.64), but the first two terms in the arguments of the delta functions. This yields

$$h(t) = -\frac{c}{2\pi r_D r_s}\int_{\infty} \mathrm{d}^3 r'\, \Delta n(\boldsymbol{r}')\delta''(p - \boldsymbol{r}' \cdot \hat{\boldsymbol{n}}) \tag{5.67}$$

where

$$p = -c(t - t_0) + r_s + r_D, \tag{5.68}$$

$$\hat{\boldsymbol{n}} = \hat{\boldsymbol{r}}_s + \hat{\boldsymbol{r}}_D. \tag{5.69}$$

Neglecting a distortion that arises because $\hat{\boldsymbol{n}}$ is not really a unit vector, we can write

$$h(t) = \text{const.} \cdot \left[\frac{\partial^2}{\partial p^2} \mathscr{R}_3\{\Delta n(\boldsymbol{r})\} \right]. \tag{5.70}$$

This result has a simple physical interpretation. All points $\boldsymbol{r}'$ that have the same propagation delay (from source to scattering point to detector) contribute to $h(t)$ for the same t. (See Fig. 12.) If r_s and r_D are large, these points lie on a surface that is approximately planar and normal to $\hat{\boldsymbol{r}}_s + \hat{\boldsymbol{r}}_D$. One can think of points on this plane forming a mirror. The occurrence of $\partial^2/\partial p^2$ in (5.70) is also understandable. If $\Delta n(\boldsymbol{r})$ is either invariant or linearly varying in the direction $\hat{\boldsymbol{n}}$, there is no reflection from this "mirror".

Another way to understand (5.70) is to take its temporal Fourier transform, yielding the standard Born-approximation result (WOLF [1969]; DÄNDLIKER and WEISS [1970])

$$H(\nu) = \text{const.}\ \nu^2 \int_\infty \mathrm{d}^3 r'\ \Delta n(\boldsymbol{r}')\exp[(2\pi i\nu/c)(\hat{\boldsymbol{n}}\cdot\boldsymbol{r}')]. \tag{5.71}$$

The factor ν^2 corresponds to the second derivative, and the integral is recognized as a 3D *spatial* Fourier transform, with spatial frequency

$$\boldsymbol{\sigma} = \nu\hat{\boldsymbol{n}}/c.$$

Thus, a monochromatic source of frequency ν at $\boldsymbol{r}_s$ and a detector at $\boldsymbol{r}_D$ give information about one point in the 3D Fourier transform of $\Delta n(\boldsymbol{r})$. Varying ν, but keeping $\boldsymbol{r}_s$ and $\boldsymbol{r}_D$ fixed, gives information about other spatial frequencies along a line in direction $\hat{\boldsymbol{n}}$ in Fourier space. The temporal impulse contains all temporal frequencies, and hence maps out this entire line in one measurement. By the central-slice theorem, this information is equivalent to $\lambda_{\hat{n}}(p)$ for all p and fixed $\hat{\boldsymbol{n}}$.

Very similar results have been obtained by LACOURT, VIENOT and GOEDGEBUER [1976] for the temporal impulse response of a 2D diffracting aperture.

5.6. COMPTON SCATTERING

The inelastic scattering of X-rays by electrons was first discovered by FLORANCE [1910] and GRAY [1913]. However, the effect is now associated with the name of A. H. COMPTON [1923a,b; 1926], who carried out a series of very precise experiments and gave a theoretical explanation of the main features of the observations. A good account of the early history is given by STUEWER and COOPER [1977].

An elementary derivation of the energy loss by Compton scattering considers the electron to be free and initially at rest (see, for example, TIPLER [1969]). The incident X-ray photon has energy E_0. After scattering through an angle θ, it has a reduced energy E, given by

$$\frac{1}{E} = \frac{1}{E_0} + \frac{1}{mc^2}(1 - \cos\theta), \tag{5.72}$$

with m being the rest mass of the electron and c being the speed of light. The important point for the present discussion is that there is a unique relationship between E and θ; the photons scattered through one particular angle θ have a discrete energy E.

The situation is very different when the electrons are initially not at rest but have some distribution of momenta, as they must when they are bound to an atom. Then the scattered photons, even for fixed θ, have a distribution of energies, and analysis of this energy spectrum can give information about the momentum distribution of the scattering electrons. This information is of fundamental importance because the probability density function for the electron momentum $\boldsymbol{P}$ is given by

$$n(\boldsymbol{P}) = \chi^*(\boldsymbol{P})\chi(\boldsymbol{P}), \tag{5.73}$$

where $\chi(\boldsymbol{P})$ is the electron wavefunction in the momentum representation, which is just the 3D Fourier transform of the usual configuration-space wavefunction $\psi(\boldsymbol{r})$. Thus, measurement of the Compton spectrum provides a direct check on theoretical wavefunctions (WILLIAMS [1977]).

A simple classical analysis that shows the relationship between $n(\boldsymbol{P})$ and the Compton spectrum was given by DUMOND [1929, 1933]. See also STUEWER and COOPER [1977] and PLATZMAN and TZOAR [1977]. The analysis is just a nonrelativistic application of the principles of conservation of energy and momentum. The photon momentum before scattering is $\hbar\boldsymbol{k}_0$, where $\boldsymbol{k}_0$ is the wavevector. The momentum after scattering is $\hbar\boldsymbol{k}$. The photon energy before and after scattering is $\hbar ck_0$ and $\hbar ck$, respectively. The electron momentum after scattering is $\boldsymbol{P} + \Delta\boldsymbol{P}$, where, by conservation of momentum,

$$\Delta\boldsymbol{P} = \hbar(\boldsymbol{k} - \boldsymbol{k}_0) \equiv -\hbar\Delta\boldsymbol{k}. \tag{5.74}$$

The energy gained by the electron, and hence, lost by the photon, is

$$\Delta E = E_0 - E = \frac{|\boldsymbol{P} + \Delta \boldsymbol{P}|^2}{2m} - \frac{P^2}{2m} = \frac{|\Delta \boldsymbol{P}|^2}{2m} + \frac{\boldsymbol{P} \cdot \Delta \boldsymbol{P}}{m}$$

$$= \frac{\hbar^2}{2m}(k^2 + k_0^2 - 2\boldsymbol{k} \cdot \boldsymbol{k}_0) - \frac{\hbar}{m}\boldsymbol{P} \cdot \Delta \boldsymbol{k}$$

$$= \frac{E^2}{2mc^2} + \frac{E_0^2}{2mc^2} - \frac{EE_0}{mc^2}\cos\theta - \frac{\hbar}{m}\boldsymbol{P} \cdot \Delta \boldsymbol{k}. \tag{5.75}$$

This expression is valid for nonrelativistic electron energies such that $\Delta E \ll mc^2$. Hence,

$$\Delta E \approx \frac{E_0^2}{mc^2}(1 - \cos\theta) - \frac{\hbar}{m}\boldsymbol{P} \cdot \Delta \boldsymbol{k}. \tag{5.76}$$

The first term agrees with the usual expression for the Compton shift, (5.72), within the nonrelativistic approximation. The second term is a Doppler shift present because the scattering electron is in motion. Note that only the component of $\boldsymbol{P}$ in the direction $\Delta \boldsymbol{k}$ influences ΔE.

Equation (5.76) was derived for a fixed $\boldsymbol{P}$; if there is a distribution in $\boldsymbol{P}$, the scattered flux for a particular ΔE is given by

$$I(\Delta E) = \text{const.} \int_\infty \mathrm{d}^3 P\, n(\boldsymbol{P})\, \delta\left[\Delta E - (\Delta E)_0 + \frac{\hbar}{m}\boldsymbol{P} \cdot \Delta \boldsymbol{k}\right], \tag{5.77}$$

where

$$(\Delta E)_0 = \frac{E_0^2}{mc^2}(1 - \cos\theta). \tag{5.78}$$

Equation (5.77) has the structure of a 3D Radon transform *in momentum space*. The usual parameters $\hat{\boldsymbol{n}}$ and p are given by

$$\hat{\boldsymbol{n}} = -\Delta \boldsymbol{k}/|\Delta \boldsymbol{k}|, \tag{5.79}$$

$$p = \frac{m}{\hbar|\Delta \boldsymbol{k}|}[\Delta E - (\Delta E)_0]. \tag{5.80}$$

Hence, spectra taken at various scattering angles can be used to map out the full 3D Radon transform of $n(\boldsymbol{P})$. Some approximation to the inverse Radon transform can then be used to reconstruct $n(\boldsymbol{P})$ (MIJNARENDS [1977]).

A closely related technique that also measures $\mathscr{R}_3\{n(\boldsymbol{P})\}$ is positron annihilation (HAUTOJÄRVI [1979]; WEST [1974]).

An interesting mathematical point arises when Compton scattering or positron annihilation studies are carried out in oriented single crystals. Then it is convenient to use a spherical harmonic decomposition akin to (4.63), but with suitable linear combinations of spherical harmonics that reflect the symmetry of the crystal. These so-called lattice harmonics (MIJNARENDS [1967, 1977]) allow reasonable reconstructions of $n(\boldsymbol{P})$ from a relatively small number of points in Radon space.

5.7. MISCELLANEOUS APPLICATIONS

Many applications of the Radon transform to astronomy have been reported. Indeed, perhaps the first experimental implementation of the inverse Radon transform was the 1936 work of V. AMBARTSUMIAN, as reported by CORMACK [1982]. Ambartsumian examined the problem of deducing the 3D distribution of velocities of stars from measurements of doppler shifts, which are sensitive only to the radial component of velocity, and obtained a reconstruction of the velocity distribution projected on the galactic plane. BRACEWELL [1974] reviews many other astronomical applications, including strip-scan radio-astronomy, lunar occultation of radio sources, radar measurements of lunar roughness, and estimation of the density of globular clusters. ALTSCHULER and PERRY [1974] suggest 3D modeling of the solar corona, and the attenuated Radon transform could be applicable to solar limb darkening.

Another early application was to statistics. CRAMÉR and WOLD [1936] used the mD Radon transform to prove various theorems about mD probability distributions, using known results for 1D distributions.

Since the 1950s, there has been interest in optical computers for performing the inverse Radon transform. KORENBLYUM, TETEL'BAUM and TYUTIN [1958] designed an optical reconstruction system for medical CT, and the extensive work since then is reviewed by GMITRO, GREIVENKAMP, SWINDELL, BARRETT, CHIU and GORDON [1980]. One recent development is the application of optical Radon transformers to data-processing problems that initially do not involve projections at all (BARRETT [1982b, 1983]). Here the idea is to use the Radon transform as a pre-processing step to reduce 2D or 3D data sets to a series of more manageable 1D data sets.

Electron microscopy is another important application area. Here the main problem is that it is experimentally difficult to get a complete data set, and the theoretical work is aimed at the ill-posed problem of reconstruction from incomplete projections.

This survey certainly does not exhaust the possibility applications of Radon theory. The proceedings of the Brookhaven (MARR [1974]) and Stanford (GORDON [1975]) conferences should be perused to gain a better impression of the richness of this field.

Acknowledgements

The writing of this chapter was begun while the author was on sabbatical leave in Erlangen, W. Germany. The hospitality and encouragement of Adolf Lohmann and his group and the financial support of the Alexander von Humboldt foundation are gratefully acknowledged. The manuscript was capably typed by Debbie Spargur and critically read by Art Gmitro, Lee Giles, Anne Clough, William Swindell, and Kyle Voss, to all of whom I am indebted. Financial support was also provided by the National Cancer Institute under grant no. CA-23417.

References

ABRAGAM, A., 1961, *The principles of Nuclear Magnetism* (Oxford University Press, London).

ALTSCHULER, M. S. and R. M. PERRY, 1974, The Three-dimensional Solar Corona, in: *Techniques of Three-dimensional Reconstruction*, Proc. Int. Workshop at Brookhaven Nat. Lab., ed. R. B. Marr (Brookhaven, Upton, LI).

ANGER, H. O., 1958, Scintillation Camera, Rev. Sci. Instrum. **29,** 27.

ANGER, H. O., 1964, Scintillation Camera with Multichannel Collimators, J. Nucl. Med. **5,** 515.

BARRETT, H. H., 1982a, Dipole-sheet Transform, J. Opt. Soc. Amer. **72,** 468.

BARRETT, H. H., 1982b, Optical Processing in Radon Space, Opt. Lett. **7,** 248.

BARRETT, H. H., 1983, Three-dimensional Image Reconstruction from Planar Projections, with Application to Optical Data Processing, in: *Transformations in Optical Signal Processing*, eds. W. T. Rhodes, J. R. Fienup and B. E. A. Saleh (S.P.I.E., Bellingham, WA).

BARRETT, H. H. and W. SWINDELL, 1981, *Radiological Imaging: Theory of Image Formation, Detection and Processing* (Academic, New York).

BELLINI, D., M. PIACENTINI, C. CAFFORIO and F. ROCCA, 1979, Compensation of Tissue Absorption in Emission Tomography, IEEE Trans. Acoust. Speech and Signal Process. **ASSP-27,** 213.

BOERNER, W.-M., 1979, Development of Physical Optics Inverse Scattering Techniques Using Radon Projection Theory, in: *Mathematical Methods and Applications of Scattering Theory*, Lecture Notes in Physics, vol. 130 (Springer, Berlin).

BOJARSKI, N. N., 1979, *N*-dimensional Fast Fourier Transform Tomography for Incomplete Information and Its Application to Inverse Scattering Theory, in: *Mathematical Methods and Applications of Scattering Theory*, Lecture Notes in Physics, vol. 130 (Springer, Berlin).

BRACEWELL, R. N., 1965, *The Fourier Transform and its Applications* (McGraw–Hill, New York).

BRACEWELL, R. N., 1974, Three-dimensional Reconstruction: An Overview, in: *Techniques of Three-dimensional Reconstruction*, Proc. Int. Workshop held at Brookhaven Nat. Lab., ed. R. B. Marr (Brookhaven, Upton, IL).

BROOKS, R. A. and G. DICHIRO, 1976, Principles of Computer Assisted Tomography (CAT) in Radiographic and Radioisotopic Imaging, Phys. Med. Biol. **21,** 689.

BROWNELL, G. L., J. A. CORREIA and R. G. ZAMENHOF, 1978, Positron Instrumentation, Rec. Adv. Nucl. Med. **5,** 1.

BUDINGER, T. F. and G. T. GULLBERG, 1977, Transverse Section Reconstruction of Gamma-ray Emitting Radionuclides in Patients, in: *Reconstruction Tomography in Diagnostic Radiology and Nuclear Medicine*, eds. M. Ter-Pogossian et al.

CHAN, C. K. and N. H. FARHAT, 1981, Frequency Swept Tomographic Imaging of Three-dimensional Perfectly Conducting Objects, IEEE Trans. Antennas & Propag. **AP-29,** 312.

CHIU, M. Y., 1980, Three-dimensional Radiographic Imaging, Ph.D. Dissertation (Univ. of Arizona, Tucson).

CHIU, M. Y., H. H. BARRETT and R. G. SIMPSON, 1980. Three-dimensional Reconstruction from Planar Projections, J. Opt. Soc. Amer. **70,** 755.

CLOUGH, A. V. and H. H. BARRETT, 1983, The Attenuated Radon and Abel Transforms, J. Opt. Soc. Amer., to be published in November issue.

COMPTON, A. H., 1923a, The Spectrum of Scattered X-rays, Phys. Rev. **22,** 409.

COMPTON, A. H., 1923b, A Quantum Theory of the Scattering of X-rays by Light Elements, Phys. Rev. **21,** 207 and 483.

COMPTON, A. H., 1926, *X-rays and Electrons* (Van Nostrand, New York).

CORMACK, A. M., 1963, Representation of a Function by Its Line Integrals, with Some Radiological Applications, J. Appl. Phys. **34,** 2722.

CORMACK, A. M., 1964, Representation of a Function by Its Line Integrals, with Some Radiological Applications, II, J. Appl. Phys. **35,** 2908.

CORMACK, A. M., 1973, Reconstruction of Densities from Their Projections with Applications in Radiological Physics, Phys. Med. Biol. **18,** 195.

CORMACK, A. M., 1982, Computed Tomography: History and Some Recent Developments, Amer. Math. Soc. short course on computed tomography, Cincinnati, Jan. 1982.

COURANT, R. and D. HILBERT, 1962, *Methods of Mathematical Physics* (Wiley, London).

CRAMÉR, H. and H. WOLD, 1936, Some Theorems on Distribution Functions, J. Lond. Math. Soc. **11,** 290.

DÄNDLIKER, R. and K. WEISS, 1970, Reconstruction of the Three-dimensional Refractive Index from Scattered Waves, Opt. Commun. **1,** 323.

DAS, Y. and W. M. BOERNER, 1978, On Radar Target Shape Estimation Using Algorithm for Reconstruction from Projections, IEEE Trans. Antennas & Propag. **AP-26,** 274.

DEANS, S. R., 1978, A Unified Radon Inversion Formula, J. Math. Phys. **19,** 2346.

DEANS, S. R., 1983, *The Radon Transform and Some of its Applications* (Wiley, New York).

DEVANEY, A. J., 1980, Inverse Source and Scattering Problems in Optics, in: *Optics in* 4 *Dimensions*, eds. L. M. Narducci and M. A. Machado (AIP, New York).

DUMOND, J. W. M., 1929, Compton Modified Line Structure and Its Relation to the Electron Theory of Solid Bodies, Phys. Rev. **33,** 643.

DUMOND, J. W. M., 1933, The Linear Momenta of Electrons in Atoms and in Solid Bodies as Revealed by X-ray Scattering, Rev. Med. Phys. **5,** 1.

EIN-GAL, M., 1975, The Shadow Transform: An Approach to Cross-sectional Imaging, Ph.D. Dissertation (Stanford Univ.).

FARRAR, T. C. and E. D. BECKER, 1971, *Pulse and Fourier Transform NMR – Introduction to Theory and Methods* (Academic, New York).

FLORANCE, D. C. H., 1910, Primary and Secondary Gamma Rays, Phil. Mag. **20,** 921.

GASKILL, J. D., 1978, *Linear Systems, Fourier Transforms and Optics* (Wiley, New York).

GEL'FAND, I. M., M. I. GRAEV and N. YA. VILENKIN, 1966a, *Integral Geometry and Representation Theory* (Academic, New York).

GEL'FAND, I. M., M. I. GRAEV and N. YA. VILENKIN, 1966b, *Generalized Functions*, vol. 5 (Academic, New York), pp. 1–74.

GINDI, G. R., J. ARENDT, H. H. BARRETT, M. Y. CHIU, A. ERVIN, C. L. GILES, M. A. KUJOORY, E. L. MILLER and R. G. SIMPSON, 1982, Imaging with Rotating-Slit Apertures and Rotating Collimators, Med. Phys. **9,** 324.

GMITRO, A. F., J. E. GREIVENKAMP, W. SWINDELL, H. H. BARRETT, M. Y. CHIU and S. K. GORDON, 1980, Optical Computers for Reconstructing Objects from Their X-ray Projections, Opt. Eng. **19,** 260.

GORDON, R., ed., 1975, *Image Processing for 2-D and 3-D Reconstruction from Projections*, Proc. Symp. held at Stanford Univ., Aug. 4–7, 1975 (Optical Society of America, Washington).

GORDON, R., G. T. HERMAN and S. A. JOHNSON, 1975, Image Reconstruction from Projections, Scientific American, Oct., 56.

GRADSHTEYN, I. S. and I. M. RYZHIK, 1980, *Table of Integrals, Series, and Products* (Academic, New York).

GRAY, J. A., 1913, The Scattering and Absorption of the Rays of Radium, Phil. Mag. **26,** 611.

GULLBERG, G. T., 1979, The Attenuated Radon Transform: Theory and Application in Medicine and Biology, Ph.D. Thesis (Univ. of California, Berkeley).

HANSEN, E. W., 1981, Theory of Circular Harmonic Image Reconstruction, J. Opt. Soc. Amer. **71,** 304.

HAUTOJÄRVI, P., ed., 1979, *Positrons in Solids*, Topics in Current Physics, vol. 12 (Springer, Berlin).

HAWKINS, W., 1982, The Mathematics of Computed Tomography, Ph.D. dissertation (Univ. of Arizona, Tucson).

HELGASON, S., 1965, The Radon Transform on Euclidean Spaces, Compact Two-point Homogeneous Spaces and Grassmann Manifolds, Acta Math. **113,** 153.

HELGASON, S., 1980, *The Radon Transform* (Birkhäuser, Boston).

HERMAN, G. T., 1980, *Image Reconstruction from Projections* (Academic, New York).

HERMAN, G. T. and F. NATTERER, eds., 1981, *Mathematical Aspects of Computerized Tomography*, Lecture Notes in Medical Informatics, vol. 8 (Springer, Berlin).

JACKSON, J. D., 1975, *Classical Electrodynamics*, 2nd Ed. (Wiley, New York).

JASZCZAK, R., R. E. COLEMAN and C. B. LIM, 1980, SPECT: Single Photon Emission Computed Tomography, IEEE Trans. Nucl. Sci. **NS-27,** 1137.

JOHN, F., 1955, *Plane Waves and Spherical Means Applied to Partial Differential Equations* (Interscience, New York).

KAK, A. C., 1979, Computerized Tomography with X-ray, Emission, and Ultrasound Sources, Proc. IEEE **67,** 1245.

KARSTAEDT, N., R. L. WITCOFSKI and C. L. PARTAIN, 1981, eds., *Proc. Int. Symp. on NMR Imaging* (Bowman Gray School of Medicine Press).

KAUFMAN, L., L. E. CROOKS and A. R. MARGULIS, 1981, eds., *Nuclear Magnetic Resonance Imaging in Medicine* (Igaku–Shoin, Tokyo).

KENNAUGH, E. M. and D. L. MOFFATT, 1965, Transient and Impulse Response Approximations, Proc. IEEE **53,** 893.

KEYES, W. I., 1975, The Fan-beam Gamma Camera, Phys. Med. Biol. **20,** 489.

KORENBLYUM, B. I., S. I. TETEL'BAUM and A. A. TYUTIN, 1958, About One Scheme of Tomography, Izv. Vyschikh Uchebnykh Zavedenii-Radiofizika **1,** 151. (In Russian; English translation available from H. H. Barrett.)

LACOURT, A., J. C. VIENOT and J. P. GOEDGEBUER, 1976, Reassessing Basic Landmarks in Space–Time Optics, Opt. Commun. **19,** 68.

LAX, P. D. and R. S. PHILLIPS, 1967, *Scattering Theory* (Academic, New York).

LEWIS, R. M., 1969, Physical Optics Inverse Diffraction, IEEE Trans. Antennas & Propag. **AP-24,** 276.

LEWITT, R. M. and R. H. T. BATES, 1978, Image Reconstruction from Projections (Parts I–IV), Optik **50,** 19.

LIGHTHILL, M. J., 1962, *Fourier Analysis and Generalized Functions* (Cambridge University Press).

LINDGREN, A. G. and P. A. RATTEY, 1981, The Inverse Discrete Radon Transform with Applications to Tomographic Imaging Using Projection Data, in: *Advances in Electronics and Electron Physics*, vol. 56, ed. C. Marton (Academic, New York).

LUDWIG, D., 1966, The Radon Transform on Euclidean Space, Comm. Pure Appl. Math. **19,** 49.

MAGER, R. D. and N. BLEISTEIN, 1978, An Examination of the Limited Aperture Problem of Physical Optics Inverse Scattering, IEEE Trans. Antennas & Propag. **AP-26,** 695.

MANSFIELD, P. and P. G. MORRIS, 1982, *NMR Imaging in Biomedicine* (Academic, Cambridge).

MARR, R. B., ed., 1974, *Techniques of Three-dimensional Reconstruction*, Proc. Int. Workshop held at Brookhaven National Lab. (Brookhaven, Upton, LI).

MARR, R. B., C.-N. CHEN and P. C. LAUTERBUR, 1981, On Two Approaches to 3D Reconstruction in NMR Zeugmatography, in: *Mathematical Aspects of Computerized Tomography*, Lecture Notes in Medical Informatics, vol. 8, eds. G. T. Herman and F. Natterer (Springer, Berlin).

MIJNARENDS, P. E., 1967, Determination of Anisotropic Momentum Distributions in Positron Annihilation, Phys. Rev. **160,** 512.

MIJNARENDS, P. E., 1977, Reconstruction of Three-dimensional Distributions, in:*Compton Scattering*, ed. B. Williams (McGraw–Hill, New York) ch. 10.

MIJNARENDS, P. E., 1979, Electron Momentum Densities in Metals and Alloys, in: *Positrons in Solids*, ed. P. Hautojärvi (Springer, Berlin) ch. 2.

MORSE, P. M. and H. FESHBACH, 1953, *Methods of Theoretical Physics* (McGraw–Hill, New York).

PLATZMAN, P. and N. TZOAR, 1977, Theory, in: *Compton Scattering*, ed. B. Williams (McGraw–Hill, New York) ch. 2.

PYKETT, I. L., 1982, NMR Imaging in Medicine, Scientific American **246,** 78.

PYKETT, I. L., J. H. NEWHOUSE, F. S. BUONANNO, T. G. BRADY, M. R. GOLDMAN, J. P. KISTLER and G. M. POHOST, eds., 1982, *Principles of N.M.R. Imaging* **143,** 157.

RADON, J., 1917, Über die Bestimmung von Funktionen durch ihre Integralwerte längs gewisser Mannigfaltigkeiten, Ber. Saechs. Akad. Wiss. (Leipzig) **69,** 262.

RADULOVIC, P. T. and C. M. VEST, 1975, Direct Three-dimensional Reconstruction, in: *Image Reconstruction for 2-D and 3-D Reconstruction from Projections*, ed. R. Gordon (Digest of Technical Papers, Stanford, CA, Aug. 4–7, 1975).

ROCKMORE, A. G., R. V. DENTON and B. FRIEDLANDER, 1979, Direct Three-Dimensional Image Reconstruction, IEEE Trans. Antennas & Propag. **AP-26,** 274.

SHEPP, L. A., 1980, Computerized Tomography and Nuclear Magnetic Resonance, J. Comput. Assist. Tomog. **4,** 94.

SHEPP, L. A. and J. B. KRUSKAL, 1978, Computerized Tomography: the New Medical X-ray Technology, Amer. Math. Monthly **85,** 420.

SLICHTER, C. P., 1963, *Principles of Magnetic Resonance* (North-Holland, Amsterdam).

SMITH, K. T., D. C. SOLOMON and S. L. WAGNER, 1977, .Practical and Mathematical Aspects of the Problem of Reconstructing Objects from Radiographs, Bull. Amer. Math. Soc. **83,** 1227.

STEWART, A. T., 1957, Momentum Distribution of Metallic Electrons by Positron Annihilation, Can. J. Phys. **35,** 168.

STUEWER, R. H. and M. J. COOPER, 1977, History, in: *Compton Scattering*, ed. B. Williams (McGraw–Hill, New York) ch. 1.

SWINDELL, W. and H. H. BARRETT, 1977, Computerized Tomography: Taking Sectional X-rays, Physics Today **30,** 32.

TIPLER, P. A., 1969, *Foundations of Modern Physics* (Worth Publ., New York).

TRETIAK, O. J. and P. DELANEY, 1977, The Exponential Convolution Algorithm for Emission Computed Axial Tomography, in: *Information Processing in Medical Imaging* (Biomedical Computing Technology Information Center, Nashville, TN).

TRETIAK, O. J. and C. METZ, 1980, The Exponential Radon Transform, SIAM J. Appl. Math. **39,** 341.

VAN DER POL, B. and TH. J. WEIJERS, 1934, Tchebycheff Polynomials and Their Relation to Circular Functions, Bessel Functions and Lissajous Figures, Physica **1,** 78.

VERLY, J. G., 1981, Circular and Extended Circular Harmonic Transforms and Their Relevance to Image Reconstruction from Line Integrals, J. Opt. Soc. Amer. **71,** 825.

VEST, C. M. and D. G. STEEL, 1978, Reconstruction of Spherically Symmetric Objects from Slit-imaged Emission: Application to Spatially Resolved Spectroscopy, Opt. Lett. **3,** 54.

WEST, R. N., 1974, *Positron Studies of Condensed Matter* (Taylor and Francis, London).

WILLIAMS, B., ed., 1977, *Compton Scattering* (McGraw–Hill, New York).

WOLF, E., 1969, Three-dimensional Structure Determination of Semi-transparent Objects from Holographic Data, Opt. Commun. **1,** 153.

YOUNG, J. D., 1976, Radar Imaging from Ramp Response Signatures, IEEE Trans. Antennas & Propag. **AP-24,** 276.

E. WOLF, PROGRESS IN OPTICS XXI
© ELSEVIER SCIENCE PUBLISHERS B.V. 1984

IV

ZONE PLATE CODED IMAGING: THEORY AND APPLICATIONS

BY

NATALE M. CEGLIO

University of California, Lawrence Livermore National Laboratory, Livermore, CA 94550, U.S.A.

and

DONALD W. SWEENEY*

School of Mechanical Engineering, Purdue University, West Lafayette, IN 47907, U.S.A.

* Current address: Sandia National Laboratories, Livermore, CA 94550, U.S.A.

CONTENTS

This chapter is dedicated to the memory of N. M. Ceglio's aunt, Sera Sciacca.

§ 1. Introduction

Sources of short wavelength radiations such as X-rays and sub-atomic particles are not easily imaged because of difficulties in achieving ray bending by the conventional methods of reflection, refraction and diffraction. In such cases, coded imaging (CI) techniques can play an important role in determining source location and distribution. In the twenty years since they were first proposed (MERTZ and YOUNG [1961]), coded techniques have made significant imaging contributions in the fields of X-ray astronomy (DICKE [1968], YOUNG [1963]), nuclear medicine (BARRETT [1972], BARRETT, WILSON, DE MEESTER and SCHARFMAN [1973], ROGERS, JONES and BEIERWALTES [1973]), nuclear engineering (ROSE, JACOBS and KENNEY [1975], ROSE [1976]), and inertial confinement fusion (CEGLIO and COLEMAN [1977], CEGLIO and LARSEN [1980], CEGLIO, ATTWOOD and LARSEN [1982], BRUNOL, SAUNEUF and GEX [1979], FENIMORE, CANNON, VAN HULSTEYN and LEE [1979]). Although taking varied forms in its application, CI, when reduced to basics, is a two-step process. In the first step, source information is recorded or encoded by geometrical shadowcasting through a coded aperture (no ray bending is involved). In the second step, image reconstruction or decoding is achieved using a numerical or optical procedure matched to the coded aperture design. The rich variety of forms in which CI techniques are manifested arises from the many different aperture designs and decoding methods that are used. Aperture designs include the Fresnel zone plate (MERTZ [1965]), non-redundant pinhole array (DICKE [1968], ABLES [1968]), uniformly redundant array (FENIMORE and CANNON [1978]), single annular ring (WALTON [1973], BRUNOL, SAUNEUF and GEX [1979]), as well as others (BARRETT, STONER, WILSON and DE MEESTER [1974]). Image decoding can be accomplished optically using coherent image reconstruction (MERTZ [1965], BARRETT and HORRIGAN [1973]) or incoherent analog methods (DICKE [1968], GASKILL, WHITEHEAD and GRAY [1972]). Digital image decoding may also be employed using either deconvolution (SINGLETON, RANSON and MITTRA [1976]) or matched filter (FENIMORE and CANNON [1978], BROWN [1973]) techniques.

In this paper, we provide a comprehensive mathematical framework for a specific CI technique, zone plate coded imaging (ZPCI), in which the coded aperture is a Fresnel zone plate and the image decoding method is coherent optical reconstruction. We also present an overview of applications of zone plate coded imaging in laser fusion experiments. Because coded apertures offer a significant improvement (five orders of magnitude is typical) in radiation collection over pinhole optics of equivalent resolution, and because laser fusion targets are small, ZPCI has been particularly well-suited for imaging low level X-ray and particle emissions from these experiments.

§§ 2–4 are a presentation of the linear theory of ZPCI, neglecting all nonlinear effects which arise from and tend to complicate the analysis of optically reconstructed images. § 2 develops the mathematical formalism which serves as the basis for both the linear and nonlinear analysis. §§ 3 and 4 present two distinct but complementary ZPCI analyses of extended source distributions. In § 3, an impulse response analysis treats the source as a weighted accumulation of discrete point sources. In § 4, an eigenfunction analysis is provided in which the extended source is treated as a superposition of Gaussian–Laguerre polynomials, the eigenfunctions of the Fresnel transform. Both analyses provide expressions for an intuitive understanding of the matters of planar and tomographic resolution and coded imaging diffraction efficiency in all orders. § 5 is a discussion of a number of specific issues of practical importance in coded imaging applications: quantum noise effects, nonlinear recording, and the spectral content of a reconstructed X-ray image. § 6 concludes with an illustrated survey of the use of ZPCI for the microscopy of laser imploded fusion targets. A compilation of ZPCI data is presented, including: 3.5 MeV alpha particle images of the region of thermonuclear burn, and multispectral X-rays images covering the spectral range from 3–30 keV.

§ 2. Linear Mathematical Formulation

In this section, the mathematical foundations of zone plate coded imaging are developed. The imaging geometry for this two-step process is illustrated schematically in Fig. 1. An incoherent radiation source casts a geometrical shadow through an aperture having the form of a Fresnel zone plate, thereby producing a showgraph or coded image. The coded image is a convolution of the source distribution with the coded aperture transmission function. As such, it is a complicated shadow pattern which does not resemble the source. The recorded shadowgraph is processed to produce an optical transparency which

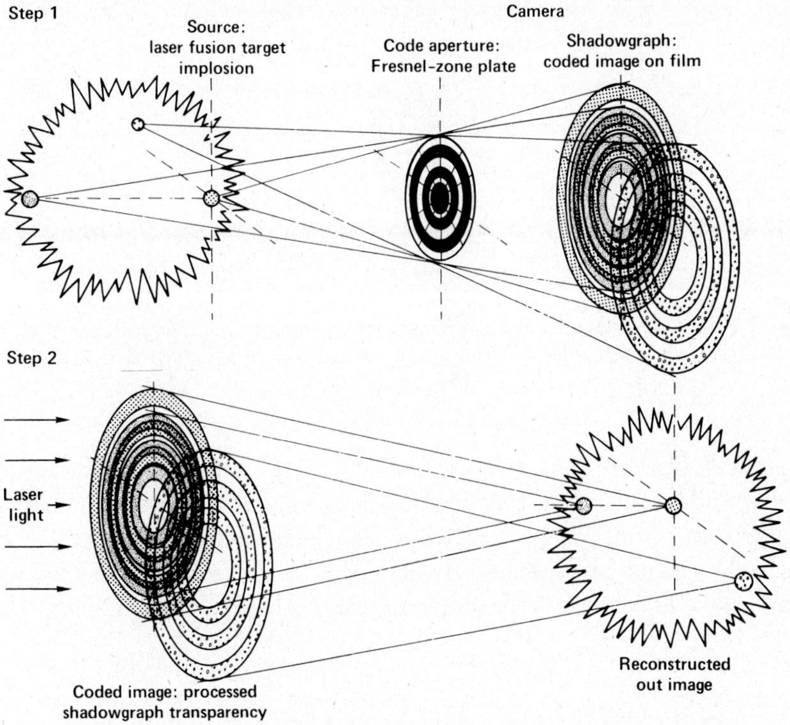

Fig. 1. Imaging geometry for the two-step ZPCI technique. In step 1, radiation emitted from the source casts a shadow pattern through a Fresnel zone plate aperture onto recording film. Each point in the source casts a separate zone plate shadow, and each shadow, by its size and position, uniquely characterizes the position of its associated source point. In step 2, the processed shadowgraph is illuminated with a low power, visible laser beam. Each zone plate shadow focuses the incident laser light to a diffraction-limited image spot. The image of the source is scaled and inverted.

when coherently illuminated produces an optical image of the original radiation source. The two-step process may be viewed as follows: The coherent diffraction process in step two performs a Fresnel transform on the shadowgraph pattern. This serves to precisely unfold the source–aperture convolution achieved when the source casts its shadow through the Fresnel zone plate aperture in step one.

Figure 2 shows the two-dimensional coordinate systems of the source, coded aperture, and coded image. This introductory analysis follows closely the early work of BARRETT and HORRIGAN [1973], and so uses their notation for convenience. A planar source distribution $f(\boldsymbol{r})$ is considered here for simplicity,

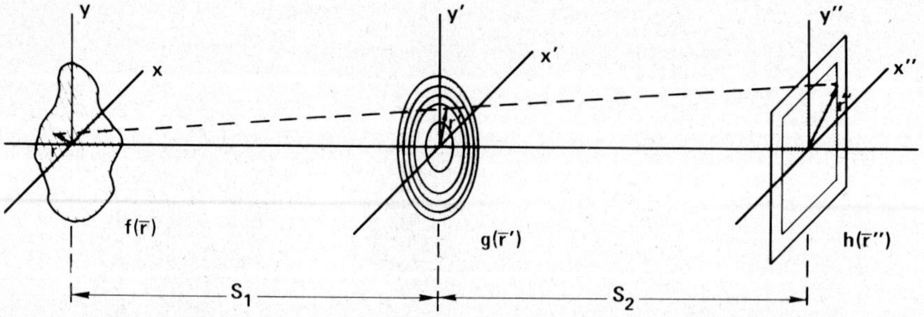

Fig. 2. Coordinate systems for the coded aperture shadowgraph recording process. Radiation from a source point at $\boldsymbol{r}$ passes through the coded aperture at $\boldsymbol{r}'$ and is recorded in the shadowgraph at $\boldsymbol{r}''$.

but this discussion can easily be generalized to three-dimensional distributions. We consider radiation originating at a source point $\boldsymbol{r}$, passing through the aperture at $\boldsymbol{r}'$, and being recorded in the shadowgraph plane at $\boldsymbol{r}''$. Since diffraction effects are negligible during coded image recording, the radiation travels in a straight line from source to shadowgraph and the 2D vectors are related by

$$\boldsymbol{r}' = a\boldsymbol{r}'' + b\boldsymbol{r}, \tag{2.1}$$

where

$$a = \frac{s_1}{s_1 + s_2}; \qquad b = \frac{s_2}{s_1 + s_2}.$$

The shadowgraph distribution $h(\boldsymbol{r}'')$ is the accumulation of all the source quanta which successfully pass through the coded aperture $g(\boldsymbol{r}')$ and as such, may be written

$$h(\boldsymbol{r}'') = \frac{1}{4\pi(s_1 + s_2)^2} \iint f(\boldsymbol{r})\, g(a\boldsymbol{r}'' + b\boldsymbol{r})\, \mathrm{d}^2\boldsymbol{r}. \tag{2.2}$$

Equation (2.2) holds true for any coded aperture function, $g(\boldsymbol{r}')$. In this analysis, our interest centers on the Fresnel zone plate coded aperture, a series of alternating transparent and opaque equi-area, annular rings. Its transmission function may be represented by the Fourier series expansion

$$g(\boldsymbol{r}') = \left(\frac{1}{2} + \frac{1}{\pi \mathrm{i}} \sum_{m=-\infty\,(odd)}^{\infty} \frac{1}{m} \exp\left(-\mathrm{i}\pi m \left(\frac{r'}{r_1}\right)^2\right)\right) \mathrm{Circ}\left(\frac{\mathrm{r}'}{\mathrm{r}_\mathrm{N}}\right). \tag{2.3}$$

In this expansion r_1 and r_N are the outer radii of the first (central) and Nth (outermost) zones, respectively. The Circ function accounts for the finite zone plate size (GOODMAN [1968]). Substituting eq. (2.3) into eq. (2.2) yields the complete expression for the zone plate coded image

$$h(\boldsymbol{r}'') = h_0 + \sum_{m=-\infty\,(\text{odd})}^{\infty} h_m(\boldsymbol{r}''). \tag{2.4}$$

We have, for convenience, separated the individual orders in the Fourier representation of the zone plate aperture. The contribution of the mth order term to the shadowgraph exposure is

$$h_m(\boldsymbol{r}'') = \frac{1}{4\pi^2(s_1+s_2)^2 mi} \iint f(\boldsymbol{r}) \exp\left[-im\pi\left(\frac{a\boldsymbol{r}''+b\boldsymbol{r}}{r_1}\right)^2\right] \times \operatorname{Circ}\left(\frac{|a\boldsymbol{r}''+b\boldsymbol{r}|}{r_N}\right) \mathrm{d}^2\boldsymbol{r}. \tag{2.5}$$

The bias or d.c. contribution is

$$h_0 = \frac{1}{8\pi(s_1+s_2)^2} \iint f(\boldsymbol{r}) \operatorname{Circ}\left(\frac{|a\boldsymbol{r}''+b\boldsymbol{r}|}{r_N}\right) \mathrm{d}^2\boldsymbol{r}. \tag{2.6}$$

Optical reconstruction of the zone plate coded image is accomplished by coherent illumination of the shadowgraph transparency (as one would a hologram) and viewing the Fresnel diffraction pattern in the appropriate plane. It is convenient at this point in the analysis to incorporate the assumption of linear shadowgraph processing. In particular, it is assumed that the shadowgraph transparency is processed such that its *amplitude* transmission function is linearly proportional to $h(\boldsymbol{r}'')$. The neglect of all nonlinear effects associated with the coded imaging process allows the development of a mathematically simplified linear theory of ZPCI. It is important to note, however, that the linear theory developed here provides a quantitatively accurate description of ZPCI capabilities and limitations. The nonlinear effects neglected here are considered in § 5.2.

Assuming linear shadowgraph processing, the reconstructed optical amplitude distribution an axial distance d beyond the uniformly illuminated shadowgraph is (GOODMAN [1968] ch. 4)

$$U(\boldsymbol{r}_2) = \frac{\exp\left(i2\pi\dfrac{d}{\lambda}\right)}{i\lambda d} \iint h(\boldsymbol{r}'') \exp\left(i\frac{\pi}{\lambda d}(\boldsymbol{r}_2-\boldsymbol{r}'')^2\right) \mathrm{d}^2\boldsymbol{r}, \tag{2.7}$$

where λ is the reconstruction wavelength, and $\boldsymbol{r}_2$ is the position vector in the reconstruction plane. Substituting eq. (2.4) into eq. (2.7) yields

$$U(\boldsymbol{r}_2) = U_0 + \sum_{m=-\infty\,(\text{odd})}^{\infty} U_m(\boldsymbol{r}_2). \tag{2.8}$$

again $U(\boldsymbol{r}_2)$ is separated into a d.c. contribution U_0, and a summation over higher order contributions U_m. The d.c. contribution

$$U_0 = K_0 \iint h_0 \exp\left(\frac{\mathrm{i}\pi}{\lambda d}(\boldsymbol{r}_2 - \boldsymbol{r}'')^2\right) \mathrm{d}^2\boldsymbol{r}'', \tag{2.9}$$

is to a good approximation uniform over the region of interest in the reconstruction plane. The mth order contribution to the reconstructed image is

$$U_m(\boldsymbol{r}_2) = K \iint \mathrm{d}^2\boldsymbol{r}'' \iint \mathrm{d}^2\boldsymbol{r}\, f(\boldsymbol{r}) \exp\left(-\mathrm{i}\frac{\pi m}{r_1^2}(a\boldsymbol{r}'' + b\boldsymbol{r})^2\right)$$
$$\times \exp\left(\mathrm{i}\frac{\pi}{\lambda d}(\boldsymbol{r}_2 - \boldsymbol{r}'')^2\right) \mathrm{Circ}\left(\frac{|a\boldsymbol{r}'' + b\boldsymbol{r}|}{r_N}\right). \tag{2.10}$$

Each point in the source distribution casts a shadow which, according to eq. (2.10), launches a paraxial spherical wave focused at an axial location

$$d_m = \frac{r_1^2}{a^2 m\lambda} \tag{2.11}$$

If eq. (2.10) is evaluated only in this focal plane, then after algebraic manipulation it can be written in the simplified form

$$U_m(\boldsymbol{r}_2) = K_m \iint f\left(-\frac{s_1}{s_2}\boldsymbol{r}^*\right) S_m(\boldsymbol{r}_2 - \boldsymbol{r}^*)\, \mathrm{d}^2\boldsymbol{r}^*, \tag{2.12}$$

where

$$K_m = \frac{Na^2}{2\pi s_1 s_2} \exp\left(\mathrm{i}\frac{2\pi r_1^2}{ma^2\lambda^2}\right); \tag{2.13a}$$

$$S_m(\boldsymbol{r}_2 - \boldsymbol{r}^*) = \frac{J_1(|\pi a(\boldsymbol{r}_2 - \boldsymbol{r}^*)|/(\Delta r/m))}{|\pi a(\boldsymbol{r}_2 - \boldsymbol{r}^*)|/(\Delta r/m)} \exp\left(\mathrm{i}\frac{m\pi a^2}{r_1^2}(\boldsymbol{r}_2 - \boldsymbol{r}^*)^2\right); \tag{2.13b}$$

$$\boldsymbol{r}^* = -\frac{s_2}{s_1}\boldsymbol{r}; \tag{2.13c}$$

$$N = \left(\frac{r_N}{r_1}\right)^2, \qquad \text{number of coded aperture zones;} \tag{2.13d}$$

$$\Delta r = \frac{r_1^2}{2r_N}, \qquad \text{width of the outermost zone.} \tag{2.13e}$$

Equation (2.12) relates the optically reconstructed image to the scaled and inverted source distribution, f. The scaled source distribution is convolved with S_m, the mth order impulse response for the ZPCI technique. It is important to note that eq. (2.12) relates the source *intensity* distribution, f, to the mth order reconstructed *amplitude* distribution U_m, whereas in the laboratory the reconstructed *intensity* distribution $|U_m|^2$ is typically measured. While eqs. (2.11) and (2.12) are the key reconstruction equations, eq. (2.8) reminds us that at the reconstruction plane a distance d_m from the shadowgraph there is not only the focused image contribution from U_m, but also the out-of-focus contributions from all the other U_i. The out-of-focus light forms a non-uniform background, commonly referred to as ghosts, artefacts, or side-lobe effects (CANNON and FENIMORE [1980], WILSON, BARRETT, DEMEESTER and FARMELANT [1973]). The degree to which the non-uniform background interferes with the perception or interpretation of the focused image depends on the order number of the focused image and on relative source and zone plate scale sizes. A detailed discussion of such matters is deferred to § 4.4. Until then, we concentrate on analyzing the characteristics of the mth order reconstructed image noting that there do exist situations for which the non-uniform background effects are completely negligible in optical reconstructions.

§ 3. Impulse Response Analysis of ZPCI

Although historically proposed as a method for locating discrete point sources in the X-ray sky (MERTZ and YOUNG [1961]), recent interest in ZPCI has centered on its usefulness for the investigation of extended radiation source distributions in the laboratory. Having formulated the image reconstruction equation as an impulse-response convolution integral, it is mathematically convenient to model the extended radiation source as a collection of adjacent, discrete point sources,

$$f(\mathbf{r}) = \sum_i A_i \delta(\mathbf{r} - \mathbf{R}_i). \tag{3.1}$$

This methodology allows direct evaluation of the planar and tomographic resolution capabilities of ZPCI.

3.1. PLANAR RESOLUTION

There is a great deal of arbitrariness in the choice of a planar or transverse resolution criterion for optically reconstructed coded images. Since image reconstruction is a coherent process, the standard Rayleigh resolution criterion is not quite appropriate. Nevertheless, we wish to define a resolution standard similar to the Rayleigh criterion so that comparison may be easily made with other incoherent imaging techniques. We, therefore, choose a point source pair, and calculate the reconstructed intensity distribution as a function of point pair separation. The point pair is considered marginally resolved when the reconstructed intensity distribution has a saddle to peak ratio of 0.735, similar to the Rayleigh criterion for incoherent sources. Consider a pair of point sources located at positions $\pm \boldsymbol{x}$,

$$f(\boldsymbol{r}) = \delta(\boldsymbol{r} + \boldsymbol{x}) + \delta(\boldsymbol{r} - \boldsymbol{x}). \tag{3.2}$$

Substitution into eq. (2.12) yields an mth order reconstructed field distribution

$$U_m(\boldsymbol{r}_2) = K_m\left(S_m\left(\boldsymbol{r}_2 - \frac{s_2}{s_1}\boldsymbol{x}\right) + S_m\left(\boldsymbol{r}_2 + \frac{s_2}{s_1}\boldsymbol{x}\right)\right).$$

The reconstructed intensity distribution $I_m \propto |U_m|^2$ contains a cross product term that involves the phase difference between the shadows of the two point sources. For real sources, the phase difference is zero. The resulting intensity distribution is then,

$$I_m(\boldsymbol{r}_2) = K'_m\left(\frac{J_1(x^-)}{x^-} + \frac{J_1(x^+)}{x^+}\right)^2. \tag{3.3}$$

For notational convenience, we defined the scaled position variables,

$$x^- \equiv \left|\boldsymbol{r}_2 - \frac{s_2}{s_1}\boldsymbol{x}\right|\frac{\pi a m}{\Delta r}, \qquad x^+ \equiv \left|\boldsymbol{r}_2 + \frac{s_2}{s_1}\boldsymbol{x}\right|\frac{\pi a m}{\Delta r}.$$

The saddle point and peaks of $I(\boldsymbol{r}_2)$ occur at $\boldsymbol{r}_2 = 0$ and $\pm (s_2/s_1)\boldsymbol{x}$, respectively. The saddle/peak intensity ratio is

$$\frac{I_s}{I_p} = \frac{4(J_1(x_0)/x_0)^2}{(J_1(2x_0)/2x_0 + 1/2)^2},$$

where

$$x_0 = \frac{s_2}{s_1}\frac{\pi a m}{\Delta r}|\boldsymbol{x}|.$$

Setting $I_s/I_p = 0.735$ yields a transcendental equation with the solution $x_0 = 2.574$. This corresponds to a marginally resolved point pair separation, $2|\boldsymbol{x}|$, equal to the mth order planar resolution δ_m

$$\delta_m = 1.64\left(\frac{s_1 + s_2}{s_2}\right)\frac{\Delta r}{m}. \qquad (3.4)$$

Equation (3.4) is an important statement about the resolution capability of ZPCI. Note that according to eq. (3.4) resolution capability improves with reconstruction in higher order. This is demonstrated experimentally in Fig. 3 in which a coded image of a point source pair nominally separated by 6 μm is reconstructed in first, third and fifth order. In this case the zone plate coded

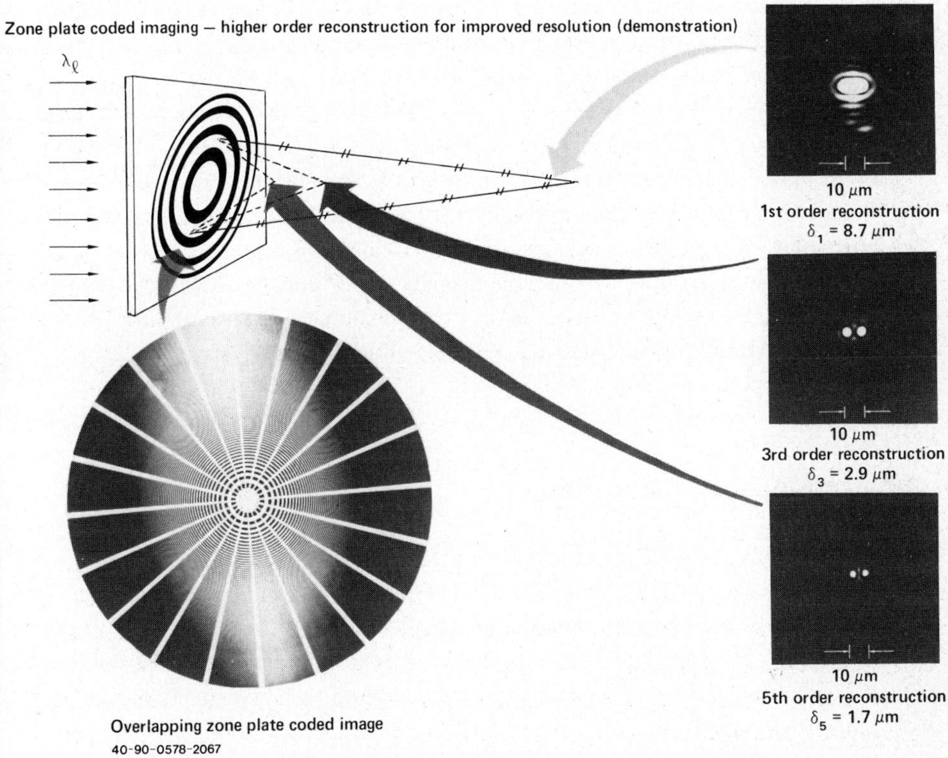

Fig. 3. Image reconstruction in higher order for improved resolution. The coded image of a point source pair nominally separated by 6 μm is reconstructed in first, third, and fifth order. In first order the resolution is insufficient to resolve the points. However, in third and fifth order the point pair is clearly resolved.

aperture has a minimuml zone width, $\Delta r = 5.3\ \mu m$, and $(s_1 + s_2)/s_2 \approx 1$, so that nominal first-order resolution is $\delta_1 \approx 8.7\ \mu m$, insufficient to resolve the point source pair. In third order, however, the nominal resolution is $\delta_3 \approx 2.9\ \mu m$ and the point pair is clearly resolved, while in fifth order, $\delta_5 = 1.7\ \mu m$, the point pair is even better resolved. Resolution improvement in higher order reconstruction has been previously noted (GUR and FORSYTH [1978]). It is possible because high spatial frequency information is recorded in the sharp edge definition of the zone plate shadowgraph. There are, however, practical considerations which limit the usefulness of image reconstruction in higher order. Ultimately, the higher order resolution for optical reconstruction is limited by spherical aberration effects resulting from the fact that the geometrical zone plate produces only a paraxial approximation to a spherical wavefront (YOUNG [1972]). Also, nonlinear effects which are neglected here will be shown to be more severe in higher-order image reconstructions than in first order. Also, the short wavelength requirement for geometrical shadowcasting becomes more stringent for reconstruction in higher order, since the high spatial frequency information resides in the shadowgraph edge definition which can be seriously compromised by diffraction blurring during shadowgraph recording. In addition to the above, there are effects arising primarily from continuous source distributions, which limit the usefulness of image reconstruction in higher order. In particular, non-uniform background effects which are negligible for the point sources illustrated here become increasingly important for continuous source distributions as image reconstruction order is increased. Discussion of these practical limitations is deferred to the eigenfunction analysis of continuous source distributions in § 4.

3.2. TOMOGRAPHIC RESOLUTION

In what follows, the tomographic or three-dimensional imaging capability of ZPCI is assessed. The establishment of a general tomographic resolution criterion is complicated by the fact that successful resolution of three-dimensional image detail is strongly dependent on source intensity distribution. For example, two source points separated in depth by a given distance can be more easily resolved if they are of equal intensity, than if one is significantly more intense than the other. As a result, the following should be viewed as the development of a figure of merit rather than a definitive criterion for the tomographic resolution of ZPCI.

The tomographic capability of the ZPCI technique can be understood intuitively by considering two on-axis point sources at unequal distances from the zone plate aperture. The point sources will cast two overlapping zone plate shadows of unequal size because of their different source to zone plate distances. When coherently reconstructed, the zone plate shadows will focus the incident light to a pair of axially separated diffraction-limited spots.

The tomographic analysis again follows the simple point-source-pair model. The source points lie on the axis of the zone plate aperture, are separated by a distance $\Delta s_1 \ll s_1$, and have intensity ratio ψ (i.e., $\psi = 1$ implies equal source intensities). The overlapping zone plate shadows reconstruct in mth order to a pair of image points axially separated by

$$|\Delta d_m| = 2d_m \left(\frac{s_2}{s_1 + s_2}\right) \frac{\Delta s_1}{s_1}, \tag{3.5}$$

where d_m is given by eq. (2.11). In order to evaluate the resolution of the axially separated point-pair image we calculate the axial distribution of the reconstructed field in the vicinity of the mth order images. This is achieved by summing the axial field distributions from each of the reconstructed point images. In doing this, LOMMEL's [1885] analysis for the field distribution in the vicinity of the focal spot of a thin converging lens is appropriately applied (BORN and WOLF [1975]). Squaring the reconstructed field distribution yields the axial intensity distribution in the vicinity of the mth order reconstructed images. The axial intensity plots shown in Fig. 4 were computed in this rigorous manner (CEGLIO [1976]).

Without displaying the algebraic detail involved in the rigorous calculation, the basic results for tomographic resolution can be obtained from the expression for the axial intensity distribution in the vicinity of a single point image (BORN and WOLF [1975]),

$$\frac{I(\alpha_m)}{I_0} = \left(\frac{\sin(\alpha_m/4)}{\alpha_m/4}\right)^2,$$

where

$$\alpha_m = 4\pi m N \left(\frac{s_2}{s_1 + s_2}\right)\left(\frac{\Delta s_1}{s_1}\right),$$

is a dimensionless tomographic parameter.

A measure of the tomographic capability of the ZPCI process is the distance in object space to the first null in the axial intensity distribution. This occurs

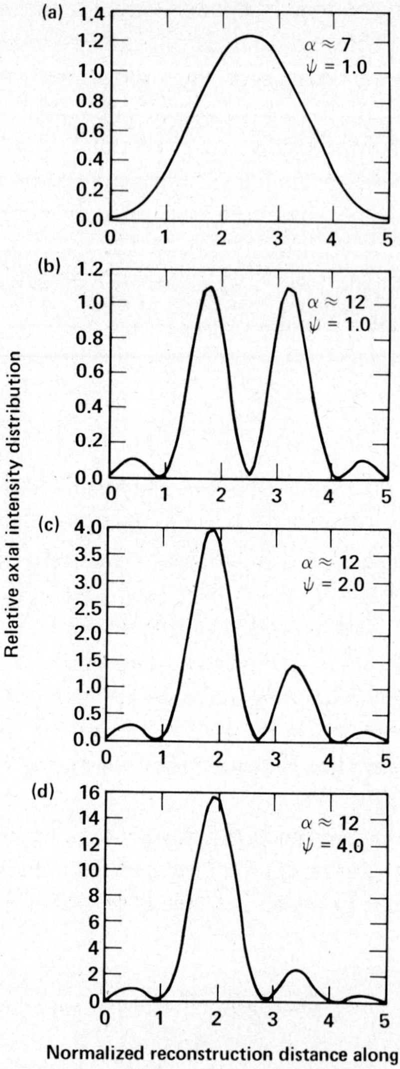

Fig. 4. Tomographic resolution calculations. The axial distribution of reconstructed *intensity* for an axially separated point source pair is plotted for different values of the tomographic parameter α and the intensity ratio ψ. At $\alpha \approx 4\pi$, $\psi = 1$ (b) the point pair is clearly resolved, leading to eq. (3.6). At large ψ (d) the image of the weak point is lost in the sidelobes of the strong point.

when $\alpha_m = \pm 4\pi$; or when the object space separation is

$$\Delta_m = \frac{s_1}{Nm} \frac{(s_1 + s_2)}{s_2}. \tag{3.6}$$

Equation (3.6) is useful as a tomographic figure of merit. Δ_m is proportional to s_1 and inversely proportional to N. Improved tomographic resolution can be achieved using a large (N) zone plate placed close to the source. Equation (3.6) also predicts improved tomographic resolution for reconstruction in higher order, however, when considering continuous source distributions, this conclusion is subject to the same limitations mentioned earlier in the discussion of planar resolution. This matter is treated again in § 4.2.

Figure 4 shows a series of plots of $I(\psi)$ at various α_m values for point source pairs of varying relative intensities ($\psi = 1, 2$ and 4). We note that for point source pairs of unequal relative intensity ($\psi > 1$) tomographic separation is more difficult. As pointed out earlier, source intensity maps into the reconstructed amplitude field, so that the intensity ratio of the optically reconstructed image peaks goes as ψ^2. The tomographic resolution capability is thereby limited ($\psi \geq 4$ seems sufficient) by the difficulty in discriminating between the axial sidelobes of the strong image point and the peak value of the weak image point.

§ 4. Eigenfunction Analysis of Continuous Source Distributions

In the previous sections, ZPCI was characterized by its impulse response. The image of an extended source was represented by eq. (2.12) as the convolution of the scaled source distribution with the impulse response. In this section, we pursue a fundamentally different but equally valid approach where continuous source imaging is analyzed directly. A contribution of the following analysis is the additional insight it provides to ZPCI. It is shown, for example, that ZPCI can be described by equations of the same form as the familiar Gaussian laser beam propagation equations. Many interesting quantitative results follow. The analysis clearly shows the following features:

(a) Image resolution, both planar and tomographic.

(b) The manner in which spatial information is stored as phase and contrast modulation in a coded image. (For a given object the information for progressively higher orders is stored in the shadowgraph over progressively smaller regions centered about the middle of the shadowgraph.)

(c) That conventional concepts of zone plate diffraction efficiency are inapplicable to the optical reconstruction of coded images of continuous sources. (For example, for an optimally recorded Gaussian source, only about 5% of the incident energy is diffracted into the first order image; the relative energy diffracted into progressively higher orders decreases with order number to the fourth power.)

(d) Most importantly, that linear ZPCI will only be successful (i.e., good image signal to noise ratio and negligible non-uniform background) if the characteristic size of the object is small compared to the radius of the first zone of the zone plate (cf. eq. (4.9)).

The analysis here is based on the precept that the shadow casting process may be mathematically represented as a Fresnel transform. Subsequent optical reconstruction may also be represented as a Fresnel transform. The eigenfunctions of the Fresnel transform are well known from stable laser oscillator theory (SIEGMAN [1971]). The Hermite–Gaussian polynomials in cartesian coordinates and the Gaussian–Laguerre polynomials in polar coordinates form complete, orthogonal sets that retain their functional form when Fresnel transformed. When a coded imaging source is represented as a superposition of terms in these series, then all of the integrals in the recording and reconstruction process can be analytically evaluated. The reconstructed field for any order can be easily found in any plane.

In polar coordinates, any general source distribution may be represented by the series expansion

$$f(r, \phi) = \sum_{p=0}^{\infty} \sum_{l=-\infty}^{\infty} c_p^{(l)} f_p^{(l)}(r, \phi), \tag{4.1}$$

where $c_p^{(l)}$ are the series coefficients and

$$f_p^{(l)}(r, \phi) = \left(\frac{r\sqrt{2}}{\sigma}\right)^l L_p^l\left(2\,\frac{r^2}{\sigma^2}\right) \exp\left(\frac{-r^2}{\sigma^2} - il\phi\right);$$

$L_p^{(l)}$ are the associated Laguerre polynomials. The eigenfunction analysis is most easily illustrated using the first series term $p = 0$, $l = 0$, which is a simple two-dimensional Gaussian. The analysis for higher order terms follows in similar fashion. Consider then the Gaussian source distribution

$$f(r, \phi) = f(r) = \exp(-r^2/\sigma^2), \tag{4.2}$$

where

$$\int_0^{2\pi} \int_0^{\infty} f(r, \phi) r \,\mathrm{d}r \,\mathrm{d}\phi = \pi\sigma^2.$$

It follows from eqs. (2.2) and (2.3), that the shadowgraph exposure is

$$h(r'', \phi'') = \frac{\pi\sigma^2}{2} + \frac{1}{i\pi} \sum_{m=-\infty\,(\text{odd})}^{\infty} \frac{1}{m} \int_0^{2\pi} \int_0^{\infty} \exp\left(\frac{-r^2}{\sigma^2}\right) \times \exp\left(\frac{-im\pi}{r_1^2}(br + ar'')^2\right) r\,\mathrm{d}r\mathrm{d}\phi, \tag{4.3}$$

where the $4\pi(s_1 + s_2)^{-2}$ factor and the Circ function have been suppressed for simplicity. The second exponential term in eq. (4.3) is the Fresnel kernel. The integral in eq. (4.3) can be evaluated by completing the square. A fair amount of algebra follows that is not repeated here (see SIEGMAN [1971]). The solution can be put in the form

$$h(r'') = \frac{\pi\sigma^2}{2} + \sum_{m=-\infty\,(\text{odd})}^{\infty} \frac{1}{m\pi\rho_m} \exp(i\psi_m) \exp\left[-i\pi\left(\frac{1}{R_m} - \frac{i}{\pi\omega_m^2}\right)(r'')^2\right]. \tag{4.4}$$

Each summation term in eq. (4.4) is in the form of a Gaussian spherical wave. Wave parameters like R_m and ω_m are functions of the coded imaging system geometry and source size; specifically,

$$R_m = f_m\left[1 + \left(\frac{\pi\omega_0^2}{f_m}\right)^2\right], \tag{4.5a}$$

$$\omega_m = \omega_0\left[1 + \left(\frac{f_m}{\pi\omega_0^2}\right)^2\right]^{1/2}, \tag{4.5b}$$

$$\psi_m = \tan^{-1}(f_m/\pi\omega_0^2), \tag{4.5c}$$

$$\rho_m = \frac{mb^2}{r_1^2}\left(\frac{\omega_m}{\omega_0}\right), \tag{4.5d}$$

$$f_m = r_1^2/ma^2, \tag{4.5e}$$

$$\omega_0^2 = (\sigma b/a)^2. \tag{4.5f}$$

The equations above can each be associated with classical Gaussian beam propagation equations. In fact, eqs. (4.4) and (4.5) can be obtained from the Gaussian beam equations by making the parameter substitution

$$\left(\frac{\pi\omega_0^2}{f_m}\right)_{\text{ZPCI}} \leftrightarrow \left(\frac{\pi\omega_0^2}{\lambda L}\right)_{\text{optical}}. \tag{4.6}$$

The dimensionless optical parameter on the right is the so-called Fresnel number. The far field of a Gaussian beam is reached when this parameter becomes much smaller than unity. Similarly, we define the ZPCI Fresnel number, F_m. The ZPCI Fresnel number has an interesting geometrical interpretation. If the zoneplate is projected back onto the object plane using a point source located on the axis in the shadowgraph plane, then the ZPCI Fresnel number is equal to $m\pi$ times the number of zones covered by the object. It is shown below that F_m is an important figure of merit for characterizing the shadowgraph recording process. In order to produce reconstructed ZPCI images of high S/N with negligible non-uniform background effects, it is important that F_m be less than unity. Image quality improves as F_m is decreased.

For linear recording the amplitude transmittance of the shadowgraph is linearly proportional to the exposure. If this shadowgraph transparency is illuminated with a unit amplitude plane wave, the field just after the transparency is simply equal to the transmittance. The scalar optical field a distance L beyond the shadowgraph is described by the Fresnel diffraction integral. However, it is not necessary to explicitly evaluate the Fresnel diffraction integral because we note that $h(\mathbf{r}'')$, and in the linear case $t(\mathbf{r}'')$, is simply a sum of Gaussian-spherical wavefronts. The propagation characteristics of such wavefronts are well known.

Equations (4.4) and (4.5) are then the key results of this analysis; they represent a simple functional form for the shadowgraph exposure for a continuous object. We can gain a greater physical appreciation of eq. (4.4) by writing it in the form

$$h(r'') = \frac{\pi\sigma^2}{2} + \frac{2}{\pi} \sum_{m=1,3,5}^{\infty} \frac{1}{m\rho_m} \exp\left(\frac{-(r'')^2}{\omega_m^2}\right) \sin\left(\frac{\pi}{R_m}(r'')^2 - \left(\psi_m - \frac{\pi}{2}\right)\right). \tag{4.7}$$

When the shadowgraph is optically reconstructed each term in the summation of eq. (4.7) yields the $\pm m$ diffracted orders. The sinusoidal term provides the high frequency "grating" modulation that diffracts the reconstruction beam into the $\pm m$ orders. The spatial distribution of the grating fringe visibility is determined by the multiplicative Gaussian envelope term.

In order to optimize S/N in the reconstructed image, it is necessary to maximize the diffracted image component relative to the undiffracted (d.c.) component. This is achieved by maximizing the width and amplitude parameters, W_m and A_m, which are defined as follows:

$$W_m = \frac{\text{Gaussian envelope width}}{\text{Grating modulation period}} \text{ (in } r''^2 \text{ space)} \sim \frac{\omega_m^2}{R_m},$$

and
$$A_m \equiv \frac{\text{Gaussian envelope amplitude}}{\text{D.C. component amplitude}} \sim \frac{1}{m\rho_m \sigma^2}.$$

W_m is a measure of the number of grating periods contributing to the reconstructed image, and A_m measures the relative amplitudes of the image and background contributions. The parametric dependencies of W_m and A_m are indicated above. They may also be written as explicit functions of the ZPCI Fresnel number, F_m:

$$W_m \sim \frac{1}{F_m}, \tag{4.8a}$$

$$A_m \sim \frac{1}{mF_m}\left(\frac{1}{(1 + 1/F_m^2)^{1/2}}\right). \tag{4.8b}$$

W_m varies inversely with F_m, as does A_m for large values of F_m. For F_m of order unity, A_m increases slowly with decreasing F_m values.

A useful guide for experimental design is to require that the ZPCI Fresnel number be less than unity. This reduces to the requirement

$$\frac{\sigma}{r_1} < \frac{1}{b(m\pi)^{1/2}} \tag{4.9}$$

for good image S/N. When interpreted to apply to generalized continuous sources, eq. (4.9) requires that the characteristic large scale dimension of the source be less than the diameter of the central zone of the coded aperture.

The S/N condition expressed in eq. (4.9) is a quantitative expression of the "information compression" requirement for a S/N advantage in coded imaging techniques. This is discussed in § 5.1. The condition $F_m \ll 1$ is physically equivalent to the condition that the shadowgraph be in the far field of the Gaussian wave waist (i.e., reconstructed image). Therefore, as F_m is decreased, a higher degree of information compression is achieved, because the shadowgraph recedes further into the far field of the reconstructed image (i.e., the Gaussian waist).

From either eq. (4.4) or eq. (4.7) we see that when the shadowgraph is reconstructed each term launches a Gaussian spherical wavefront. Each Gaussian wavefront is characterized by a waist location and size. The ZPCI image of interest is located at the waist of the corresponding Gaussian wavefront. The Gaussian waists will form at locations, $L = d_m$, such that the optical and ZPCI Fresnel numbers of eq. (4.6) are equal. The waists are located at

$$d_m = \frac{f_m}{\lambda} = \frac{r_1^2}{a^2 \lambda m} \qquad m = \pm 1, \pm 3, \cdots. \tag{4.10}$$

The spot size at the Gaussian waist, and hence, the ZPCI magnification, is the same for all m,

$$\sigma_{\text{image}} = \omega_0 = \frac{b}{a} \sigma = \frac{s_2}{s_1} \sigma. \tag{4.11}$$

Equations (4.10) and (4.11) are identical to eqs. (2.11) and (2.13c) from the impulse response analysis.

4.1. PLANAR RESOLUTION

On reconstruction, if the shadowgraph is to launch a Gaussian wavefront then the entire Gaussian envelope (see eq. (4.7)) associated with the wavefront must be encoded in the shadow plane. Clearly, the zoneplate and the shadowgraph detection plane must be large enough so the envelope will not be truncated. We can use this requirement to define planar resolution. Planar resolution may, therefore, be defined as the size of the smallest Gaussian source $\sigma = \delta_m$ that can be encoded with a given size zoneplate without significant truncation of the Gaussian envelope. If we require that the contrast envelope decrease by a factor e^{-2} at the edge of the zoneplate shadow, then the resolution can be obtained by manipulation of eqs. (4.5) and (4.7). The result is

$$\delta_m = \left(\frac{s_1 + s_2}{s_2}\right) \frac{\Delta r}{m}, \tag{4.12}$$

where again Δr is the width of the outermost zone. Note the functional identity with eq. (3.4) for resolution obtained using point sources. The difference in the numerical coefficients of eqs. (3.4) and (4.12) is of no significance. The coefficient values result from the flexibility in the definition of resolution.

4.2. TOMOGRAPHIC RESOLUTION

The tomographic resolution using the eigenfunction analysis reveals the dependence of the axial resolution on the source intensity distribution. Since a Gaussian object intensity distribution reconstructs a Gaussian-spherical wavefront, the optical Rayleigh range, d_R, associated with the reconstructed Gaussian wavefront is a measure of axial structure. Using the ZPCI notation, the Rayleigh range is

$$d_R = \frac{\pi\sigma^2}{\lambda}\left(\frac{b}{a}\right)^2.$$

The corresponding displacement in object space may be defined as the tomographic resolution

$$\Delta_m = \frac{\pi m}{2}\left(\frac{\sigma}{r_1}\right)^2 \frac{s_1 s_2}{(s_1 + s_2)}. \tag{4.13}$$

For comparison purposes, eq. (4.13) can be manipulated to form a result similar to eq. (3.6) by considering a limiting case. Let the Gaussian object have a width σ equal to the planar resolution for the first order; then use eq. (4.12) to eliminate σ in eq. (4.13). After some algebraic manipulation, the tomographic resolution is

$$\Delta_m = \frac{m}{N}\left(\frac{s_1 + s_2}{s_2}\right)s_1. \tag{4.14}$$

This result is functionally the same as eq. (3.6) obtained for a pair of point sources except that m appears in the numerator rather than the denominator. Rather than improve with order number, the tomographic resolution in this case gets worse with order number. The presence of the order number in the numerator is easily described physically. For $\sigma \gg \delta_1$, all reconstructed Gaussian orders have the same waist size and, hence, the same Rayleigh range. (This statement is not true for point sources, i.e., $\sigma \ll \delta_1$. So, while eq. (3.6) is valid for point sources, eq. (4.14) represents the tomographic capabilities of ZPCI for finite size, continuous sources.) In image space, then, there is no dependence of the tomographic resolution on order number. However, these equal distances in image space are scaled by m in mapping to object space coordinates (see eq. (3.5)). This introduces the order number in the numerator of eq. (4.14).

4.3. IMAGE INTENSITY

It is easy to calculate the total power and the image intensity in each order. If the shadowgraph transparency is processed so that the bias transmittance is 0.5, then the Gaussian amplitude modulation distribution for the mth order is

$$|U_m| = \frac{1}{\pi^2 \sigma^2 m \rho_m} \exp\left(\frac{-r''^2}{\omega_m^2}\right)$$
$$= \frac{1}{m\pi(F_m^2+1)^{1/2}} \exp\left(\frac{-F_m^2}{F_m^2+1}\left(\frac{r''}{\omega_0}\right)^2\right). \tag{4.15}$$

The power into the mth order for a plane wave reconstruction of irradiance I_0 is

$$P_m = I_0 \int_0^{2\pi} \int_0^{\infty} |U_m|^2 r'' \, dr'' \, d\theta = I_0 \left(\frac{\omega_0^2}{2\pi m^2 F_m^2}\right)$$
$$= I_0 \frac{r_1^4}{2a^2 b^2 \pi^3 m^4 \sigma^2}. \tag{4.16}$$

The power into each order decreases as m^{-4} for a continuous source, not as m^{-2} as it does for a point object. The power also varies as σ^{-2}.

The m^{-4} dependence of diffracted power can be easily explained. As the order increases, the carrier contrast decreases by m^{-1}. In addition, the width of the Gaussian envelope of the diffractive region of the shadowgraph decreases by m^{-1} (for $F_m \ll 1$). So then, the overall area under the square of the Gaussian envelope decreases with order number as m^{-4}.

The maximum image intensity for the Gaussian image will be

$$I_{max} = \frac{2P_m}{\pi \omega_0^2},$$

then from eq. (4.16), we obtain the interesting result,

$$I_{max} = I_0 \left(\frac{1}{\pi m F_m}\right)^2 = I_0 \left(\frac{r_1}{m\pi b\sigma}\right)^4,$$

indicating that I_{max} varies as $m^{-4} (r_1/\sigma)^4$.

4.4. BACKGROUND CONTRIBUTIONS FOR CONTINUOUS SOURCES

When a shadowgraph is reconstructed with a planar, coherent beam, all of the Fresnel orders (in and out of focus) contribute to the total amplitude in any plane. In the plane corresponding to the mth order focus, for example, other order terms represent undesirable contributions that are referred to as non-uniform background.

The presence of these background terms has been used as an argument for using coded aperture techniques other than the Fresnel zone plate (CANNON and FENIMORE [1980]). The impulse response of ZPCI contains undesirable sidelobes due to the contributions of the out-of-focus orders. Other coded apertures, such as uniformly redundant arrays, have impulse responses that do not exhibit such sidelobes (FENIMORE and CANNON [1978]). However, published images (CEGLIO, ATTWOOD and LARSEN [1982]) of small, continuous sources using the ZPCI technique suggest that the sidelobes are not as deleterious as has been suggested. In this section, we show quantitatively that if one considers continuous sources (in this case a simple Gaussian object having $F_m < 1$), the background contributions to the first order reconstructed image are insignificant. In higher order focal planes, however, the background terms can easily become significant.

We have presented all of the equations necessary to calculate the total optical field in any plane for a Gaussian source distribution. (For specific details, see Appendix A.) The spot size, absolute phase distribution, and amplitude associated with each order can be determined. For example, in the case of unit illumination and 0.5 bias transmittance, the contribution of the mth order in the nth order focal plane is (see eq. (4.4))

$$U_{mn}(r_2) = \left(\frac{2P_m}{\pi}\right)^{1/2} \frac{1}{\omega_{mn}} \exp\left(-\frac{r_2^2}{\omega_{mn}^2}\right) \exp\left[-\mathrm{i}\left(\frac{\pi}{R_{mn}} r_2^2 - \psi_{mn}\right)\right]. \quad (4.17)$$

Manipulation of the Gaussian beam equations predicts

$$P_m = \frac{r_1^4}{2m^4\pi^3 a^2 b^2 \sigma^2}, \quad (4.18a)$$

$$\omega_{mn} = \omega_0 \left[1 + \left(\frac{f_m}{\pi\omega_0^2}\right)^2 \left(\frac{m-n}{n}\right)^2\right]^{1/2}, \quad (4.18b)$$

$$R_{mn} = f_m \left(\frac{m-n}{n}\right)\left[1 + \left(\frac{\pi\omega_0^2}{f_m}\right)^2 \left(\frac{n}{m-n}\right)^2\right], \quad (4.18c)$$

$$\psi_{mn} = \tan^{-1}\left[\left(\frac{f_m}{\pi\omega_0^2}\right)\left(\frac{m-n}{n}\right)\right]. \tag{4.18d}$$

The objective then is to manipulate the above equations to characterize the effect of these background terms.

Near the optical axis the image term is in phase quadrature with all other orders and the d.c. term (when $F_m \ll 1$). Away from the optical axis, the phase structure of the background terms can in principal allow them to phase modulate the image term. A sufficient condition to avoid such background effects is to have the image amplitude much larger than the background contributions. In addition, if the background terms are kept in approximate phase quadrature with the image term, their interference with the image integrity will be minimized. This will be achieved if the image is contained within a region sufficiently close to the optical axis, a condition equivalent to that of eq. (4.9).

The ratio of the maximum amplitude (i.e., on the optical axis) of the nth order image term to that of the mth order background term is

$$\frac{U_{nn}(0)}{U_{mn}(0)} = \frac{m^2}{n^2}\left[1 + \left(\frac{f_n}{\pi\omega_0^2}\right)^2\left(\frac{n-m}{n}\right)^2\right]^{1/2}. \tag{4.19}$$

Several interesting observations follow from this equation: The first order amplitude is at least nine times that of the third order in the first order focal plane. In the third order focal plane, however, the first order background term may exceed the amplitude of the third order (unless F_n is very much less than unity). Secondly, the strongest non-uniform background contribution in the first order (real) image plane is that due to the negative first order (virtual) image term. Finally, note the importance of the ZPCI Fresnel number in eq. (4.19). The relative background contributions become more significant as F_n becomes larger.

Restricting attention to the first order image, the third order background contribution can be the most harmful, because it has the most rapidly varying phase structure (see eqs. (4.17) and (4.18c)). For positions away from the optical axis the third order term is the first to move out of phase quadrature with the first order image. (In fact, it is the radial variation of the phase structure of these out-of-focus background terms that produces the sequence of dark rings which are noticed surrounding the first order image in the optical reconstruction.) As mentioned above, in order to be unaffected by these radial phase variations, the first order image must be small enough that it fits within the central region, where the third order phase is nearly constant. This is equivalent to the requirement that the separation between the first and third order image planes be greater than the Rayleigh range associated with the Gaussian beams, that is,

$$(d_1 - d_3) = \frac{2r_1^2}{3\lambda a^2} > \frac{\pi\omega_0^2}{\lambda}. \qquad (4.20)$$

Simple manipulation of this equation produces the same imaging condition as eq. (4.9); that is, again we find that the ZPCI Fresnel number must be less than one.

§ 5. Additional Topics in ZPCI

5.1. QUANTUM NOISE CONSIDERATIONS

In this section we discuss the basic physical processes which determine the signal to noise ratio (SNR) for an optically reconstructed zone plate image. Our approach avoids mathematical complexity, and strives for an intuitive appreciation of those quantum noise processes which are unique to two-step coded imaging techniques. We build on and generalize the basic results established in the pioneering work by BARRETT and DEMEESTER [1967].

Coded imaging techniques are arduous, indirect methods for measuring the spatial distribution of a source. A primary justification for the use of coded methods is that they are less wasteful of radiation than direct methods (e.g. pinhole imaging), and therefore, present the possibility of an enhanced SNR. A coded aperture collects radiation over a large solid angle without sacrificing spatial resolution. The natural expectation of this increased collection efficiency is that it will lead to improved statistics and enhanced SNR. That this is *often* true accounts for the interest in coded imaging methods; that it is *not always* true provides the motivation for our discussion.

The quantum noise limitations of coded aperture methods are illustrated in the following comparative example: An unknown source distribution casts a shadow through a pinhole aperture, and then (for an equal time) through a zone plate coded aperture of equivalent resolution. From each of these shadowgraphs we wish to reconstruct the original source. There are many more quanta recorded in the zone plate shadow than in the pinhole shadow. However, each recorded quantum in the pinhole image carries with it more information about source location. Each image point, when projected back through the pinhole, delineates a narrow cone in space from which it could have come. The narrower the cone the more information about source location. In the zone plate shadow each recorded quantum, when projected back through the coded aperture, delineates a large region of space from which it could have come. So, although there are many more quanta in the zone plate shadow, each carries with it much less information about source location.

Similar back-projection arguments may be used in discussing quantum noise in a "reconstructed" image. In a pinhole image only those quanta recorded within a resolution element contribute to the noise at a given point. However, in a zone plate shadowgraph *all* the recorded quanta contribute to the noise at any given point in the reconstructed image. Therefore, unfavorable effects such as increased source size and phenomena which reduce contrast can quickly erode the SNR advantage provided by coded aperture methods. As source size increases, each new point adds quanta which contribute noise but no signal to the reconstruction of all *other* image points. Contrast reducing phenomena add quanta which not only add to the noise, but also reduce signal intensity. So then, increased collection efficiency does not by itself guarantee increased SNR in coded aperture imaging.

The fundamental source of noise in coded imaging is the quantization of the emissions recorded in the shadowgraph. For example, a point source of limited strength does not generate a spatially continuous zone plate shadow. Individual quanta from the source interact with the detector to produce discrete, randomly positioned responses (e.g., spots on film) within the envelope of the zone plate pattern. Figure 5 shows such a "spotty" zone plate shadowgraph, made using a point source of 6 MeV protons (STONE and CEGLIO [1983]).

An imperfectly delineated zone plate coded image suffers from a spatially distributed shot noise, that is, random fluctuations in the shadowgraph transmission, $t(r'')$. In the optical reconstruction of the coded image, statistical fluctuations in the shadowgraph transmission lead to fluctuations in the reconstructed field, $U(\boldsymbol{r}_2)$. The expected value of the field, $\langle U(\boldsymbol{r}_2)\rangle$, is the reconstructed image (i.e. signal), and its variance ($\sigma_U^2 = \langle |U|^2\rangle - \langle U\rangle^2$) is the average noise intensity.

The effects of coherence are important in determining the dependence of SNR on the mean and variance of the statistically varying reconstructed field. For the purpose of illustration, we divide the amplitude transmission of the coded image into a deterministic component, that would be obtained if the shadowgraph distribution were continuous, and a random component, that accounts completely for the quantization of the emission process. Since the reconstruction process is linear, the optical fields resulting from each of these terms can be found independently. Evaluation of the SNR must account for interference between the deterministic image (i.e. signal) and the coherent noise,

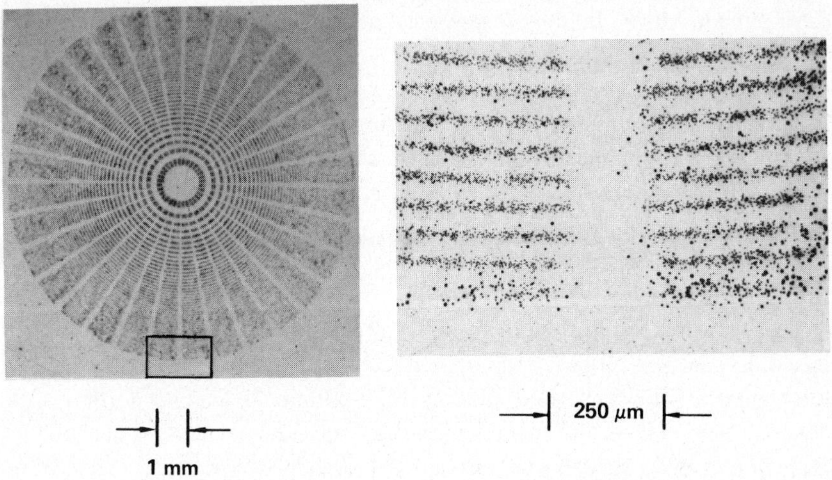

Fig. 5. Spatial shot noise in a zone plate coded image. This shadowgraph of a point source of 6 MeV protons illustrates the quantum nature of the recorded coded image.

and the fact that the observable is the reconstructed intensity, not the field. Appropriate accounting for these effects leads to the expression (BARRETT and DEMEESTER [1974], GOODMAN [1967]),

$$\mathrm{SNR} = \frac{I_s/\sigma_U^2}{(1 + 2I_s/\sigma_U^2)^{1/2}}$$

where I_s, the signal intensity, is equal to $\langle U \rangle^2$. The limiting forms of this expression illustrate the importance of coherence effects in the SNR,

$$\mathrm{SNR} \simeq I_s/\sigma_U^2, \qquad \text{when } I_s/\sigma_U^2 \ll 1; \tag{5.1a}$$

$$\mathrm{SNR} \simeq \left(\frac{I_s}{2\sigma_U^2}\right)^{1/2}, \qquad \text{when } I_s/\sigma_U^2 \gg 1. \tag{5.1b}$$

When the signal is much weaker than the noise (eq. (5.1a)), cross terms (signal × noise) may be neglected relative to the random intensity variations of the noise wave itself. In this case the standard deviation of the statistically varying *intensity* distribution is simply σ_U^2, and eq. (5.1a) results. When the signal is much stronger than the noise (eq. (5.1b)), the predominant noise contribution is the interference between the weak random wave and the strong deterministic wave. In this case the standard deviation of the intensity is $(2I_s \sigma_U^2)^{1/2}$ resulting in eq. (5.1b).

There are two important quantum noise principles which allow us to relate the SNR of the reconstructed image to directly observable (and controllable) shadowgraph parameters. Simply stated these are:

P1. *All* quanta recorded in the shadowgraph contribute to the noise at every point in the reconstructed image.

P2. Only the statistically averaged modulation of the shadow pattern contributes to the signal.

These maxims follow directly from the formal quantum noise theory, and are based on a simple model of the optical reconstruction process. In this model the regular modulation of the shadowgraph pattern focusses a portion of the coherent illumination into the reconstructed image, while the statistical fluctuations in the shadowgraph transmission scatter light over a broad region of the reconstruction plane. The spectral bandwidth of the noise is presumed to be "very broad". This means that *all* shadowgraph fluctuations scatter light into the image region, or equivalently, that the spatial distribution of the noise is virtually uniform in the vicinity of the reconstructed image.

In quantitative terms **P1** means that σ_U^2 is proportional to $\langle N \rangle$, where N is the total number of quanta recorded in the entire shadowgraph. This allows us to relate σ_U^2 to the bias transmission level, t_b, of the shadowgraph. The precise form of the relation between σ_U^2 and t_b depends on the model used for the recording medium. We prefer the simple model of BARRETT and DEMEESTER [1974], in which the recording medium is a linear, positive film for which amplitude transmission is directly proportional to exposure. In this case $\sigma_U^2 \propto t_b$. The noise is directly proportional to the bias level of the shadowgraph.

Our second maxim, **P2,** relates the reconstructed image, $\langle U \rangle$, to the expected value of the shadowgraph modulation, $\langle \Delta t \rangle$. This follows directly from the linear theory in § 2. So then, signal intensity increases as $\langle \Delta t \rangle^2$, and the noise increases as t_b. The SNR then varies as,

$$\mathrm{SNR} \propto \frac{\langle \Delta N \rangle^2}{\langle N \rangle} = \frac{\langle \Delta t \rangle^2}{t_b}, \quad \text{low signal limit;} \tag{5.2a}$$

$$\mathrm{SNR} \propto \frac{\langle \Delta N \rangle}{\langle N \rangle} = \frac{\langle \Delta t \rangle}{t_b}, \quad \text{low noise limit.} \tag{5.2b}$$

It is useful to apply these insights to an analysis of Fig. 6, a rather noisy zone plate shadowgraph made using a point source of 100 keV X-rays (STONE

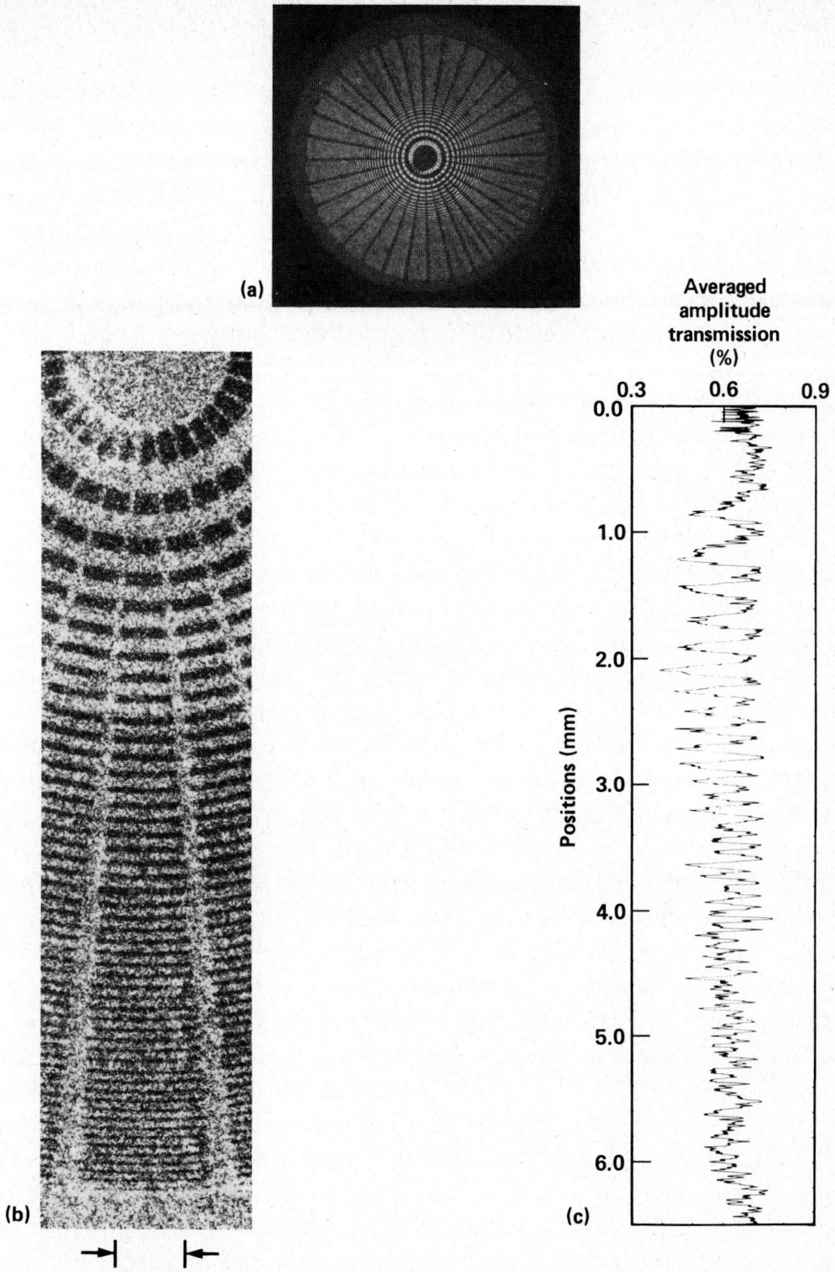

Fig. 6. (a) A 'noisy' zone plate coded image made using a point source of 100 keV X-rays. The fractional X-ray transmission through the solid zones of the coded *aperture* was 50%. (b) An optical micrograph of a radial segment of the shadowgraph. (c) A radial plot of amplitude transmission averaged over azimuthal angle. (Polarity is reversed in (a) to enhance visibility.)

[1983], STONE and CEGLIO [1983]). The fractional X-ray transmission through the solid zones of the coded *aperture* was approximately 50%, so that the coded image has significant exposure in both "transparent zone" and "solid zone" regions. According to **P1** all the recorded quanta, those in the open zone regions as well as those which leaked through the solid zones, contribute to the noise in the reconstructed image. Quanta added to an open zone region increase both $\langle \Delta t \rangle$ and t_b equally. So, while adding to the noise, they make an even greater contribution to the signal, thereby improving the overall SNR. Quanta which leak through the solid zones serve to increase t_b, while also decreasing $\langle \Delta t \rangle$, thereby rapidly degrading the overall SNR.

Figure 6c is an instructive illustration of how the coherent reconstruction field that samples the entire shadowgraph can extract a signal from an apparently noisy array of quanta. A radial plot of amplitude transmission averaged over azimuthal angle is shown. This is a good approximation to the ensemble averaged amplitude transmission function, $\langle t \rangle$. Clearly shown is $\langle \Delta t \rangle$, illustrating the existence of a strong signal in what on a small scale (Fig. 6b) would appear to be a very noisy array of quanta.

We have up to now primarily focussed on the causes of quantum noise, and the limitations it imposes on coded imaging methods. We now consider the mechanism of information compression by which coded techniques achieve SNR enhancement over pinhole methods. Information compression is defined as the area of the illuminated shadowgraph to the area of the final reconstructed image. Using the analysis from either § 3 or § 4, it follows easily that significant information compression can be achieved using coded aperture imaging when the radiation source is small. The importance of small source size can be easily understood physically. Consider the optical reconstruction of a coded image produced by a radiation source much smaller than the coded aperture. The shadowgraph modulation focusses a portion of the incident illumination into a magnified image of the source. (The amount of light focussed into the image is proportional to $(\int\int \langle \Delta t \rangle \mathscr{F} \, d^2 r)^2$, where the integral is over the shadowgraph area and $\mathscr{F}$ is the Fresnel kernel.) At the same time the coherent noise is scattered over a larger area ($\gtrsim$ shadowgraph area) in the reconstruction plane. Thus, while the total integrated energy in the deterministic signal may be comparable to or even less than the total integrated noise, the *local* SNR at the position of the reconstructed image can be quite high due to compression of the signal, but not the noise, into the small area of the image.

The SNR advantage provided by information compression can very rapidly erode with increased source size. New source points degrade the SNR because quanta from each new source add noise but no signal to the images of all other source points. In addition, increased source size reduces the information compression ratio.

To summarize the quantum noise results, we provide three guidelines or "rules or thumb" for coded imaging of laboratory sources. The guidelines are supplemented by quantitative illustrations, which clarify their meaning and limits of applicability. The guidelines are listed below, and then discussed in turn:

G1. Any process which serves to decrease the ratio of *exposure* modulation to bias level, $\langle \Delta\varepsilon \rangle / \varepsilon_b$, in the coded image will decrease the SNR of the reconstructed image.

G2. A continuous source should be limited in size to less than the diameter of the central zone of the coded aperture.

G3. Coded techniques are well suited for imaging a bright source on a dark background, but are particularly ill-suited for imaging a weak source on a bright background.

We may illustrate **G1** by considering the effect on SNR of two specific deleterious processes: (i) radiation leakage through the opaque zones of the coded aperture, and (ii) a uniform background exposure of the coded image. We compare the effect of these same phenomena on SNR in pinhole images, and draw conclusions concerning the sensitivity of the different imaging techniques to departures from ideal conditions.

We note that **G1** is stated in terms of the statistically varying *exposure*, $\varepsilon(\boldsymbol{r}'')$, in the shadowgraph, not its amplitude transmission, $t(\boldsymbol{r}'')$. This is because the precise statement of **G1** in terms of amplitude transmission depends on the model used to relate $\varepsilon(\boldsymbol{r}'')$ to $t(\boldsymbol{r}'')$ in the shadowgraph recording medium. As before, we employ the model used by BARRETT and DEMEESTER [1974]. However, our results differ in that we allow the coded *aperture* transmission to have values other than zero and one, whereas the analysis of BARRETT and DEMEESTER [1974] may be used only for binary coded apertures*. In this model there are no saturation effects, and $t(\boldsymbol{r}'')$ is proportional to $\varepsilon(\boldsymbol{r}'')$. In addition, eqs. (5.2a) and (5.2b) are valid, and $\langle \Delta t \rangle / t_b$ may replace $\langle \Delta\varepsilon \rangle / \varepsilon_b$ in the statement of **G1**.

Figure 7 shows the ensemble averaged amplitude transmission functions of coded images recorded under three different coded aperture conditions. In the

* BARRETT and DEMEESTER's [1974] representation of the amplitude transmission of the shadowgraph (their eq. (1)) is not appropriate for a coded aperture which is not strictly binary. This difficulty becomes apparent in the evaluation of the variance of the amplitude (their eqs. (25)–(28)). The use of a weight function, $Z(r)$, even if modified, does not properly account for the statistics of those quanta which are transmitted through the solid regions of the aperture.

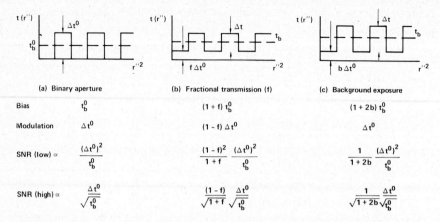

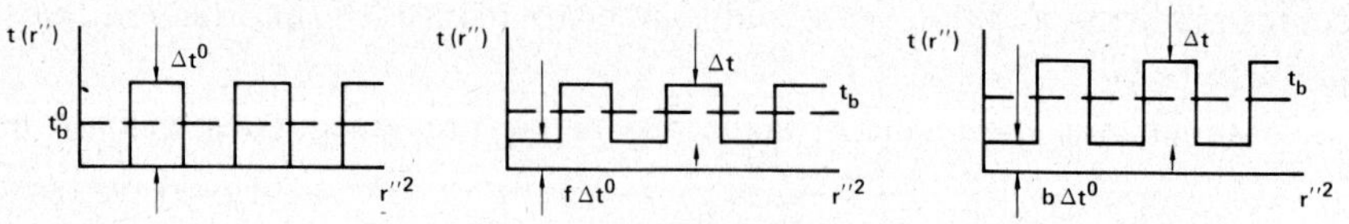

	(a) Binary aperture	(b) Fractional transmission (f)	(c) Background exposure
Bias	t_b^0	$(1+f)\, t_b^0$	$(1+2b)\, t_b^0$
Modulation	Δt^0	$(1-f)\, \Delta t^0$	Δt^0
SNR (low) $\propto$	$\frac{(\Delta t^0)^2}{t_b^0}$	$\frac{(1-f)^2}{1+f} \frac{(\Delta t^0)^2}{t_b^0}$	$\frac{1}{1+2b} \frac{(\Delta t^0)^2}{t_b^0}$
SNR (high) $\propto$	$\frac{\Delta t^0}{\sqrt{t_b^0}}$	$\frac{(1-f)}{\sqrt{1+f}} \frac{\Delta t^0}{\sqrt{t_b^0}}$	$\frac{1}{\sqrt{1+2b}} \frac{\Delta t^0}{\sqrt{t_b^0}}$

Fig. 7. Deleterious phenomena that effect SNR in coded image recording. Amplitude transmission functions and relevant quantum noise parameters are listed for coded images recorded under three different conditions. (a) Ideal conditions, a binary aperture with no background; (b) the aperture has a fractional transmission f through its solid zones; (c) the aperture is binary but there is a background exposure leading to a transmission value $b\Delta t^0$.

exposure that produced Fig. 7a the coded *aperture* was binary, that is, its solid zones were truly opaque. In Fig. 7b the coded aperture had a fractional transmission f through its solid zones. In Fig. 7c the coded *image* suffers from an additive, uniform background exposure, $b\Delta t^0$. Figure 7 also lists expressions for the transmission bias and modulation, and the relative SNR in the low signal (eq. (5.2a)) and low noise (eq. (5.2b)) limits. Note that for equal amounts of additional radiation, leakage through the solid zones is more detrimental to SNR than an additive background. Also note that the reduction in SNR due to departure from ideal conditions is greater when the SNR is low than when it is high.

Figure 8 illustrates the effects of these same deleterious phenomena in a pinhole image. In a pinhole image the signal at any "point" (i.e. within a resolution element) is proportional to the number of recorded quanta (above the background) at that "point", and the noise is proportional to the square root of the total number of quanta (including background) at that "point". Figure 8 also lists the SNR for a pinhole image in the three cases under discussion. Note that departures from ideal conditions have a greater adverse effect on coded techniques than on direct imaging methods. This is again due to the fact that in a coded image all recorded quanta contribute to the noise (recall **P1**). Whereas, in a direct image only those quanta recorded in the image region contribute to its noise. Quanta recorded outside the image region add to the

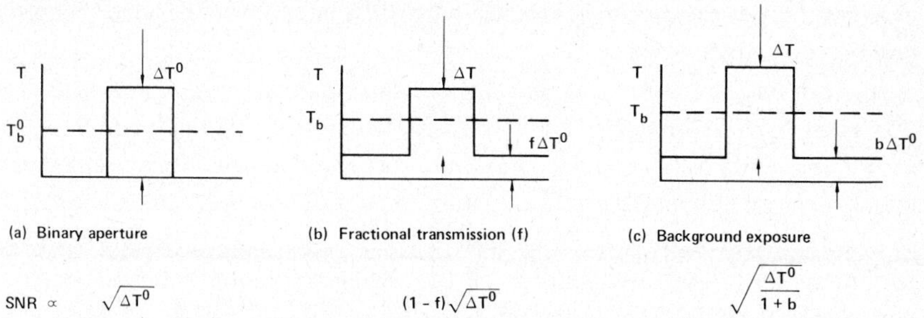

Fig. 8. Deleterious phenomena that effect SNR in pinhole image recording. Recorded pinhole image distributions and SNR's are listed for (a) ideal recording conditions; (b) a pinhole aperture with fractional transmission f through its walls; and (c) a background exposure leading to a transmission value $b\Delta T^0$.

background, and thereby reduce image modulation (i.e. signal), but they do not directly add to the noise in the image.

The source-size criterion established in **G2** derives from our earlier discussion of information compression, and the eigenfunction analysis of § 4. The $F_m < 1$ criterion of § 4 (eq. (4.9)) is in fact equivalent to **G2**. One should not infer from this guideline that ZPCI is of no value for sources larger than the diameter of the central zone. Instead, **G2** should be regarded as an experimental design criterion. It is important to match the band-pass of the coded aperture to the spatial frequency content of the source. In order to record all the spatial frequency content of the source, a coded aperture with central zone larger than the source should be used. Otherwise, quanta associated with low spatial frequencies will (because of zone blurring) contribute negligibly to the signal while contributing fully to the overall noise.

The final guideline, **G3,** is provided as a caution to those who would use coded imaging techniques in radiography. Although coded methods can be quite advantageous for imaging small, self luminous, laboratory sources, coded radiography is in general useful only under very limited conditions. The primary difficulty in radiographic applications arises from radiation which passes around the outside of the object*. This "shine around" radiation adds no useful information, since it does not pass through the object, but does add considerable noise to the shadowgraph. As a result, unless effective measures are taken to

* For radiographic information to be properly encoded in the shadowgraph, each backlit point in the object must illuminate the entire coded aperture. Since the coded aperture is typically much larger than, and close to the object, the backlighting source must be larger than the object for proper encoding of radiographic information.

eliminate “radiographic shine around”, the SNR advantage of coded methods can be very rapidly lost.

It is not easy to arrive at a unifying conclusion concerning quantum noise processes in coded imaging. ZPCI, as first proposed by MERTZ and YOUNG [1961] for stellar X-ray imaging, was quite successful in providing a significant quantum noise advantage over conventional pinhole methods. In our attempts to adapt this technique to different source distributions under different imaging conditions, we must be mindful of the stringent limitations and narrow constraints within which the quantum noise advantage may be retained.

5.2. NONLINEAR EFFECTS

To this point, the theoretical analysis of ZPCI has been strictly linear; that is, the amplitude transmission $t(\boldsymbol{r}'')$ has been presumed linearly proportional to the shadowgraph exposure $h(\boldsymbol{r}'')$. However, the direct optical reconstruction of zone plate coded images involves a number of nonlinear steps and processes which violate the linear shadowgraph recording assumption. One nonlinearity arises in the power law detection of the final reconstructed image. (The source intensity distribution is proportional to the reconstructed *field amplitude*, but the reconstructed *intensity* is recorded.) This nonlinearity, and others like it, can be easily unfolded from the final intensity. Another nonlinearity, which is not as easy to unfold, is the nonlinear relation between the shadowgraph exposure distribution $h(\boldsymbol{r}'')$ and the amplitude transmission distribution $t(\boldsymbol{r}'')$ operative in the optical reconstruction. Historically, coded images have been directly recorded on photographic emulsions or particle track detectors (FLEISCHER, PRICE and WALKER [1975], CEGLIO and BENTON [1980]), and contact printed onto high resolution photographic plates for subsequent optical reconstruction. Although it is conceivable that detailed processing of shadowgraph recording media and reconstruction plates could yield a linearized transparency, such care would be a serious practical impediment to the implementation of the ZPCI technique. It is, therefore, important to understand the extent to which information about the source distribution is compromised by nonlinear shadowgraph effects in the direct optical reconstruction of the coded images.

In the following paragraphs we establish the mathematical formalism and provide analysis of a few illustrative cases of the effects of nonlinear shadowgraph recording on optically reconstructed images. The results of this analysis indicate that, for optical reconstruction in first order, nonlinear artefacts provide negligible distortion of the reconstructed image. That is, isoemission contours in the source map into isointensity contours in the optically reconstructed image. In general, however, the relation between contour values of the reconstructed optical field and values of the source emission contours will be nonlinear. Contours of the reconstructed optical field will accurately represent isoemission contours in the source, but the contour values in the two cases will not necessarily be linearly related. Such a condition may affect estimates of FWHM values of source distributions as well as relative peak values for bipolar source distributions. However, for cases of practical interest, in which the recording nonlinearity is not particularly severe over the exposure range of the shadowgraph, the effect of nonlinear shadowgraph recording will be shown to be essentially negligible. That is, nonlinear shadowgraph recording serves only to limit the dynamic range of the ZPCI technique in cases of practical interest.

The impulse response analysis of § 3 can be used in the mathematical formulation of the problem of nonlinear shadowgraph recording. Although this approach may be somewhat tedious, it provides interesting geometrical illustrations of the nonlinear contributions in image reconstruction. We may represent any source distribution, $f(\boldsymbol{r})$, as an accumulation of emission contours, f_i, where

$$f_i(\boldsymbol{r}) = a_i \sum_j \delta(\boldsymbol{r} - \boldsymbol{R}_{ij}). \tag{5.3}$$

All points on a given emission contour have the same intensity, a_i. The complete source distribution, $f(\boldsymbol{r})$, is simply a sum over individual contours,

$$f(\boldsymbol{r}) = \sum_i a_i \sum_j \delta(\boldsymbol{r} - \boldsymbol{R}_{ij}). \tag{5.4}$$

In the limit of linear shadowgraph recording, the first order reconstructed field distribution using eq. (2.12) is

$$U(\boldsymbol{r}_2) = K_1 \sum_i a_i \sum_j S\left(\boldsymbol{r}_2 - \left[-\frac{s_2}{s_1}\boldsymbol{R}_{ij}\right]\right). \tag{5.5}$$

In the case of nonlinear shadowgraph recording the reconstructed field distribution is somewhat more complicated. As will be demonstrated in the following examples, the nonlinear first order reconstructed field distribution may be written in the form

$$U(\boldsymbol{r}_2) = K_1 \sum_i b_i \sum_j S\left(\boldsymbol{r}_2 - \left[-\frac{s_2}{s_1}\boldsymbol{R}_{ij}\right]\right) + \Gamma(\boldsymbol{r}_2), \tag{5.6}$$

where $\Gamma(\boldsymbol{r}_2)$ represents a nonlinear artefact contribution to the image. In cases of practical interest $\Gamma(\boldsymbol{r}_2)$ is negligible compared to the first term. The coefficients b_i in the first term are in general also affected by the nonlinearity, and may be written as a nonlinear function of the a_i coefficients as well as other parameters of the nonlinear recording process, i.e.,

$$b_i = f(a_1, a_2, \ldots, \alpha, \beta, \ldots), \tag{5.7}$$

where α and β represent film parameters, for example. If certain assumptions can be made about the severity of the nonlinearity and the range of the shadowgraph exposure, then the reconstructed image is in this limit a linear replica of the source distribution, that is eq. (5.7) reduces to

$$b_i \approx k a_i, \tag{5.8}$$

where k is the same constant for all contours, and $\Gamma(\boldsymbol{r}_2)$ is negligible.

In the following paragraphs we illustrate the nonlinear impulse response formalism discussed above using a point source pair

$$f(\boldsymbol{r}) = A_1 \delta(\boldsymbol{r} - \boldsymbol{R}_1) + A_2 \delta(\boldsymbol{r} - \boldsymbol{R}_2). \tag{5.9}$$

A more general distribution of source points could be analyzed in similar fashion to what follows, although it would require more complicated algebraic manipulation. Nevertheless, the same general conclusions arise from the general distribution as from the point pair illustration (CEGLIO [1976]). The primary advantage of this analysis is that it provides an intuitive insight to the source of nonlinear artefacts in zone plate coded imaging. While some quantitative conclusions about the relative importance of nonlinear artefacts can be made, the major burden of the quantitative analysis is left to computer calculations of relevant examples using the eigenfunction analysis of continuous source distributions later in this section.

The shadowgraph exposure for the two-point source is

$$h(\boldsymbol{r}'') = \frac{1}{4\pi(s_1 + s_2)^2} \sum_{i=1}^{2} \frac{A_i}{2} [1 + g^*(a\boldsymbol{r}'' + b\boldsymbol{R}_i)], \tag{5.10}$$

where g^* is a square wave pattern in $|\boldsymbol{r}|^2$ of amplitude unity and zero bias*. Equation (5.10) can be separated into d.c. and higher order components,

$$h(\boldsymbol{r}'') = h_0 + h_1(\boldsymbol{r}''), \tag{5.11}$$

* Equation (5.10) assumes a binary coded aperture with zero transmission through its solid zones. Generalizations for fractional transmission through the solid zones have been made by CEGLIO [1976].

where

$$h_0 = \frac{1}{8\pi(s_1 + s_2)^2}(A_1 + A_2) = \text{constant}, \tag{5.12a}$$

$$h_1(\boldsymbol{r}'') = \frac{1}{8\pi(s_1 + s_2)^2}\sum_{i=1}^{2} A_i g^*(a\boldsymbol{r}'' + b\boldsymbol{R}_i). \tag{5.12b}$$

At this stage of the development it becomes necessary to introduce the form of the nonlinearity relating $t(\boldsymbol{r}'')$ and $h(\boldsymbol{r}'')$. Since the thrust of the analysis will be to expand the nonlinear expression in a power series and analyze the form of the various power series contributions, the precise form of the nonlinearity is not critical at this juncture. The form of the nonlinearity will only affect the coefficients of the various power series terms. A nonlinearity appropriate for a photographically recorded X-ray exposure is the exponential (MEES [1966], YAGODA [1949]):

$$t(\boldsymbol{r}'') = exp(-\gamma h) = exp(-\gamma h_0)\, exp(-\gamma h_1). \tag{5.13}$$

This can be expanded as

$$t(\boldsymbol{r}'') = K_1 \sum_{k=0}^{\infty} \frac{(\gamma h_1)^k}{k!}(-1)^k, \tag{5.14}$$

where $K_1 = \exp(-\gamma h_0)$. Combining eqs. (5.12) and (5.14) gives

$$t(\boldsymbol{r}'') = K_1 \sum_{k=0}^{\infty} (-1)^k \frac{K_2^k}{k!} \sum_{i=1}^{2} A_i g_i^{*k}, \tag{5.15}$$

where

$$K_2 = \frac{\gamma}{8\pi(s_1 + s_2)^2}.$$

It is useful to investigate the form of the various terms of this power series expansion

$$t(\boldsymbol{r}'') = K_1 \Big[1 \qquad (k = 0)$$

$$+ (-1)K_2(A_1 g_1^* + A_2 g_2^*) \qquad (k = 1;\ \text{linear term})$$

$$+ \frac{K_2^2}{2}(A_1^2 + 2A_1A_2 g_1^* g_2^* + A_2^2) \qquad (\text{k} = 2)$$

$$+ (-1)\frac{K_2^3}{6}(A_1^3 g_1^* + 3A_1^2 A_2 g_2^* + 3A_1 A_2^2 g_1^* + A_2^3 g_2^*) \qquad (k = 3)$$

$$+ \frac{K_2^4}{24}(A_1^4 + 4A_1^3 A_2 g_1^* g_2^* + 6A_1^2 A_2^2 + 4A_1 A_2^3 g_1^* g_2^* + A_2^4) \qquad (k = 4)$$

$$+ \cdots \Bigg]. \tag{5.16}$$

In the above we have used

$$(g_i^*)^n = \begin{cases} 1 & \text{for } n \text{ even} \\ g_i^* & \text{for } n \text{ odd.} \end{cases} \tag{5.17}$$

All even values of k in the expansion contribute d.c. and $g_1^* g_2^*$ cross terms. All odd terms contribute g_i^* terms. The shadowgraph transmission for the point-source pair then reduces to an expression of the form

$$t(\boldsymbol{r}'') = \alpha_0 + \sum_{i=1}^{2} \alpha_i g_i^* + \alpha_x g_1^* g_2^*. \tag{5.18}$$

The first order reconstructed field distribution has the form

$$U(\boldsymbol{r}_2) = \sum_{i=1}^{2} \alpha_i S\left(\boldsymbol{r}_2 - \left(-\frac{s_2}{s_1}\boldsymbol{R}_i\right)\right) + \alpha_x \Gamma(g_1^* g_2^*), \tag{5.19}$$

reminiscent of eq. (5.6), where Γ is the Fresnel transform of the $g_1^* g_2^*$ cross term and the α_i are the nonlinear combinations of the A_1, A_2 source coefficients.

It is useful at this point to provide an intuitive guide to the physical origin of the nonlinear cross term $\Gamma(g_1^* g_2^*)$ as well as the nonlinear contribution to the coefficients α_i of the linear g_i^* terms. Moiré patterns produced from overlapping zone plate transparencies are extremely helpful in this discussion. The transmission through a simple zone plate transparency is $g = \frac{1}{2}(1 + g^*)$. If two such transparencies are laterally displaced and superimposed the net transmission is

$$g_1 g_2 = \tfrac{1}{4}(1 + g_1^* + g_2^* + g_1^* g_2^*). \tag{5.20}$$

This transmission function illustrates all the features of the nonlinearly recorded shadowgraph (e.g., d.c. term, linear terms, nonlinear cross terms) represented by eq. (5.18). The coefficients are of course not the same in this demonstration. Figure 9a shows the overlapping zone plate transparencies. The Moiré pattern produced by the beating of the g_1^* and g_2^* terms is the $g_1^* g_2^*$ cross term contribution. It is a zone plate pattern which has a focussing effect on the incident coherent illumination.

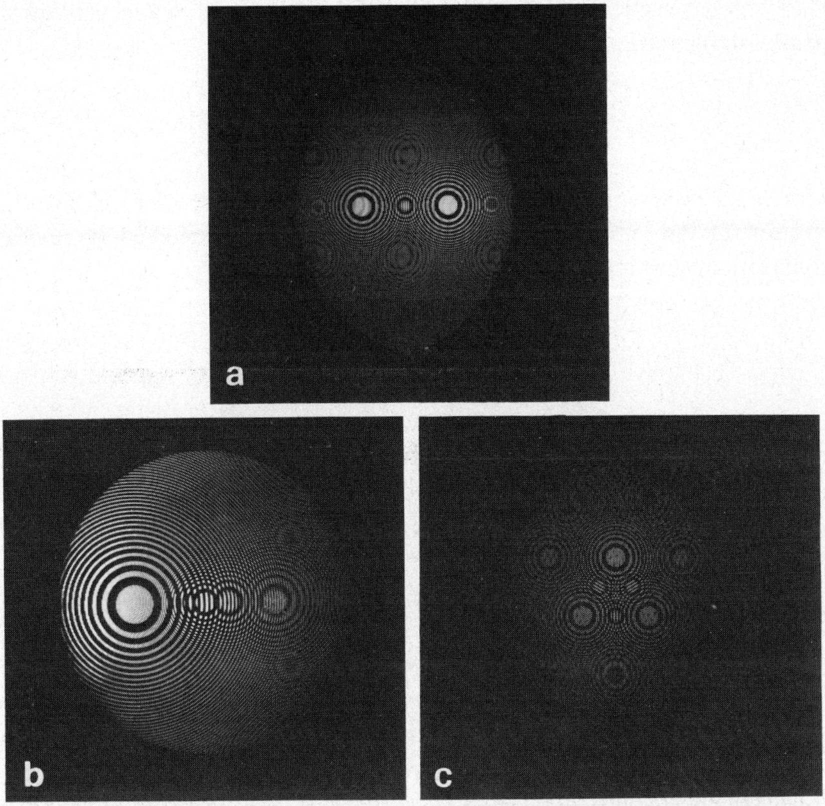

Fig. 9. Nonlinear effects in coded imaging are illustrated by the Moiré patterns produced by overlapping zone plate shadows. (a) Moiré patterns produced by the beating of two displaced zone plate shadows (two-point correlations). (b) Moiré patterns illustrate the nonlinear contributions to the linear term coefficients in eq. (5.18). (c) Moiré patterns produced by the beating of three displaced zone plate shadows (three-point correlations).

A detailed algebraic analysis of the $g_1^* g_2^*$ cross term in eq. (5.18) supports the Moiré interpretation of the nonlinear effects and provides a quantitative analysis of the nonlinear artefact term. Using eq. (2.3) we expand the $g_1^* g_2^*$ term below:

$$g_1^* g_2^* = -\frac{4}{\pi^2}\left\{\sum_{k,m} \frac{1}{km}\right.$$

$$\left.\times \exp\left(\frac{-\mathrm{i}\pi}{r_1^2}\left[(k+m)a^2\boldsymbol{r}''^2 + b^2(k\boldsymbol{R}_1^2 + m\boldsymbol{R}_2^2) + 2ab\boldsymbol{r}''\cdot(k\boldsymbol{R}_1 + m\boldsymbol{R}_2)\right]\right)\right\}. \tag{5.21}$$

Upon reconstruction this cross product term produces nonlinear artefact points (designated as (k, m)) at reconstruction distances

$$d_{k,m} = \frac{r_1^2}{(k+m)a\lambda^2} = \frac{d_1}{(k+m)} \quad (k, m \quad \text{odd}), \tag{5.22}$$

where d_1 is the first order reconstruction distance for the true image. The lateral position of the (k, m)th nonlinear artefact point is determined by similar arguments and is located at

$$\boldsymbol{r}_2(k, m) = \frac{b}{a} \frac{k\boldsymbol{R}_1 + m\boldsymbol{R}_2}{k+m} \quad (k, m \quad \text{odd}). \tag{5.23}$$

The nonlinear artefact contributions have a number of noteworthy features. For an ideal zone plate aperture there are no even values for k or m in the expansions so that there are no focussed $g_1^* g_2^*$ artefacts contributions in the d_1 reconstruction plane. Therefore all even term contributions are unfocussed and significantly reduced in intensity in the d_1 plane. Note, however, that had we considered a three-point example there would be three-point artefact correlations of the form $g_1^* g_2^* g_3^*$ that can appear in the d_1 plane (see Fig. 9c). A careful investigation of the amplitudes of such artefact terms shows them to be small compared to the amplitudes of the primary image terms (CEGLIO [1976]). This is especially true for more complicated source distributions. However, these conclusions depend on the severity of the nonlinearity encountered.

The nonlinear contributions to the α_i coefficients of the linear terms may also be identified as the beating together of zone plate patterns. A nonlinear contribution to the coefficients of g_2^* arises from the beating of g_1^* with $g_1^* g_2^*$. This is illustrated in Fig. 9b, in which a zone plate pattern, g_1^*, is superimposed on a displaced pattern of half scale, simulating $g_1^* g_2^*$. The beating of these two patterns together produces a variety of Moirés, with a distinct zone plate pattern occurring in the original g_2^* position (i.e. symmetrically located on the opposite side of $g_1^* g_2^*$ away from g_1^*).

The coefficients α_i of the primary image terms in eq. (5.18) can each be represented as a series expansion in the coefficients A_1, A_2. The first term of the series expansion is a term linear in A_i. The rest of the series represent the nonlinear contributions to the amplitude of the reconstructed primary image. The question of relevance is whether there is ever a condition in which the linear term dominates the series so that the reconstruction could be termed linear. A detailed analysis of the coefficients shows that for appropriate limits on the source level, A_i, the degree of nonlinearity, γ, and the dynamic range, the reconstruction can indeed be linear. In the following paragraphs examples of practical interest are numerically analyzed to illustrate that true linear reconstruction of source distributions can be obtained.

The point-wise discrete source distribution analysis of the nonlinear effects provides a useful intuitive insight through Moiré effects into nonlinear artefacts in coded imaging. However, further quantitative analysis of this model is not particularly useful since it is not easily generalized. In the remainder of this section, the eigenfunction analysis from § 4 is used to numerically simulate nonlinear shadowgraph recording in cases of practical interest. This approach provides confidence in the measurements of parameters such as FWHM values and peak-to-peak ratios from real reconstructed images.

The simulation procedure begins by calculating the shadowgraph exposure for any continuous source distribution that has been expanded in terms of the Gaussian–Laguerre polynomials (i.e., a generalized form of eq. (4.4)). The exposure calculations include the bias and all orders through $|m| = 19$. In the general nonlinear case, the mapping between exposure and amplitude transmittance must be known to find the shadowgraph transmittance. Fortunately, in many real applications the nonlinear relationship between exposure and transmittance is at least approximately known. For example, the response of silver halide film to X-rays is an optical density linearly proportional to the exposure. The relationships between amplitude transmittance and exposure used in the these simulations are given in table 1.

The optically reconstructed image for the nonlinear shadowgraph may be found by either expanding the final transmittance function in terms of the Gaussian–Laguerre polynomials or by numerically evaluating the Fresnel

TABLE 1

Summary of the experimental parameters used in the nonlinear film simulations presented in Figs. 10–13.

Figure number	First zone radius, r_1 (μm)	Number of zones, N	Magnification, (b/a)	Fresnel number	Image order, m	Nonlinear film model	Source type	Simulation of figure
10	316	250	1.5	0.074	1	(a)	X-ray	16a, a′
11	316	250	1.5	0.342	1	(a)	X-ray	16b, b′
12	100	100	5.2	0.028	1	(b)	alpha	19
13	158	250	3.0	0.045	1	(a)	X-ray	18a, b, c

(a) X-ray nonlinear film model: $t = k_1 \, e^{-k_2 h}$; k_1, k_2 constants.
(b) alpha particle nonlinear film model: $t = k_1 (h)^{-\gamma/2}$; k_1, γ constants.

diffraction integral. In the following examples, the Fresnel integral is numerically evaluated. To make the computations tractable, only the radial series terms from eq. (4.1) are used. In this case, the Fresnel integral can be expressed as a zero order Hankel transform. Typically, 4096 radial samples of the shadowgraph transparency are used in the numerical integration.

The results of several simulations are shown in Figs. 10–13. The important experimental parameters for each simulation are tabulated in table 1. Each figure is a simulation of a corresponding experiment presented in § 6, as indicated in table 1. The simulation parameters match the actual experiments in basic source dimensions, recording geometry, and film characteristics. The functional forms of the film nonlinearities are also given in table 1. The numerical constants required by the models were determined by microdensitometer scans of the experimentally recorded shadowgraphs. Each figure shows (a) a radial lineout of the calculated shadowgraph exposure, (b) the nonlinear relationship used to map exposure to amplitude transmittance, and (c) a comparison between the scaled first order image amplitude (i.e., the reconstructed source) and the original source. The insert in panel (a) of each figure shows the envelope of the lineout for all zoneplate zones; the larger graph in panel (a) shows the lineout for only the first 30 zones. In all four cases the ZPCI Fresnel number is much less than 1.

With the exception of Fig. 12, the imaging errors introduced by the nonlinear shadowgraph recording simulated here do not affect source widths or amplitude ratios by more than a few percent. The largest error is in Fig. 12 where the low level exposures are nonlinearly clipped by the threshold level of the film. The nonlinearity simulated in Fig. 12 introduces a 15% reduction in the apparent FWHM of the source.

These simulations provide quantitative confidence in the numerical parameters measured using ZPCI in § 6. Further, these and other simulations demonstrate that so long as the film nonlinearities are not severe, the linear reconstruction terms dominate the first order reconstructed image. In typical experiments then, extreme film nonlinearities in the shadowgraph recording process (like saturation or clipping) are avoided. The shadowgraph film is otherwise processed using normal procedures.

5.3. SPECTRAL REPONSE OF THE ZPCI TECHNIQUE

The zone plate coded imaging technique is used in a variety of important applications to image the X-ray emissions from broadband source distributions. In such applications it is important to know the spectral response of the imaging technique in order to determine the spectral content of the optically reconstructed image. Simple aperture shadowcasting techniques (e.g., pinhole camera, ZPCI) are generally represented as broadly achromatic imaging methods, requiring only that the wavelength of the radiation recorded be much smaller than the minimum feature size of the coded aperture. However, in practical applications the spectral response of these techniques may be significantly narrowed as a result of the spectral characteristics of: (i) the image recording medium, (ii) the transmission through filter materials interposed between the source and the recording medium, and (iii) the transmission through the nominally opaque regions of the shadowcasting aperture.

Because ZPCI is a two-step imaging technique involving optical reconstruction, the spectral response analysis of ZPCI is somewhat more involved than that for one-step, direct imaging methods (e.g., pinhole camera). Nevertheless, it shall be demonstrated that in the limit of linear shadowgraph recording the spectral response of the ZPCI technique is virtually the same as that for one-step imaging methods.

Consider an imaging geometry in which an X-ray source distribution $f(\boldsymbol{r}, \lambda)$ casts its shadow through an aperture $A(\boldsymbol{r}', \lambda)$, passes through an X-ray filter with transmission $\tau_F(\lambda)$, and is recorded on X-ray film with a spectral sensitivity $\sigma(\lambda)$. The spectral content of the recorded shadowgraph will be different from that of the source.

The source spectrum is defined as the spatially integrated, spectral distribution of the source emission,

$$\Omega_{\text{source}}(\lambda) = \iint f(\boldsymbol{r}, \lambda)\, \mathrm{d}^2\boldsymbol{r}. \tag{5.24}$$

The source spectrum is generally measured independently using a non-imaging detector. Our goal is to determine the spectral content of the image of the source, and thereby determine the spectral response of the two-step ZPCI technique.

The aperture transmission function may be written as

$$A(\boldsymbol{r}', \lambda) = \tau_A(\lambda) + (1 - \tau_A[\lambda])g(\boldsymbol{r}'), \tag{5.25}$$

where $\tau_A(\lambda)$ is the fractional transmission at wavelength λ of the nominally opaque portions of the aperture, and $g(\boldsymbol{r}')$ is the aperture spatial transmission function. The intensity distribution in the shadowgraph after passing through the aperture and filter may be written

$$h(\boldsymbol{r}'', \lambda) = \tau_F \tau_A \iint f(\boldsymbol{r}, \lambda)\, \mathrm{d}^2\boldsymbol{r} + \tau_F(1 - \tau_A) \iint f(\boldsymbol{r}, \lambda) g(\boldsymbol{r}')\, \mathrm{d}^2\boldsymbol{r}. \tag{5.26}$$

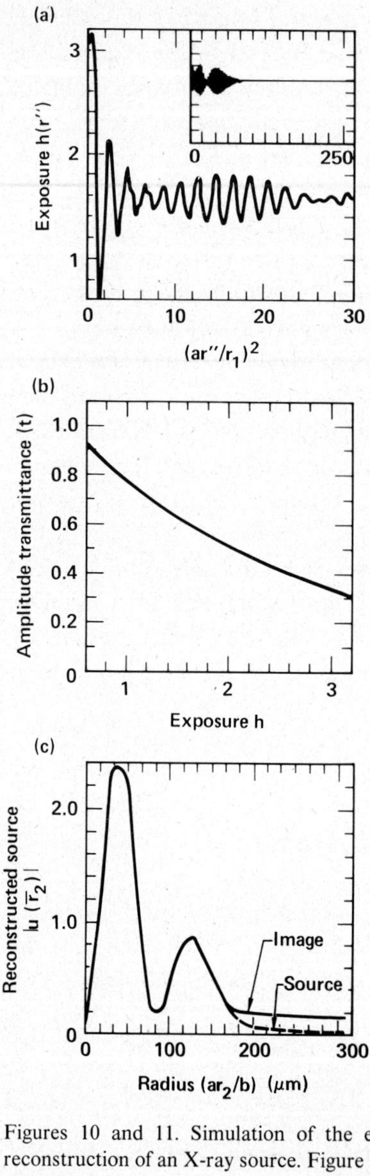

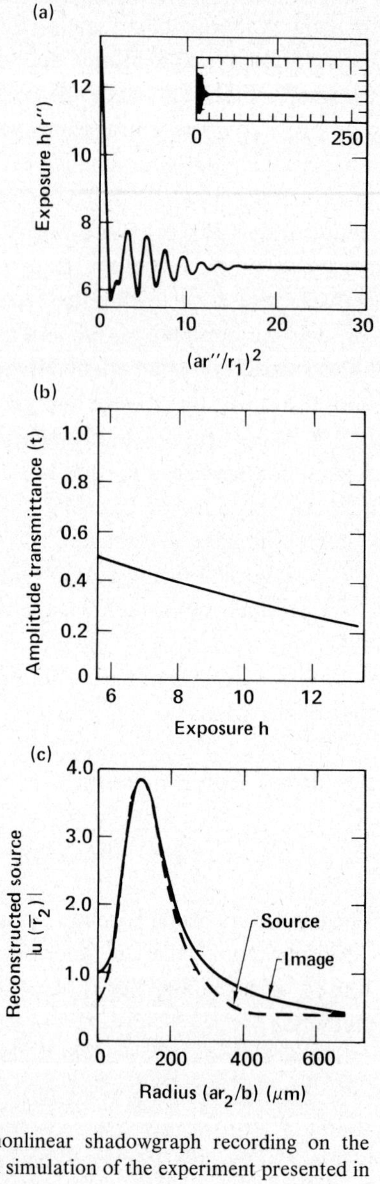

Figures 10 and 11. Simulation of the effects of nonlinear shadowgraph recording on the reconstruction of an X-ray source. Figure 10, left, is a simulation of the experiment presented in Fig. 16a, a′.

Figure 11, right, is a direct simulation of the dimensions and parameters presented in Fig. 16b, b′; the shape of the source is also quite similar to Figs. 18c, d, and 20. Note that the Fresnel number in this experiment was 0.34. This rather large value leads to the narrow, low contrast lineout envelope shown in the insert in panel (a). The other figures in this series have much smaller Fresnel numbers and correspondingly wider lineout envelopes. (See table 1 for other parameters.)

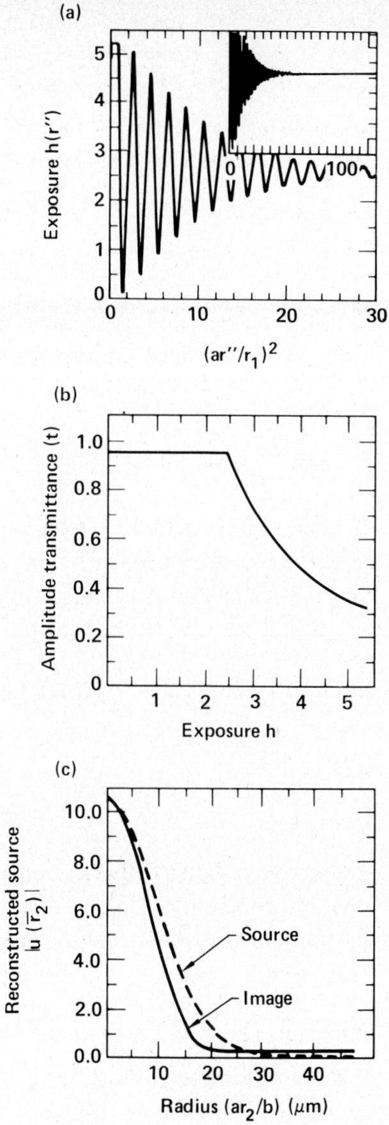

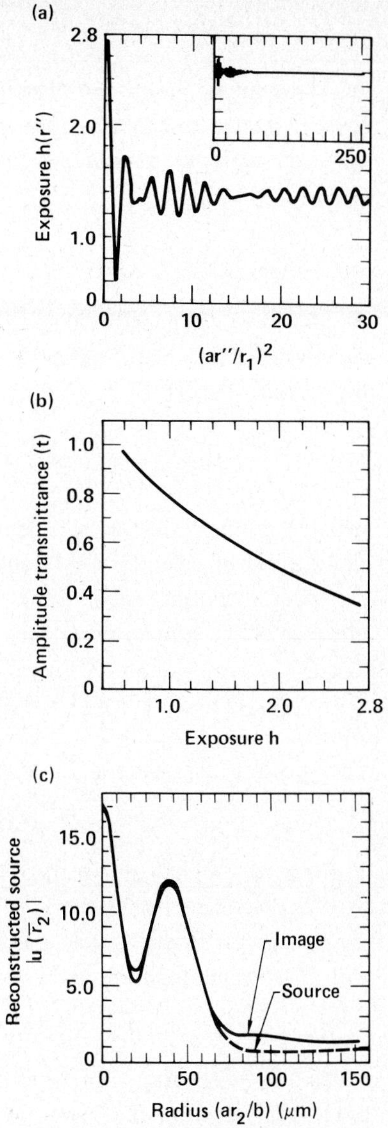

Fig. 12 (left). Simulation of the effects of nonlinear shadowgraph recording on the reconstruction of an alpha particle source. This is a simulation of the experiment presented in Fig. 19. In this case, a nuclear track detector was used to record the original shadowgraph, which was subsequently contact printed onto high resolution plates for optical reconstruction. In this simulation, we assume a sharp nonlinear film threshold effect. As a result, low level exposures are nonlinearly clipped. The effect of this nonlinearity is a narrowing of the image distribution, producing a 15% reduction in the apparent FWHM of the source. (See table 1 for other parameters.)

Fig. 13 (right). Simulation of the effects of nonlinear shadowgraph recording on the reconstruction of an X-ray source. This is a direct simulation of the experiment presented in Fig. 18a, b, c. (See table 1 for other parameters.)

In X-ray applications, the photographic density of the shadowgraph film is linearly related to exposure (MEES [1966], YAGODA [1949]). For a film with spectral sensitivity $\sigma(\lambda)$ the recorded photographic density distribution in the shadow is

$$D(\boldsymbol{r}'') = \int \sigma h(\boldsymbol{r}'', \lambda)\,\mathrm{d}\lambda. \tag{5.27}$$

Equations (5.26) and (5.27) can be combined and represented as the sum of two terms, $D(\boldsymbol{r}'') = D_0 + D_*(\boldsymbol{r}'')$. The first term is the spatially uniform bias, the second term is

$$D_*(\boldsymbol{r}'') = \iint g(\boldsymbol{r}')\left(\int \sigma\tau_F(1 - \tau_A) f(\boldsymbol{r}, \lambda)\,\mathrm{d}\lambda\right)\mathrm{d}^2\boldsymbol{r}, \tag{5.28}$$

where the inner integral term defines an effective source spatial distribution. For the case of a simple pinhole camera $D(\boldsymbol{r}'')$ is the recorded image. Therefore, the spectral content, $\Omega(\lambda)$, of the pinhole image in density units is simply that of the effective source, or

$$\Omega_{\text{pin}}(\lambda) = \iint \sigma\tau_{\text{F}}(1 - \tau_{\text{A}}) f(\boldsymbol{r}, \lambda)\,\mathrm{d}^2\boldsymbol{r} = \sigma\tau_{\text{F}}(1 - \tau_{\text{A}})\,\Omega_{\text{source}}(\lambda). \tag{5.29}$$

The spectral response for the pinhole camera is then,

$$R_{\text{pin}}(\lambda) = \sigma\tau_{\text{F}}(1 - \tau_{\text{A}}). \tag{5.30}$$

For a zone plate camera the recorded photographic density distribution is used as an optical transparency for the subsequent coherent reconstruction of the source distribution. The amplitude transmission distribution associated with the density function given in eq. (5.27) is

$$t(\boldsymbol{r}'') = \exp[-1.15 D(\boldsymbol{r}'')] = \exp[-1.15(D_0 + D_*)] \tag{5.31}$$

which may be expanded. In the limit of linear shadowgraph recording the linear term dominates the expansion, and we may write

$$t(\boldsymbol{r}'') = K(1 - 1.15 D_*), \tag{5.32}$$

where

$$K = \exp(-1.15 D_0). \tag{5.33}$$

The first order reconstructed field distribution is

$$U(\boldsymbol{r}_2) = K \iint S(\boldsymbol{r}_2 - \boldsymbol{r}^*)\left(\int \sigma\tau_{\text{F}}(1 - \tau_{\text{A}}) f\left(-\frac{a}{b}\boldsymbol{r}^*, \lambda\right)\mathrm{d}\lambda\right)\mathrm{d}^2\boldsymbol{r}^*. \tag{5.34}$$

As with eq. (5.28) the inner integral term defines an effective source spatial distribution. The associated spectral content of the reconstructed zone plate image in units of optical field amplitude is then,

$$\Omega_{\mathrm{ZPCI}}(\lambda) = \iint \sigma\tau_{\mathrm{F}}(1 - \tau_{\mathrm{A}}) f\left(-\frac{a}{b}\boldsymbol{r}^*, \lambda\right) \mathrm{d}^2\boldsymbol{r}^*. \tag{5.35}$$

The spectral response for the two step ZPCI process is, therefore,

$$R_{\mathrm{ZPCI}}(\lambda) = \sigma\tau_{\mathrm{F}}(1 - \tau_{\mathrm{A}}). \tag{5.36}$$

Comparison of eqs. (5.30) and (5.36) shows that while ZPCI is a two-step process, the spectral response of the technique is the same as for one-step imaging processes.

§ 6. Applications of ZPCI in Laser Fusion Experiments

Although first proposed for use in X-ray astronomy and later for tomography in nuclear medicine, the greatest impact of zone plate coded imaging has been in the microscopy of laser irradiated targets in the inertial confinement fusion effort. In laser fusion experiments, small D–T filled spherical microballoons are compressed to thermonuclear conditions by multiple, high intensity laser beams. Laser compressed targets are ideal sources for zone plate coded imaging. They are small (typically $\sim 10^{-2}$ cm), bright, X-ray and particle emitters with a great deal of small-scale spatial structure. Since these targets are generally much smaller than the coded apertures used to image them (i.e., eq. (4.9) is satisfied), significant S/N advantage is attained from the large collection solid angle which the coded aperture affords. In addition coded aperture techniques are readily extended to high energy X-ray and particle imaging, and they have a tomographic capability as well.

The ability to exploit the advantages provided by coded imaging methods in laser fusion experiments has been in large part due to the rapid advance in microfabrication technology allowing the fabrication of micro-Fresnel zone plate coded apertures (CEGLIO [1981], SMITH [1980]). These apertures are free-standing, gold zone plates with microscopic minimum linewidth (typically $\Delta r \sim 1$–15 μm), hundreds of zones (typically $N \sim 100$–250), and high aspect ratios (A.R. ≡ material thickness/Δr). Figure 14 shows three different micro-Fresnel zone structures in various stages of microfabrication. Figure 14a is a SEM photograph of a polymer electroplating mold used in the zone plate fabrication process. Figure 14c shows the electroplated zone plate pattern,

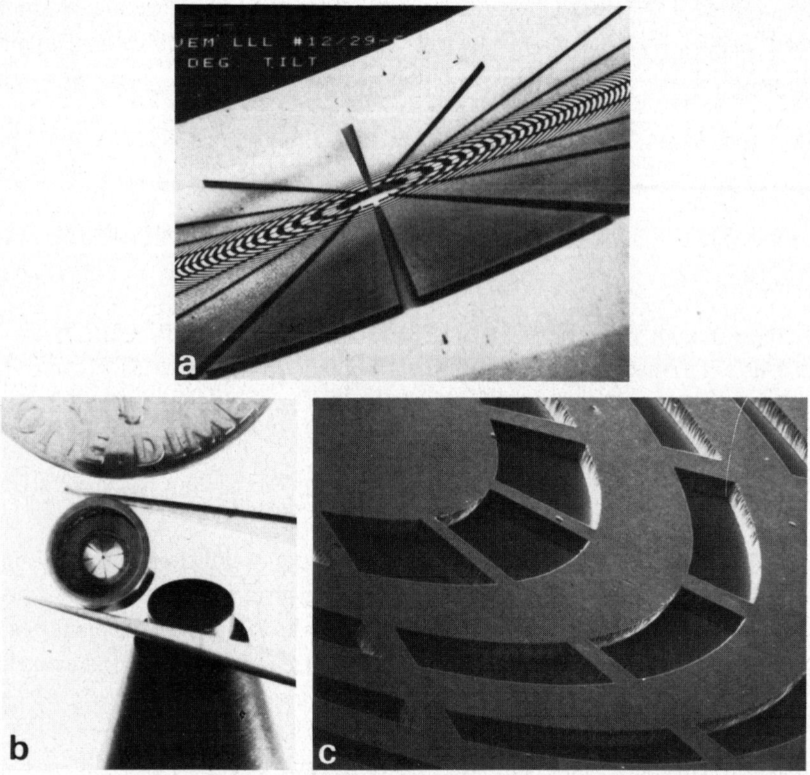

Fig. 14. Micro-Fresnel zone structures. (a) A polymer electroplating mold used in the zone plate fabrication process ($N = 250$ zones, $\Delta r = 15$ µm; $t \geqslant 100$ µm thick). (b) A free standing gold zone plate held together by radial struts ($N = 100$ zones; $\Delta r = 5$ µm; $t \geqslant 7$ µm thick). (c) An electro-polated gold zone plate pattern after the polymer mold has been removed ($N = 250$ zones; $\Delta r = 5$ µm; $t \geqslant 12$ µm thick).

and Fig. 14b shows a free-standing (held together by radial struts) gold Fresnel zone plate in its holder, ready for mounting on the nose cone of a zone plate shadow camera. Complete details of zone plate fabrication methods are provided elsewhere (Ceglio and Smith [1978], Ciarlo and Ceglio [1980], Ceglio, Stone and Hawryluk [1981]).

Over the past seven years of application of ZPCI to laser fusion experiments, there have been a number of opportunities for comparison of coded imaging results with those from traditional imaging techniques, and with calculational estimates of target performance. A number of such comparisons are illustrated in Figs. 15 and 16. Figure 15 is a comparison of a pinhole camera

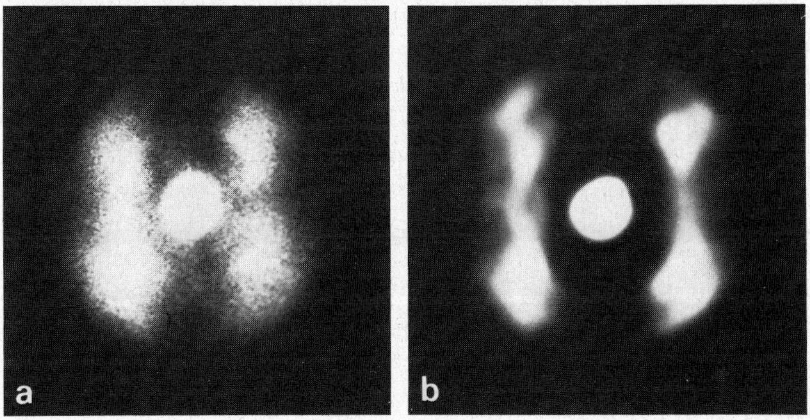

Fig. 15. Comparison of (a) pinhole camera image and (b) reconstructed zone plate image of a laser irradiated microsphere target. Pinhole image courtesy of G. CHARATIS and R. JOHNSON, KMS Fusion Inc.

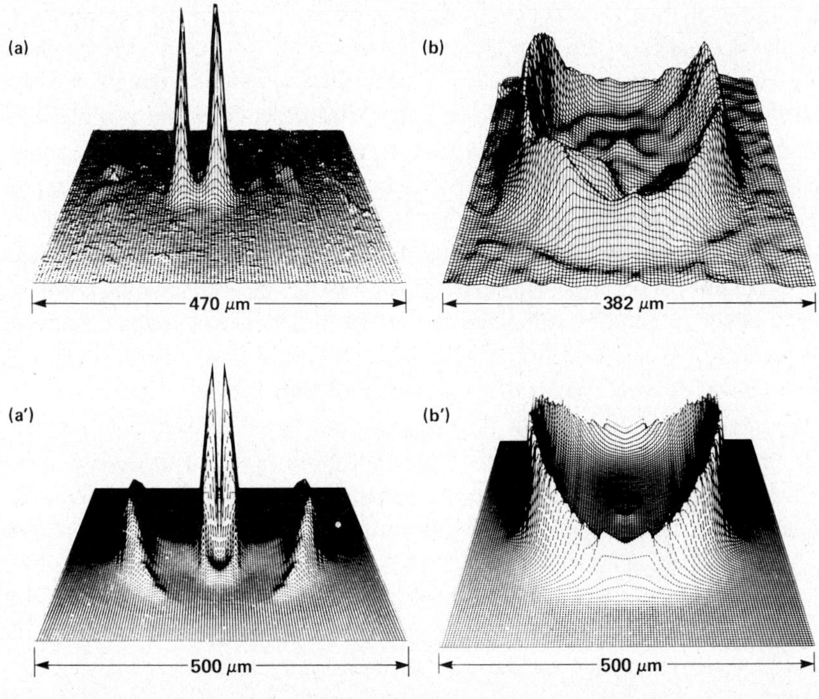

Fig. 16. Comparisons of (a, b) measured (using zone plate coded imaging) and (a′, b′) calculated X-ray images of laser irradiated fusion targets in two different spectral bands: (a) 4–7 keV; (b) 17–30 keV. In these isometric representations of the image data, time integrated source intensity is plotted vertically versus the two spatial dimensions.

image and a reconstructed zone plate image of a laser irradiated target. The target was a gold coated, spherical glass microshell filled with D–T gas. It was 86 μm in diameter and had a 1 μm thick glass wall with 2000 Å gold coating. The target was illuminated from two anti-parallel directions by a Nd-glass laser operating at λ = 1.06 μm. Laser pulse duration was 500 ps, intensity on target was 2×10^{15} W/cm^2. The X-ray images of Fig. 15 each represent a spectral band of moderate width centered at approximately 3 keV. Each has a nominal spatial resolution ~ 10 μm. The images are time integrated. The images show the directly illuminated target surface and corona lighting up in X-ray emission, as well as the final compressed core of the target after it has imploded. The Fresnel zone plate used to produce the reconstructed image in Fig. 15b was a free standing structure of 12 μm thick gold with 250 zones, and a minimum zone width of 5 μm. Image magnification in this experiment was 3 ×.

Figure 16 provides comparisons of measured (using ZPCI) and calculated X-ray images of laser irradiated fusion targets in two different spectral bands. Figure 16a, a′ compares measured and calculated images of the thermal X-ray emission in an energy band (4–7 keV). Figure 16b, b′ compares measured and calculated images of the suprathermal X-ray emission in the energy band 17–30 keV. (At these elevated X-ray energies the reduced target emission levels render the pinhole camera, with its small collection solid angle, inadequate for high S/N imaging.) The targets in Figs. 16a, a′ and 16b, b′ are similar. Each has a diameter ≃ 300 μm, and a glass wall thickness ≃ 1.5–2 μm (see CEGLIO and LARSEN [1980] for more details). The zone plate coded aperture used in these experiments was a 24 μm thick gold, free standing structure with 250 zones and 10 μm wide outermost zone. Nominal image resolution is 20 μm. The data shows good overall agreement between the measured and calculated X-ray images. However, as expected, measured images show detailed features which idealized theoretical calculations cannot fully predict.

While comparisons of zone plate images with those derived from conventional methods are reassuring, it is the application of ZPCI to measurement regimes where conventional methods are inadequate that is of greatest interest. Zone plate cameras routinely provide multispectral X-ray images of laser driven fusion targets (CEGLIO, ATTWOOD and LARSEN [1982]). On a single target shot the zone plate camera simultaneously records a sequence of distinct X-ray images (typically three to six images) in separate energy bands. In addition to X-ray applications, zone plate coded apertures have produced high resolution images of the alpha particle emissions from the thermonuclear burn of laser fusion targets (CEGLIO and COLEMAN [1977], CEGLIO [1978]). Another unique application of coded imaging methods is tomography. ZPCI

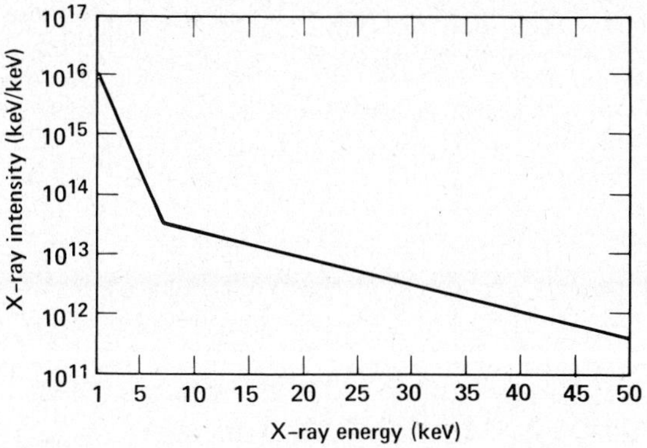

Fig. 17. Typical X-ray spectrum from a laser irradiated fusion target.

has been used, in limited cases, to provide tomographic X-ray images of laser irradiated fusion targets (CEGLIO, ATTWOOD and GEORGE [1977]). In subsequent paragraphs we explore each of these coded imaging applications in greater detail.

Zone plate coded apertures are routinely used to record time integrated X-ray emissions from laser driven fusion targets. In these experiments coded images are typically recorded in a multilayer filter–film pack. This arrangement allows simultaneous recording of as many as six distinct images in separate energy channels ranging from a few keV to over 30 keV. Energy-channel separation using a simple filter–film pack is facilitated by the rapidly falling X-ray spectrum typical of laser fusion targets. Figure 17 shows the X-ray spectrum, and table 2 provides the filter–film pack design for the image data of Fig. 18. Presented in Fig. 18 is a sequence of X-ray images simultaneously recorded by a single zone plate camera on a single target shot. Shown are the photographically recorded reconstructed image, a 2D iso-intensity contour map of the image, and the calculated (using eq. (5.36)) spectral content of the image, in five separate energy bands centered at 3, 6, 9, 14, and 20 keV. The target experiment illustrated in Fig. 18 is the same as that discussed earlier in connection with Fig. 15. The pinhole camera, also used on this shot (see Fig. 15a), was unable to provide a similar sequence of multispectral images because of its small collection solid angle (approximately 10^{-6} sr versus 2×10^{-2} sr for the zone plate camera). The image sequence of Fig. 18 covers a range in spectral flux in excess of four orders of magnitude. The ability to access image data over this broad dynamic range results from the large

TABLE 2

Filter-film pack design for the multispectral zone plate camera which provided the images of Fig. 18 Foils and films are listed in order of increasing distance from the target.

Foil or film	Specification
Protective foils [a]	125 μm beryllium
Coded aperture [b]	Gold Fresnel zone plate $N = 250$ (number of zones) $\Delta r = 5$ μm (outermost zone) $t = 12.2$ μm (thickness)
Film layer 1 [c]	50 μm beryllium Kodak type M X-ray film Kodak type R X-ray film Kodak type R X-ray film
Film layer 2	500 μm beryllium Kodak type A X-ray film Kodak type M X-ray film Kodak type R X-ray film
Film layer 3	125 μm aluminum Kodak type A X-ray film Kodak type M X-ray film
Film layer 4	750 μm aluminum Kodak type A X-ray film Kodak type M X-ray film Kodak type A X-ray film

[a] Serves to shield coded aperture from target debris.
[b] At a distance $S_1 = 2.5$ cm from the target.
[c] At a distance $S_2 = 7.5$ cm behind the coded aperture.

collection solid angle of the ZPCI technique, and the design and processing methods used for the filter–film pack.

The multispectral X-ray images of Fig. 18 illustrate how target emission features change with increasing X-ray energy. The 3 keV image is dominated by emission from two distinct target features: the directly illuminated gold corona blowing off the original surface of the microsphere; and the tightly compressed core of hot plasma at target center. At higher energies (14 and

Fig. 18. Multispectral X-ray images of a laser imploded microsphere target. Presented are the photographically recorded reconstructed image, its associated iso-emission contour map, and the calculated spectral content of the image in five separate energy bands peaked at (a) 3 keV, (b) 6 keV, (c) 9 keV, (d) 14 keV, and (e) 20 keV.

(a)

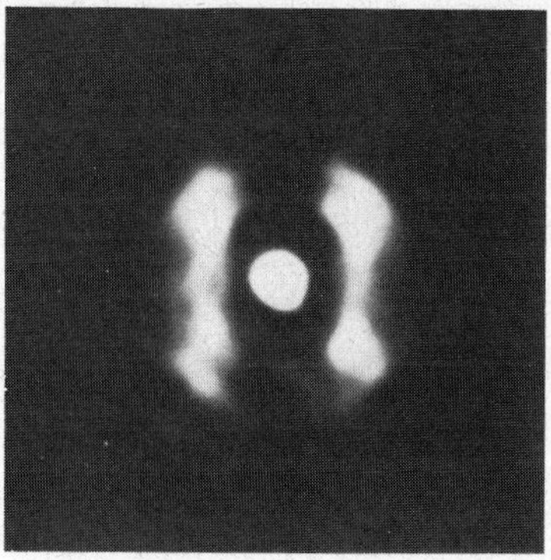

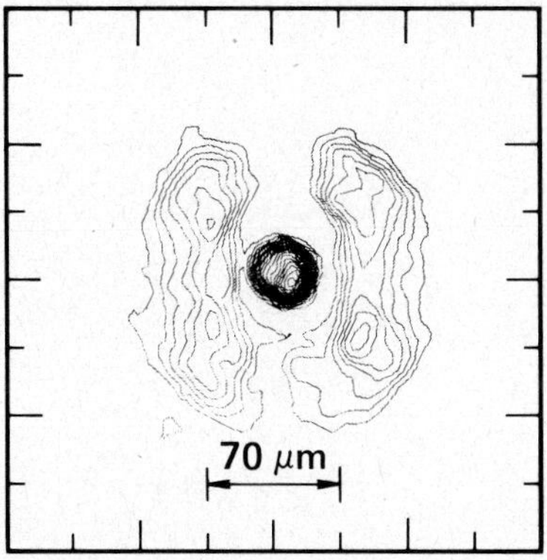
70 μm

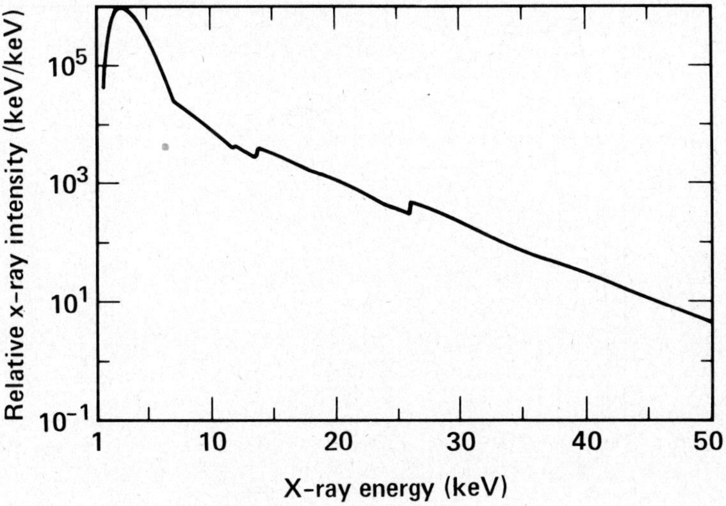
Relative x-ray intensity (keV/keV)
10^5
10^3
10^1
10^-1
1
10
20
30
40
50
X-ray energy (keV)

(b)

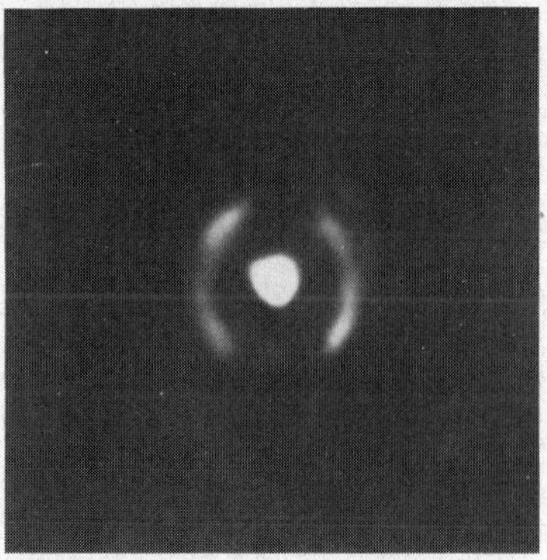

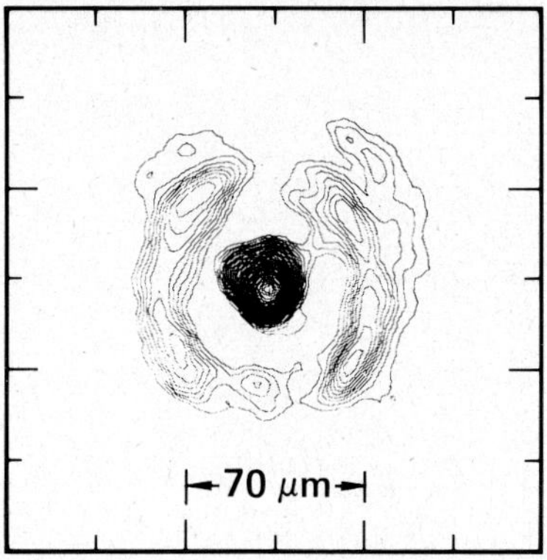
70 μm

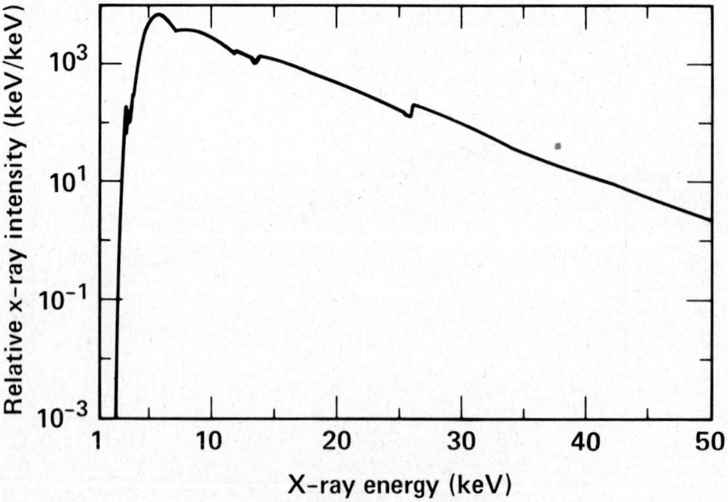
Relative x-ray intensity (keV/keV)
10³
10¹
10⁻¹
10⁻³
1
10
20
30
40
50
X-ray energy (keV)

(c)

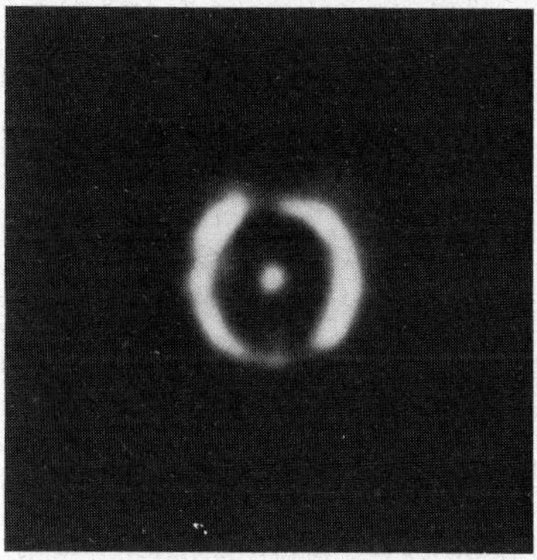

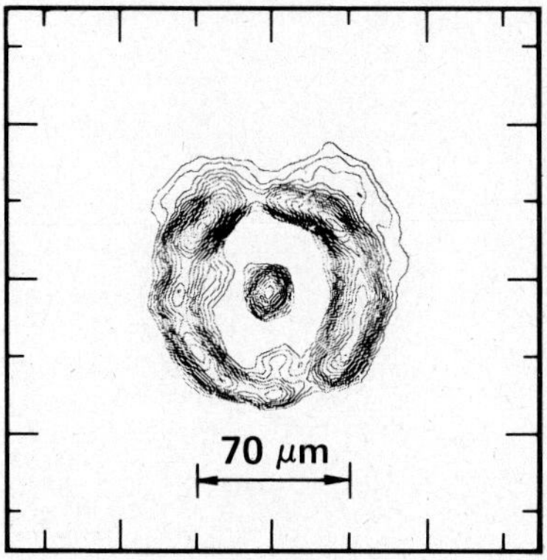
70 μm

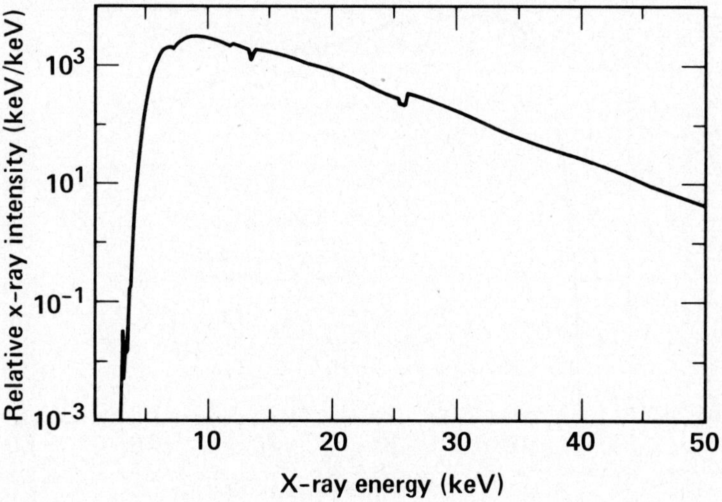
Relative x-ray intensity (keV/keV)
10^3
10^1
10^-1
10^-3
10
20
30
40
50
X-ray energy (keV)

(d)

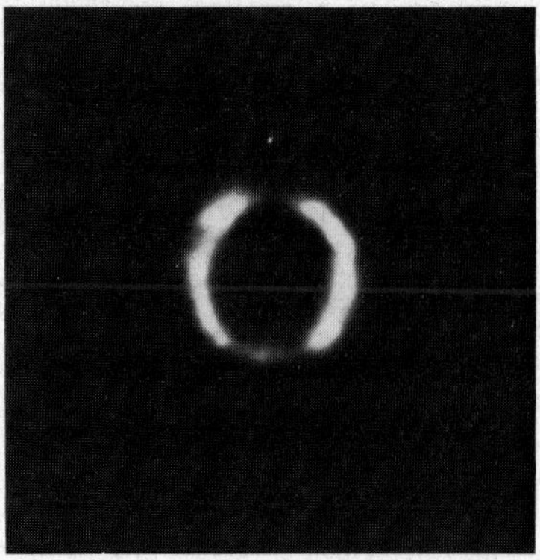

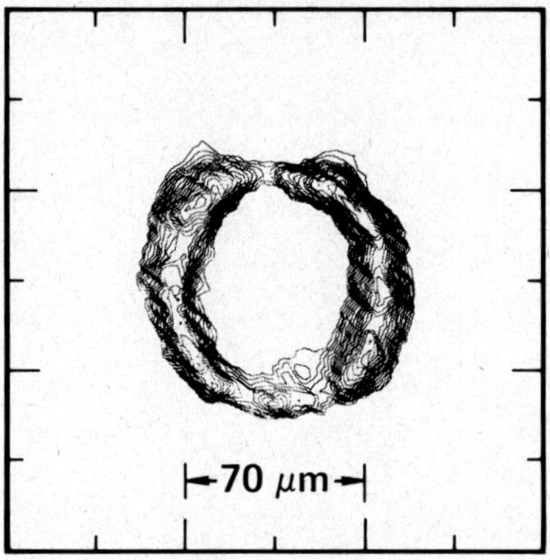
70 μm

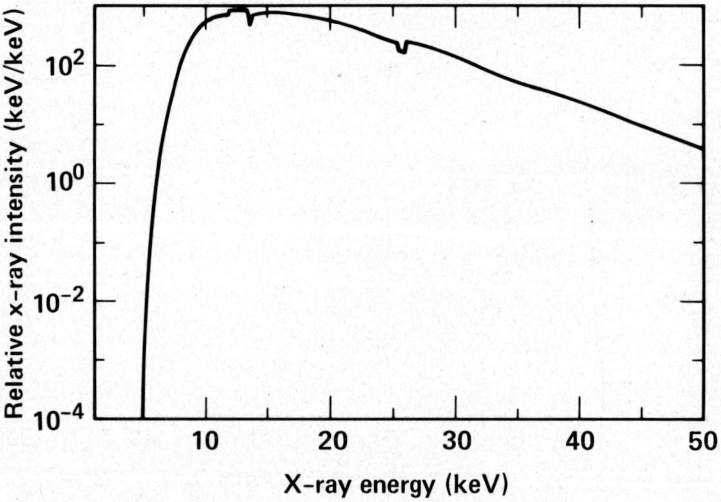
Relative x-ray intensity (keV/keV)
10^2
10^0
10^-2
10^-4
10
20
30
40
50
X-ray energy (keV)

(e)

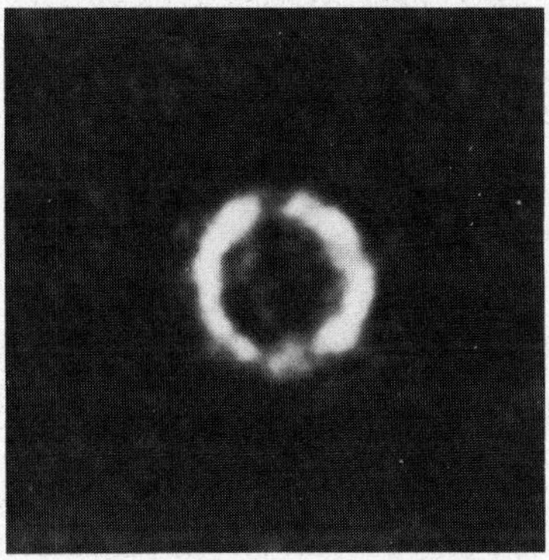

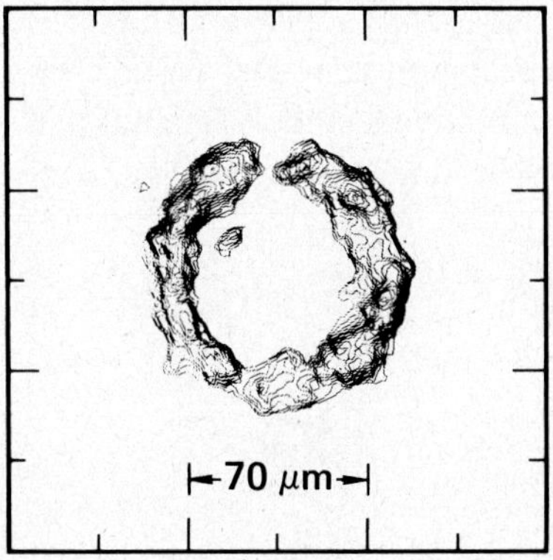
70 μm

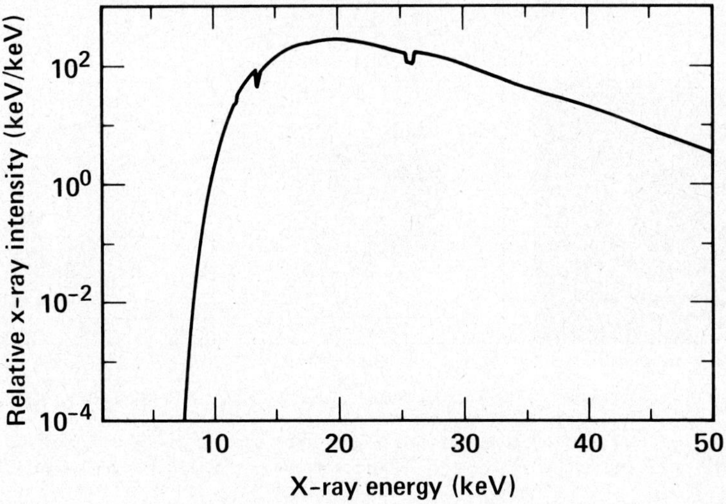
Relative x-ray intensity (keV/keV)
10^2
10^0
10^-2
10^-4
10
20
30
40
50
X-ray energy (keV)

20 keV) the dominant image feature is the suprathermal X-ray emission from the unimploded microshell. This is bremsstrahlung from suprathermal electrons interacting with the gold coated glass shell early in the compression process. The images at 6 and 9 keV show the transition from the low energy thermal emission features to the high energy suprathermal features.

Coded apertures may be used for high resolution imaging of charged particle source distributions (CEGLIO [1977]). The only imaging requirement is that the particles follow straight line trajectories from the source through the coded aperture to the detector. An important charged particle imaging application was the use of ZPCI to image the 3.5 MeV alpha particle emission from the thermonuclear (TN) burn of laser fusion targets (CEGLIO and COLEMAN [1977]). The collection efficiency of coded apertures and the small size of the TN burn region made ZPCI very well suited for this application. Other charged particle imaging methods were not quite appropriate. Electrostatic and magnetic focussing techniques, as well as conventional pinhole cameras had inadequate resolution and insufficient collection solid angle. For example, in the experiment of Fig. 19

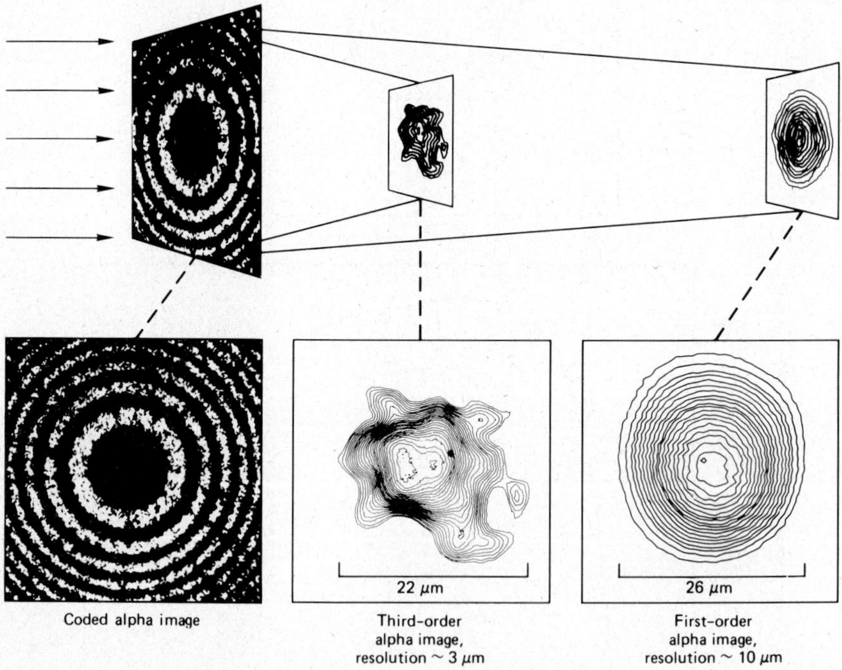

Fig. 19. Image of the thermonuclear burn in a laser fusion target, reconstructed in first and third order. The third order image, having a threefold resolution improvement, reveals spatial detail not apparent in the first order image.

the target emitted a total of 3×10^8 alpha particles. The zone plate camera collected roughly 5×10^5 alphas, and yielded a reconstructed image (in third order) with 3 μm resolution. An equivalent pinhole camera (same target distance, same resolution) would have collected only two alpha particles, a number insufficient to produce an image.

Figure 19 shows the coded alpha-particle image along with iso-intensity contour maps of the reconstructed image in first and third order. The coded alpha image in this experiment was recorded in a 6 μm thick cellulose nitrate (CN) film, a threshold-type ion track detector. The CN acted as a highly discriminating detector allowing us to record a 100 micro-joule alpha-particle signal amidst tens of joules of background radiation (X-rays, electrons, heavy ions). In this experiment the CN film was placed close behind an 8 μm thick Be filter foil. The Be stopped heavy ions from reaching the CN, while letting the alpha's through with minimal scattering. Each alpha particle incident on the film produced a microscopic damage channel, which upon subsequent etching became a pinhole through the 6 μm CN layer. The alpha-particle pinhole array (i.e., coded image) was then contact printed onto a photographic plate and optically reconstructed using a helium–neon laser, as illustrated in Fig. 19.

The optically reconstructed alpha images have good S/N, as demonstrated by the successful reconstruction in third order with threefold improvement in spatial resolution. The imaging geometry in this experiment easily satisfied the $F_m < 1$ criterion for good S/N, established in § 4. The characteristic radial scale length (σ) for the alpha source distribution was approximately 10 μm. The coded aperture used was a free standing Fresnel zone plate with 100 zones and 5 μm minimum zone width. Image magnification was 5.4 ×, and the target to zone plate distance was 1 cm.

The alpha images of Fig. 19 provided the first direct information about the thermonuclear burn geometry in laser fusion targets. They confirmed that the TN reactions did indeed occur within the compressed target core. They showed the emission volume was "egg-shaped" with some high spatial frequency structure, perhaps characteristic of non-uniformities in the laser-target illumination.

Much of the early interest in the laboratory application of coded imaging arose from its potential for tomography. In actual experiments, however, the tomographic payoff from coded aperture methods has not justified the early enthusiasm. This is perhaps not surprising considering the analysis of § 3.2, where it is demonstrated that locally intense image features can dominate the reconstructed image over many tomographic resolution lengths. Coded imaging tomography has had limited application in laser fusion experiments. This has been for at least three reasons: (1) laser fusion targets are small, placing stringent demands on tomographic resolution; (2) target emissions are fairly symmetric, so that there isn't a great deal of interesting tomographic data to be had, and (3) compressed target emissions are typically dominated by a bright compressed core at target center, which limits tomographic resolution as discussed in § 3.2.

Figure 20 shows the first ZPCI tomographic image data recorded on a laser–target interaction experiment (CEGLIO, ATTWOOD and GEORGE [1977], CEGLIO [1976]). These data are from a low energy, two-sided target illumination in which no significant compression was achieved. There is no intense compressed core of emission to limit the tomographic separation of the image data. Figure 20 shows the coded X-ray image, and iso-intensity contour maps of the reconstructed image in three planes approximately 37 μm (in source dimensions) apart. According to the resolution criterion of eq. (3.6), the image data of Fig. 20 is marginally resolved ($\Delta \sim 100$ μm from eq. (3.6), and the separation of the outer planes is 74 μm). Nevertheless, a detailed analysis indicates that

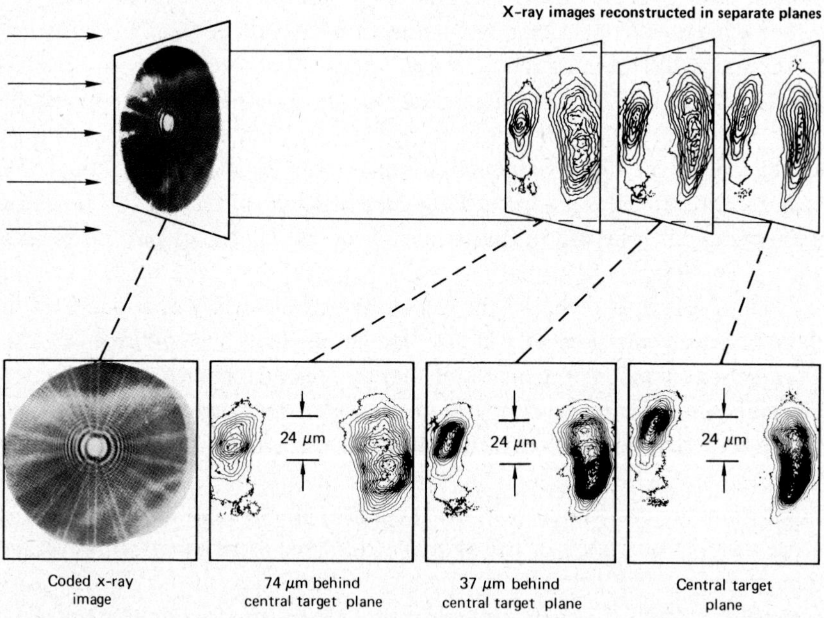

Fig. 20. Tomographic image reconstruction of the X-ray emission from a laser-irradiated, glass microsphere target. The tomographic information is retrieved by viewing the reconstructed image distribution in separate reconstruction planes. In this way the three-dimensional source can be synthesized plane by plane.

the images of Fig. 20 are indeed a tomographic representation of the laser irradiated target. However, no new information about the performance of this or any other laser fusion targets has to date been derived from coded aperture tomography.

There are a number of new directions in coded imaging applications which are worthy of review in this section. Advances in microfabrication technology are making possible the fabrication of thicker gold zone plate coded apertures (STONE [1983], STONE and CEGLIO [1983], CEGLIO, STONE and HAWRYLUK [1981]). This is extending ZPCI capabilities to higher energy X-rays and more penetrating charged particles. Figure 21 presents resolution test results using "point" sources of 100 keV X-rays and 6 MeV protons. Shown are coded images and higher order reconstructed images taken with a gold zone plate 75 μm thick with $N = 100$ zones and 30 μm minimum zone width.

Matched-filter-pair (MFP) coded apertures were recently proposed for narrowband imaging of specific line emissions from broadband laboratory X-ray sources (CEGLIO [1981]). Figure 22 illustrates this concept. The coded aperture has alternate zones made of materials having slightly different absorption edge energies (e.g., $Al_{K_{ab}} \simeq 1.56$ keV and $Si_{K_{ab}} \simeq 1.84$ keV, or $Cu_{K_{ab}} \simeq 8.98$ keV and $Ni_{K_{ab}} \simeq 8.33$ keV). Material thicknesses for the alternate zones are chosen such that all zones have equal X-ray transmission except in the narrow band of energies between the absorption edges. Under such conditions a broadband X-ray source casts a high contrast shadow only for X-ray energies between the absorption edges. This is illustrated in Fig. 23, in which the ZPCI spectral response is plotted for a copper–nickel, MFP coded aperture with an aluminum filter.

Laboratory tests using a MFP coded aperture have not yet been conducted. There are quantum noise concerns arising from the violation of guideline **G1** of § 5.1. Specifically, concern centers on the bias build-up associated with the low contrast exposure of the film from spectral components outside the narrow bandpass of the MFP coded aperture. Such a bias build-up could limit the SNR advantage of coded aperture imaging in this application.

Another recent proposal is the coded imaging spectrometer illustrated in Fig. 24. An X-ray transmission grating (with period much smaller than the

---

Fig. 21. ZPCI resolution test results for 100 keV X-rays and 6 MeV protons. The point response coded images (a) and (b) are reconstructed in higher order to demonstrate diffraction limited resolutions of approximately 4 μm and 9 μm for 100 keV X-rays and 6 MeV protons, respectively. (a′) and (b′) show isometric representations of the reconstructed point response, and linear scans ((a″) and (b″)) taken through the center of the images.

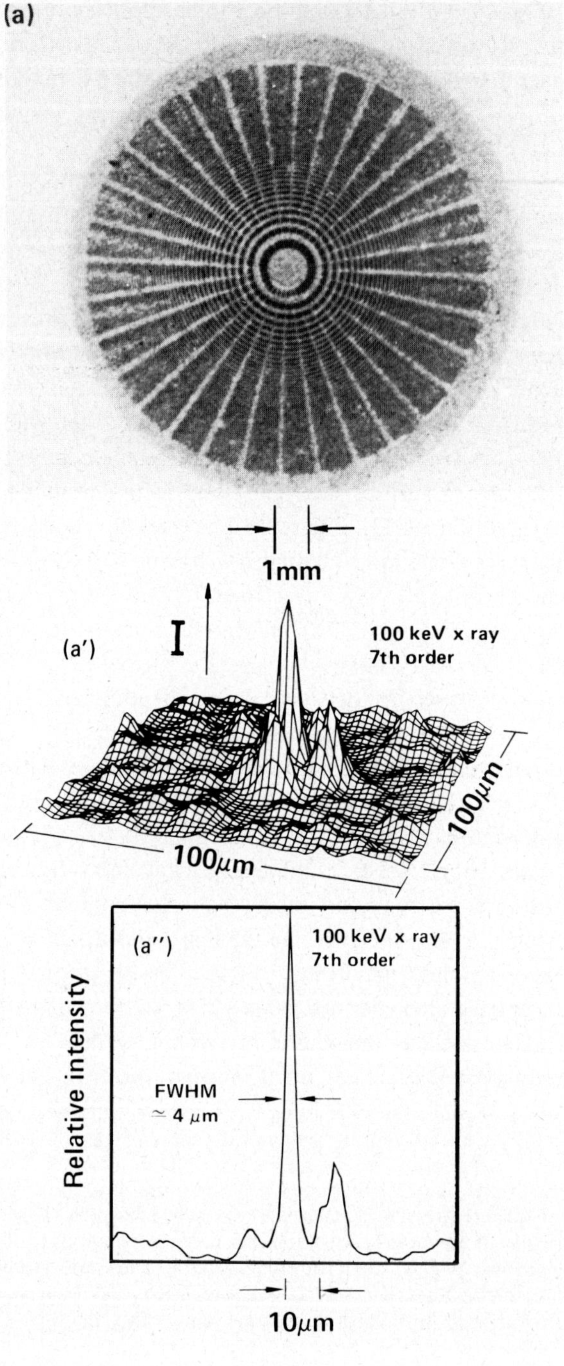
(a)
1mm
(a′)
I
100 keV x ray
7th order
100μm
100μm
(a″)
100 keV x ray
7th order
Relative intensity
FWHM
≃ 4 μm
10μm

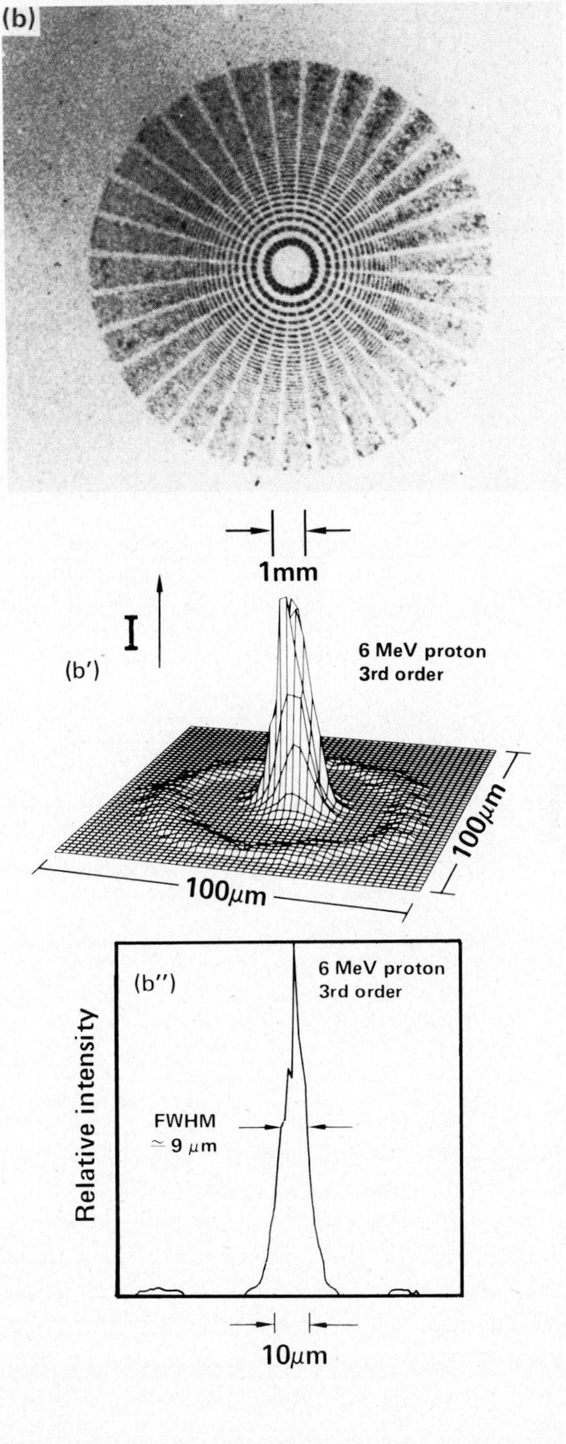
(b)
1mm
I
(b′)
6 MeV proton
3rd order
100μm
100μm
(b″)
6 MeV proton
3rd order
Relative intensity
FWHM
≃ 9 μm
10μm

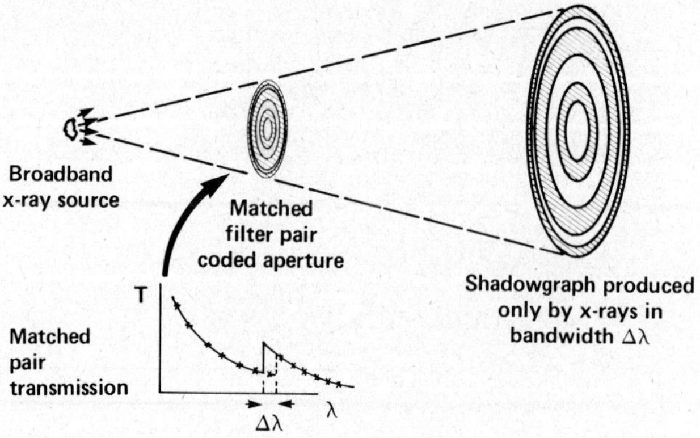

Fig. 22. Matched-filter-pair coded aperture proposed for narrowband imaging of a broadband X-ray source.

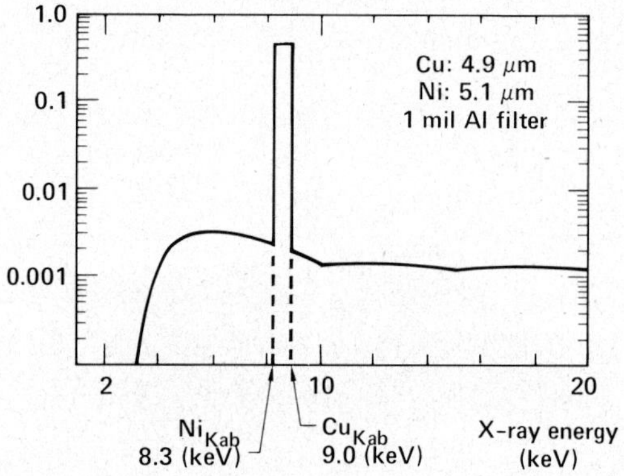

Fig. 23. Matched-filter-pair coded aperture spectral response. A "copper–nickel" zone plate coded aperture gives a narrowband spectral response between 8.3 and 9.0 keV. The ZPCI spectral response for the matched filter pair zone plate is $R = \tau_F(\tau_{Cu} - \tau_{Ni})\sigma(\lambda)$. In this figure the film response, $\sigma(\lambda)$, has been left out.

minimum zone width) is coupled with a zone plate coded aperture to produce spectrally dispersed coded images. These are optically reconstructed into a spatially resolved X-ray spectrum of the source. As with the MFP coded aperture, there are reservations concerning the SNR advantage that can be achieved with the coded imaging spectrometer. The spectral dispersion produced by the transmission grating leads to blurred zone plate shadows, which reduce the information compression ratio in image reconstruction. The

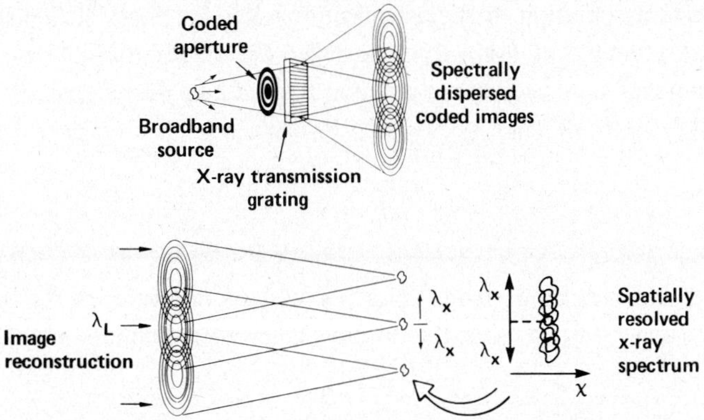

Fig. 24. Coded Imaging Spectrometer. A coded aperture serves as the entrance slit of a transmission grating spectrometer. It yields spectrally dispersed coded images, which may be reconstructed into a spatially resolved X-ray spectrum.

additional spectral information is acquired at the expense of the overall image SNR.

Acknowledgements

In the course of the past seven years, many individuals have contributed to the advances in coded imaging which are reported here. The authors are in particular indebted to Dr. David Attwood for his continued encouragement and support of our work. Experimental measurements were materially aided by the skilled contributions of Gary Stone, Gary Howe and the LLNL Technical Photography Group. Advances in coded aperture fabrication benefited from the contributions of Dino Ciarlo and Prof. Henry Smith's microfabrication groups both at Lincoln Laboratory and at the Massachusetts Institute of Technology. Work performed under the auspices of the U.S. Department of Energy by the Lawrence Livermore National Laboratory under contract number W-7405-Eng-48.

Appendix A. Gaussian Beam Propagation and ZPCI Equations

In order to derive the reconstructed optical field in any plane for a Gaussian source distribution (i.e., eqs. (4.18)–(4.19)), it is important to make the appropriate association of these equations with the classical Gaussian beam propagation equations (see SIEGMAN [1971]). From eq. (4.4) we note that when the linear shadowgraph is reconstructed it launches a series of Gaussian spherical wavefronts each converging to a waist of size ω_0 at a distance

$$d_m = \frac{f_m}{\lambda}. \tag{A.1}$$

from the shadowgraph plane. The radius of curvature of the mth order Gaussian spherical wavefront as it emerges from the shadowgraph is

$$R = \frac{R_m}{\lambda}. \tag{A.2}$$

The width of the mth order Gaussian wavefront at the shadowgraph plane is simply ω_m. So then, eqs. (4.5a – f) represent the parameters for a spherical Gaussian wavefront at a distance

$$Z = \frac{f_m}{\lambda}$$

from its waist. The classical expression for the radius of curvature of the Gaussian spherical wavefront a distance Z from its waist is

$$R(Z) = Z + \left(\frac{\pi\omega_0^2}{\lambda}\right)^2 \frac{1}{Z}. \tag{A.3}$$

If in eq. (A.3) we let $Z = f_m/\lambda$, the expression reduces to eq. (A.2). So then, in order to determine the wavefront characteristics at a distance L behind the shadowgraph one must evaluate the classical expressions at a distance

$$Z = \frac{f_m}{\lambda} - L \tag{A.4}$$

from the waist. If, for example, L is chosen equal to the reconstruction distance for the nth order wavefront,

$$L = \frac{f_n}{\lambda}, \tag{A.5}$$

then applying eqs. (A.4) and (A.5) in eq. (A.3) we get eq. (4.18c) of § 4.

References

ABLES, J. G., 1968, Proc. Astron. Soc. Aust. **4,** 1972.

BARRETT, H. H., 1972, J. Nucl. Med. **13,** 382.

BARRETT, H. H. and G. D. DEMEESTER, 1974, Appl. Opt. **13,** 1100.

BARRETT, H. H. and F. A. HORRIGAN, 1973, Appl. Opt. **12,** 2686.

BARRETT, H. H., W. W. STONER, D. T. WILSON and G. D. DEMEESTER, 1974, Opt. Eng. **13,** 539.

BARRETT, H. H., D. T. WILSON, G. D. DEMEESTER and H. SCHARFMAN, 1973, Opt. Eng. **12,** 8.

BORN, M. and E. WOLF, 1975, Principles of Optics (Pergamon Press, Oxford).

BROWN, C. M., 1973, J. Appl. Phys. **45,** 1806.

BRUNOL, J., R. SAUNEUF and J. P. GEX, 1979, Opt. Commun. **31,** 129.

CANNON, T. M. and E. E. FENIMORE, 1980, Opt. Eng. **19,** 283.

CEGLIO, N. M., 1976, Ph.D. Thesis (Massachusetts Institute of Technology).

CEGLIO, N. M., 1977, J. Appl. Phys. **48,** 1563.

CEGLIO, N. M., 1978, Energy and Technology Review, LLNL UCRL-52000-78-1, 1.

CEGLIO, N. M., 1981, in: Low Energy X-Ray Diagnostics, eds. D. T. Attwood and B. L. Henke (AIP, New York) p. 210.

CEGLIO, N. M., D. T. ATTWOOD and E. V. GEORGE, 1977, J. Appl. Phys. **48,** 1566.

CEGLIO, N. M., D. T. ATTWOOD and J. T. LARSEN, 1982, Phys. Rev. **A25,** 2351.

CEGLIO, N. M. and E. V. BENTON, 1980, LLNL-UCRL-82550-Rev. 1.

CEGLIO, N. M. and L. W. COLEMAN, 1977, Phys. Rev. Lett. **39,** 20.

CEGLIO, N. M. and J. T. LARSEN, 1980, Phys. Rev. Lett. **44,** 579.

CEGLIO, N. M. and H. I. SMITH, 1978, Rev. Sci. Instrum. **49,** 15.

CEGLIO, N. M., G. F. STONE and A. M. HAWRYLUK, 1981, J. Vac. Sci. & Technol. **19,** 886.

CIARLO, D. R. and N. M.CEGLIO, 1980, Proc. S.P.I.E. **221,** 64.

DICKE, R. H., Astrophys. J. **153,** L101.

FENIMORE, E. E. and T. M. CANNON, 1978, Appl. Opt. **17,** 337.

FENIMORE, E. E., T. M. CANNON, D. B. VAN HULSTEYN and P. LEE, 1979, Appl. Opt. **18,** 945.

FLEISCHER, R. L., P. B. PRICE and R. M. WALKER, 1975, Nuclear Tracks in Solids (Univ. of California Press, Berkeley).

GASKILL, J. D., F. R. WHITEHEAD and J. E. GRAY, 1972, Proc. S.P.I.E. **35,** 193.

GOODMAN, J. W., 1967, in: Modern Optics, ed. J. Fox (Polytechnic Press, Brooklyn, NY).

GOODMAN, J. W., 1968, Introduction to Fourier Optics (McGraw–Hill, San Francisco).

GUR, J. and J. FORSYTH, 1978, Appl. Opt. **17,** 1.

LOMMEL, E., 1885, Abh. Bayer Akad. **15,** Abth. 2, 233.

MEES, C. E. K., 1966, Theory of the Photographic Process, 3rd. Ed. (MacMillan, New York).

MERTZ, L., 1965, Transformations in Optics (Wiley, New York) ch. 3.

MERTZ, L. and N. O. YOUNG, 1961, in: Proc. Int. Conf. on Optical Instruments and Techniques (Chapman and Hall, London) p. 305.

ROGERS, W. L., L. W. JONES and W. H. BEIERWALTES, 1973, Opt. Eng. **12,** 13.

ROSE, P. J., 1976, Ph.D Thesis (Pennsylvania State Univ.).

ROSE, P. J., A. M. JACOBS and E. S. KENNEY, 1975, Nucl. Technol. **26,** 101.

SIEGMAN, A. E., 1971, An Introduction to Lasers and Masers (McGraw–Hill, New York).

SINGLETON, R. M., P. L. RANSON and R. MITTRA, 1976, IEEE Trans. Biomed. Eng. **BME-23,** 239.

SMITH, H. I., 1980, Proc. Microcircuit Engineering '80, Int. Conf. on Microlithography, ed. R. P. Kramer (Delft Univ. Press, Amsterdam) p. 463.

STONE, G. F., 1983, M.S. Thesis (Rochester Inst. of Technology).

STONE, G. F. and N. M. CEGLIO, 1983, Proc. 12th Int. Conf. on Solid State Nuclear Track Detectors, Mexico, in press.

STONE, G. F. and N. M. CEGLIO, 1983, J. Vac. Sci. & Technol., in press.
WALTON, P. W., 1973, J. Nucl. Med. **14,** 861.
WILSON, D. T., H. H. BARRETT, G. D. DEMEESTER and M. H. FARMELANT, 1973, Opt. Eng. **12,** 133.
YAGODA, H., 1949, Radioactive Measurements with Nuclear Emulsions (Wiley, New York).
YOUNG, M., 1963, Sky and Telescope **25,** 8.
YOUNG, M., 1972, J. Opt. Soc. Amer. **62,** 972.

E. WOLF, PROGRESS IN OPTICS XXI
© ELSEVIER SCIENCE PUBLISHERS B.V. 1984

V

FLUCTUATIONS, INSTABILITIES AND CHAOS IN THE LASER-DRIVEN NONLINEAR RING CAVITY*

BY

JOHN C. ENGLUND, ROBERT R. SNAPP and WILLIAM C. SCHIEVE

Center for Studies in Statistical Mechanics, The University of Texas at Austin
Austin, TX 78712, U.S.A.

* This work was partially supported by the R. A. Welch Foundation.

CONTENTS

§ 1. Introduction

Physics, chemistry, engineering, and even the social sciences have lately seen a remarkable growth of interest in far-from-equilibrium phenomena. Particular attention has been focused upon the bifurcation of steady states, the growth of fluctuations, the onset of chaos, and the appearance of temporal–spatial structure (NICOLIS and PRIGOGINE [1977], HAKEN [1978], SCHIEVE and ALLEN [1981], SWINNEY and GOLLUB [1981], TOMITA [1982] and HELLEMAN [1980, 1982]). Such phenomena are fundamentally *nonlinear*. On the macroscopic level this nonlinearity appears in the rate equations of spatially homogeneous systems and in the partial differential equations of systems with important spatial inhomogeneities. The remarkable feature is that despite the vast physical and mathematical scope – nonlinear equations being highly individual – there are universal properties.

The laser-like systems of modern optics exhibit many of these properties in an especially uncluttered setting where their essential features are not hidden by irrelevant complexity. This fact, given early emphasis by Haken (HAKEN [1975e, 1978]) has gained prominence with the recent appearance of new theoretical results for several such systems (BOWDEN, CIFTAN and ROBL [1981], ABRAHAM and SMITH [1982] and ORAEVSKII [1981]). Among these are the driven nonlinear cavity and the laser with saturable absorber, which exhibit hysteresis and other properties analogous to those of first-order phase transitions. In particular, the ring cavity with a nonlinear medium shows the universal approach to chaotic behavior argued by FEIGENBAUM [1978, 1979, 1980] in a general context, suggested in this instance by IKEDA [1979] and IKEDA, DAIDO and AKIMOTO [1980] as a mechanism for optical chaos (SNAPP, CARMICHAEL and SCHIEVE [1981]). This important system is the focus of the present work*. First in § 2 we will treat the stationary response of the nonlinear ring cavity to an injected field utilizing the Maxwell–Bloch equations for a medium of two-level atoms first considered by BONIFACIO

* For a recent review of similar phenomena in the injected nonlinear Fabry–Perot cavity, see ABRAHAM and SMITH [1982].

and LUGIATO [1978b]. The mean-field approximation is formulated and the hysteresis behavior analogous to a first-order phase transition is obtained (BONIFACIO and LUGIATO [1976, 1978a]). In § 3 we turn to the *onset of instability* from these stationary solutions by means of a linear stability analysis. This treatment is closely related to the early work on the unidirectional ring laser of RISKEN and NUMMEDAL [1968a,b]. The stability is here analyzed in a general way and simplifying limits are discussed. The instabilities leading to hysteresis and also self-pulsing from the upper branch in the bistable region are evidenced (BONIFACIO, GRONCHI and LUGIATO [1979b])* . Another instability, in the limit of fast transverse atomic relaxation (IKEDA [1979], CARMICHAEL, SNAPP and SCHIEVE [1982]), is discussed. This instability leads to period-doubling bifurcations and to aperiodic (chaotic) motion, to be discussed in § 6.

In the next sections we turn to the time-dependent behavior of the nonlinear ring cavity. Section 4 describes the deterministic switching in the bistable region utilizing the Maxwell–Bloch equations. Attention is focused on slowing near the marginal points. Section 5 treats the effects of internal noise upon the hysteresis behavior – both stationary and time-dependent. Both the microscopic and mesoscopic (phenomenological-stochastic) points of view are taken. Comments are made upon various approximations made in obtaining a Fokker–Planck equation from the fully quantum master equation (LAX [1968], HAKEN [1970]). The Fokker–Planck equation for absorptive bistability is obtained in the good-cavity limit by the method of GORDON [1967] and HAKEN [1975b]. The fluctuation dynamics are then formulated from a simple phenomenological point of view and the same Fokker–Planck equation is obtained (BULSARA, SCHIEVE, and GRAGG [1978]). This point of view is similar to that taken for the laser by RISKEN [1970]. The steady-state solutions of the Fokker–Planck equation are discussed; the eigenvalues (and thus time dependence) of the Fokker–Planck operator are approximated analytically and numerically (ENGLUND, SCHIEVE, ZUREK and GRAGG [1981]). The validity of the "classic" KRAMERS [1940] approximation for the lowest nonzero eigenvalue near the marginal point is discussed. Also the eigenvalues at the critical point are examined.

In the final section, § 6, we discuss how deterministic chaos may arise from instabilities present in several laser-related systems. In particular, we focus on the sequence of instabilities present in the nonlinear ring cavity with fast transverse relaxation, discussed in § 3. The universal properties seen in period-doubling sequences (FEIGENBAUM [1978]) are shown.

* See also a review by these authors in BOWDEN, CIFTAN and ROBL [1981].

§ 2. Semiclassical Theory and Steady-State Response

In this section, we formulate the dynamics and present the steady-state response of the driven ring cavity, by means of the Maxwell–Bloch equations for two-level atoms coupled to the radiation field. The first theoretical study of a driven nonlinear cavity was given for a Fabry–Perot resonator by SZÖKE, DANEU, GOLDHAR and KURNIT [1969], who used a rate equation to model the dynamics of the intracavity medium. The earliest treatments using Maxwell–Bloch equations were given by SPENCER and LAMB [1972] for the Fabry–Perot cavity, and by BONIFACIO and LUGIATO [1978b] for the ring cavity. In the following discussion, we begin by deriving the latter formulation and reviewing the associated steady-state results; we then introduce the "mean-field" approximation, which simplifies, somewhat, the theoretical description, and conclude with some remarks on qualitative aspects of the steady states.

A schematic representation of the driven ring cavity is given in Fig. 1, with arrows indicating the direction of propagation. We adopt a "folded" cylindrical coordinate system (r, ϕ, z), where the z-axis is along the optical path. The monochromatic field $\boldsymbol{E}_i$ is incident upon a mirror of reflectivity R and transmissivity $T \equiv 1 - R$. The nonlinear (saturable) intracavity medium is confined between the points $z = 0$ and $z = L$, and is located a distance l from the output mirror, which we assume to be identical to the input mirror, and through which the transmitted field $\boldsymbol{E}_t$ emerges. We suppose that all other mirrors are perfectly reflecting and that the round-trip optical path of the internal field $\boldsymbol{E}$ is of length $\mathscr{L}$.

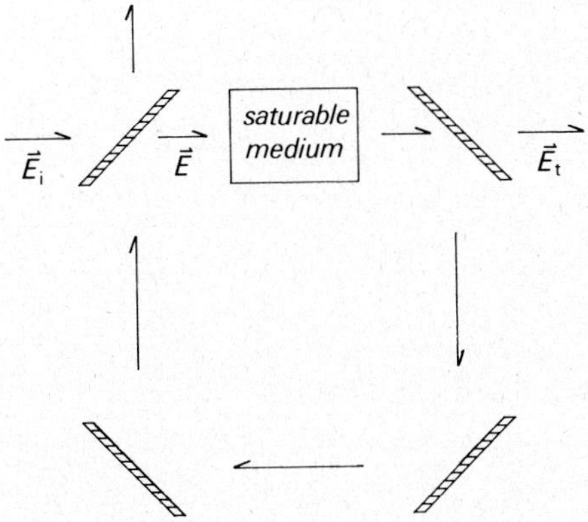

Fig. 1. Schematic representation of the nonlinear ring cavity.

The fields $\boldsymbol{E}_{\rm i}$, $\boldsymbol{E}_{\rm t}$ and $\boldsymbol{E}$ are taken to be plane-polarized, and we assume that they, and all functions describing the medium, are symmetric about the z-axis, so that we may henceforth neglect all ϕ-dependence. Then, defining the field magnitudes,

$$E_{\rm i}(r, z, t) \equiv |\boldsymbol{E}_{\rm i}|, \qquad E_{\rm t}(r, z, t) \equiv |\boldsymbol{E}_{\rm t}|, \qquad E(r, z, t) \equiv |\boldsymbol{E}|,$$

and supposing that transverse field variations are negligible over a wavelength of the incident field (i.e., $\partial E(r, z, t)/\partial r \ll \partial E(r, z, t)/\partial z$), Maxwell's equations, in SI units, give

$$\left(\frac{\partial^2}{\partial z^2} - \frac{1}{c^2}\frac{\partial^2}{\partial t^2}\right) E(r, z, t) = \frac{1}{\varepsilon_0 c^2}\frac{\partial^2}{\partial t^2} P(r, z, t) \tag{2.1}$$

(LAMB [1964]), with the (lossless) medium described by the polarization magnitude $P(r, z, t)$. Here, ε_0 is the permittivity of the vacuum and c is the speed of light. (BALLAGH, COOPER, HAMILTON, SANDLE and WARRINGTON [1981] have treated the case of a Gaussian cavity-mode structure.)

We suppose that the medium consists of an ensemble of $\mathcal{N}$ two-level atoms, each as pictured in Fig. 2. Emission and absorption between the excited (a) and ground (b) states occur at a frequency $\omega_{\rm A}$, which may vary from atom to atom (inhomogeneous broadening). The Weisskopf–Wigner decay rate of the excited atom to the unamplified field modes is denoted by $\gamma_\downarrow$. Finally, we have included an excitation rate $\gamma_\uparrow$ so that we might discuss the laser limit of the dynamics in § 3; for the passive medium of Fig. 1, we shall take $\gamma_\uparrow = 0$.

With these definitions, and assuming an electric dipole interaction of the atoms with the internal field $E(r, z, t)$, we may describe the semiclassical dynamics of the atomic density matrix

$$\begin{pmatrix} \rho_{\rm aa}(r, z, \omega_{\rm A}, t) & \rho_{\rm ab}(r, z, \omega_{\rm A}, t) \\ \rho_{\rm ba}(r, z, \omega_{\rm A}, t) & \rho_{\rm bb}(r, z, \omega_{\rm A}, t) \end{pmatrix}$$

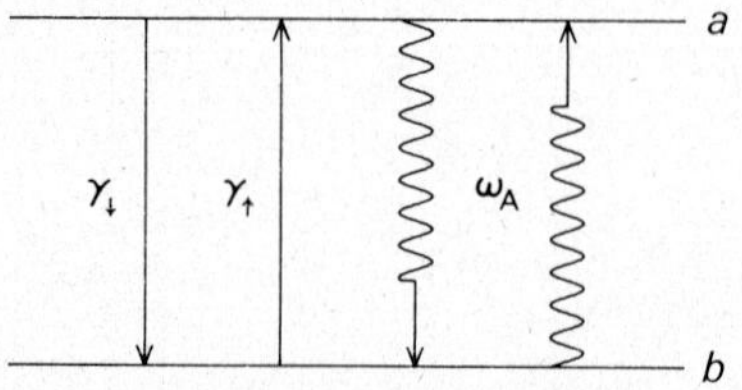

Fig. 2. The two-level atom. The wavy arrows represent the cavity-enhanced transitions.

by the optical Bloch equations (HAKEN [1970], LOUISELL [1973], SARGENT, SCULLY and LAMB [1974]):

$$\frac{\partial \rho_{ab}(r, z, \omega_A, t)}{\partial t} = -(i\omega_A + \gamma_\perp)\rho_{ab}(r, z, \omega_A, t) - \left(\frac{i\mu}{\hbar}\right) E(r, z, t)[\rho_{aa}(r, z, \omega_A, t) - \rho_{bb}(r, z, \omega_A, t)], \quad (2.2a)$$

$$\frac{\partial \rho_{aa}(r, z, \omega_A, t)}{\partial t} = -\gamma_\downarrow \rho_{aa}(r, z, \omega_A, t) + \gamma_\uparrow \rho_{bb}(r, z, \omega_A, t) - \left(\frac{i\mu}{\hbar}\right) E(r, z, t)[\rho_{ab}(r, z, \omega_A, t) - \text{c.c.}], \quad (2.2b)$$

and

$$\frac{\partial \rho_{bb}(r, z, \omega_A, t)}{\partial t} = \gamma_\downarrow \rho_{aa}(r, z, \omega_A, t) - \gamma_\uparrow \rho_{bb}(r, z, \omega_A, t) + \left(\frac{i\mu}{\hbar}\right) E(r, z, t)[\rho_{ab}(r, z, \omega_A, t) - \text{c.c.}]. \quad (2.2c)$$

Here, μ is the projection along the direction of polarization of the electric dipole matrix element between levels (a) and (b). The atomic dipole decay constant is represented by

$$\gamma_\perp = \tfrac{1}{2}\gamma_\parallel + \gamma_{ph},$$

where $\frac{1}{2}\gamma_\parallel \equiv \frac{1}{2}(\gamma_\uparrow + \gamma_\downarrow)$ is the radiative contribution and γ_{ph} is the contribution of "dephasing" processes, such as elastic collisions.

We suppose the atoms to be identical, except for possible differences in ω_A because of, for example, atomic motion or crystal imperfections. Then, the polarization matrix is related to the density matrix by

$$P(r, z, t) = \mu \int_{-\infty}^{\infty} d\omega_A\, W(\omega_A)[\rho_{ab}(r, z, \omega_A, t) + \text{c.c.}], \quad (2.3)$$

where $W(\omega_A)$ is the normalized distribution of atomic frequencies ω_A.

Equations (2.1)–(2.3), along with the boundary conditions

$$E(0, t) = \sqrt{T}\, E_i(0, t) + RE\left(L, t - \frac{\mathscr{L} - L}{c}\right) \quad (2.4a)$$

and

$$E_t(l, t) = \sqrt{T}\, E\left(L, t - \frac{l}{c}\right), \tag{2.4b}$$

now describe the dynamics of the driven ring cavity in full. In proceeding to their solution, however, it is advantageous to apply the slowly-varying-amplitude approximation. Specifically, we decompose the field $E(r, z, t)$ and the polarization $P(r, z, t)$ by factoring out the running-wave phasor $e^{i(kz - \omega t)}$, where k is the wave number of the driving field E_i and $\omega = kc$ is its frequency. Thus,

$$E(r, z, t) \equiv w(r)[\mathscr{E}(z, t)\, e^{i(kz - \omega t)} + \text{c.c.}] \tag{2.5a}$$

and

$$P(r, z, t) \equiv \mu[\mathscr{P}(r, z, t)\, e^{i(kz - \omega t)} + \text{c.c.}], \tag{2.5b}$$

where we have also factored the transverse (e.g., Gaussian) mode function $w(r)$ from $E(r, z, t)$. We define $\mathscr{E}_i$ and $\mathscr{E}_t(z, t)$ similarly:

$$E_i(r, z, t) \equiv w(r)[\mathscr{E}_i(z, t)\, e^{i(kz - \omega t)} + \text{c.c.}] \tag{2.5c}$$

and

$$E_t(r, z, t) \equiv w(r)[\mathscr{E}_t(z, t)\, e^{i(kz - \omega t)} + \text{c.c.}]. \tag{2.5d}$$

Then, if we assume the amplitudes $\mathscr{E}(z, t)$ and $\mathscr{P}(z, t)$ to be slowly varying with respect to $e^{i(kz - \omega t)}$, eq. (2.1) becomes

$$\frac{\partial \mathscr{E}(z, t)}{\partial t} + c\frac{\partial \mathscr{E}(z, t)}{\partial z} = \frac{i\omega\mu}{2\varepsilon_0} \frac{\int_0^\infty r\, dr\, w^*(r) \mathscr{P}(r, z, t)}{\int_0^\infty r\, dr\, |w(r)|^2}. \tag{2.6}$$

Introducing the definitions

$$\mathscr{P}(r, z, \omega_A, t)\, e^{i(kz - \omega t)} \equiv \rho_{ab}(r, z, \omega_A, t)$$

and

$$\mathscr{D}(r, z, \omega_A, t) \equiv \rho_{aa}(r, z, \omega_A, t) - \rho_{bb}(r, z, \omega_A, t),$$

we also have

$$\frac{\partial \mathscr{P}(r, z, \omega_A, t)}{\partial t} = -\gamma_\perp(1 + i\delta_A)\mathscr{P}(r, z, \omega_A, t) - \left(\frac{i\mu}{\hbar}\right) w(r) \mathscr{E}(z, t) \mathscr{D}(r, z, \omega_A, t) \tag{2.7a}$$

and

$$\frac{\partial \mathscr{D}(r, z, \omega_A, t)}{\partial t} = \gamma_{\|} [\mathscr{D}_0 - \mathscr{D}(r, z, \omega_A, t)]$$

$$+ 2\left(\frac{i\mu}{\hbar}\right) w(r)[\mathscr{E}(z, t)\, \mathscr{P}^*(r, z, \omega_A, t) - \text{c.c.}], \qquad (2.7b)$$

with

$$\mathscr{P}(r, z, t) = \int_{-\infty}^{\infty} d\omega_A\, W(\omega_A) \mathscr{P}(r, z, \omega_A, t) \qquad (2.8)$$

and

$$\mathscr{D}_0 \equiv \frac{(\gamma_\uparrow - \gamma_\downarrow)}{(\gamma_\uparrow + \gamma_\downarrow)} \mathscr{N},$$

the steady-state inversion of the undriven cavity. We have made the rotating-wave approximation in ignoring the rapidly varying factors $e^{\pm 2i(kz - \omega t)}$ in eq. (2.7b), and have introduced the atomic detuning parameter

$$\delta_A \equiv (\omega_A - \omega)/\gamma_\perp .$$

Defining the cavity detuning parameter

$$\delta_c \equiv (\omega_c - \omega)/\kappa,$$

where ω_c is the frequency of the cavity mode nearest resonance and

$$\kappa \equiv cT/\mathscr{L}$$

is the empty-cavity decay constant for the field, we may rewrite the boundary conditions (2.4a,b) as

$$\mathscr{E}(0, t) = \sqrt{T}\, \mathscr{E}_i + R\, e^{-i\delta_c T}\, \mathscr{E}\left(L, t - \frac{\mathscr{L} - L}{c}\right) \qquad (2.9a)$$

and

$$\mathscr{E}_t(l, t) = \sqrt{T}\, \mathscr{E}\left(L, t - \frac{l}{c}\right). \qquad (2.9b)$$

For the most part, we shall neglect transverse field variations and let $w(r) \equiv 1$. Under this limit, the steady-state response of the ring-cavity model represented by eqs. (2.6)–(2.9) has been obtained for homogeneous broadening by BONIFACIO, LUGIATO and GRONCHI [1979], ROY and ZUBAIRY [1980] and CARMICHAEL and HERMANN [1980]. With $\gamma_{\uparrow} = 0$, they find the parametric relations, in terms of the dummy variable s,

$$x^2 = \left(\frac{1+\delta_A^2}{s^2-1}\right)(\alpha L - \ln s^2)$$

and

$$y^2 = \frac{x^2}{T^2}[s^2 + R^2 - 2Rs\cos(\delta_A \ln s - \delta_c T)], \tag{2.10b}$$

for the normalized incident and transmitted fields

$$x \equiv \frac{2\mu \mathscr{E}_t(0)}{\hbar(\gamma_\perp \gamma_\| T)^{1/2}} \quad \text{and} \quad y \equiv \frac{2\mu \mathscr{E}_i}{\hbar(\gamma_\perp \gamma_\| T)^{1/2}}. \tag{2.11}$$

In eq. (2.10a),

$$\alpha \equiv \frac{\mu^2 \omega \mathscr{N}}{\hbar c \varepsilon_0 \gamma_\perp (1+\delta_A^2)} \tag{2.12}$$

is the linear absorption coefficient. N.B.: BONIFACIO and LUGIATO [1978b] define an absorption coefficient α that is one half that used here. Below saturation, their α gives a decay in field amplitude $|\mathscr{E}(z)| = |\mathscr{E}(0)|e^{-\alpha z}$, while eq. (2.12) gives a decay in intensity $|\mathscr{E}(z)|^2 = |\mathscr{E}(0)|^2 e^{-\alpha z}$. This explains certain apparent discrepancies between our presentations (as in the definition of C, below, eq. (2.20)). In the absence of detuning ($\delta_A = \delta_c = 0$), eqs. (2.10a,b) further simplify to

$$\alpha L = \ln\left[1 + T\left(\frac{y}{x} - 1\right)\right]^2 + x^2\left\{\left[1 + T\left(\frac{y}{x} - 1\right)\right]^2 - 1\right\} \tag{2.13}$$

(BONIFACIO and LUGIATO [1978b]), representing pure absorption. These results have been generalized to a model with inhomogeneous broadening, where $W(\omega_A)$ is a Lorentzian function, by GRONCHI and LUGIATO [1980].

Before proceeding to a description of the response curves generated by eqs. (2.10–2.13), we will formulate the *mean-field approximation* to eqs. (2.6–2.9). This facilitates a more realistic treatment of inhomogeneous broadening and of transverse field variations; it also simplifies greatly the analysis of deterministic switching and, when appropriately generalized, leads to a tractable quantum-statistical model of the driven ring cavity.

We begin by developing eqs. (2.9a,b) in power series about $(z, t) = (0, t)$ and (l, t), respectively:

$$\mathscr{E}(0, t) = \sqrt{T}\,\mathscr{E}_{\mathrm{i}} + R\,\mathrm{e}^{-\mathrm{i}\delta_{\mathrm{c}}T}\left[\mathscr{E}(0, t) + L\frac{\partial\mathscr{E}(0, t)}{\partial z} - \frac{(\mathscr{L} - L)}{c}\frac{\partial\mathscr{E}(0, t)}{\partial t} + \dots\right] \tag{2.14a}$$

and

$$\mathscr{E}_{\mathrm{t}}(l, t) = \sqrt{T}\left[\mathscr{E}(l, t) + (L - l)\frac{\partial\mathscr{E}(l, t)}{\partial z} - \frac{l}{c}\frac{\partial\mathscr{E}(l, t)}{\partial t} + \dots\right]. \tag{2.14b}$$

In addition we use the steady-state absorption length α^{-1} to define the dimensionless coordinates $\tilde{z} = z\alpha$ and $\tilde{t} = t\alpha c$, giving

$$\mathscr{E}(0, t) = \sqrt{T}\mathscr{E}_{\mathrm{i}} + R\,\mathrm{e}^{-\mathrm{i}\delta_{\mathrm{c}}T}\left[\mathscr{E}(0, t) + \alpha L\frac{\partial\mathscr{E}(0, t)}{\partial\tilde{z}} - \alpha(\mathscr{L} - L)\frac{\partial\mathscr{E}(0, t)}{\partial\tilde{t}} + \dots\right] \tag{2.15a}$$

and

$$\mathscr{E}_{\mathrm{t}}(l, t) = \sqrt{T}\left[\mathscr{E}(l, t) + \alpha(L - l)\frac{\partial\mathscr{E}(l, t)}{\partial\tilde{z}} - \alpha l\frac{\partial\mathscr{E}(l, t)}{\partial\tilde{t}} + \dots\right]. \tag{2.15b}$$

Then, assuming that the derivatives $\partial^n\mathscr{E}/\partial\tilde{z}^n$ and $\partial^n\mathscr{E}/\partial\tilde{t}^n$ remain finite, the mean-field limit

$$T \to 0, \tag{2.16a}$$

with

$$\alpha L = O(T), \tag{2.16b}$$

$$\kappa = O(1), \tag{2.16c}$$

$$\delta_{\mathrm{c}} = O(1), \tag{2.16d}$$

$$\frac{L}{\mathscr{L}} = \lambda - O(T), \tag{2.16e}$$

reduces eqs. (2.15a,b) to

$$c\frac{\partial \mathscr{E}(0,t)}{\partial z} = \left(\frac{1}{\lambda} - 1\right)\frac{\partial \mathscr{E}(0,t)}{\partial t} - \frac{\kappa}{\lambda}[\mathscr{E}_{\mathrm{i}}/\sqrt{T} - (1 + \mathrm{i}\delta_{\mathrm{c}})\mathscr{E}(0,t)] \quad (2.17a)$$

and

$$\mathscr{E}_{\mathrm{t}}(l,t) = \sqrt{T}\,\mathscr{E}(l,t), \quad (2.17b)$$

where we have let the cavity-enhanced fields $\mathscr{E}_{\mathrm{i}}/\sqrt{\mathrm{T}}$ and $\mathscr{E}_{\mathrm{t}}(0,t)/\sqrt{\mathrm{T}}$ remain finite in the limit. Combining eq. (2.17a) with eq. (2.6), we obtain

$$\frac{\partial \mathscr{E}(t)}{\partial t} = \kappa[\mathscr{E}_{\mathrm{i}}/\sqrt{T} - (1 + \mathrm{i}\delta_{\mathrm{c}})\mathscr{E}(t)] + \frac{\mathrm{i}\omega\mu\lambda}{2\varepsilon_0}\frac{\displaystyle\int_0^\infty r\,\mathrm{d}r\,w^*(r)\mathscr{P}(r,t)}{\displaystyle\int_0^\infty r\,\mathrm{d}r\,|w(r)|^2}, \quad (2.18)$$

with $\mathscr{E}(t) \equiv \mathscr{E}(0,t)$ and $\mathscr{P}(r,t) \equiv \mathscr{P}(r,0,t)$. Equations (2.18), (2.7)–(2.8) and (2.17b) comprise the time-dependent mean-field approximation to the Maxwell–Bloch equations. Clearly, condition (2.16a) eliminates all spatial dependence caused by the presence of the mirrors, while (2.16b) removes the spatial effects associated with the presence of the medium. The mean-field theory is therefore, equivalent to a single-mode theory.

Bonifacio, Lugiato and Gronchi [1979] have obtained the mean-field limit in the steady state (see also Bonifacio and Lugiato [1978b] and Roy and Zubairy [1980]) without introducing the "fill factor" λ of eq. (2.16e). Equations (2.18) and (2.7 – 2.8) also may be derived by generalizing the single-mode quantum-statistical model of the laser (see § 5) and then taking the semiclassical limit of the Heisenberg equations (Bonifacio and Lugiato [1978d], Hassan, Drummond and Walls [1978]).

We now reconsider, under the mean-field approximation, the steady-state response of the ring cavity. In the place of eqs. (2.10a,b), we find

$$y = x\left\{\left[1 + \frac{2C(\delta_{\mathrm{A}}^2 + 1)}{\delta_{\mathrm{A}}^2 + 1 + x^2}\right] + \mathrm{i}\left[\delta_{\mathrm{c}} - \frac{2C\delta_{\mathrm{A}}(\delta_{\mathrm{A}}^2 + 1)}{\delta_{\mathrm{A}}^2 + 1 + x^2}\right]\right\} \quad (2.19)$$

(Bonifacio and Lugiato [1978d], Hassan, Drummond and Walls [1978]), where

$$C \equiv \frac{\alpha L}{4T} \quad (2.20)$$

remains finite in the limits (2.16a,b) and (2.15). Equation (2.19) has the absorptive limit ($\delta_A = \delta_c = 0$)

$$y = x\left[1 + \frac{2C}{1 + x^2}\right] \tag{2.21}$$

(SZÖKE, DANEU, GOLDHAR and KURNIT [1969], BONIFACIO and LUGIATO [1976]). (It is of interest that ASQUINI and CASAGRANDE [1981] have shown that eqs. (2.13) and (2.21) describe the driven *bidirectional* ring cavity, also.) Taking the dispersive limit (for $\delta_c = 0$),

$$\delta_A \gg 1 \tag{2.22a}$$

and

$$x^2 \ll 1, \tag{2.22b}$$

gives the response

$$y = x\left\{1 - 2C\delta_A\left[1 - \frac{x^2}{\delta_A^2 + 1}\right]\right\} \tag{2.23}$$

(GIBBS, MCCALL and VENKATESAN [1976, 1978], SCHWENDIMANN [1979]). These mean-field results have been generalized to treat inhomogeneous broadening by HASSAN, DRUMMOND and WALLS [1978] and by BONIFACIO and LUGIATO [1978d], who use Gaussian and Lorentzian functions, respectively, for $W(\omega_A)$. In addition, DRUMMOND [1981a,b] has considered a Gaussian transverse mode function $w(r)$.

At this point, a word should be said concerning the mean-field approximation introduced by BONIFACIO and LUGIATO [1976, 1978a] in describing the dynamics of the driven *Fabry–Perot* resonator (see also LUGIATO and BONIFACIO [1978]). This approximation leads to Maxwell–Bloch equations that are precisely those given above for the ring cavity; however, one must introduce an arbitrary factorization, in addition to the limits (2.16a–e), that weakens the validity of those equations. To clarify this, we give a brief treatment for the Fabry–Perot cavity.

First, we apply the slowly-varying-amplitude approximation for standing wave fields,

$$E(z, t) = \sqrt{2}\sin kz[\mathscr{E}(z, t)\,e^{-i\omega t} + \text{c.c.}] \tag{2.24a}$$

and

$$P(z, t) = \mu\sqrt{2}\sin kz[\mathscr{P}(z, t)\,e^{-i\omega t} + \text{c.c.}]. \tag{2.24b}$$

Then, neglecting transverse field variations and broadening inhomogeneities, for simplicity, we obtain the Maxwell–Bloch equations

$$\frac{\partial \mathscr{E}(t)}{\partial t} = \kappa[\mathscr{E}_i/\sqrt{T} - (1 + i\delta_c)\mathscr{E}(t)] + \frac{i\omega\mu}{\varepsilon_0}\frac{1}{L}\int_0^L dz \sin^2 kz\, \mathscr{P}(z, t), \tag{2.25a}$$

$$\frac{\partial \mathscr{P}(z, t)}{\partial t} = -\gamma_\perp(1 + i\delta_A)\mathscr{P}(z, t) - \left(\frac{i\mu}{\hbar}\right)\mathscr{E}(t)\mathscr{D}(z, t) \tag{2.25b}$$

and

$$\frac{\partial \mathscr{D}(z, t)}{\partial t} = \gamma_\parallel[\mathscr{D}_0 - \mathscr{D}(z, t)] + 4\left(\frac{i\mu}{\hbar}\right)\sin^2 kz[\mathscr{E}(t)\mathscr{P}^*(z, t) - \text{c.c.}], \tag{2.25c}$$

where we have applied the limits (2.16a–e).

The functions $\mathscr{P}(z, t)$ and $\mathscr{D}(z, t)$ now may be Fourier-analyzed; this leads to an infinite hierarchy of Bloch equations (CARMICHAEL [1980]), a situation vastly more complicated than with the ring cavity, in which all z-dependence is eliminated. To simplify matters BONIFACIO and LUGIATO [1978a] average eqs. (2.25b,c) over the length L of the medium, then *factor* all averages of products, i.e.,

$$\frac{1}{L}\int_0^L dz\, F(z)G(z) \rightarrow \left(\frac{1}{L}\int_0^L dz\, F(z)\right)\left(\frac{1}{L}\int_0^L dz\, G(z)\right), \tag{2.26}$$

where $F(z)$ and $G(z)$ are any of the z-dependent terms. In essence, this replaces the factors of $\sin^2 kz$ in eqs. (2.25a,c) by 1/2. It is easy to see that this reduces eqs. (2.25a–c) to those previously given for the ring cavity. This neglect of z-dependence is also intrinsic to the rate-equation model of SZÖKE, DANEU, GOLDHAR and KURNIT [1969], who introduced the limits (2.16a,b), as well.

Although the z-dependence of eqs. (2.25a–c) poses complications in treating the dynamics of the Fabry–Perot device, the steady state is readily obtainable, and it is therefore possible to judge the validity of assuming (2.26). In particular, SPENCER and LAMB [1972] have shown that the steady-state response for a non-dispersive medium is

$$y = x + \frac{2C}{x}\left[1 - \frac{1}{(1+2x^2)^{1/2}}\right]. \tag{2.27}$$

Carmichael [1980], Hermann [1980], McCall and Gibbs [1980], Agrawal and Carmichael [1980] and Carmichael and Agrawal [1981] have made detailed comparisons of the Fabry–Perot's steady-state characteristics with and without (2.26). As one would expect, since the factors of $\sin^2 kz$ occur in the interaction terms of eqs. (2.25a,c), the differences are greatest for large amplitudes x. Similar studies have been carried out by Meystre [1978], Abraham, Bullough and Hassan [1979], Roy and Zubairy [1980] and Abraham, Hassan and Bullough [1980], who, however, use a truncated hierarchy of Bloch equations in their analyses. From these numerous investigations, it appears preferable to avoid applying (2.26). However, we should emphasize that it greatly simplifies the analysis of the Fabry–Perot cavity dynamics, while maintaining an essentially correct qualitative description of them in many cases.

Finally, we discuss some qualitative aspects of the steady-state response of the ring cavity. In Fig. 3, we have used eq. (2.13) to plot x versus y for several values of C, with T fixed. The response curve is single-valued below the critical

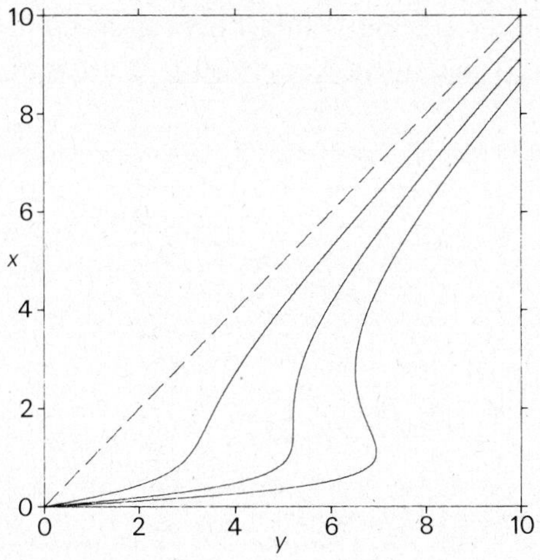

Fig. 3. The onset of bistability for a purely absorbing medium ($\delta_A = \delta_c = 0$). The three solid response curves are plotted from eq. (2.13) in the text, with $C = 2$, 4.11 and 6, respectively from left to right, and $T = 0.05$; for the value $C \approx 4.11$, the curve contains a critical point. The dashed line represents the response of the empty cavity.

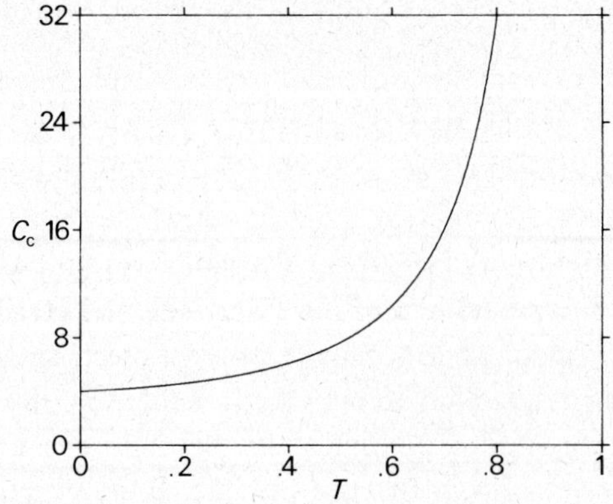

Fig. 4. The critical value of C, C_c, is plotted versus the transmission coefficient T of the cavity, for $\delta_A = \delta_c = 0$; the region above the line is the bistable regime. As $T \to 0$, C_c approaches the mean-field value 4, while C_c diverges as $T \to 1$.

value $C = C_c$ and triple-valued above it. As we shall see in § 3, however, all loci with $dy/dx < 0$ are unstable; therefore, for $C > 4$, we have the possibility of a *bistable response* and *hysteresis* (see § 4). As T is varied, the critical value C_c also changes, as shown in Fig. 4, approaching 4 in the mean-field limit, $T \to 0$ (BONIFACIO and LUGIATO [1976]). A comparison between the exact (eq. (2.13)) and the mean-field (eq. (2.21)) response is presented in Fig. 5 for a case in which the mean-field condition (2.16b) is not satisfied. For similar comparisons, see BONIFACIO and LUGIATO [1978b], BONIFACIO, LUGIATO and

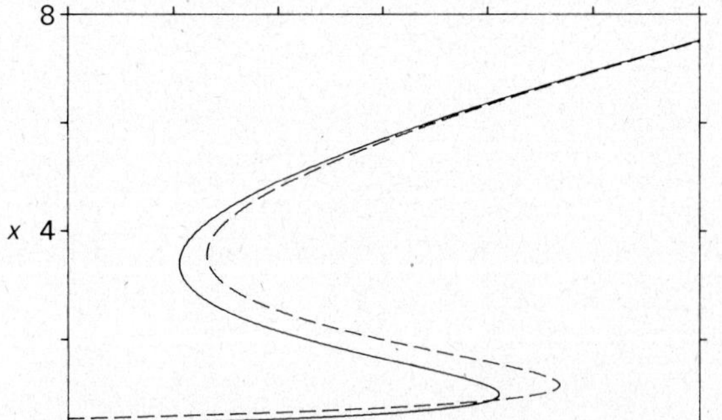

Fig. 5. A comparison of the response curves obtained from eq. (2.13) (solid curve) and the mean-field eq. (2.21) (dashed curve), with $\alpha L = 1.6$ and $T = 0.05$ ($C = 8$); the medium is purely absorbing ($\delta_A = \delta_c = 0$). Note that the condition, eq. (2.16b) is not satisfied.

GRONCHI [1979], ROY and ZUBAIRY [1980] and GRONCHI and LUGIATO [1980].

The purely absorptive cavity can have at most a triple-valued response; however, FELBER and MARBURGER [1976] have shown that dispersion can lead to a multi-valued response of higher order. (Their analysis, for a Fabry–Perot cavity filled with a Kerr medium, has been repeated for the ring cavity by IKEDA [1979].) Roughly speaking, the response curve acquires a new kink whenever $\alpha L \delta_A$ is increased by 2π; the intensity-dependent phase shifts thereby encounter an additional cavity resonance. This behavior, pictured in Fig. 6, is not encountered in the mean-field eq. (2.19) which has at most three roots. (See also MARBURGER and FELBER [1978] and CARMICHAEL and HERMANN [1980].) GRONCHI and LUGIATO [1980] have shown that inhomogeneous broadening tends to eliminate this multi-valued response; BONIFACIO and LUGIATO [1978d] and HASSAN, DRUMMOND and WALLS [1978] have also demonstrated that large inhomogeneous broadening may eliminate purely absorptive bistability, while still allowing bistability with dispersion. As in the purely absorptive case, C must be greater than 4 for a bistable response (BONIFACIO, LUGIATO and GRONCHI [1979]); general conditions have been given by HASSAN, DRUMMOND and WALLS [1978], AGRAWAL and CARMICHAEL [1979] and BONIFACIO, GRONCHI and LUGIATO [1979a], in the mean-field approximation.

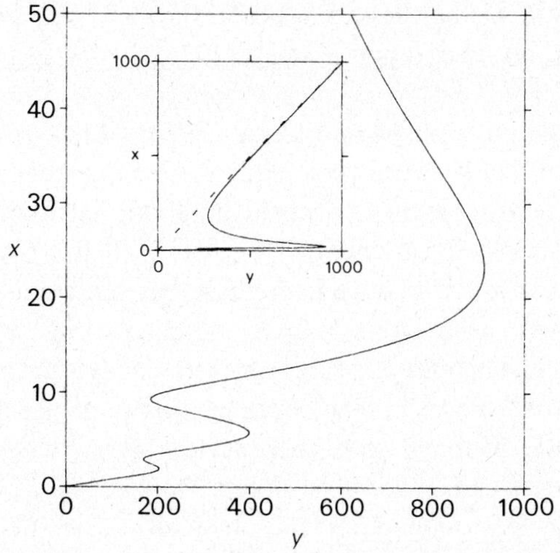

Fig. 6. Multivalued response for a dispersive medium, obtained from eq. (2.10) with $\alpha L = 4$, $T = 0.05$, $\delta_A = -3\pi$ and $\delta_c = 0$. The dashed line in the inset is the empty-cavity response curve.

The bistability threshold, i.e., the value of y for which x is triple-valued (C fixed), may be raised or lowered by dispersion (HASSAN, DRUMMOND and WALLS [1978], WILLIS and DAY [1979] and AGRAWAL and CARMICHAEL [1979]). Calculations by DRUMMOND [1981a,b] indicate that the threshold is raised when inhomogeneous broadening and/or transverse spatial variations are included. Finally, WILLIS and DAY [1979], AGRAWAL and CARMICHAEL [1979], BONIFACIO, GRONCHI and LUGIATO [1979a] and ROY and ZUBAIRY [1980] have shown that an increase in dispersion decreases the size of the hysteresis cycle.

§ 3. The Onset of Instability of the Stationary States

Numerous linear stability analyses of the stationary states of the nonlinear unidirectional ring cavity and related systems have appeared in the literature. Most have been based upon the semiclassical study of a homogeneously broadened, unidirectional ring laser of RISKEN and NUMMEDAL [1968a,b], where it is shown that certain nonresonant cavity modes appear if the pump parameter exceeds a critical value. In addition, their numerical calculations indicate that as the stationary (cw) solution loses its stability the cavity field amplitude undergoes self-pulsing (oscillatory limit cycles). The threshold condition for this instability was simultaneously derived in a quantum mechanical description by GRAHAM and HAKEN [1968]. Further calculations by HAKEN and OHNO [1976] (cf. OHNO and HAKEN [1976]), using a generalized mode analysis (HAKEN [1975c,d]), analytically confirm Risken and Nummedal's numerically derived pulse shape and phase velocity. More recently, GERBER and BÜTTIKER [1979] have discovered additional instabilities in this system if the phase difference between the cavity field amplitude and atomic polarization is nonzero. Numerically integrating the semiclassical laser equations, MAYR, RISKEN and VOLLMER [1981] have found that these latter instabilities lead to periodic and chaotic behavior. MANDEL and AGRAWAL [1982] have also found self-pulsing instabilities in a ring laser with counter propagating modes.

BONIFACIO and LUGIATO [1978c] were the first to analyze the stability of the stationary states of the system under review: a ring cavity containing a homogeneously broadened, saturable atomic medium with a unidirectional, longitudinally injected, monochromatic field. By extending Risken and Nummedal's analysis to this system they derived an analogous instability condition in the limit of pure absorption. In particular they found a finite region of instability on the upper steady-state branch in the mean-field limit. Numerical integrations of the Maxwell–Bloch equations (BONIFACIO, GRONCHI and LUGIATO [1979b]) indicate that within this region of parameter space two types of dynamical behavior occur: either the cavity field amplitude decays to the corresponding steady state on the lower branch ("precipitation"), or it is attracted to a rapidly oscillating limit cycle (self-pulsing), as is the case for the ring laser. These results were reviewed and numerically extended away from the mean-field limit by GRONCHI, BENZA, LUGIATO, MEYSTRE and SARGENT [1981]. Several further investigations (BENZA and LUGIATO [1979b, 1982], BENZA, LUGIATO and MEYSTRE [1980], and LUGIATO, BENZA, NARDUCCI and FARINA [1981]) employing generalized mode analyses – in the manner of HAKEN [1975c,d] and HAKEN and OHNO [1976] – have shed greater light on the bifurcations between the stationary and self-pulsing states in the absorptive limit. CASAGRANDE, LUGIATO and ASQUINI [1980] found a similar instability in a coherently driven nonlinear Fabry–Perot cavity, but with a higher threshold than the ring cavity.

The scarcity of stability analyses of the nonlinear ring cavity with nonzero dispersion contrasts sharply against the amount of research in the limit of pure absorption. This is not surprising considering the accompanying complexities in the Maxwell–Bloch equations. As we shall later see, it is not possible in this case to derive, in closed form, a characteristic equation for the eigenvalues of the linearized equations. Thus, most analyses of the dispersive system have incorporated simplifying limits. For example, LUGIATO [1980b] extracted a characteristic equation by using a perturbation expansion in the mean-field limit. Similarly, by applying the limit of infinite transverse broadening (IKEDA [1979]), IKEDA, DAIDO and AKIMOTO [1980] derived a characteristic equation for a ring cavity containing a Kerr medium (the limit of infinite dispersion). CARMICHAEL, SNAPP and SCHIEVE [1982] generalized these results by considering a medium with finite dispersion. In the following, we shall elucidate the necessity of these simplifying limits by analyzing the stability of this system in a generalized framework.

The stationary states $\mathscr{E}_s(z)$, $\mathscr{P}_s(z)$ and $\mathscr{D}_s(z)$ of the Maxwell–Bloch equations for a homogeneously broadened, two-level atomic medium [eqs. (2.6) and (2.7a,b) with $w(r) \equiv 1$] satisfy

$$\frac{\mathrm{d}\mathscr{E}_s(z)}{\mathrm{d}z} = -\frac{\alpha}{2}(1 - \mathrm{i}\delta_A)\frac{\mathscr{E}_s(z)}{1 + |\mathscr{E}_s(z)|^2/\mathscr{E}_s^2}, \tag{3.1a}$$

$$\mathscr{P}_s(z) = \frac{\mathscr{N}}{2}\left(\frac{\gamma_\parallel}{\gamma_\perp}\right)^{1/2} \frac{\mathrm{i} + \delta_A}{(1 + \delta_A^2)^{1/2}} \frac{1}{\mathscr{E}_s} \frac{\mathscr{E}_s(z)}{1 + |\mathscr{E}_s(z)|^2/\mathscr{E}_s^2} \tag{3.1b}$$

and

$$\mathscr{D}_s(z) = -\frac{\mathscr{N}}{1 + |\mathscr{E}_s(z)|^2/\mathscr{E}_s^2}, \tag{3.1c}$$

where $\mathscr{E}_s = [\hbar(\gamma_\perp \gamma_\parallel)^{1/2}/2\mu](1 + \delta_A^2)^{1/2}$ is the nonresonant saturation amplitude. The electric field amplitudes must also satisfy the ring cavity boundary conditions, eqs. (2.9a,b). In particular, eq. (2.9a) implies that the stationary field amplitude satisfies

$$\mathscr{E}_s(0) = \sqrt{T}\,\mathscr{E}_i + R\,\mathscr{E}_s(L)\,e^{-i\phi}, \tag{3.2}$$

where $\phi = \delta_c T$. We now assume that in the vicinity of a stationary state the dynamical variables may be expressed as a superposition of normal modes with exponential time dependences (see for example CHANDRASEKHAR [1961]). In particular, we let

$$\mathscr{E}(z, t) = \mathscr{E}_s(z)\left\{1 + \sum_\lambda \mathscr{A}_{\mathscr{E},\lambda}(z)\,e^{\lambda t}\right\} \tag{3.3a}$$

and

$$\mathscr{E}^*(z, t) = \mathscr{E}_s^*(z)\left\{1 + \sum_\lambda \mathscr{A}_{\mathscr{E}^*,\lambda}(z)\,e^{\lambda t}\right\} \tag{3.3b}$$

with similar equations for $\mathscr{P}(z, t)$, $\mathscr{P}^*(z, t)$ and $\mathscr{D}(z, t)$, where the eigenvalues λ are yet to be determined. Equating $\mathscr{E}(z, t)$ with the conjugate of $\mathscr{E}^*(z, t)$, etc., yields the mode amplitude relationships

$$\mathscr{A}_{\mathscr{E}^*,\lambda}(z) = \mathscr{A}^*_{\mathscr{E},\lambda^*}(z), \tag{3.4a}$$

$$\mathscr{A}_{\mathscr{P}^*,\lambda}(z) = \mathscr{A}^*_{\mathscr{P},\lambda^*}(z), \tag{3.4b}$$

and

$$\mathscr{A}_{\mathscr{D},\lambda}(z) = \mathscr{A}^*_{\mathscr{D},\lambda^*}(z). \tag{3.4c}$$

Implicit in this assumption is that the system is close enough to a stationary state so that the moduli of the above amplitudes are much less than unity. After substituting eqs. (3.3a,b) etc. into the Maxwell–Bloch equations for a homogeneously broadened medium, and neglecting all higher ordered terms, the mode amplitudes are found to satisfy

$$\frac{d}{dz}\begin{pmatrix}\mathscr{A}_{\mathscr{E},\lambda}(z)\\ \mathscr{A}_{\mathscr{E}^*,\lambda}(z)\end{pmatrix} = \begin{pmatrix}\mathscr{M}_{1,\lambda}(z) & \mathscr{M}_{2,\lambda}(z)\\ \mathscr{M}^*_{2,\lambda^*}(z) & \mathscr{M}^*_{1,\lambda^*}(z)\end{pmatrix}\begin{pmatrix}\mathscr{A}_{\mathscr{E},\lambda}(z)\\ \mathscr{A}_{\mathscr{E}^*,\lambda}(z)\end{pmatrix}, \tag{3.5a}$$

$$\begin{pmatrix} \mathscr{A}_{\mathscr{P},\lambda}(z) \\ \mathscr{A}_{\mathscr{P}^*,\lambda}(z) \end{pmatrix} = \begin{pmatrix} \mathscr{M}_{3,\lambda}(z) & \mathscr{M}_{4,\lambda}(z) \\ \mathscr{M}^*_{4,\lambda^*}(z) & \mathscr{M}^*_{3,\lambda^*}(z) \end{pmatrix} \begin{pmatrix} \mathscr{A}_{\mathscr{E},\lambda}(z) \\ \mathscr{A}_{\mathscr{E}^*,\lambda}(z) \end{pmatrix}, \tag{3.5b}$$

and

$$\mathscr{A}_{\mathscr{D},\lambda}(z) = \mathscr{M}_{5,\lambda}(z)\mathscr{A}_{\mathscr{E},\lambda}(z) + \mathscr{M}^*_{5,\lambda^*}(z)\mathscr{A}_{\mathscr{E}^*,\lambda}(z), \tag{3.5c}$$

where

$$\mathscr{M}_{1,\lambda}(z) = -\frac{\lambda}{c} + \frac{\alpha}{2}\frac{1 - \mathrm{i}\delta_{\mathrm{A}}}{1 + |\mathscr{E}_{\mathrm{s}}(z)|^2/\mathscr{E}_{\mathrm{s}}^2}[1 - \mathscr{M}_{3,\lambda}(z)], \tag{3.6a}$$

$$\mathscr{M}_{2,\lambda}(z) = -\frac{\alpha}{2}\frac{1 - \mathrm{i}\delta_{\mathrm{A}}}{1 + |\mathscr{E}_{\mathrm{s}}(z)|^2/\mathscr{E}_{\mathrm{s}}^2}\mathscr{M}_{4,\lambda}(z), \tag{3.6b}$$

$$\mathscr{M}_{3,\lambda}(z) = \frac{\gamma_\perp(1 + \mathrm{i}\delta_{\mathrm{A}})}{\lambda + \gamma_\perp(1 + \mathrm{i}\delta_{\mathrm{A}})}[1 + \mathscr{M}_{5,\lambda}(z)], \tag{3.6c}$$

$$\mathscr{M}_{4,\lambda}(z) = \frac{\gamma_\perp(1 + \mathrm{i}\delta_{\mathrm{A}})}{\lambda + \gamma_\perp(1 + \mathrm{i}\delta_{\mathrm{A}})}\mathscr{M}^*_{5,\lambda^*}(z), \tag{3.6d}$$

and

$$\mathscr{M}_{5,\lambda}(z) = \frac{-\frac{1}{2}\gamma_\parallel(1 + \mathrm{i}\delta_{\mathrm{A}})[\lambda + \gamma_\perp(1 - \mathrm{i}\delta_{\mathrm{A}})](\lambda + 2\gamma_\perp)}{(\lambda + \gamma_\parallel)[(\lambda + \gamma_\perp)^2 + (\gamma_\perp\delta_{\mathrm{A}})^2] + \gamma_\perp\gamma_\parallel(\lambda + \gamma_\perp)(1 + \delta_{\mathrm{A}}^2)|\mathscr{E}_{\mathrm{s}}(z)|^2/\mathscr{E}_{\mathrm{s}}^2}\,\frac{|\mathscr{E}_{\mathrm{s}}(z)|^2}{\mathscr{E}_{\mathrm{s}}^2} \tag{3.6e}$$

The boundary condition, eq. (2.9a), and its conjugate become

$$\mathscr{E}_{\mathrm{s}}(0)\,\mathscr{A}_{\mathscr{E},\lambda}(0) = R\,\mathscr{E}_{\mathrm{s}}(L)\mathscr{A}_{\mathscr{E},\lambda}(L)\exp\left[-\left(\lambda\frac{\mathscr{L} - L}{c} + \mathrm{i}\phi\right)\right] \tag{3.7a}$$

$$\mathscr{E}^*_{\mathrm{s}}(0)\,\mathscr{A}_{\mathscr{E}^*,\lambda}(0) = R\,\mathscr{E}^*_{\mathrm{s}}(L)\mathscr{A}_{\mathscr{E}^*,\lambda}(L)\exp\left[-\left(\lambda\frac{\mathscr{L} - L}{c} - \mathrm{i}\phi\right)\right]. \tag{3.7b}$$

In principle, the stability of a given stationary state, satisfying equations (3.1a) and (3.2), may now be tested with eq. (3.7a,b) and the integrated solution of eq. (3.5a).* After some algebraic manipulations, a characteristic equation

* A similar approach is taken by CARMICHAEL [1983b] after assuming that (3.5a) may be integrated.

might be obtained, from which the admissible values of λ may be derived. If the real parts of all the eigenvalues are negative, then the stationary state is asymptotically stable; if one or more are zero, but all nonpositive, then it is marginally stable; and if one or more are positive, the state is unstable. The feasibility of this method relies upon the explicit integrability of eq. (3.5a). However, because the matrix therein is not diagonalizable, a general stability analysis of these stationary states does not appear to be possible. Characteristic equations, however, can still be obtained in several restricted cases: (i) the empty-cavity limit ($\alpha L \to 0$), (ii) the limit of pure absorption ($\delta_A = \phi = 0$), (iii) the mean-field limit of BONIFACIO and LUGIATO [1978a] and (iv) the limit of fast transverse atomic relaxation ($\gamma_\perp \to \infty$). We shall now examine how eqs. (3.5a) reduce in each of these cases.

In the empty-cavity limit, αL tends to zero, so that $\mathscr{M}_{1,\lambda}(z) = -\lambda/c$ and $\mathscr{M}_{2,\lambda}(z) = 0$. Then, eq. (3.5a) may be integrated and a characteristic equation is obtained with the roots

$$\lambda_n^{(\pm)} = \frac{c}{\mathscr{L}} \ln R + \mathrm{i}(2\pi n \pm \phi), \qquad n = 0, \pm 1, \pm 2, \pm 3, \ldots . \tag{3.8}$$

Since all $\lambda_n^{(\pm)}$ have negative real parts, the empty-cavity modes are asymptotically stable.

Less trivial results are realized in the limit of pure absorption. After letting $\delta_A = 0$, the matrix in eq. (3.5a) is symmetric. Equation (3.5a) then has the solution

$$\begin{pmatrix} \mathscr{A}_{\mathscr{E},\lambda}(z) \\ \mathscr{A}_{\mathscr{E}^*,\lambda}(z) \end{pmatrix} = \exp\left[\int_0^z \mathscr{M}_{1,\lambda}(z')\,\mathrm{d}z'\right]$$

$$\times \begin{pmatrix} \cosh \int_0^z \mathscr{M}_{2,\lambda}(z')\,\mathrm{d}z' & \sinh \int_0^z \mathscr{M}_{2,\lambda}(z')\,\mathrm{d}z' \\ \sinh \int_0^z \mathscr{M}_{2,\lambda}(z')\,\mathrm{d}z' & \cosh \int_0^z \mathscr{M}_{2,\lambda}(z')\,\mathrm{d}z' \end{pmatrix} \begin{pmatrix} \mathscr{A}_{\mathscr{E},\lambda}(0) \\ \mathscr{A}_{\mathscr{E}^*,\lambda}(0) \end{pmatrix} \tag{3.9}$$

which we evaluate at $z = L$. The two resulting equations with equations (3.7a,b) comprise a system of four homogeneous linear equations. The existence of nontrivial solutions requires that the determinant of the coefficient matrix vanish. Hence, one obtains a characteristic equation with roots specified by

$$\lambda_n^{(+)} = \frac{c}{\mathscr{L}} \ln R - 2\pi \mathrm{i} n \frac{c}{\mathscr{L}} - \frac{\alpha}{2} \frac{c}{\mathscr{L}} \int_0^L \Gamma(\lambda_n^{(+)}, |\mathscr{E}_s(z)|^2)\,\mathrm{d}z, \tag{3.10a}$$

$$\lambda_n^{(-)} = \frac{c}{\mathscr{L}} \ln R - 2\pi i n \frac{c}{\mathscr{L}} - \frac{\alpha}{\mathscr{L}} \frac{\gamma_\perp}{\lambda_n^{(-)} + \gamma_\perp} \int_0^L \frac{dz}{1 + |\mathscr{E}_s(z)|^2/\mathscr{E}_s^2}, \tag{3.10b}$$

where

$$\Gamma(\lambda, |\mathscr{E}_s(z)|^2) = \frac{\gamma_\perp}{1 + |\mathscr{E}_s(z)|^2/\mathscr{E}_s^2} \frac{\gamma_\parallel [1 - |\mathscr{E}_s(z)|^2/\mathscr{E}_s^2] + \lambda}{(\lambda + \gamma_\perp)(\lambda + \gamma_\parallel) + \gamma_\perp \gamma_\parallel |\mathscr{E}_s(z)|^2/\mathscr{E}_s^2}. \tag{3.11}$$

Equation (3.10a) agrees with the calculations of BONIFACIO and LUGIATO [1978c] and is cast in the form of equation (24) of GRONCHI, BENZA, LUGIATO, MEYSTRE and SARGENT [1981]*. To our knowledge, the existence of eigenvalues satisfying equation (3.10b) has not been discussed. The integrals in eqs. (3.10a,b) may be trivially evaluated, from which we obtain,

$$\lambda_n^{(+)} = \frac{c}{\mathscr{L}} \ln R - 2\pi i n \frac{c}{\mathscr{L}} - \frac{\gamma_\perp}{\lambda_n^{(+)} + \gamma_\perp} \frac{c}{\mathscr{L}} \ln \frac{\mathscr{E}_s(0)}{\mathscr{E}_s(L)}$$
$$+ \frac{1}{2} \frac{c}{\mathscr{L}} \left(1 + \frac{\gamma_\perp}{\lambda_n^{(+)} + \gamma_\perp}\right) \ln \left(\frac{(\lambda_n^{(+)} + \gamma_\perp)(\lambda_n^{(+)} + \gamma_\parallel) + \gamma_\perp \gamma_\parallel |\mathscr{E}_s(0)|^2/\mathscr{E}_s^2}{(\lambda_n^{(+)} + \gamma_\perp)(\lambda_n^{(+)} + \gamma_\parallel) + \gamma_\perp \gamma_\parallel |\mathscr{E}_s(L)|^2/\mathscr{E}_s^2} \right), \tag{3.12a}$$

$$\lambda_n^{(-)} = \frac{c}{\mathscr{L}} \ln R - 2\pi i n \frac{c}{\mathscr{L}} - \frac{\gamma_\perp}{\lambda_n^{(-)} + \gamma_\perp} \frac{c}{\mathscr{L}} \ln \frac{\mathscr{E}_s(0)}{\mathscr{E}_s(L)}. \tag{3.12b}$$

Equation (3.12b) is a quadratic equation in $\lambda_n^{(-)}$; it can be shown, by an extension of the Hurwitz criterion (see MARDEN [1949]), that its roots always lie in the left half of the complex plane. Therefore, the presence of an instability must occur through the transcendental equation (3.12a). By numerical iteration of eq. (3.10a) GRONCHI, BENZA, LUGIATO, MEYSTRE and SARGENT [1981] have shown that equation (3.12a) can possess roots with positive real parts. BONIFACIO and LUGIATO [1978c] have examined eq. (3.10a) in the mean-field limit (cf. eqs. (2.16) and (2.20)). Under these circumstances the slowly varying cavity field amplitude, $\mathscr{E}_s(z)$, is independent of z. Letting y and x denote the

* There is a typographical error in formula (21) of the cited work. It should read like eq. (3.11).

normalized injected and transmitted field amplitudes (cf. eqs. (2.11)), eqs. (3.10a,b) become*

$$\lambda_n^3 + c_{2,n}\lambda_n^2 + c_{1,n}\lambda_n + c_{0,n} = 0, \tag{3.13a}$$

$$\lambda_n^2 + d_{1,n}\lambda_n + d_{0,n} = 0, \tag{3.13b}$$

where

$$c_{2,n} = \kappa + \gamma_\perp + \gamma_\parallel + \mathrm{i}\alpha_n, \tag{3.14a}$$

$$c_{1,n} = \gamma_\parallel(\gamma_\perp + \kappa) + \gamma_\perp\gamma_\parallel x^2 + \kappa\gamma_\perp\frac{y}{x} + (\gamma_\perp + \gamma_\parallel)\mathrm{i}\alpha_n, \tag{3.14b}$$

$$c_{0,n} = \kappa\gamma_\perp\gamma_\parallel\left[2x^2 + (1 - x^2)\frac{y}{x}\right] + \gamma_\perp\gamma_\parallel(1 + x^2)\mathrm{i}\alpha_n, \tag{3.14c}$$

$$d_{1,n} = \kappa + \gamma_\perp + \mathrm{i}\alpha_n, \tag{3.14d}$$

and

$$d_{0,n} = \gamma_\perp\left(\kappa\frac{y}{x} + \mathrm{i}\alpha_n\right). \tag{3.14e}$$

The parameter κ is the cavity decay constant, and $\alpha_n = 2\pi nc/\mathscr{L}$ is the nth cavity mode frequency. We have also made use of the absorptive mean-field limit steady-state relationship, eq. (2.21). Note that in the limit $y \to 0$ eqs. (3.13a,b) reduce to the characteristic equations derived by RISKEN and NUMMEDAL [1968a,b] for the ring laser; their laser pump parameter equals the cavity field intensity (above threshold). Although this association is unphysical (i.e., if $y = 0$, the stationary state $x = 0$ is globally stable below threshold), it exemplifies the equivalence of the mean-field and single-mode approximations and suggests that the self-pulsing instabilities, found in each system, stem from a more general instability.

The cubic eq. (3.13a) is in principle exactly soluble; however, because of the algebraic complexity of its roots, it is useful to approximate their values by a perturbation analysis. After expanding λ_n in terms of successive powers of T, i.e. $\lambda_n = \lambda_n^{(0)} + T\lambda_n^{(1)} + T^2\lambda_n^{(2)} + \ldots$, and assuming the good-cavity limit, $\kappa = O(T) \ll \gamma_\perp, \gamma_\parallel$, GRONCHI, BENZA, LUGIATO, MEYSTRE and SARGENT [1981] obtain

$$\lambda_{n,1} = -\mathrm{i}\alpha_n - \kappa\left[1 + \frac{2C\gamma_\perp}{1 + x^2}\,\frac{\gamma_\parallel(1 - x^2) - \mathrm{i}\alpha_n}{(\gamma_\perp - \mathrm{i}\alpha_n)(\gamma_\parallel - \mathrm{i}\alpha_n) + \gamma_\perp\gamma_\parallel x^2}\right] + O(T^2) \tag{3.15}$$

* We assume that $(\lambda_n + \gamma_\perp)(\lambda_n + \gamma_\parallel) + \gamma_\perp\gamma_\parallel x^2 \neq 0$.

and

$$\lambda_{n,\frac{2}{3}} = -\tfrac{1}{2}\{\gamma_\perp + \gamma_\| \pm [(\gamma_\perp - \gamma_\|)^2 - 4\gamma_\perp\gamma_\| x^2]^{1/2}\} + O(T). \tag{3.16}$$

CARMICHAEL [1983a,b] notes that eq. (3.16) is not valid in the limit $\kappa \approx \gamma_\perp \approx \gamma_\|$. However, in the good-cavity limit, for sufficiently small T, it is clear that the real parts of $\lambda_{n,2}$ and $\lambda_{n,3}$ are negative. Hence, the condition for an instability to occur is that $\lambda_{n,1}$ has a positive real part, or

$$1 + \frac{2C\gamma_\perp}{1+x^2}\,\frac{\gamma_\|(1-x^2)[\gamma_\perp\gamma_\|(1+x^2) - \alpha_n^2] + (\gamma_\perp + \gamma_\|)\alpha_n^2}{[\gamma_\perp\gamma_\|(1+x^2) - \alpha_n^2]^2 + (\gamma_\perp + \gamma_\|)^2\alpha_n^2} < 0. \tag{3.17}$$

For the on-resonance cavity mode ($n = 0$) the above reduces to

$$1 + 2C\frac{1-x^2}{(1+x^2)^2} = \frac{\mathrm{d}y}{\mathrm{d}x} < 0. \tag{3.18}$$

Hence, all regions of negative slope in the steady-state diagram are unstable.

Either by a direct analysis of eq. (3.17), or by applying Risken and Nummedal's analysis to eq. (3.13a) one can show that the stationary state is unstable if

$$R \geqslant 0 \tag{3.19}$$

and

$$S \pm R^{1/2} \geqslant 0, \tag{3.20}$$

where R (not to be confused with the reflectivity of the first and second cavity mirrors) and S are given by

$$R = \gamma_\perp^2\gamma_\|^2 x^4\left(1 - \frac{y}{x}\right)^2 - 2\gamma_\perp\gamma_\| x^2\left[3\gamma_\|^2 + 4\gamma_\perp\gamma_\| + \frac{y}{x}(3\gamma_\perp^2 - \gamma_\|^2) - \gamma_\perp^2\left(\frac{y}{x}\right)^2\right] + \left(\gamma_\|^2 - \gamma_\perp^2\frac{y}{x}\right)^2 \tag{3.21}$$

and

$$S = \gamma_\|(3\gamma_\perp x^2 - \gamma_\|) - \frac{y}{x}\gamma_\perp(\gamma_\perp + \gamma_\| x^2), \tag{3.22}$$

provided that there exists a cavity mode such that

$$\lambda_{\mathrm{i,min}} < |\lambda_{\mathrm{i},n}| < \lambda_{\mathrm{i,max}}, \tag{3.23}$$

where

$$\lambda_{\mathrm{i},\substack{\max\\\min}} = \frac{1}{\sqrt{2}}(S \pm R^{1/2})^{1/2} \tag{3.24}$$

and $\lambda_{\mathrm{i},n}$ denotes the imaginary part of $\lambda_{1,n}$ (BONIFACIO, GRONCHI and LUGIATO [1979b]). Because of conditions (3.23) only a finite number of cavity modes are involved in this instability. For sufficiently small T, $\lambda_{\mathrm{i},n} \simeq -\alpha_n$, hence, (3.23) may be satisfied by choosing the roundtrip cavity length $\mathscr{L}$ to be sufficiently large. Since $(1 + x^2)R$ may be expressed as a cubic in x^2 with a negative leading coefficient, condition (3.19) imposes an upper bound on x^2. In particular, if $\gamma_\perp = \gamma_\parallel = \gamma$, then $R = 4\gamma^4(C^2 - 4x^2)$, hence, (3.19) reduces to $x \leqslant C/2$. The other inequality (3.20) places a lower bound on x^2, excluding this instability from the lower steady-state branch.

The bounds on $\lambda_{\mathrm{i},n}$ in (3.23) become

$$\lambda_{\mathrm{i},\substack{\max\\\min}} = \gamma[x^2 - C - 1 \pm (C^2 - 4x^2)^{1/2}]^{1/2} < \omega_{\mathrm{R}}, \tag{3.25}$$

where $\omega_{\mathrm{R}} = \gamma x$ is the Rabi frequency.

By numerically integrating the Maxwell–Bloch equations, BONIFACIO, GRONCHI and LUGIATO [1979b] and GRONCHI, BENZA, LUGIATO, MEYSTRE and SARGENT [1981] confirmed that this instability occurs only on the upper steady-state branch in the "bistable" region and that within these bounds two kinds of behavior are prevalent: either the electric field precipitates to the corresponding steady state on the lower branch, or it undergoes self-pulsing – much like the solutions of RISKEN and NUMMEDAL [1968a,b] for the ring laser.

LUGIATO [1980b] has extended this mean-field analysis by considering the case of nonzero dispersion. Neglecting all terms with derivatives in z (i.e. those on the left-hand side of eq. (3.5a)), he finds that if C is chosen large enough, then the domain of the positive-slope instability broadens beyond the bistable region. However, in the limit of pure dispersion ($\Delta \gg 1$) this domain "practically vanishes".

We now return to the generalized stability equations, (3.5) through (3.7), and consider the limit of fast transverse atomic relaxation, $\gamma_\perp \gg \gamma_\parallel, \kappa$. Following IKEDA [1979], we adiabatically eliminate the polarization; thus the Maxwell–Bloch equations become

$$\mathscr{E}(z, t) = \mathscr{E}\left(0, t - \frac{z}{c}\right) \exp[-\tfrac{1}{2}\alpha z(1 - \mathrm{i}\delta_{\mathrm{A}})\Phi(z, t)] \tag{3.26a}$$

and

$$\frac{1}{\gamma_{\|}}\frac{\partial}{\partial t}\Phi(z,t) = 1 - \Phi(z,t) - \frac{\left|\mathscr{E}\left(0, t-\frac{z}{c}\right)\right|^2}{\mathscr{E}_{\mathrm{s}}^2}\,\frac{1}{\alpha z}\left[1 - \exp(-\alpha z\Phi(z,t))\right], \tag{3.26b}$$

where we have introduced

$$-\mathscr{N}\Phi(z,t) = \frac{1}{z}\int_0^z \mathrm{d}z'\,\mathscr{D}\left(z', t - \frac{z-z'}{c}\right) \tag{3.27}$$

as the average atomic inversion over the path of a wave front in the retarded frame. Applying (3.26a), the boundary condition (2.9a) may be written

$$\mathscr{E}(0,t) = \sqrt{T}\,\mathscr{E}_{\mathrm{i}} + R\,\mathscr{E}\left(0, t - \frac{\mathscr{L}}{c}\right)\exp\left[-\tfrac{1}{2}\alpha L(1-\mathrm{i}\delta_{\mathrm{A}})\,\Phi\left(L, t - \frac{\mathscr{L}-L}{c}\right)\right]. \tag{3.28}$$

The matrix elements in eq. (3.5a) reduce to

$$\mathscr{M}_{1,\lambda}(z) = -\lambda/c + \Gamma_\lambda(z) \tag{3.29a}$$

and

$$\mathscr{M}_{2,\lambda} = \Gamma_\lambda(z), \tag{3.29b}$$

where

$$\Gamma_\lambda(z) = \frac{\alpha}{2}\,\frac{1-\mathrm{i}\delta_{\mathrm{A}}}{1+|\mathscr{E}_{\mathrm{s}}(z)|^2/\mathscr{E}_{\mathrm{s}}^2}\left[1 + \frac{\lambda}{\gamma_{\|}} + \frac{|\mathscr{E}_{\mathrm{s}}(z)|^2}{\mathscr{E}_{\mathrm{s}}^2}\right]^{-1}\frac{|\mathscr{E}_{\mathrm{s}}(z)|^2}{\mathscr{E}_{\mathrm{s}}^2}. \tag{3.30}$$

Eq. (3.5a) may now be integrated, and we obtain

$$\begin{pmatrix}\mathscr{A}_{\mathscr{E},\lambda}(z)\\ \mathscr{A}_{\mathscr{E}^*,\lambda}(z)\end{pmatrix} = \mathrm{e}^{-\frac{\lambda}{c}z}\begin{pmatrix}1+\Lambda_\lambda(z) & \Lambda_\lambda(z)\\ \Lambda^*_{\lambda^*}(z) & 1+\Lambda^*_{\lambda^*}(z)\end{pmatrix}\begin{pmatrix}\mathscr{A}_{\mathscr{E},\lambda}(0)\\ \mathscr{A}_{\mathscr{E}^*,\lambda}(0)\end{pmatrix}, \tag{3.31}$$

where

$$\Lambda_\lambda(z) = \tfrac{1}{2}(1-\mathrm{i}\delta_{\mathrm{A}})\left[1 + \frac{\lambda}{\gamma_{\|}} + \frac{|\mathscr{E}_{\mathrm{s}}(z)|^2}{\mathscr{E}_{\mathrm{s}}^2}\right]^{-1}\frac{|\mathscr{E}_{\mathrm{s}}(0)|^2 - |\mathscr{E}_{\mathrm{s}}(z)|^2}{\mathscr{E}_{\mathrm{s}}^2}. \tag{3.32}$$

Applying the boundary conditions for the mode amplitudes, eqs. (3.7a,b), we obtain the characteristic equation reported by CARMICHAEL, SNAPP and SCHIEVE [1982]:

$$1 - 2B\mathscr{S}\mathrm{e}^{-\lambda\frac{\mathscr{L}}{c}} + B^2\mathrm{e}^{-2\lambda\frac{\mathscr{L}}{c}} = 0, \tag{3.33}$$

where

$$B = R\,\mathrm{e}^{-\alpha L \Phi_s(L)} \frac{\left[1 + \frac{\lambda}{\gamma_\parallel} + \frac{|\mathscr{E}_s(0)|^2}{\mathscr{E}_s^2}\right]^{1/2}}{\left[1 + \frac{\lambda}{\gamma_\parallel} + \frac{|\mathscr{E}_s(0)|^2}{\mathscr{E}_s^2}\mathrm{e}^{-\alpha L \Phi_s(L)}\right]^{1/2}}, \tag{3.34a}$$

$$\begin{aligned}\mathscr{S} = &\left\{\left(1 + \frac{\lambda}{\gamma_\parallel} + \frac{1}{2}\frac{|\mathscr{E}_s(0)|^2}{\mathscr{E}_s^2}[1 + \mathrm{e}^{-\alpha L \Phi_s(L)}]\right)\cos\left[\frac{\alpha L \delta_A}{2}\Phi_s(L) - \phi\right]\right.\\ &\left. + \frac{\delta_A}{2}\frac{|\mathscr{E}_s(0)|^2}{\mathscr{E}_s^2}[1 - \mathrm{e}^{-\alpha L \Phi_s(L)}]\sin\left[\frac{\alpha L \delta_A}{2}\Phi_s(L) - \phi\right]\right\}\\ &\times\left[1 + \frac{\lambda}{\gamma_\parallel} + \frac{|\mathscr{E}_s(0)|^2}{\mathscr{E}_s^2}\right]^{-1/2}\left[1 + \frac{\lambda}{\gamma_\parallel} + \frac{|\mathscr{E}_s(0)|^2}{\mathscr{E}_s^2}\mathrm{e}^{-\alpha L \Phi_s(L)}\right]^{-1/2},\end{aligned} \tag{3.34b}$$

and

$$-\mathscr{N}\Phi_s(L) = \frac{1}{L}\int_0^L \mathscr{D}_s(z)\,\mathrm{d}z \tag{3.35}$$

is the steady state of expression (3.27) evaluated at $z = L$.

It is easy to confirm that eq. (3.33) conforms with the previous results. For example, in the empty-cavity limit the roots of the above satisfy eq. (3.8). Similarly, in the limit of pure absorption ($\delta_A, \phi \to 0$), its roots satisfy eq. (3.12a,b) if the limit $\gamma_\perp \to \infty$ is taken in each. However, if the mean-field limit is now taken, the self-pulsing instability no longer occurs. This is also apparent from condition (3.17) if the limit $\gamma_\perp \to \infty$ is taken with $\mathrm{d}y/\mathrm{d}x$ positive. Thus, it appears that the absorptive self-pulsing instability is stabilized by large pressure broadening (cf. LUGIATO, ASQUINI and NARDUCCI [1982]).

As in the absorptive limit, an explicit determination of the eigenvalues in eq. (3.33) is impossible, hence, any positive-slope instability must be found by either numerical methods after additional limits are imposed. One can show, however, that regions of negative slope on steady-state diagrams of $|\mathscr{E}_t|$ versus

$|\mathscr{E}_i|$ are unstable, as one might expect from degree theory considerations (cf. SATTINGER [1973]). From the steady states of eqs. (3.26) and (3.28), the characteristic eq. (3.33) with $\lambda = 0$ may be rewritten as

$$\left[1 + \frac{|\mathscr{E}_s(0)|^2}{\mathscr{E}_s^2}\right] e^{-\alpha L \Phi_s(L)} \frac{|\mathscr{E}_i|}{|\mathscr{E}_t|} \frac{d|\mathscr{E}_i|}{d|\mathscr{E}_t|}$$

$$= \left[1 + \frac{|\mathscr{E}_s(0)|^2}{\mathscr{E}_s^2} e^{-\alpha L \Phi_s(L)}\right] \frac{|\mathscr{E}_i|}{|\mathscr{E}_s(0)|} \frac{d|\mathscr{E}_i|}{d|\mathscr{E}_s(0)|} = 0. \qquad (3.36)$$

Hence, the turning points, characterized by $d|\mathscr{E}_i|/d|\mathscr{E}_s(0)| = 0$, are marginally stable. A perturbation analysis near such turning point yields the desired result (CARMICHAEL, SNAPP and SCHIEVE [1982]).

Rather than analyze the stability of the positive-sloped branches numerically, we impose the simplifying limit of an instantaneous medium response time, $\gamma_{\parallel} \mathscr{L}/c \to \infty$. Then, eqs. (3.26b) and (3.28) are transformed into the IKEDA [1979] map:

$$\mathscr{E}(0, t + (n+1)\tau) =$$

$$= \mathscr{E}_i \sqrt{T} + R\mathscr{E}(0, t + n\tau) \exp\left[-\frac{\alpha L}{2}(1 - i\delta_A)\Phi_n\right] \exp(-i\phi), \qquad (3.37a)$$

$$\Phi_n - 1 = \frac{1}{\alpha L}[1 - e^{-\alpha L \Phi_n}] \frac{|\mathscr{E}(0, t + n\tau)|^2}{\mathscr{E}_s^2}, \qquad (3.37b)$$

where $0 \leqslant t < \tau = \mathscr{L}/c$. Although this limit is singular and unphysical, the map possesses many of the qualitative features exhibited by the delay-differential system, eqs. (3.26) and (3.28). For example, IKEDA [1979] has found that the iterates of (3.37) undergo a complicated bifurcation sequence as $\mathscr{E}_i$ is slightly varied (cf. § 6).

Applying the limit $\gamma_{\parallel} \mathscr{L}/c \to \infty$ to eq. (3.33), we find

$$\lambda_n^{(\pm)} = -\frac{c}{\mathscr{L}} \ln \frac{1}{B} [\mathscr{S} + (\mathscr{S}^2 - 1)^{1/2}] \pm 2\pi i n \frac{c}{\mathscr{L}}, \qquad n = 0, 1, 2, \ldots, \qquad (3.38)$$

where now

$$B = R\, e^{-\alpha L \Phi_s(L)/2} \frac{[1 + |\mathscr{E}_s(0)|^2]^{1/2}}{[1 + e^{-\alpha L \Phi_s(L)} |\mathscr{E}_s(0)|^2/\mathscr{E}_s^2]^{1/2}} \qquad (3.39a)$$

and

$$\mathscr{S} = \left\{ \left(1 + \frac{1}{2} \frac{|\mathscr{E}_s(0)|^2}{\mathscr{E}_s^2} [1 + e^{-\alpha L \Phi_s(L)}] \right) \cos\left[\frac{\alpha L \delta_A}{2} \Phi_s(L) - \phi \right] + \frac{\delta_A}{2} \frac{|\mathscr{E}_s(0)|^2}{\mathscr{E}_s^2} [1 - e^{-\alpha L \Phi_s(L)}] \sin\left[\frac{\alpha L \delta_A}{2} \Phi_s(L) - \phi \right] \right\} \times \left[1 + \frac{|\mathscr{E}_s(0)|^2}{\mathscr{E}_s^2} \right]^{-1/2} \left[1 + \frac{|\mathscr{E}_s(0)|^2}{\mathscr{E}_s^2} e^{-\alpha L \Phi_s(L)} \right]^{-1/2}. \tag{3.39b}$$

Further analysis reveals that steady states are *stable* if

$$2B|\mathscr{S}| < 1 + B^2. \tag{3.40}$$

This may also be obtained by requiring that the eigenvalues of the Jacobian matrix of the Ikeda map have moduli less than unity. Using eq. (3.36), the steady states of the map are *unstable* if either

$$\mathscr{S} > \frac{1 + B^2}{2B} \quad \Leftrightarrow \quad \frac{d|\mathscr{E}_i|}{d|\mathscr{E}_s(0)|} < 0 \tag{3.41a}$$

or

$$\mathscr{S} < -\frac{1 + B^2}{2B} \quad \Leftrightarrow \quad \frac{|\mathscr{E}_i|}{|\mathscr{E}_s(0)|} \frac{d|\mathscr{E}_i|}{d|\mathscr{E}_s(0)|} > 2(1 + B^2). \tag{3.41b}$$

Condition (3.41a) is a restatement of instability in regions of negative slope, while (3.41b) is evidence of the positive-slope instability identified by IKEDA [1979].

Note that, unlike the self-pulsing instability of Bonifacio and Lugiato, this one may occur on the lowest branch in the steady-state diagram. Also, adjacent to each turning point is a neighborhood of positive slope in which the steady state is stable. Further, since all the eigenvalues $\lambda_n^{(\pm)}$ share the same real part, the onset of instability may involve every cavity mode. Finally, the Ikeda instability is primarily dispersive. If $\delta_A = 0$, then

$$\mathscr{S} = \frac{1 + \frac{1}{2}(1 + e^{-\alpha L \Phi_s(L)})|\mathscr{E}_s(0)|^2/\mathscr{E}_s^2}{[(1 + |\mathscr{E}_s(0)|^2/\mathscr{E}_s^2)(1 + e^{-\alpha L \Phi_s(L)}|\mathscr{E}_s(0)|^2/\mathscr{E}_s^2)]^{1/2}} \cos\phi. \tag{3.42}$$

Thus, for example, if $|\phi| \leqslant \pi/2$, the instability condition (3.14b) is never satisfied. Although CARMICHAEL [1981b] has found parameter values in the absorptive limit for which (3.41b) is met, the steady state only loses its stability to a solution with period 2τ. No higher-order bifurcations are observed.

IKEDA, DAIDO and AKIMOTO [1980] have analyzed the stability of a similar system: a unidirectional ring cavity containing a Kerr medium. Note that the collection of two-level atoms considered in § 2 behaves like a Kerr liquid if the limit of large pressure broadening is taken with the dispersive limit:

$$\delta_A \gg 1 \qquad \text{and} \qquad |\mathscr{E}(z,t)|^2/\mathscr{E}_s^2 \ll 1,$$

such that

$$\frac{\alpha L \delta_A}{2} \frac{|\mathscr{E}(z,t)|^2}{\mathscr{E}_s^2} \approx 2\pi \qquad \text{and} \qquad \frac{\alpha L}{2} \frac{|\mathscr{E}(z,t)|^2}{\mathscr{E}_s^2} \ll 1.$$

That is, the medium's inversion remains well below saturation and the nonlinear phase shift is of much greater significance than the nonlinear absorption. Under this double limit, the Maxwell–Bloch and boundary equations, i.e. (3.26) and (3.28), become

$$\hat{\mathscr{E}}(z,t) = \hat{\mathscr{E}}(0, t - z/c) \exp[-\tfrac{1}{2}\alpha z(1 - i\delta_A)] \exp[i\psi(z,t)], \tag{3.43a}$$

$$\frac{1}{\gamma_\parallel} \frac{\partial}{\partial t} \psi(z,t) = -\psi(z,t) + \frac{z}{L} \operatorname{sgn}(n_2) \left| \hat{\mathscr{E}}\left(0, t - \frac{z}{c}\right) \right|^2, \tag{3.43b}$$

and

$$\hat{\mathscr{E}}(0,t) = \hat{\mathscr{E}}_i + B\hat{\mathscr{E}}\left(0, t - \frac{\mathscr{L}}{c}\right) \exp\left\{i\left[\psi\left(L, t - \frac{\mathscr{L} - L}{c}\right) - \psi_0\right]\right\}, \tag{3.43c}$$

where we have introduced

$$\psi(z,t) = \frac{\alpha L \delta_A}{2} [\Phi(z,t) - 1].$$

Here, $\psi_0 = \phi - \alpha L \delta_A/2$ is the phase shift (including linear dispersion) per round trip, and $n_2 = -(\alpha\delta_A/2k)(1 + \delta_A^2)^{-1} \mathscr{E}_s^{-2}$ is the nonlinear part of the refractive index. For convenience, we have adopted the dimensionless field units of IKEDA, DAIDO and AKIMOTO [1980], i.e.

$$\hat{\mathscr{E}}_{\rm i} = [\tfrac{1}{2}T\alpha L|\delta_{\rm A}|(\alpha L)^{-1}(1-{\rm e}^{-\alpha L})]^{1/2}\mathscr{E}_{\rm i}$$
$$= [TkL|n_2|(\alpha L)^{-1}(1-{\rm e}^{-\alpha L})]^{1/2}E_{\rm i} \qquad (3.44a)$$

and

$$\hat{\mathscr{E}} = [\tfrac{1}{2}\alpha L|\delta_{\rm A}|(\alpha L)^{-1}(1-{\rm e}^{-\alpha L})]^{1/2}\mathscr{E}$$
$$= [kL|n_2|(\alpha L)^{-1}(1-{\rm e}^{-\alpha L})]^{1/2}E. \qquad (3.44b)$$

The parameters found in the characteristic equation (3.34) simplify accordingly:

$$B = R\,{\rm e}^{-\alpha L/2} \qquad (3.45a)$$

and

$$\mathscr{S} = \cos(|\hat{\mathscr{E}}_{\rm s}(0)|^2 - \psi_0) - \frac{|\hat{\mathscr{E}}_{\rm s}(0)|^2}{1+\lambda/\gamma_{\parallel}}\sin(|\hat{\mathscr{E}}_{\rm s}(0)|^2 - \psi_0). \qquad (3.45b)$$

In the limit of instantaneous medium response, $\gamma_{\parallel}\mathscr{L}/c \to \infty$, the matter and boundary equations (3.43b,c) can be incorporated into the complex valued map

$$\hat{\mathscr{E}}(0, t+(n+1)\tau) = \hat{\mathscr{E}}_{\rm i} + B\hat{\mathscr{E}}(0, t+n\tau)\exp\{{\rm i}[{\rm sgn}(n_2)|\hat{\mathscr{E}}(0, t+n\tau)|^2 - \psi_0]\}. \qquad (3.46)$$

Unlike Ikeda's former map, this one has the particular advantage of not depending on a transcendental quantity. Given $\hat{\mathscr{E}}_{\rm i} = |\hat{\mathscr{E}}_{\rm i}|{\rm e}^{{\rm i}\phi_{\rm i}}$, the fixed points of (3.46), $\hat{\mathscr{E}}_{\rm ss} = |\hat{\mathscr{E}}_{\rm ss}|{\rm e}^{{\rm i}\phi_{\rm ss}}$, are determined by

$$|\hat{\mathscr{E}}_{\rm i}| = |\hat{\mathscr{E}}_{\rm ss}|[1 + B^2 - 2B\cos(|\hat{\mathscr{E}}_{\rm ss}|^2 - \psi_0)]^{1/2} \qquad (3.47a)$$

and

$$\phi_{\rm ss} - \phi_{\rm i} = {\rm sgn}(n_2)\arctan\frac{B\sin(|\hat{\mathscr{E}}_{\rm ss}|^2 - \psi_0)}{1 - B\cos(|\hat{\mathscr{E}}_{\rm ss}|^2 - \psi_0)}. \qquad (3.47b)$$

As before, in the limit $\gamma_{\parallel}\mathscr{L}/c \to \infty$, $\mathscr{S}$ loses its "dependence" on λ, further simplifying the stability criterion (3.40).

Although a characteristic equation of the fixed points of the map (3.46) does appear in the paper by IKEDA, DAIDO and AKIMOTO [1980], a more complete stability analysis may be found in CARMICHAEL, SNAPP and SCHIEVE [1982]. Note that the steady states are related to three control parameters B, ψ_0 and $|\hat{\mathscr{E}}_{\rm i}|$. Because of the existence of multiple steady states corresponding to the same set of control parameters, it is impossible to sketch a complete steady

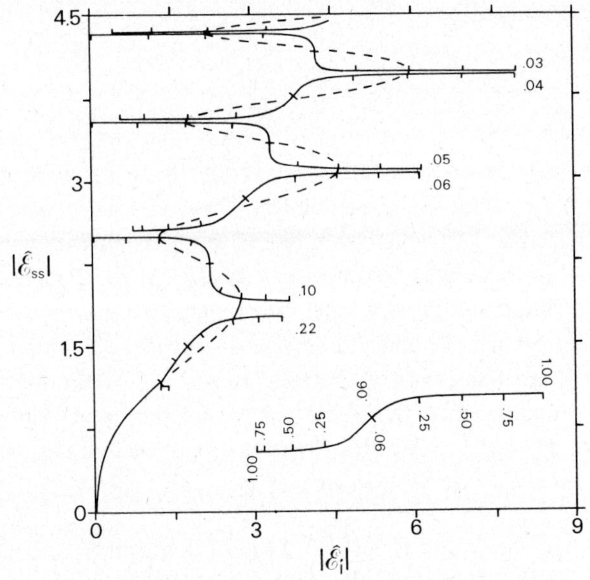

Fig. 7. Stability diagram for the stationary states of equations (3.46) with $B = 0.5$ and $\psi_0 = 0.0$. Solid curves represent stable fixed points, broken curves correspond to unstable fixed points. To generalize this diagram for other values of B, curvilinear axes are provided. See the text for a description.

state diagram. We shall thus limit our illustration of the stable and unstable regions by considering only one or two parameters at a time.

In Fig. 7 we plot $|\hat{\mathscr{E}}_{ss}|$ versus $|\hat{\mathscr{E}}_i|$ for $B = 0.5$ and $\psi_0 = 0$. The curvilinear axes are included so that the reader may visualize the corresponding curves for different values of B. Each of these axes has nine divisions: the value of B corresponding to the central division of each axis is indicated by the number to its right, the remaining division labels are identical to those in the inset.

For $\psi_0 = 0$ and any choice of $B = \bar{B}$, the curve $|\hat{\mathscr{E}}_{ss}|$ versus $|\hat{\mathscr{E}}_i|$ winds around, crossing each axis at $\bar{B}$. Numbering from the bottom, the even numbered axes contain the turning points for each branch $(\mathrm{d}|\hat{\mathscr{E}}_i|/\mathrm{d}|\hat{\mathscr{E}}_s(0)| = 0)$; between these points the steady states are unstable according to condition (3.41a). The regions between crossings of the remaining axes contain the unstable steady states described by (3.41b). All other points represent stable fixed points. If $\bar{B}$ is less than the value of a central division, that axis is ignored.

Next, we consider the domain of stability subject to variations in ψ_0 as well as in B. The boundaries, or marginal points, between the stable and unstable regions are determined by

$$\mathscr{S} = \mp \frac{1 + B^2}{2B}. \tag{3.48}$$

(The upper sign characterizes the onset of the Ikeda instability; the lower corresponds to the negative-slope instability.) Now viewing $\mathscr{S}$ as a function of $|\hat{\mathscr{E}}_{ss}|$ parametrized by ψ_0, we find that one of these instabilities will exist if $\mathscr{S}$ exceeds $(1 + B^2)/2B$ or falls below $-(1 + B^2)2B$. At the very least $\mathscr{S}$ must achieve one of these values as a local extremum. Therefore we describe the onset of an instability by requiring that (3.48) possess a double root, thus simultaneously satisfying the equation $\mathrm{d}\mathscr{S}/\mathrm{d}|\hat{\mathscr{E}}_{ss}| = 0$. Denoting the coordinates of these double roots $(B, \psi_0^{(+)}, |\hat{\mathscr{E}}_{ss}^{(+)}|)$ for the upper sign $(-)$ in eq. (3.48), and $(B, \psi_0^{(-)}, |\hat{\mathscr{E}}_{ss}^{(-)}|)$ for the lower $(+)$ sign, we find they lie on curves defined by

$$|\hat{\mathscr{E}}_{ss}^{(\pm)}|^4 = -2 + \frac{1}{2}\left[\frac{1 + B^2}{2B}\right]^2 \left\{1 + \left[1 + 8\left(\frac{2B}{1 + B^2}\right)\right]^{1/2}\right\}, \tag{3.49a}$$

$$\psi_0^{(+)} = (|\hat{\mathscr{E}}_{ss}^{(+)}|^2 + \Psi - \pi) \bmod 2\pi, \tag{3.49b}$$

$$\psi_0^{(-)} = (|\hat{\mathscr{E}}_{ss}^{(-)}|^2 + \Psi) \bmod 2\pi, \tag{3.49c}$$

and

$$\Psi = \arccos\left[\frac{1 + B^2}{2B}\,\frac{2}{2 + |\hat{\mathscr{E}}_{ss}^{(\pm)}|^4}\right], \tag{3.49d}$$

where $-\pi < \psi_0^{(\pm)} < \pi$ and $0 \leqslant \Psi < \pi/2$. Using the steady-state relationships (3.47a), we obtain the corresponding incident intensities:

$$|\hat{\mathscr{E}}_{i}^{(+)}|^2 = (1 + B^2)|\hat{\mathscr{E}}_{ss}^{(+)}|^2 \frac{4 + |\hat{\mathscr{E}}_{ss}^{(+)}|^4}{2 + |\hat{\mathscr{E}}_{ss}^{(+)}|^4} \tag{3.50a}$$

and

$$|\hat{\mathscr{E}}_{i}^{(-)}|^2 = (1 + B^2)|\hat{\mathscr{E}}_{ss}^{(-)}|^2 \frac{|\hat{\mathscr{E}}_{ss}^{(-)}|^4}{2 + |\hat{\mathscr{E}}_{ss}^{(-)}|^4}. \tag{3.50b}$$

Finally we consider the region of stability subject to variations in detuning $\psi_0^{(\pm)}$ and $|\hat{\mathscr{E}}_{i}^{(\pm)}|$. In Fig. 8 the shaded regions correspond to stable fixed points on the lower three steady-state branches. Note that the lowest branch is either

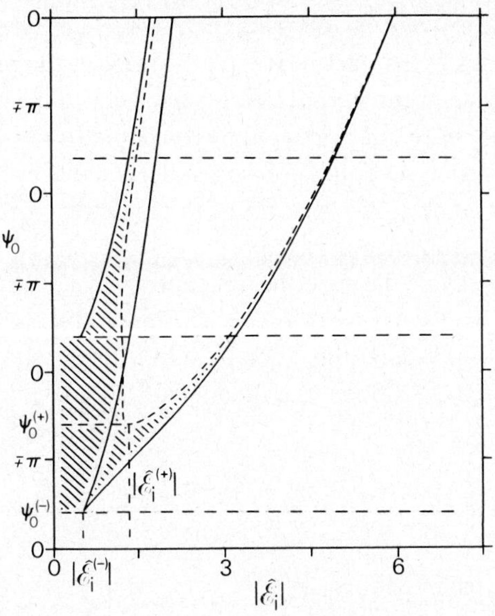

Fig. 8. Stability diagram for the stationary states with $B = 0.5$. Solid curves denote points of marginal stability described by $(|\hat{\mathscr{E}}_i^{(-)}|, \psi^{(-)})$ using equations (3.49) and (3.50). Broken lines denote the marginal points $(|\hat{\mathscr{E}}_i^{(+)}|, \psi^{(+)})$. Shaded regions represent stable fixed points. The horizontal dashed lines demarcate the three lower steady-state branches.

completely stable or partially unstable, while every upper branch is partially unstable. In particular:

(1) If $0 \leqslant \psi_0^{(-)} < \pi$, then the lowest branch is completely stable if either $\psi_0^{(-)} < \psi_0 < \pi$ or $-\pi < \psi_s < \psi^{(+)}$, and partially unstable if $\psi_0^{(-)} < \psi_0 < \psi_0^{(+)}$.

(2) If $-\pi < \psi_0^{(-)} \leqslant 0$, then the lowest branch is completely stable if $\psi_0^{(-)} < \psi_0 < \psi_0^{(+)}$ and partially unstable if $\psi_0^{(+)} \leqslant \psi_0 < \pi$ or $-\pi < \psi_0 < \psi_0^{(-)}$.

In § 6 we shall give a detailed description of the bifurcation sequences exhibited by the map, eq. (3.46). Now, however, we return to the mean-field, absorptive limit and consider how the presence of two simultaneously stable stationary states may influence the dynamic behavior of this system.

§ 4. Bistable Operation and Deterministic Switching

The analysis in § 3 indicates that the ring cavity may exhibit a bistable response, in accordance with the early experimental findings of GIBBS, MCCALL and VENKATESAN [1976], who used a Fabry–Perot cavity in their investigations. The slowing of the dynamic response (BISCHOFBERGER and SHEN [1978, 1979], GRANT and KIMBLE [1983]) near the marginal points of the response curve, and the relevance of this behavior to the practical problem of optical memory, have motivated numerous treatments of the switching dynamics. In the present section, we describe the bistable operation of the ring cavity and briefly outline results pertaining to the deterministic dynamics; the role of noise will be considered in § 5.

We begin by simplifying the mean-field model of eqs. (2.7)–(2.8), (2.17b) and (2.18) – we limit our discussion to the case of pure absorption ($\delta_A = \delta_c = 0$) and, furthermore, introduce the "good-cavity" limit, $\kappa \ll \gamma_\perp, \gamma_\parallel$ (WILLIS [1977]). Under these conditions, the functions $\mathscr{P}(t)$ and $\mathscr{D}(t)$ follow the field adiabatically, with

$$\frac{\partial \mathscr{P}(t)}{\partial t} \approx 0, \qquad \frac{\partial \mathscr{D}(t)}{\partial t} \approx 0,$$

and we obtain the mean-field equation

$$\frac{1}{\kappa}\frac{\mathrm{d}x}{\mathrm{d}t} = y - x - \frac{2Cx}{1+x^2} \tag{4.1}$$

(BONIFACIO, GRONCHI and LUGIATO [1978]). The steady-state condition for this case has been given in eq. (2.21), and an example is depicted in Fig. 9. There, x_1, x_2 and x_3 correspond to the three steady-state responses to the (scaled) injected field y_0. We will assume that the points at x_1 and x_3 are stable;

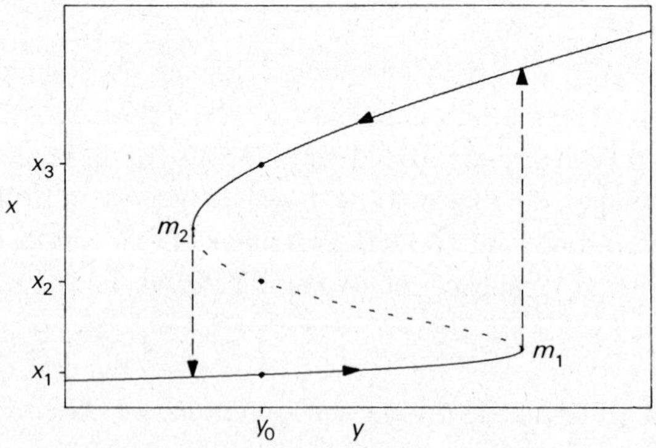

Fig. 9. Schematic of the hysteresis cycle for a bistable response. The values x_1, x_2 and x_3 are the steady states corresponding to $y = y_0$, within the bistable region. The dashed portion of the response curve is the locus of the unstable states x_2. For adiabatic operation, switching occurs at the marginal points m_1 and m_2.

the solutions (x_2, y) were shown to be unstable in § 3, and their locus is indicated by the dashed part of the S-shaped curve. The points m_1 and m_2 are the marginal points of the response curve, and bound the bistable regime. As y is increased, the point at x_1 along the lower branch approaches m_1. Upon reaching m_1, the response jumps to the upper (x_3) branch. If y is now reduced, the point at x_3 moves down along the upper branch until reaching m_2, at which time it jumps to the lower (x_1) branch. The hysteresis cycle corresponding to this behavior is indicated in the figure by the dashed arrows at the marginal points and is illustrated by the photograph in Fig. 10.

An alternate expression for the dynamics may be obtained from eq. (4.1) by introducing the "potential"

$$V(r, \phi) \equiv \tfrac{1}{2}[r^2 + y^2 - 2ry\cos\phi + 2C\ln(1 + r^2)], \qquad (4.2)$$

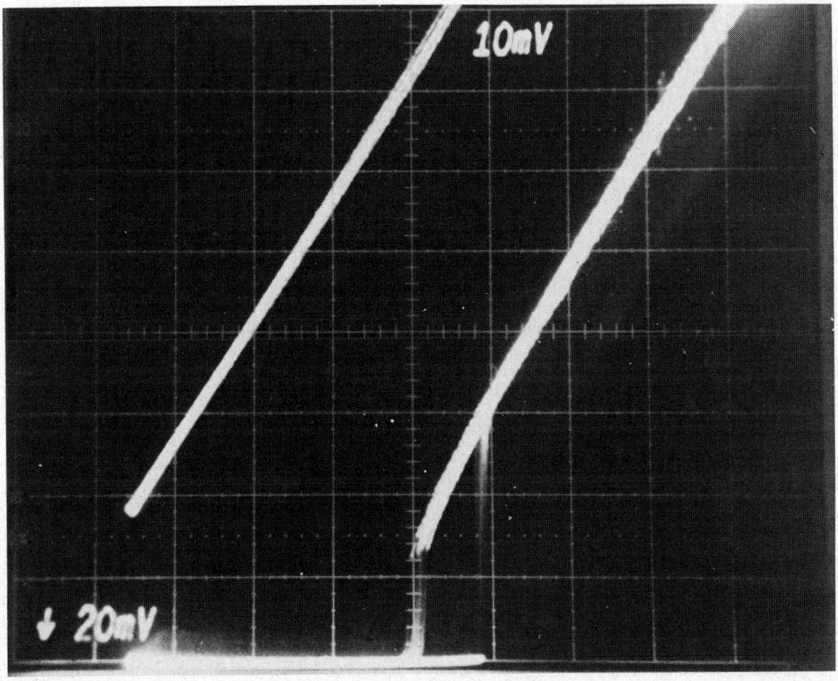

Fig. 10. Hysteresis-cycle behavior of absorptive bistability, observed in experiments with intracavity atomic beams that have been optically prepumped to behave as two-level atoms. The horizontal axis records input power to the cavity; the vertical axis is the output power. At the marginal point m_1 (cf. Fig. 9) the input power is 350–400 μW. The experiment is further described by GRANT and KIMBLE [1982, 1983], who furnished the photograph. The diagonal line to the left is the empty-cavity response.

in terms of which

$$\frac{1}{\kappa}\frac{\mathrm{d}r}{\mathrm{d}t} = -\frac{\partial V(r, \phi)}{\partial r} \tag{4.3a}$$

and

$$\frac{1}{\kappa}\frac{\mathrm{d}\phi}{\mathrm{d}t} = -\frac{1}{r^2}\frac{\partial V(r, \phi)}{\partial \phi}, \tag{4.3b}$$

where we have let

$$x \equiv r\,\mathrm{e}^{\mathrm{i}\phi}.$$

(N.B. The field variables r and ϕ should not be confused with the spatial variables r and ϕ of § 2.) We have chosen the phase of y equal to zero. Equations (4.3a,b) may be thought of as kinetic equations for an over-damped point particle moving in the potential $V(r, \phi)$, and with a damping constant κ^{-1}. The steady-state solutions (x, y) of eq. (4.1) now correspond to the extrema of $V(r, \phi)$, which is depicted in Fig. 11 for the values C and y_0 of Fig. 9. The stable points at x_1 and x_3 are now represented by the potential minima, which are separated by the maximum at x_2. In the context of eqs. (4.3a,b), these extrema are isolated from one another. With the introduction of noise (§ 5), transitions between the potential wells will be possible; however, the times involved in this *stochastic* switching are typically quite large (unless artificially enhanced) in comparison with laboratory time scales, so that the hysteresis behavior of Fig. 9 is preserved.

The problem of *deterministic* switching is illustrated in Fig. 12. In contrast to the situation depicted in Fig. 9, where we increased y_0 adiabatically in following the hysteresis cycle, we now increase y_0 to y_1 in a time much faster than the response time of the cavity. In addition, we choose y_1 outside the bistable regime, by an amount Δy. Otherwise, the system would relax to the lower branch. Then, it is possible, with eq. (4.1), to follow the evolution of the initial point to the upper branch (up-switching). The problem of down-switching is exactly the reverse: one then decreases y_1 rapidly to a point below the bistable regime and follows the point to the lower branch. BONIFACIO and MEYSTRE [1978, 1979] and HOPF and MEYSTRE [1979] have provided numerical analyses of this problem both in the "good-cavity" and "bad-cavity" ($\kappa \gg \gamma_\perp, \gamma_\parallel$) limits. Equation (4.1) was solved analytically by BENZA and LUGIATO [1979a], and MANDEL and ERNEUX [1982b] and ERNEUX and

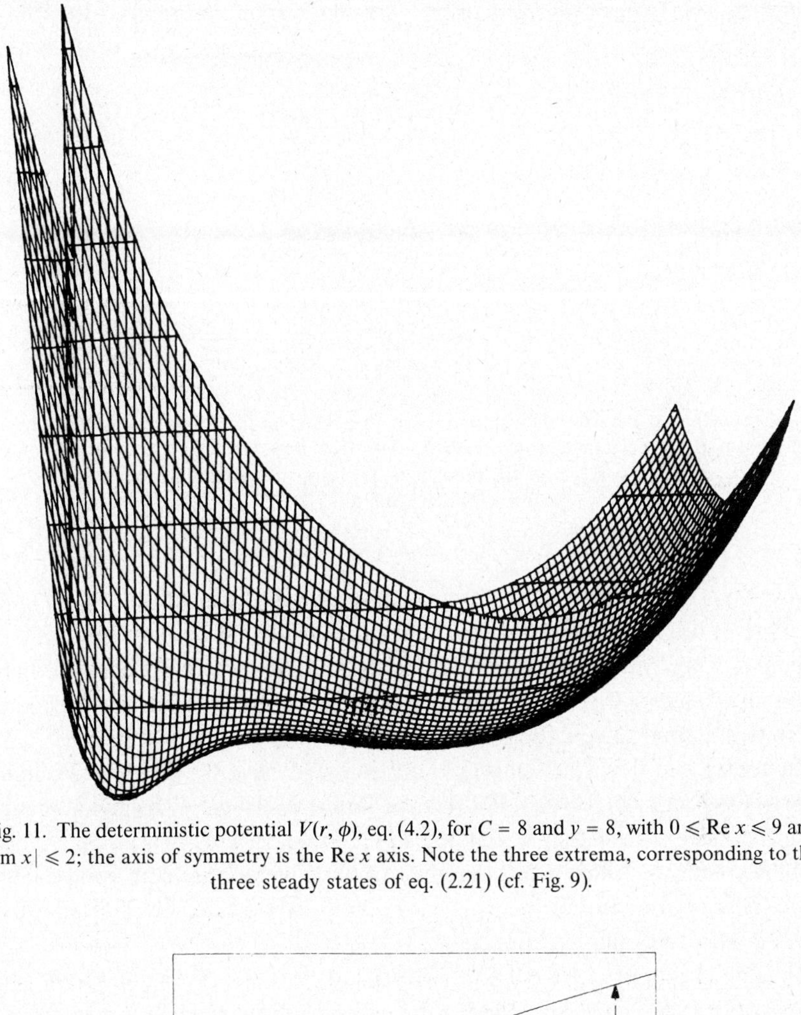

Fig. 11. The deterministic potential $V(r, \phi)$, eq. (4.2), for $C = 8$ and $y = 8$, with $0 \leqslant \mathrm{Re}\, x \leqslant 9$ and $|\mathrm{Im}\, x| \leqslant 2$; the axis of symmetry is the $\mathrm{Re}\, x$ axis. Note the three extrema, corresponding to the three steady states of eq. (2.21) (cf. Fig. 9).

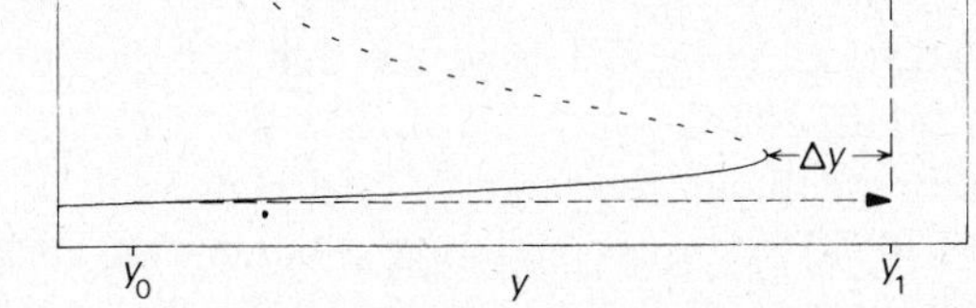

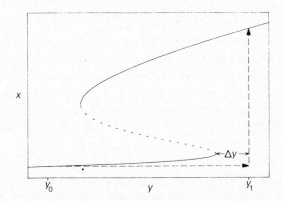

Fig. 12. Schematic of deterministic up-switching. The injected field is increased suddenly from y_0 to a value y_1 of magnitude Δy above the rightmost marginal point.

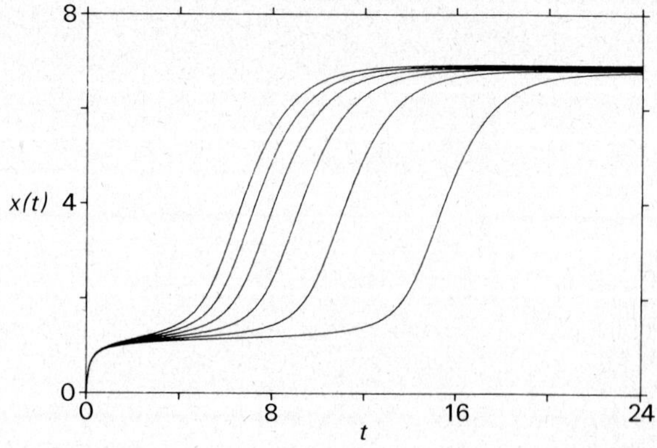

Fig. 13. Switching dynamics obtained from the mean-field eq. (4.1). The six curves, respectively from right to left, are obtained using $\Delta y = 0.02, 0.04, 0.06, 0.08, 0.10$ and 0.12, with $y_0 = 0$ and $C = 8$ (cf. Fig. 12). Note the prominent slowing near the marginal point at $x = (7 - 4\sqrt{2})^{1/2} \approx 1.16$.

MANDEL [1983] have given analytic results for large C and arbitrary κ, $\gamma_\perp$ and $\gamma_\parallel$. One of the more interesting of the features exhibited by these studies is a regime of *marginal slowing* near the points m_1 and m_2, which depends strongly upon the value of Δy. Once the marginal points are passed, the response exhibits a comparatively rapid evolution to its steady state. See Fig. 13.

In contrast to the rather smooth transition evident in the good-cavity limit, BONIFACIO and MEYSTRE [1978] have found a "ringing" behavior in the approach to the steady-state for the bad-cavity limit. This is associated with Rabi oscillations, whose effects are lost in the cavity-dominated good-cavity limit. This is also true of the $\gamma_\parallel \ll \kappa$, $\gamma_\perp$ limit investigated by DRUMMOND [1982], who gives numerical results.

Finally, HOPF, MEYSTRE, DRUMMOND and WALLS [1979] and HOPF and MEYSTRE [1980] have investigated the effect of dispersion on the switching dynamics. In that case, the two-dimensional nature of the fields x and y is of significance. In fact, one-dimensional representations, such as Fig. 6, can be misleading – switching may be induced even within the (apparently) bistable regimes. Furthermore, switching may accompany rapid changes in the phase of the incident field (phase switching) as well as in its amplitude. Similar results are reported by SELLONI, QUATTROPANI, SCHWENDIMANN and BALTES [1981], using the anharmonic oscillator model proposed by SELLONI and SCHWENDIMANN [1979]. Transverse mode structure and diffraction effects are included in the analysis of phase switching by TAI, GIBBS and MOLONEY

[1982]; by not applying the mean-field approximation, they also show that Ikeda instabilities (see § 3) may be avoided with a proper choice of parameters.

§ 5. Noise-Driven Transitions for Bistable Operation

As we have remarked in the previous section, the introduction of noise will destroy the hysteresis-cycle behavior of the bistable steady-state response. Here, we present two descriptions of noise phenomena for absorptive bistability, one based purely upon quantum-statistical effects, the other based upon (classical) statistical fluctuations in the atomic density. Both treatments lead to Fokker–Planck equations, and we show that these are identical under certain simplifying assumptions; this allows us to study a single equation. We present and discuss results for both the steady-state operation and the approach to the steady state (switching). Finally, the critical dynamics are briefly considered.

The single-mode quantum-statistical treatment of the coherently driven ring-cavity system may be based upon the weak-coupling master equation (HAKEN [1970] and LOUISELL [1973])

$$i\hbar \frac{d\hat{\rho}(t)}{dt} = \hat{\mathscr{L}}\hat{\rho}(t), \tag{5.1}$$

where $\hat{\rho}(t)$ is the von Neumann density operator, and where the dissipative Liouvillian $\hat{\mathscr{L}}$ is defined by

$$\hat{\mathscr{L}} \equiv \hat{\mathscr{L}}_{AF} + i\hbar(\hat{\Lambda}_A + \hat{\Lambda}_F), \tag{5.2}$$

with

$$\hat{\mathscr{L}}_{AF}\hat{\rho} \equiv i\hbar g[(\hat{a}^\dagger \hat{\Sigma}^- - \hat{a}\hat{\Sigma}^+), \hat{\rho}], \tag{5.3a}$$

$$\hat{\Lambda}_A\hat{\rho} \equiv \frac{\mathscr{N}}{L}\int_0^L dz\{\tfrac{1}{2}\gamma_\parallel([\hat{\sigma}^-(z)\hat{\rho}, \hat{\sigma}^+(z)] + [\hat{\sigma}^-(z), \hat{\rho}\hat{\sigma}^+(z)]) + \tfrac{1}{4}(\gamma_\perp - \tfrac{1}{2}\gamma_\parallel)([\hat{\sigma}_z(z)\hat{\rho}, \hat{\sigma}_z(z)] + [\hat{\sigma}_z(z), \hat{\rho}\hat{\sigma}_z(z)])\} \tag{5.3b}$$

and

$$\hat{\Lambda}_F\hat{\rho} \equiv \kappa\{[(\hat{a} - \alpha_0)\hat{\rho}, (\hat{a}^\dagger - \alpha_0^*)] + [(\hat{a} - \alpha_0), \hat{\rho}(\hat{a}^\dagger - \alpha_0^*)]\} \tag{5.3c}$$

(BONIFACIO and LUGIATO [1978d], BONIFACIO, GRONCHI and LUGIATO [1978], HASSAN, DRUMMOND and WALLS [1978]). Here, for simplicity, we have neglected transverse field variations, and have assumed that the coherent driving field and the (homogeneously broadened) medium are in resonance with a cavity mode, so that dispersive effects are absent. (We have thus avoided the possibility of the Ikeda instability.) The slowly-varying-amplitude and the rotating-wave approximations have also been made in obtaining (5.3), which is in an interaction representation. Finally, we have neglected the influence of thermal photons on the statistics. The medium is represented by the collective operators $\hat{\Sigma}^{\pm}$ and $\hat{\Sigma}_z$, defined by

$$\hat{\Sigma}^{\pm} \equiv \frac{\mathcal{N}}{L} \int_0^L \mathrm{d}z\, \hat{\sigma}^{\pm}(z)\, \mathrm{e}^{\pm ikz} \tag{5.4a}$$

and

$$\hat{\Sigma}_z \equiv \frac{\mathcal{N}}{L} \int_0^L \mathrm{d}z\, \hat{\sigma}_z(z) \tag{5.4b}$$

in terms of the Pauli operators $\hat{\sigma}^{\pm}(z)$ and $\hat{\sigma}_z(z)$ satisfying

$$[\hat{\sigma}^{+}(z), \hat{\sigma}^{-}(z')] = \hat{\sigma}_z(z)\delta(z - z') \tag{5.5a}$$

and

$$[\hat{\sigma}_z(z), \hat{\sigma}^{\pm}(z')] = \pm 2\hat{\sigma}^{\pm}(z)\delta(z - z'), \tag{5.5b}$$

so that

$$[\hat{\Sigma}^{+}, \hat{\Sigma}^{-}] = \hat{\Sigma}_z \tag{5.6a}$$

and

$$[\hat{\Sigma}_z, \hat{\Sigma}^{\pm}] = \pm 2\hat{\Sigma}^{\pm}. \tag{5.6b}$$

The quantized internal field is described by the Bose operators $\hat{a}^{\dagger}$ and $\hat{a}$, with

$$[\hat{a}, \hat{a}^{\dagger}] = 1, \tag{5.7}$$

and the injected field by the (complex) amplitude α_0. Finally, g is the quantum mechanical coupling constant

$$g \equiv \mu \sqrt{\frac{\omega}{2\varepsilon_0 \hbar}}. \tag{5.8}$$

One now may verify that the Heisenberg equations for $\hat{a}^\dagger$, $\hat{\Sigma}^\dagger$ and $\hat{\Sigma}$ imply the correspondence

$$\mathscr{E}(t) \leftrightarrow \sqrt{\frac{\hbar\omega}{2\varepsilon_0}}\, a^\dagger(t), \tag{5.9a}$$

$$\mathscr{P}(t) \leftrightarrow \mathrm{i}\, \hat{\Sigma}^{\pm}(t) \tag{5.9b}$$

and

$$\mathscr{D}(t) \leftrightarrow \hat{\Sigma}_z(t) \tag{5.9c}$$

with the mean-field semiclassical approach of § 2.

To facilitate the solution of eqs. (5.1)–(5.3) one may now introduce a c-number representation for $\hat{\rho}(t)$. The general case has been treated by several authors – GRONCHI and LUGIATO [1978], AGARWAL, NARDUCCI, FENG and GILMORE [1980], DRUMMOND and WALLS [1981], LUGIATO, CASAGRANDE and PIZZUTO [1982] – who obtain multi-dimensional Fokker–Planck equations in a variety of representations. In all these cases, however, we should note that the derivation of the Fokker–Planck equation involves a system-size expansion, in $1/\mathcal{N}$, that motivates the neglect of derivatives of order higher than second. Such a procedure has been criticized by VAN KAMPEN [1969]; its validity in approximating the statistics of the ring-cavity steady-state has been demonstrated, however, by BONIFACIO, GRONCHI and LUGIATO [1978]. (See also HORSTHEMKE and BRENIG [1977] and HORSTHEMKE, MALEK-MANSOUR and BRENIG [1977].) This strongly suggests that the Fokker–Planck approximation is justified for obtaining the lower-order eigenvalues and eigenfunctions of $\hat{\rho}(t)$.

Rather than present the general case, we shall confine our Fokker–Planck treatment to the good-cavity limit $\kappa \ll \gamma_\perp, \gamma_\parallel$; this, and the bad-cavity limit $\kappa \gg \gamma_\perp, \gamma_\parallel$, were first discussed in the context of optical bistability by WILLIS [1977]. (The bad-cavity case has been treated by LUGIATO [1979], AGARWAL, NARDUCCI, FENG and GILMORE [1980] and ZARDECKI [1981]; the limit $\gamma_\parallel \gg \gamma_\perp, \kappa$, which allows a particularly simple formulation of the population dynamics, has been analyzed by DRUMMOND [1982].)

Assuming $\kappa \ll \gamma_\perp, \gamma_\parallel$, we may adiabatically eliminate the atomic variables from the master equation, using the method of GORDON [1967] and HAKEN [1975b]. We let

$$\hat{\rho}_{\mathrm{F}}(t) \equiv \mathrm{Tr}_{\mathrm{A}}\, \hat{\rho}(t),$$

where Tr_{A} indicates the trace over the atomic subspace, and represent $\hat{\rho}_{\mathrm{F}}(t)$ by the quasiprobability density $P(\alpha, t)$ of GLAUBER [1963a,b] and SUDARSHAN

[1963]. Then, adiabatic elimination yields (GRONCHI and LUGIATO [1978], LUGIATO [1979], DRUMMOND and WALLS [1981], LUGIATO, CASAGRANDE and PIZZUTO [1982])

$$\frac{1}{\kappa}\frac{\partial P(r,\phi,t)}{\partial t} = \left\{ -\left[\frac{1}{r}\frac{\partial}{\partial r} rA_r(r,\phi) + \frac{1}{r}\frac{\partial}{\partial \phi}A_\phi(r,\phi)\right] + \frac{1}{2}\left[\frac{1}{r}\frac{\partial}{\partial r} r\frac{\partial}{\partial r}B_{rr}(r) + \frac{1}{r^2}\frac{\partial^2}{\partial \phi^2}B_{\phi\phi}(r)\right]\right\} P(r,\phi,t), \tag{5.10a}$$

with

$$\int_0^\infty r\,\mathrm{d}r \int_0^{2\pi} \mathrm{d}\phi\, P(r,\phi,t) = 1, \tag{5.10b}$$

as the Fokker–Planck approximation to the system's dynamics, where

$$\alpha \equiv \sqrt{n_s}\, r\, \mathrm{e}^{\mathrm{i}\phi}, \qquad \alpha_0 \equiv \sqrt{n_s}\, y;$$

here,

$$n_s \equiv \gamma_\perp \gamma_\parallel / 4g^2$$

is the photon number at saturation. The variable ϕ, thus, represents the phase difference between the injected and the internal fields. The drift functions are

$$A_r(r,\phi) \equiv y\cos\phi - r - \frac{2Cr}{1+r^2} \tag{5.11a}$$

and

$$A_\phi(r,\phi) \equiv -\frac{y}{r}\sin\phi, \tag{5.11b}$$

so we regain the semiclassical dynamics by letting

$$\frac{1}{\kappa}\frac{\mathrm{d}r}{\mathrm{d}t} = A_r(r,\phi) \qquad \text{and} \qquad \frac{1}{\kappa}\frac{\mathrm{d}\phi}{\mathrm{d}t} = A_\phi(r,\phi),$$

in agreement with eqs. (4.3a,b), with $x = r\mathrm{e}^{\mathrm{i}\phi}$. (It should be noted that terms of order $1/n_s \ll 1$ have been neglected in eqs. (5.11a,b).) The diffusion functions of eq. (5.10) are

$$B_{rr}(r) \equiv \frac{C}{n_s}\left[\frac{2r^2}{(1+r^2)^2} - \frac{(1+2f)r^2}{(1+r^2)^3}\right] \tag{5.12a}$$

and

$$B_{\phi\phi}(r) \equiv \frac{C}{n_s} \frac{2r^2}{(1+r^2)}, \tag{5.12b}$$

where

$$f \equiv \frac{\gamma_\parallel}{2\gamma_\perp}.$$

The small-signal limit of this Fokker–Planck equation has been derived by DRUMMOND and WALLS [1980], using the third-order polarization.

The results of the preceding paragraph are in accordance with other quantum-statistical treatments, except for the diffusion variable $B_{rr}(r)$ of eq. (5.12a). In eliminating the atomic variables, WILLIS [1978] and WILLIS and DAY [1979] replace them by their semiclassical stationary values (as a function of r). This approach neglects the feedback of field fluctuations through the medium, and has a significant effect on $B_{rr}(r)$ and, hence, the photon statistics (cf. CASAGRANDE and LUGIATO [1976].) The early treatment of BONIFACIO, GRONCHI and LUGIATO [1978] produced the relation

$$B_{rr}(r) = \frac{C}{n_s} \frac{r^2}{(1+r^2)^2}. \tag{5.13}$$

The difference between this and eq. (5.12a) is attributable to two sources: the method of adiabatic elimination and the treatment of atomic correlations. First, the adiabatic elimination technique used in obtaining (5.13) is not the same as HAKEN's [1975b], which, briefly, consists in setting time derivatives of the conditional expectations

$$\frac{\mathrm{Tr}_A\, \hat{\sigma}^\pm \hat{\rho}}{\mathrm{Tr}_A\, \hat{\rho}} \quad \text{and} \quad \frac{\mathrm{Tr}_A\, \hat{\sigma}^+ \hat{\sigma}^- \hat{\rho}}{\mathrm{Tr}_A\, \hat{\rho}}$$

equal to zero; eq. (5.13) was found using

$$\frac{\partial}{\partial t} \mathrm{Tr}_A\, \hat{\sigma}^\pm \hat{\rho} = \frac{\partial}{\partial t} \mathrm{Tr}_A\, \hat{\sigma}^+ \hat{\sigma}^- \hat{\rho} = 0,$$

instead. Second, all terms arising from atomic correlations were neglected in finding eq. (5.13), but were retained in deriving (5.12a). ENGLUND [1984] has shown the following: if Haken's method of adiabatic elimination is used, while still neglecting the correlation terms, one obtains a $B_{rr}(r)$ that is twice that given by eq. (5.13), so the first (positive) term of (5.12a) is recovered; and, if one, in addition, *factors* the correlation terms (cf. WEIDLICH, RISKEN and HAKEN [1967]) instead of neglecting them, one then obtains exactly eq. (5.12a). Therefore, within the Fokker–Planck approximation, atomic correlations are negligible in the (usual) sense that correlation terms may be factored.

It is apparent from eq. (5.12a) that $B_{rr}(r)$ may become negative for small mean intensities, $\overline{r^2} \ll 1$, if $f > 1/2$ ($\gamma_{\parallel} > \gamma_{\perp}$). Such behavior indicates the presence of non-semiclassical effects, such as squeezed states and photon antibunching (DRUMMOND and WALLS [1980, 1981], CASAGRANDE and LUGIATO [1980], LUGIATO, CASAGRANDE and PIZZUTO [1982]). Since eqs. (5.10)–(5.12) are not suitable for describing these effects analytically, two alternate approaches have been developed. GRONCHI and LUGIATO [1978] have derived a Fokker–Planck equation in terms of a symmetrically ordered Wigner distribution. In contrast to eq. (5.12a), the diffusion function of this Fokker–Planck equation is positive definite. On the other hand, DRUMMOND and WALLS [1981] have used the generalized Glauber distribution of DRUMMOND and GARDINER [1980], which also leads to a positive definite $B_{rr}(r)$.

We now introduce two additional approximations to our treatment of quantum fluctuations. First, we restrict ourselves to cases where the negative contribution to $B_{rr}(r)$ in eq. (5.12a) is negligible:

$$B_{rr}(r) \simeq 2q \frac{r^2}{(1+r^2)^2},$$

with

$$q \equiv \frac{C}{n_s}. \tag{5.14}$$

As mentioned above, this approximation is equivalent to neglecting atomic correlation terms. (See WILLIS [1983] for a further discussion of its validity.) The resulting Fokker–Planck equation is still rather intractable, however; even its steady-state solution is unattainable analytically, since it lacks *detailed balance*. GRAHAM and SCHENZLE [1981a,b] have formulated a perturbation approach for such cases and have applied it to a model of dispersive optical bistability. Here, we simply assume that phase fluctuations have a negligible effect on the photon statistics of the ring cavity, especially in bistable operation, in which large fluctuations in amplitude occur. If we consider only the phase-locked value $\phi = 0$, this amounts to setting $P(r, \phi, t) = \delta(\phi)P(r, t)$, with eqs. (5.10)–(5.12) becoming

$$\frac{1}{\kappa}\frac{\partial P(x,t)}{\partial t} = \left\{-\frac{\partial}{\partial x}A(x) + \frac{1}{2}\frac{\partial^2}{\partial x^2}B(x)\right\}P(x,t), \tag{5.15a}$$

with

$$\int_0^\infty \mathrm{d}x\, P(x, t) = 1, \tag{5.15b}$$

$$A(x) \equiv y - x - \frac{2Cx}{1 + x^2}, \tag{5.16}$$

and

$$B(x) \equiv \frac{2qx^2}{(1 + x^2)^2}, \tag{5.17}$$

in terms of the (scaled) field x.

At this point, it is appropriate that we consider alternative formulations of the fluctuation dynamics. These *mesoscopic* (stochastic) approaches begin with the Maxwell–Bloch equations, then introduce noise sources to obtain Langevin equations, along with their corresponding Fokker–Planck equations. The earliest such model, by BULSARA, SCHIEVE and GRAGG [1978], is of particular interest here, as it leads precisely to eqs. (5.15)–(5.17). (See also SCHENZLE and BRAND [1978], ARECCHI and POLITI [1979], ZARDECKI [1980] and GRAGG [1981].) To begin with, we assume that the good-cavity eq. (4.1) is valid, but we replace C by a stochastic process C_t such that $C_t - C$ is a δ-correlated white noise:

$$\langle (C_t - C) \rangle = 0 \qquad \text{and} \qquad \langle (C_t - C)(C_{t'} - C) \rangle = \sigma^2 \delta(t - t'),$$

with all higher cumulants vanishing. (MOORE [1983] has investigated the case where C_t has a finite correlation time.) The quantity σ^2 is thus the variance of C_t. Since $C \propto \mathcal{N}$, we may view this approach as treating density fluctuations in the medium. With this replacement, we obtain the stochastic differential (Langevin) equation

$$\mathrm{d}x_t = \left\{ y - x_t - \frac{2Cx_t}{1 + x_t^2} \right\} \mathrm{d}t + \frac{2\sigma x_t}{1 + x_t^2} \mathrm{d}W_t, \tag{5.18}$$

where W_t is the Wiener process and $x(t)$ has been replaced by the stochastic process x_t. Equation (5.18), as it stands, is meaningless, since the *multiplicative* noise term leads to an infinity of associated Fokker–Planck equations, according to which definition of stochastic calculus one adopts. In particular, the Itô and Stratonovich rules produce Fokker–Planck equations whose drift terms differ by a term of order σ^2. However, VAN KAMPEN [1981] has demonstrated that the validity of Langevin equations cannot be established to the precision needed to distinguish between the Itô and Stratonovich interpretations. Adopting the Itô interpretation, we find exactly eqs. (5.15–17), with

$$\sigma^2 = \tfrac{1}{2}q.$$

The drift term resulting from the Stratonovich interpretation, therefore, differs only by a term of order $1/n_s$; as previously mentioned, such terms were ignored in the quantum-statistical derivation, as well. More general models of population fluctuations have been given by SCHENZLE and BRAND [1978], KONDO, MABUCHI and HASEGAWA [1980], HASEGAWA, NAKAGOMI, MABUCHI and KONDO [1980] and ZARDECKI [1980]. The effect of fluctuations in the injected field has been studied by SCHENZLE and BRAND [1978] and GRAHAM and SCHENZLE [1981a,b]; it is clear that substituting a stochastic process y_t for y in eq. (4.1) leads, in that case, to an *additive* noise term. A numerical simulation has been presented by CHROSTOWSKI and ZARDECKI [1979]. Finally, WILLIS [1983] has investigated the effect of frequency fluctuations ("jitter").

We now turn to the stationary statistical properties of the driven ring cavity. Equations (5.15)–(5.17) yield the steady-state probability density (BULSARA, GRAGG and SCHIEVE [1978], BONIFACIO, GRONCHI and LUGIATO [1978])

$$P(x, \infty) = \frac{1}{N}\frac{1}{B(x)} \exp\left\{\frac{-u(x)}{q}\right\}, \tag{5.19a}$$

where

$$\begin{aligned} u(x) &\equiv -2q \int \mathrm{d}x \left[\frac{A(x)}{B(x)}\right] \\ &= \frac{x^4}{4} - \frac{x^3 y}{3} + (1+C)x^2 - 2xy + (1+2C)\ln x + \frac{y}{x} \end{aligned} \tag{5.19b}$$

and N is the normalization integral

$$N \equiv \int_0^\infty \mathrm{d}x\, P(x, \infty).$$

From eqs. (5.15a) and (5.19b), one can show that

$$\frac{1}{\kappa}\frac{\mathrm{d}\langle x\rangle}{\mathrm{d}t} = -\left\langle \frac{B(x)}{2q}\frac{\mathrm{d}u(x)}{\mathrm{d}x}\right\rangle. \tag{5.20}$$

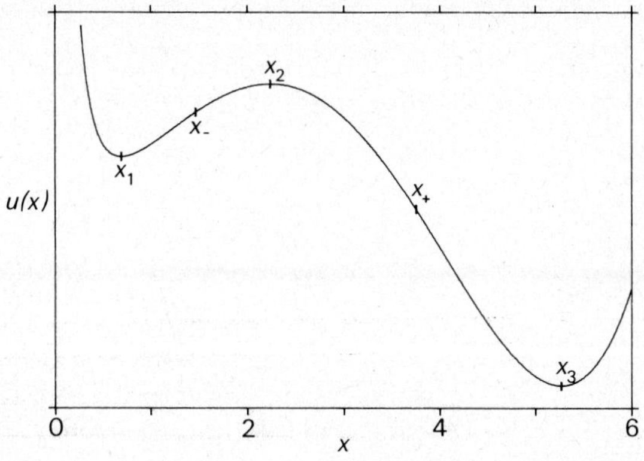

Fig. 14. The potential $u(x)$ for $C = 8$ and $y = 8.2$. The extrema x_1, x_2 and x_3 are the semiclassical steady states; the points x_- and x_+ define the coarse-graining of the probabilities $P^{(1)}$ and $P^{(3)}$ in the Kramers theory of switching.

In the semiclassical limit, where

$$\frac{1}{\kappa}\frac{\mathrm{d}x}{\mathrm{d}t} = -\frac{B(x)}{2q}\frac{\mathrm{d}u(x)}{\mathrm{d}x},$$

the potential $u(x)$ plays a role qualitatively similar to that played by $V(x)$ of § 4: i.e., its gradient indicates the direction of evolution of $x(t)$ (since $B(x) \geqslant 0$), while its extrema correspond to the semiclassical stationary states. (Note that $B(x)/2q$ is independent of q.) A plot of $u(x)$ versus x is pictured in Fig. 14 for an example of bistable operation.

The first two moments of the photon distribution are given by

$$\langle n \rangle = n_s \langle x^2 \rangle$$

and

$$\langle n^2 \rangle = n_s^2 \langle x^4 \rangle + n_s \langle x^2 \rangle,$$

where n is the number of photons. We have plotted $\langle x \rangle$ and the standard deviation,

$$\delta x \equiv (\langle x^2 \rangle - \langle x \rangle^2)^{1/2},$$

versus y in Fig. 15 for $C = 8$ and $q = 1$; the semiclassical steady-state values of x are indicated by the dashed curves. The introduction of noise sources clearly destroys the absolute stability of the semiclassical solutions. Instead, the

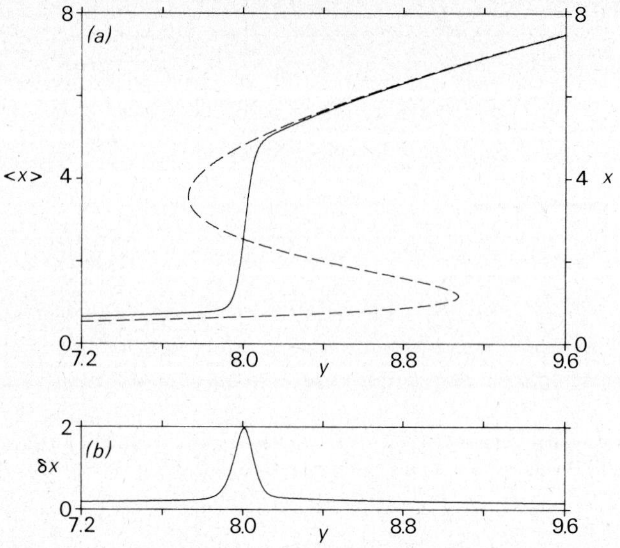

Fig. 15. Statistical results for absorptive bistability, with $C = 8$ and $q = 1$: (a) the first moment of x, $\langle x\rangle$, is plotted as a solid line against the dashed mean-field steady-state response; (b) the standard deviation of x, δx, shows an increase in fluctuations in the transition region exhibited in (a). As $q \to 0$, the peak narrows about the value $y_c \approx 0.798$.

quantity $\langle x\rangle$ exhibits the behavior of a first-order phase transition. For decreasing q, the transition region narrows about the critical value y_c for which

$$U(x_1) = U(x_3), \tag{5.21}$$

where we define

$$U(x) \equiv u(x) - q \ln B(x). \tag{5.22a}$$

Equation (5.19a) then becomes

$$P(x, \infty) = \frac{1}{N} \exp\left\{\frac{-U(x)}{q}\right\}. \tag{5.22b}$$

Therefore, $U(x)$ and q play the roles of a free energy (potential) and a temperature, respectively. (The analogy with equilibrium thermodynamics is purely formal, of course.) Because of the multiplicative nature of the noise source, eq. (5.21) does not correspond to the Maxwell construction, however. Finally, Fig. 15b illustrates the enhanced fluctuations in the transition region about y_c.

The description offered in the preceding paragraph is complementary, rather than contradictory, to that presented in § 2. While eqs. (5.19) represent the true stationary state, laboratory time scales may be such that the metastable states x_1, for $y > y_c$, and x_3, for $y < y_c$, can be considered to be absolutely stable, and hysteresis curves will be observed. On such time scales, then, one may use linearization about x_1 and x_3 to analyze phenomena associated with these metastable states. (Of course, these phenomena will then also exhibit hysteresis.) Thus, LUGIATO, FARINA and NARDUCCI [1980] and FARINA, NARDUCCI, YUAN and LUGIATO [1980, 1981] have described the local relaxation of the field statistics; the absorption spectrum has been treated by HASSAN and WALLS [1978], by NARDUCCI, GILMORE, FENG and AGARWAL [1979] and by TEWARI [1980]. Linear response theory has been used to obtain spectra of the fluorescent and transmitted light, which we briefly describe. Along the low-intensity (x_1) branch the fluorescence spectrum is single-peaked, while along the high-intensity (x_3) branch it exhibits the Stark sidebands characteristic of ordinary resonance fluorescence (MOLLOW [1969]). These results were given by LUGIATO [1980a] and CARMICHAEL [1981a], who showed that earlier treatments contained incorrect assumptions regarding atomic correlations (cf. BONIFACIO and LUGIATO [1976, 1978a,e], CARMICHAEL and WALLS [1977], LUGIATO and BONIFACIO [1978], AGARWAL, NARDUCCI, FENG and GILMORE [1978] and LUGIATO [1979]). For the bad-cavity case, the spectrum of the transmitted light has a similar behavior, except for drastic line narrowing which occurs as the marginal points are approached. (See BONIFACIO and LUGIATO [1978f], NARDUCCI, GILMORE, FENG and AGARWAL [1978], AGARWAL, NARDUCCI, GILMORE and FENG [1978], LUGIATO [1979] and AGARWAL and TEWARI [1980]). This line narrowing is also found in the good-cavity limit; however, the spectrum is at most double-peaked, and exhibits no Stark sidebands. (See BONIFACIO and LUGIATO [1978f], DRUMMOND and WALLS [1980, 1981] and CASAGRANDE and LUGIATO [1980].) In this limit, DRUMMOND and WALLS [1980, 1981] and CASAGRANDE and LUGIATO [1980] have demonstrated the possibility of photon antibunching.

For sufficiently long time scales, linear analysis is inapplicable: the metastable states eventually decay, and the situation illustrated in Fig. 9 gives way to that of Fig. 15a. Analogously, the spectra also undergo first-order transitions at y_c. As our final topic, we consider the switching times associated with the decay of the metastable states. These times will indicate *slowing* in the vicinity of y_c.

To begin our treatment, we make an eigenfunction expansion of $P(x, t)$,

$$P(x, t) = \sum_{n=0}^{\infty} c_n P_n(x)\, e^{-\lambda_n \kappa t}, \tag{5.23a}$$

where we have assumed a discrete spectrum of eigenvalues λ_n. One may show that the eigenfunctions $P_n(x)$ are normalized according to

$$\int_0^\infty \mathrm{d}x \frac{P_m(x) P_n(x)}{P_0(x)} = \delta_{m,n}, \tag{5.23b}$$

so that

$$c_n = \int_0^\infty \mathrm{d}x \frac{P(x, 0) P_n(x)}{P_0(x)}. \tag{5.23c}$$

Here, we have identified $P_0(x)$ with the stationary solution $P(x, \infty)$, so that $\lambda_0 \equiv 0$. In the absence of noise, the two wells of Fig. 14 are essentially isolated and the spectrum of each well may be considered separately. The lowest-order approximation yields the relations

$$\lambda_n^{(1)} = -A'(x_1)n \tag{5.24a}$$

and

$$\lambda_n^{(3)} = -A'(x_3)n \tag{5.24b}$$

for the respective wells. These spectra, then, are simply multiples of the local relaxation rates $-A'(x_1)$ and $-A'(x_3)$ of the two wells. If noise is introduced, this degeneracy is removed and the two zero eigenvalues of eqs. (5.24) split into $\lambda_0(=0)$ and $\lambda_1(\neq 0)$. The potential wells are no longer isolated, so that transitions may occur between them. However, these "tunnelling" transitions are typically characterized by time scales much greater than those related to local relaxation within the individual wells. KRAMERS [1940], CHANDRASEKHAR [1943] and LANDAUER and SWANSON [1961] have adopted an approach that takes advantage of this separation of time scales. Accordingly, one assumes that $P(x, t)$ maintains local equilibrium within each well separately and that all time dependence is involved in the stochastic diffusion between the two wells. The evolution is then governed by the lowest two eigenvalues only: λ_0, which represents the steady state, and λ_1, which represents the switching rate and which vanishes as $q \to 0$. Specifically, one finds (ENGLUND, SCHIEVE, ZUREK and GRAGG [1981], GRAGG [1981], SCHIEVE, ENGLUND, GRAGG and SNAPP [1981])

$$\lambda_1 = \frac{1}{MNP^{(1)}(\infty)P^{(3)}(\infty)}, \tag{5.25a}$$

where

$$M \equiv 2 \int_{x_-}^{x_+} \mathrm{d}x \exp\left(\frac{u(x)}{q}\right), \tag{5.25b}$$

$$P^{(1)}(t) \equiv \int_0^{x_-} \mathrm{d}x\, P(x, t), \tag{5.25c}$$

and

$$P^{(3)}(t) \equiv \int_{x_+}^{\infty} \mathrm{d}x\, P(x, t). \tag{5.25d}$$

Here the values x_- and x_+, shown in Fig. 14, determine the regions $(0, x_-)$ and (x_+, ∞) within which $P(x, t)$ maintains a local equilibrium, so that $P^{(1)}(t)$ and $P^{(3)}(t)$ represent the total probabilities within these respective regions. Equation (5.25) agrees with the WKB estimate derived by CAROLI, CAROLI and ROULET [1979], which justifies our identifying it with the λ_1 of eq. (5.23a).

We may also treat the switching between the x_1 and x_3 potential wells by means of a mean-first-passage-time analysis (WEISS [1966]). Let $T(b|a)$ represent the mean time of first passage from $x = a$ to $x = b$. Then, using approximations valid within the switching regime, one may show that

$$\frac{1}{T(x_3|x_1)} + \frac{1}{T(x_1|x_3)} = \lambda_1, \tag{5.26}$$

where

$$T(x_3|x_1) = MNP^{(1)}(\infty) \tag{5.27a}$$

and

$$T(x_1|x_3) = MNP^{(3)}(\infty). \tag{5.27b}$$

Equations (5.26,27) are therefore in agreement with the Kramers estimate of eq. (5.25). In addition, eq. (5.26) clearly divides the switching rate λ_1 into contributions from both forward and backward diffusion. The first term dominates for $y > y_c$, while the second term dominates for $y < y_c$; however, for $y \approx y_c$, both terms must be considered, since the potential wells are of approximately equal depth. More generally, $T(b|a)$ is given by the exact equation

$$T(b|a) = 2 \int_a^b \mathrm{d}x \exp\left(\frac{u(x)}{q}\right) \int_0^x \mathrm{d}x' \exp\left(\frac{-U(x')}{q}\right). \tag{5.28}$$

Finally, we may simplify eqs. (5.25a–d) by using the quadratic approximations to $u(x)$ and $U(x)$. This gives

$$\lambda_1 = \frac{1}{4\pi q}[-u''(x_2)]^{1/2} \exp\left(\frac{-u(x_2)}{q}\right)\left\{[U''(x_1)]^{1/2} \exp\left(\frac{U(x_1)}{q}\right) + [U''(x_3)]^{1/2} \exp\left(\frac{U(x_3)}{q}\right)\right\}, \tag{5.29}$$

the familiar activation energy formula.

A detailed study of these switching-rate estimates has been given by ENGLUND, SCHIEVE, ZUREK and GRAGG [1981]. (See also GRAGG [1981] and ENGLUND, SCHIEVE, GRAGG and ZUREK [1982].) A graphical comparison is presented in Fig. 16 for the parameters $C = 8$ and $q = 1$, where, in addition to the analytic estimates, we have plotted the results of a numerical variational calculation of the eigenvalue spectrum. Curve ***a*** is the result of eqs. (5.25a–d) and curve ***b*** is the quadratic approximation, eq. (5.29). (We have also

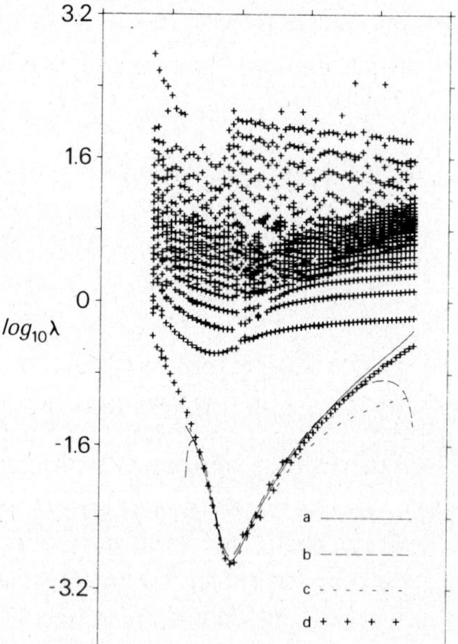

Fig. 16. Approximate eigenvalues of the absorptive bistable ring cavity for $C = 8$ and $q = 1$: (a) eq. (5.25a–d); (b) eq. (5.29); (c) eq. (5.26); (d) variational calculation.

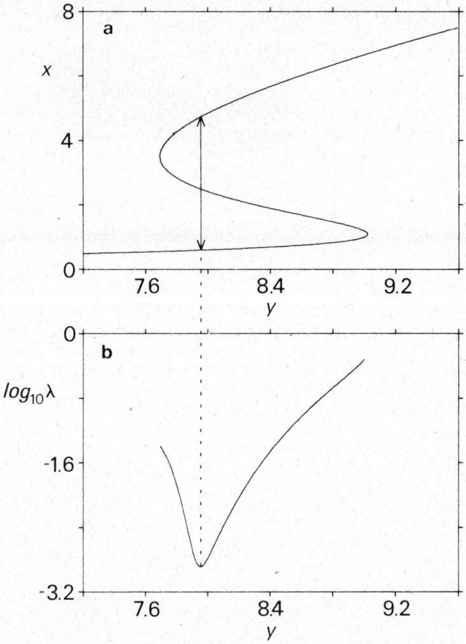

Fig. 17. Comparison of (a) the locations of the extrema of $P(x, \infty)$ with (b) result **a** of Fig. 16 $(C = 8, q = 1)$.

plotted curve ***a*** in Fig. 17a,b in order to show its relation to the bistable region.) The mean-first-passage-time estimate of eqs. (5.26,27) is plotted as curve ***c***. Finally, the variational algorithm, based upon Ritz's method and with a 45-dimensional basis, furnished the points in the curves ***d***. (SCHENZLE and BRAND [1979] have also given variational estimates for λ_1, with other parameters.) The results of Kramer's theory (curve ***a***) are in excellent agreement with λ_1 of the variational calculation, while the quadratic approximation (curve ***b***) is nearly as good, though failing drastically near the marginal points. The mean-first-passage-time estimate (curve ***c***) agrees over a somewhat more limited range of y. This clearly indicates that the higher eigenvalues ($n \geqslant 2$) play an increasingly important role in the switching dynamics as they become comparable near the marginal points. Cases ***a–d*** are again compared in fig. 18, where $q = 0.4$, and the agreement is even more striking. And, while the decrease in q has decreased the minimum switching rate by some three orders of magnitude, the upper ($n \geqslant 2$) eigenvalues have been scarcely affected. This, once again, highlights the differing roles played by the eigenvalues in the system dynamics.

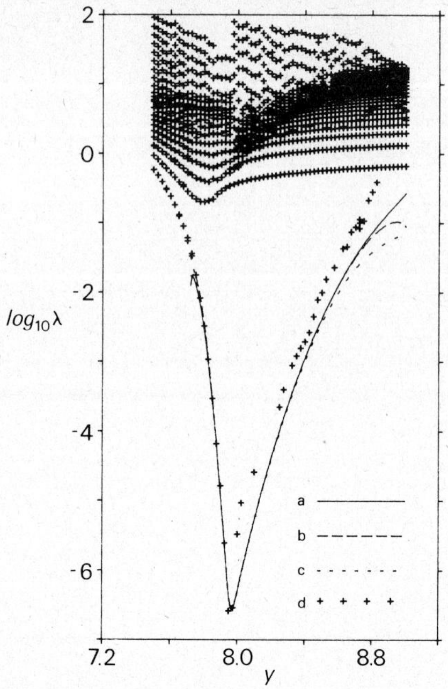

Fig. 18. Approximate eigenvalues of the absorptive bistable ring cavity for $C = 8$ and $q = 0.4$: (a) eq. (5.25a–d); (b) eq. (5.29); (c) eq. (5.26); (d) variational calculation.

Another mean-first-passage-time analysis has been conducted by FARINA, NARDUCCI, YUAN and LUGIATO [1980, 1981]. BONIFACIO, LUGIATO, FARINA and NARDUCCI [1981] have found a factor-of-two discrepancy between the mean-first-passage-time and the Kramers estimates. This is due to an error in their calculations, however. HANGGI, BULSARA and JANDA [1980] have calculated λ_1 as a continued fraction for the parameters used in Figs. 16–18; their estimates differ considerably from those presented here, indicating that more iterations are required in the continued-fraction calculation (see ENGLUND, SCHIEVE, ZUREK and GRAGG [1981] for a more detailed comparison). ZARDECKI [1980] has generated time-dependent field moments, using a Monte Carlo scheme to evaluate a path integral formulation of the Fokker–Planck dynamics, and DELLEDONNE, RICHTER and ROSS[1981] have calculated the time correlation function within the bistable region.

Finally, we consider the eigenvalues for the critical response curve, $C = 4$. Away from the critical point, the lowest-order approximation is valid for the lower eigenvalues, so

$$\lambda_n \simeq -A'(x_1)n. \tag{5.30}$$

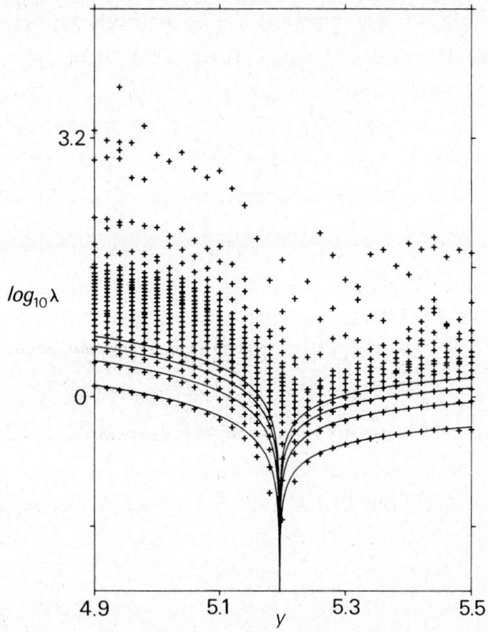

Fig. 19. Approximate eigenvalues of the absorptive ring cavity at the semiclassical mean-field critical point, $C = 4$ $(q = 0.01)$: the solid lines are the first four nonzero eigenvalues obtained from the lowest-order approximation, eq. (5.30), while the crosses are the result of a variational calculation.

This is compared with the results of a variational calculation in Fig. 19. As one might expect, the lowest eigenvalue estimates are in the closest agreement; both the lowest-order approximation and the variational results become poorer for higher n. The approximation used in deriving eq. (5.30) also fails near the critical point, since the linear coefficient of $A(x)$, $A'(x_1)$, vanishes. Instead, we have, to lowest order in x,

$$A(x) \simeq \frac{-(x-\sqrt{3})^3}{4} \qquad \text{and} \qquad B(x) \simeq \frac{3q}{8}$$

at the critical point. The quasi-WKB treatment of KUBO, MATSUO and KITAHARA [1973] leads to the approximation

$$\lambda_n \simeq \frac{\pi^{3/4}}{4}\left[\frac{\Gamma(\frac{5}{3})}{\Gamma(\frac{7}{6})}\right]^{3/2}\left[\frac{3q}{8}\right]^{1/2} n^{3/2}. \tag{5.31}$$

(This treatment follows the standard WKB method, with the exception that a low-order potential term is dropped; one may show analytically that the effect of its neglect decreases with increasing n.) While the eigenvalues no longer converge to zero, as predicted by eq. (5.30), their separation is proportional to $q^{1/2}$, which is typically very small. For this reason, the qualitative picture afforded by Fig. 19 is essentially correct, and the dramatic dip in the spectrum near the critical point $(x, y) = (\sqrt{3}, 3\sqrt{3})$ characterizes the phenomenon of *critical slowing*.

§ 6. Higher-Order Instabilities and the Transition to Chaos

It is now widely known that most dissipative, nonlinear differential systems exhibit intricate bifurcation sequences as a system parameter is varied. In particular, if the system has three or more degrees of freedom* this bifurcation sequence may include chaotic (i.e. nonperiodic) states. In this case, its phase-space trajectories may tend to a *strange attractor* (RUELLE and TAKENS [1971]): an attracting set of trajectories, about which almost all small perturbations grow exponentially (on the average). This property is termed "sensitive dependence on initial conditions" (RUELLE [1979]) and causes any small external influence (i.e. noise) acting on the system, or uncertainty in its initial state, to be amplified through time to a macroscopic scale. Even if the system is deterministic, the mixing of trajectories in a strange attractor prevents an exact prediction of any future state of the system (cf. SHAW [1981]). A classic illustration of such phenomena is found in the Lorenz system (LORENZ [1963]): a three-mode approximation of a horizontal fluid layer heated from below.

Recognizing that strange attractors are structurally stable states, and that they are generic in dissipative nonlinear systems, RUELLE and TAKENS [1971] proposed that they are the mechanism for the onset of turbulence in fluids. Because of their potential physical significance, interest in strange attractors and chaos has rapidly developed in many fields (cf. ECKMANN [1981], HELLEMAN [1980], OTT [1981], RABINOVICH [1978] and SWINNEY and GOLLUB [1981]).

* From the Poincaré–Bendixon Theorem (cf. ABRAHAM and MARSDEN [1978]) an autonomous differential system with two degrees of freedom can only approach stationary, periodic, or quasiperiodic behavior. This restriction is relaxed if the system is subject to an external, time-dependent (e.g. periodic) force.

In this section we shall discuss the appearance of strange attractors and transitions to chaotic behavior in quantum optical systems within the semi-classical and deterministic framework of §§ 2, 3 and 4. After reviewing several recent observations of chaos in single-mode lasers and related systems, we shall then focus on the time-dependent behavior of the nonlinear ring cavity. In particular, we shall describe how a period-doubling bifurcation sequence* generated by the dispersive map, eqs. (3.46), exhibits properties universal to a large class of nonlinear maps. Finally, we show how the recent numerical and experimental simulations of a related system support the existence of these universal scaling properties in a nonlinear ring cavity with a finite medium response time.

The earliest report of chaos in a quantum optical system may be traced back to GRASIUK and ORAEVSKII [1964] who observed "automodulation" in the numerically integrated equations for a single-mode laser. Later, HAKEN [1975a] showed that this model is equivalent to the Lorenz system, by a simple change of variables (cf. ORAEVSKII [1981]). GIBBON and MCGUINNESS [1980, 1982] found that this analogy is a fundamental outcome of the universality of the Lorenz equations to a certain class of dissipative systems. Appropriately, within this analogy, the laser pump parameter corresponds to the Rayleigh number as the principal bifurcation parameter. Thus the first instability in the Lorenz system appears in the single-mode laser if the pump parameter λ exceeds the critical value

$$\lambda_c = \frac{(\kappa + \gamma_\perp)(\kappa + \gamma_\perp + \gamma_\parallel)}{\gamma_\perp(\kappa - \gamma_\perp - \gamma_\parallel)},$$

provided that $\kappa > \gamma_\perp + \gamma_\parallel$. Unfortunately, this latter requirement removes this instability from the present reach of experimental confirmation. Further discussion may be found in HAKEN [1966], RISKEN, SCHMID and WEIDLICH [1966], RISKEN and NUMMEDAL [1968b] and GRAHAM [1976].

DEGEORGIO and LUGIATO [1980] have reported a similar analogy between the instabilities exhibited by a laser with a saturable absorber (LUGIATO, MANDEL, DEMBINSKI and KOSSAKOWSKI [1978]) and a truncated mode expansion of a two-component fluid layer heated from below (VELARDE and ANTORANZ [1979]). VELARDE, ANTORANZ [1980, 1981] and colleagues have

* By this we mean that as a bifurcation parameter (e.g. $\mathscr{E}_i$) is varied monotonically, an infinite sequence of instabilities is encountered. With each instability, the previous limit cycle loses its stability to one of twice its period (cf. MAY [1976] and GROSSMANN and THOMAE [1977]).

found numerical evidence for coexisting stable limit cycles (a higher-order bistability) and a strange attractor (cf. ANTORANZ, BONILLA, GEA and VELARDE [1982] and ANTORANZ, GEA and VELARDE [1981]).

Chaotic instabilities are also evident in single-mode lasers driven by sinusoidally modulated external parameters. Since only two degrees of freedom are then necessary, the atomic variables may be adiabatically eliminated in the good-cavity limit ($\kappa \ll \gamma_\perp, \gamma_\parallel$). In this manner, YAMADA and GRAHAM [1980] have numerically observed complicated sequences of higher-order bifurcations leading to chaos in a single-mode laser with a modulated injected field. There the modulation amplitude and frequency is varied, the latter chosen near the cavity decay constant. If the injected field is not modulated (cf. SPENCER and LAMB [1972] and LUGIATO [1978]), only a Hopf bifurcation (self-pulsing instability) is observed. Similarly SCHOLZ, YAMADA, BRAND and GRAHAM [1981] have observed periodic and chaotic behavior in a single-mode laser with a constant injected field and sinusoidally modulated pump parameter. Within a chaotic region, the cavity field amplitude may additionally display intermittency (cf. MANNEVILLE and POMEAU [1980] and POMEAU and MANNEVILLE [1980]). ARECCHI, MEUCCI, PUCCIONI and TREDICCE [1982] have presented qualitative experimental evidence of the existence of a bifurcation sequence leading to chaotic behavior in a CO_2 laser that was sinusoidally modulated by an intracavity electro-optical device. Periodic and chaotic instabilities have also been investigated in a Fabry–Perot cavity with an injected field of constant amplitude (FIRTH [1981], ABRAHAM, FIRTH and CARR [1982] and FIRTH and WRIGHT [1982]).

The existence of instabilities in inhomogeneously broadened lasers has created considerable interest following CASPERSON'S [1978, 1980, 1981] discovery of coherent pulsations in a cw low pressure xenon laser. In more recent experiments MAEDA and ABRAHAM [1982] and ABRAHAM, CHYBA, COLEMAN, GIOGGIA, HALAS, HOFFER, LIU, MAEDA and WESSON [1983] have observed period-two and broad-band power spectra in a ring laser. An example of this chaotic power spectrum is shown in Fig. 20.

Other devices that may exhibit periodic to chaotic transitions include a Kerr medium with a spatially dependent (sinusoidal) linear refractive index (WINFUL and COOPERMAN [1982]); an extension of a model of optical tristability (SAVAGE, CARMICHAEL and WALLS [1982] and CARMICHAEL, SAVAGE and WALLS [1983], cf. KITANO, YABUZAKI and OGAWA [1981]); and a non-resonant ring cavity, containing a nonlinear crystal which is subject to coherently driven, second-harmonic generation (SAVAGE and WALLS [1983], cf. SILBERBERG and BAR JOSEPH [1982]). In the latter system, MANDEL and ERNEUX

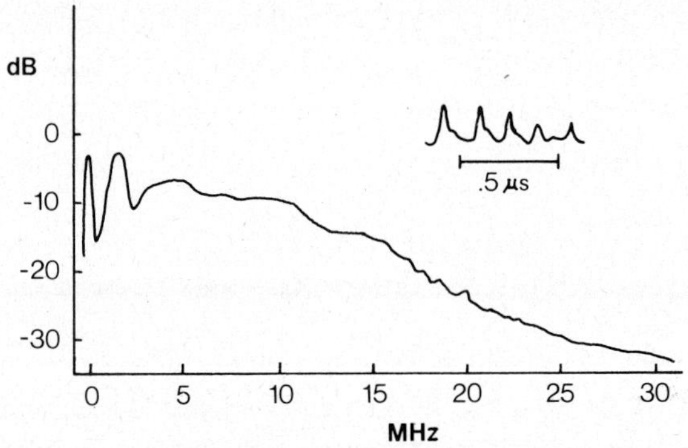

Fig. 20. The power spectrum of an inhomogeneously broadened unidirectional ring laser showing fully developed chaos. The insert shows the time series. Here ^{136}Xe is at 20 mTorr, $\kappa = 1 \times 10^{-6}\ \mathrm{s}^{-1}$ and the resolution is 100 kHz. This graph was kindly supplied by L. E. URACH, S. LIU and N. B. ABRAHAM.

[1982a] have found that if the cavity detuning is zero, then only a Hopf bifurcation occurs.

We might also mention the interesting preliminary results of LUGIATO and MILANI [1983] for a homogeneously broadened ring laser with spherical mirrors in the mean field approximation. The chaotic instabilities predicted in the plane transverse field approximation disappear with a Gaussian profile. Another indication of the sensitivity of chaotic behavior to approximation is due to MILONNI, ACKERHALT and GALBRAITH [1983]. They have found that chaotic behavior in the N-atom Jaynes–Cummings model *disappears* with the assumption of the rotating-wave approximation.

When viewed in its full generality, as a delay-differential system, the time-dependent nonlinear ring cavity with a homogeneously broadened atomic medium is a formidable numerical problem. Thus it is desirable to seek simplifying approximations, such as those discussed in §§ 2 and 3. In § 3 we mentioned that in the absorptive limit, the nonlinear ring cavity posesses a self-pulsing instability. So far, there have not been any reports of a chaotic instability in this limit (GRONCHI, BENZA, LUGIATO, MEYSTRE and SARGENT [1981]). We shall therefore focus our attention on dispersive systems.

If the mean-field limit is simultaneously taken with the limit of infinite transverse broadening, the Maxwell–Bloch and boundary equations become

$$\frac{1}{\kappa}\dot{x} = y - (1 + \mathrm{i}\delta_{\mathrm{c}})x - 2C(1 - \mathrm{i}\delta_{\mathrm{A}})x\eta, \tag{6.1a}$$

$$\frac{1}{\gamma_{\parallel}}\dot{\eta} = 1 - \eta - \frac{|x|^2}{1+\delta_{\mathrm{A}}^2}\eta, \tag{6.1b}$$

where

$$\eta(t) \equiv \Phi\left(L, t - \frac{\mathscr{L} - L}{c}\right).$$

Since x is complex valued, it is possible for these equations to exhibit chaotic behavior. LUGIATO, NARDUCCI, BANDY and PENNISE [1982] have numerically confirmed this by discovering period-doubling bifurcation sequences leading to chaos about the upper steady state branch. However, an unrealistic value of C is required for chaos (NARDUCCI, BANDY, PENNISE and LUGIATO [1983]). Similar results were previously obtained with a Kerr medium (equations (6.1), taken in the dispersive limit) by IKEDA and AKIMOTO [1982]. However, because the mean-field approximation neglects spatial effects, many significant qualitative features, such as multistability, are obscured by this limit.

This shortcoming is somewhat alleviated, in a tractable setting, by the maps derived by Ikeda and co-workers in the limit of instantaneous medium response, equations (3.37) and (3.47). Numerical iteration of these reveals period-doubling bifurcation sequences that lead to chaos as the amplitude of the injected field is varied (IKEDA [1979] and IKEDA, DAIDO and AKIMOTO [1980]). SNAPP, CARMICHAEL and SCHIEVE [1981] have presented a detailed description of the first bifurcation sequence displayed by equation (3.46) (cf. CARMICHAEL, SNAPP and SCHIEVE [1982]). In Fig. 21 we reproduce their illustration of the bifurcation sequence present on the lowest branch of the stationary state diagram $|\hat{\mathscr{E}}_{\mathrm{ss}}|$ versus $|\hat{\mathscr{E}}_{\mathrm{i}}|$ for $B = 0.5$ and $\psi_0 = 0$. We note the following:

(i) If $0 \leqslant |\hat{\mathscr{E}}_{\mathrm{i}}| < |\hat{\mathscr{E}}_{\mathrm{i}}^{\mathrm{a}}| \approx 1.25$, the stationary state of the map is stable (cf. Fig. 7).

(ii) As $|\hat{\mathscr{E}}_{\mathrm{i}}|$ is increased above $|\hat{\mathscr{E}}_{\mathrm{i}}^{\mathrm{a}}|$, the iterates of the map alternately fall on the upper and lower branches emerging from the stationary state, which is now unstable by condition (3.41b). This "two-cycle" is stable, for increasing $|\hat{\mathscr{E}}_{\mathrm{i}}|$ if $|\hat{\mathscr{E}}_{\mathrm{i}}^{\mathrm{a}}| < |\hat{\mathscr{E}}_{\mathrm{i}}| < |\hat{\mathscr{E}}_{\mathrm{i}}^{\mathrm{d}}| \approx 1.51$.

(iii) Within the interval $|\hat{\mathscr{E}}_{\mathrm{i}}^{\mathrm{b}}| < |\hat{\mathscr{E}}_{\mathrm{i}}| < |\hat{\mathscr{E}}_{\mathrm{i}}^{\mathrm{d}}|$, there are two coexisting stable attractors. First-order transitions with hysteresis take place between the two, analogous to those described in Fig. 10. Following the arrows, as $|\hat{\mathscr{E}}_{\mathrm{i}}|$ is increased above $|\hat{\mathscr{E}}_{\mathrm{i}}^{\mathrm{d}}|$, the two-cycle discontinuously switches "up" to the indicated four-cycle. If $|\hat{\mathscr{E}}_{\mathrm{i}}|$ is decreased, then the four-cycle converges to a

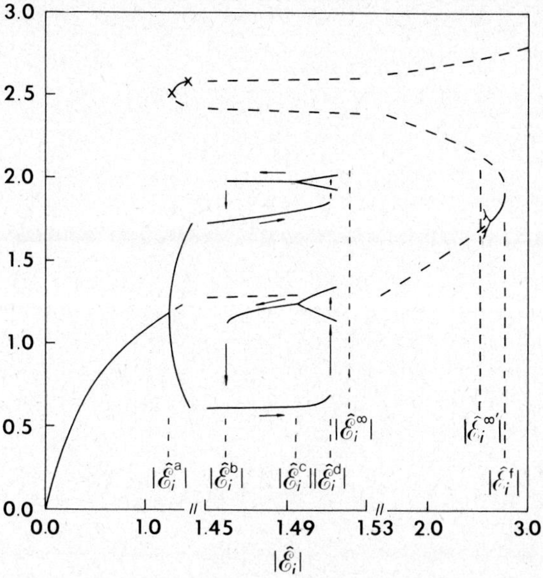

Fig. 21. Stable cycles bifurcating from the lowest branch of fixed points for the complex valued map (eq. (3.46)) for $B = 0.5$ and $\psi_0 = 0$. Beyond $|\hat{\mathscr{E}}_i|$ the unstable fixed points follow the dashed curve. Stable cycles are represented by solid lines, with $|\hat{\mathscr{E}}_n|$ evaluated at each point in the cycle.

different two-cycle at $|\hat{\mathscr{E}}_i^c| \approx 1.49$, and switches "down" to the original two-cycle at $|\hat{\mathscr{E}}_i^b| \approx 1.46$.

(iv) Above $|\hat{\mathscr{E}}_i^d|$, an increase in $|\hat{\mathscr{E}}_i|$ results in higher-order period doublings. This sequence converges "geometrically" to an accumulation point $|\hat{\mathscr{E}}_i^{\infty}| \approx 1.52$. Values of $|\hat{\mathscr{E}}_i^n|$ corresponding to the most stable cycles of periods 2×2^n are listed in table 1.

TABLE 1

Numerical results for the bifurcation velocity from the period-doubling sequence beginning along the upper branch of the hysteresis loop in Fig. 21.

2×2^n-cycle order n	$\lvert\mathscr{E}_i^n\rvert$	δ_n
2	1.517765 ± 10^{-5}	
3	$1.5202309745 \pm 1.35 \times 10^{-8}$	$4.575 \pm 2.8 \times 10^{-2}$
4	$1.52076999821 \pm 1 \times 10^{-11}$	$4.64575 \pm 1.2 \times 10^{-4}$
5	$1.52088602333 \pm 1 \times 10^{-11}$	$4.6642346 \pm 4.2 \times 10^{-6}$
6	$1.52091089882 \pm 2 \times 10^{-11}$	$4.668131 \pm 2.0 \times 10^{-5}$
7	$1.52091622761 \pm 1 \times 10^{-11}$	

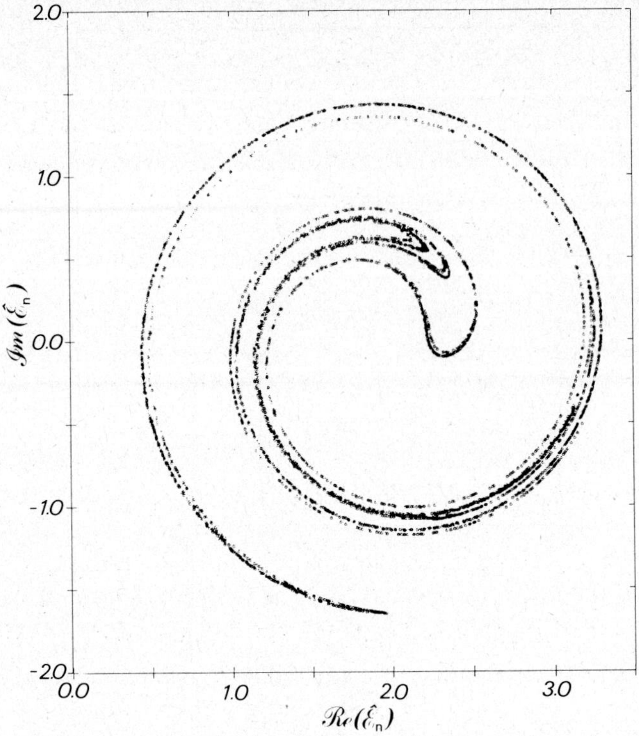

Fig. 22. A sequence of 5000 iterates of the map (eq. (3.46)) in the chaotic regime for $B = 0.5$, $\psi_0 = 0$, and $\mathscr{E}_i = 2$. The self-similarity in the spirals and the positive characteristic exponent suggest the presence of a strange attractor.

(v) Within the interval $|\hat{\mathscr{E}}_i^{\infty}| < |\hat{\mathscr{E}}_i| < |\hat{\mathscr{E}}_i^{\infty'}| \approx 2.52$, chaotic states exist with interspersed stable period-doubling bifurcation sequences. Stable period-doubling sequences with fundamentals of 12 (i.e. 12×2^n, $n = 0, 1, 2\ldots$), 10, 8, 7, 5 and 3 have been identified. In Fig. 22 we illustrate a chaotic sequence of iterates corresponding to $|\hat{\mathscr{E}}_i| = 2$. Note the local "self-similarity" near the center of the spiral attractor. The characteristic exponents of this attractor, $\lambda_+ = 0.691$ and $\lambda_- = -2.08$, were measured by an algorithm outlined by BENETTIN, GALGANI, GIORGILLI and STRELCYN [1980].

(vi) If $|\hat{\mathscr{E}}_i^{\infty'}| < |\hat{\mathscr{E}}_i| < |\hat{\mathscr{E}}_i^{e}| \approx 2.62$, the chaotic regime terminates as a new sequence of bifurcations appears. Now, however, as $|\hat{\mathscr{E}}_i|$ is increased the periods of the stable point-cycles undergo successive halvings until the stable stationary state, satisfying condition (3.40), is obtained.

(vii) If $|\hat{\mathscr{E}}_i^{e}| < |\hat{\mathscr{E}}_i| < |\hat{\mathscr{E}}_i^{f}| \simeq 2.79$ the stationary state remains stable. If $|\hat{\mathscr{E}}_i|$ exceeds $|\hat{\mathscr{E}}_i^{f}|$, then $|\mathscr{E}_{ss}|$ switches to the next higher branch.

One striking feature of these period-doubling bifurcation sequences is their apparent geometric convergence to finite accumulation points. After analyzing similar sequences in a variety of one-dimensional maps*, FEIGENBAUM [1978, 1979, 1980] proposed a universal scaling relationship among a large class of nonlinear maps. The central results of this theory are presented below for the map $x_{n+1} = f(\lambda, x_n)$ having a unique locally quadratic maximum at x_{M}. Here, λ denotes the bifurcation parameter.

(I) We first define the ratio

$$\delta_n = \frac{\lambda_n - \lambda_{n-1}}{\lambda_{n+1} - \lambda_n}, \tag{6.2}$$

where λ_n denotes the value of the bifurcation parameter corresponding to the most stable period of $m \times 2^n$ (here m is the fundamental period of the sequence). Then the bifurcation velocity, $\delta = \lim_{n \to \infty} \delta_n = 4.6692106\ldots$, is a universal constant.

(II) Without any loss of generality we assume $m = 1$. According to a theorem of JULIA [1918] (cf. METROPOLIS, STEIN and STEIN [1973]), the point x_{M} is an element of any most stable point-cycle. Hence, we characterize the kth element of the most stable point-cycle of period 2^n by

$$x_k^n = f^{(k)}(\lambda_n, x_{\mathrm{M}}) \qquad (k = 1, 2, \ldots, 2^n - 1) \tag{6.3a}$$

and

$$x_{\mathrm{M}} = f^{(0)}(\lambda_n, x_{\mathrm{M}}). \tag{6.3b}$$

(We assume that $x_k^n \neq x_{\mathrm{M}}$ for $k = 1, 2, 3, \ldots, 2^n - 1$.) By the continuous nature of the period doubling, the element of the point cycle nearest to x_k^n is $f^{(2^{n-1})}(\lambda_n, x_k^n)$: the point one half period through the cycle. We thus define the distance between these nearest pairs by

$$d_n(k) = x_k^n - f^{(2^{n-1})}(\lambda_n, x_k^n). \tag{6.4}$$

Feigenbaum conjectures that the scaling parameter

$$\alpha_n \equiv -\frac{d_n(0)}{d_{n+1}(0)} \tag{6.5}$$

* The simplest one exhibiting these properties is the one-dimensional quadratic map: $x_{n+1} = \lambda x_n(1 - x_n)$.

tends to the universal value $\alpha = \lim_{n\to\infty} \alpha_n = 2.502907875\ldots$. Similarly,

$$\lim_{n\to\infty}\left[-\frac{d_n(1)}{d_{n+1}(1)}\right] = -\alpha^2. \tag{6.6}$$

Here, $d_n(0)$ is the distance between x_M and its nearest neighbor, and $d_n(1)$ is the distance between the pair of points farthest to the right of x_M.

More generally, Feigenbaum defines the universal scaling function

$$\sigma(t) \equiv \lim_{n\to\infty} \sigma_n(t_n(k)), \tag{6.7a}$$

where

$$\sigma_n(t_n(k)) \equiv \frac{d_{n+1}(t_n(k))}{d_n(t_n(k))} \tag{6.7b}$$

and

$$t_n(k) = k/2^n \qquad (k = 0, 1, 2, \ldots). \tag{6.7c}$$

Note that $\sigma(t)$ has the properties:

(i) $\sigma(t)$ has a discontinuity at every rational number,

(ii) $\sigma(t + 1/2) = -\sigma(t)$,

(iii) $\lim_{\varepsilon\to 0^-} \sigma(0 + \varepsilon) = -\alpha^{-1}$,

(iv) $\lim_{\varepsilon\to 0^+} \sigma(0 + \varepsilon) = \alpha^{-2}$.

Although Feigenbaum formalized these universal quantities in terms of one-dimensional maps he did conjecture that they are extendable to higher-dimensional, area-contracting maps. A proof of this conjecture is outlined by COLLET, ECKMANN and KOCH [1981] (cf. COLLET and ECKMANN [1980] and COLLET, ECKMANN and LANDFORD [1980]). This latter point is important for our application as Ikeda's maps are essentially two-dimensional.

SNAPP, CARMICHAEL and SCHIEVE [1981] verified that equations (3.37) and (3.46) do exhibit these universal properties. Their results are presented in tables 1 and 2, and Fig. 23.

We emphasize that these maps are only approximations to the Maxwell–Bloch and boundary equations, and being such one may question if these period-doubling bifurcations are actually present in the nonlinear ring cavity. This question is even more significant in the light of the singular nature of the limit of instantaneous medium response. However, the physical applicability of the map, in the stable regime, is substantiated by recent numerical and experimental simulations of a related model.

TABLE 2
Numerical results for the bifurcation velocity from a period-doubling sequence in eqs. (3.37) with $\alpha L = 4.0$, $\alpha L|\delta_A| = 12\pi$, $\theta = 0$, $R = 0.95$.

4×2^n-cycle order n	$\lvert \mathscr{E}_i^n \rvert$	δ_n
3	$1.160714002 \pm 4 \times 10^{-9}$	
4	$1.1607460999 \pm 4 \times 10^{-10}$	$4.6528 \pm 6 \times 10^{-4}$
5	$1.16075299849 \pm 9 \times 10^{-11}$	$4.6659 \pm 4 \times 10^{-4}$
6	$1.1607544766991 \pm 2 \times 10^{-11}$	$4.6683 \pm 4 \times 10^{-4}$
7	$1.160754793704 \pm 4 \times 10^{-12}$	$4.6690 \pm 4 \times 10^{-4}$
8	$1.1607548615374 \pm 9 \times 10^{-13}$	

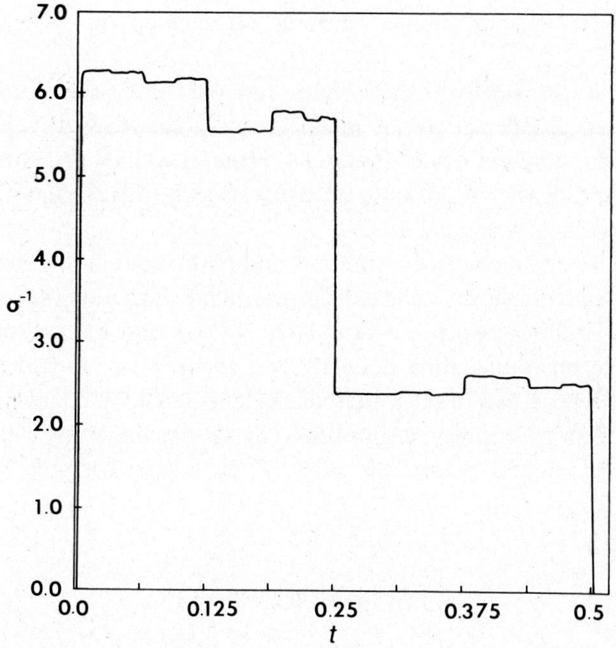

Fig. 23. Feigenbaum's universal scaling function is approximated by continuously interpolating $\sigma_7^{-1}(t_7(\kappa))$, for the period doubling sequence in Table 2.

In the limit $B \ll 1$, IKEDA, DAIDO and AKIMOTO [1980] show that the delay-differential system, equations (3.43), simplifies to

$$\frac{1}{\gamma_{\parallel}} \frac{\mathrm{d}\Psi(t)}{\mathrm{d}t} = -\Psi(t) + f(t-\tau), \tag{6.8}$$

where

$$\Psi(t) = \Psi(L, t) \tag{6.9}$$

and

$$f(t) = \operatorname{sgn}(n_2)|\hat{\mathscr{E}}_i|^2\{1 + 2B\cos[\Psi(t) - \Psi_0]\}. \tag{6.10}$$

IKEDA, KONDO and AKIMOTO [1982] give numerical evidence of period-doubling bifurcation sequences as $|\hat{\mathscr{E}}_i|$ is increased. However, beyond the first accumulation point, significant departures from the map are noted. As $|\hat{\mathscr{E}}_i|$ is increased further, successive odd harmonics of the fundamental "square wave", i.e. $\omega_n = (2^n - 1)/2\tau$, $n = 1, 2, 3, \ldots$, begin to dominate the integrated solution in the chaotic regime. Longer delay times allow higher harmonics to appear. These higher harmonics cause a "breaking-up" of the square wave solution.

Qualitative agreement with these higher harmonic bifurcations may be found in the numerical calculations of MURINA and ROZANOV [1981] and data collected from a hybrid device by GIBBS, HOPF, KAPLAN and SHOEMAKER [1981], HOPF, KAPLAN, GIBBS and SHOEMAKER [1982], and DERSTINE, GIBBS, HOPF and KAPLAN [1982].

Numerical simulations by GAO, YUAN and NARDUCCI [1983] indicate that self-pulsing and chaotic instabilities exist even in the short round-trip-time limit, i.e. $c/\mathscr{L} \gg \gamma_\parallel$. NAKATSUKA, ASAKA, ITOH, IKEDA and MATSUOKA [1983] observed the onset of period doubling and chaos in an all-optical system. However, because they used a pulsed incident field they could not resolve higher-order periods. Hence, their results only agree qualitatively with the map. Similarly they cannot detect the onset of the higher-harmonic bifurcation sequences.

Acknowledgements

We would like to acknowledge past collaboration with Professor H. J. Carmichael and Dr. R. F. Gragg which led to many results and perspectives contained in this review.

References

ABRAHAM, E. and S. D. SMITH, 1982, Rep. Prog. Phys. **45**, 815.
ABRAHAM, E., R. K. BULLOUGH and S. S. HASSAN, 1979, Opt. Commun. **29**, 109.

ABRAHAM, E., S. S. HASSAN and R. K. BULLOUGH, 1980, Opt. Commun. **33,** 93.
ABRAHAM, E., W. J. FIRTH and J. CARR, 1982, Phys. Lett. **91A,** 47.
ABRAHAM, N. B., T. CHYBA, M. COLEMAN, R. S. GIOGGIA, N. J. HALAS, L. M. HOFFER, S. N. LIU, M. MAEDA and J. C. WESSON, 1983, in: Laser Physics, eds. J. Harvey and D. Walls (Springer, New York).
ABRAHAM, R. and J. MARSDEN, 1978, Foundations of Mechanics (Benjamin, Reading, MA).
AGARWAL, G. S. and S. P. TEWARI, 1980, Phys. Rev. **A21,** 1638.
AGARWAL, G. S., L. M. NARDUCCI, D. H. FENG and R. GILMORE, 1978, in: Coherence and Quantum Optics IV, eds. L. Mandel and E. Wolf (Plenum, New York).
AGARWAL, G. S., L. M. NARDUCCI, R. GILMORE and D. H. FENG, 1978, Phys. Rev. **A18,** 620.
AGARWAL, G. S., L. M. NARDUCCI, D. H. FENG and R. GILMORE, 1980, Phys. Rev. **A21,** 1029.
AGRAWAL, G. P. and H. J. CARMICHAEL, 1979, Phys. Rev. **A19,** 2074.
AGRAWAL, G. P. and H. J. CARMICHAEL, 1980, Opt. Acta **27,** 651.
ANTORANZ, J. C., J. GEA and M. G. VELARDE, 1981, Phys. Rev. Lett. **47,** 1895.
ANTORANZ, J. C., L. L. BONILLA, J. GEA and M. G. VELARDE, 1982, Phys. Rev. Lett. **49,** 35.
ARECCHI, F. T. and A. POLITI, 1979, Opt. Commun. **29,** 361.
ARECCHI, F. T., R. MEUCCI, G. PUCCIONI and J. TREDICCE, 1982, Phys. Rev. Lett. **25,** 1217.
ASQUINI, M. L. and F. CASAGRANDE, 1981, Z. Phys. **B44,** 233.
BALLAGH, R. J., J. COOPER, M. W. HAMILTON, W. J. SANDLE and D. M. WARRINGTON, 1981, Opt. Commun. **37,** 143.
BENETTIN, G., L. GALGANI, A. GIORGILLI and J.-M. STRELCYN, 1980, Meccanica **15,** 9.
BENZA, V. and L. A. LUGIATO, 1979a, Lett. Nuovo Cim. **26,** 405.
BENZA, V. and L. A. LUGIATO, 1979b, Z. Phys. **B35,** 383.
BENZA, V. and L. A. LUGIATO, 1982, Z. Phys. **B47,** 79.
BENZA, V., L. A. LUGIATO and P. MEYSTRE, 1980, Opt. Commun. **33,** 113.
BISCHOFBERGER, T. and Y. R. SHEN, 1978, Appl. Phys. Lett. **32,** 156.
BISCHOFBERGER, T. and Y. R. SHEN, 1979, Phys. Rev. **A19,** 1169.
BONIFACIO, R. and L. A. LUGIATO, 1976, Opt. Commun. **19,** 172.
BONIFACIO, R. and L. A. LUGIATO, 1978a, Phys. Rev. **A18,** 1129.
BONIFACIO, R. and L. A. LUGIATO, 1978b, Lett. Nuovo Cim. **21,** 505.
BONIFACIO, R. and L. A. LUGIATO, 1978c, Lett. Nuovo Cim. **21,** 510.
BONIFACIO, R. and L. A. LUGIATO, 1978d, Lett. Nuovo Cim. **21,** 517.
BONIFACIO, R. and L. A. LUGIATO, 1978e, in: Coherence and Quantum Optics IV, eds. L. Mandel and E. Wolf (Plenum, New York).
BONIFACIO, R. and L. A. LUGIATO, 1978f, Phys. Rev. Lett. **40,** 1023; erratum, 1978, Phys. Rev. Lett. **40,** 1538.
BONIFACIO, R. and P. MEYSTRE, 1978, Opt. Commun. **27,** 147.
BONIFACIO, R. and P. MEYSTRE, 1979, Opt. Commun. **29,** 131.
BONIFACIO, R., M. GRONCHI and L. A. LUGIATO, 1978, Phys. Rev. **A18,** 2266.
BONIFACIO, R., M. GRONCHI and L. A. LUGIATO, 1979a, Nuovo Cim. **53B,** 311.
BONIFACIO, R., M. GRONCHI and L. A. LUGIATO, 1979b, Opt. Commun. **30,** 129.
BONIFACIO, R., L. A. LUGIATO and M. GRONCHI, 1979, in: Laser Spectroscopy IV, eds. H. Walther and K. W. Rothe (Springer, Berlin).
BONIFACIO, R., L. A. LUGIATO, J. D. FARINA and L. M. NARDUCCI, 1981, IEEE J. Quantum Electron. **QE-17,** 357.
BOWDEN, C. M., M. CIFTAN and H. R. ROBL, eds., 1981, Optical Bistability (Plenum, New York).
BULSARA, A. R., W. C. SCHIEVE and R. F. GRAGG, 1978, Phys. Lett. **68A,** 294.
CARMICHAEL, H. J., 1980, Opt. Acta **27,** 147.
CARMICHAEL, H. J., 1981a, Z. Phys. **B42,** 183.
CARMICHAEL, H. J., 1981b, Personal Communication (cf. CARMICHAEL, SNAPP and SCHIEVE [1982]).

CARMICHAEL, H. J., 1983a, Phys. Rev. **A28,** 480.

CARMICHAEL, H. J., 1983b, in: Laser Physics, eds. J. D. Harvey and D. F. Walls (Springer, New York).

CARMICHAEL, H. J. and G. P. AGRAWAL, 1981, in: Optical Bistability, eds. C. M. Bowden, M. Ciftan and H. R. Robl (Plenum, New York).

CARMICHAEL, H. J. and J. A. HERMANN, 1980, Z. Phys. **B38,** 365.

CARMICHAEL, H. J. and D. F. WALLS, 1977, J. Phys. **B10,** L685.

CARMICHAEL, H. J., R. R. SNAPP and W. C. SCHIEVE, 1982, Phys. Rev. **A26,** 3408.

CARMICHAEL, H. J., C. M. SAVAGE and D. F. WALLS, 1983, Phys. Rev. Lett. **50,** 163.

CAROLI, B., C. CAROLI and B. ROULET, 1979, J. Stat, Phys. **21,** 415.

CASAGRANDE, F. and L. A. LUGIATO, 1976, Phys. Rev. **A14,** 778.

CASAGRANDE, F. and L. A. LUGIATO, 1980, Nuovo Cim. **55B,** 173.

CASAGRANDE, F., L. A. LUGIATO and M. L. ASQUINI, 1980, Opt. Commun. **32,** 492.

CASPERSON, L. W., 1978, IEEE J. Quantum Electron. **QE-14,** 756.

CASPERSON, L. W., 1980, Phys. Rev. **A21,** 911.

CASPERSON, L. W., 1981, Phys. Rev. **A23,** 248. [See also: L. W. Casperson, 1983, in: Laser Physics, eds. J. D. Harvey and D. F. Walls (Springer, New York).]

CHANDRASEKHAR, S., 1943, Rev. Mod. Phys. **15,** 1.

CHANDRASEKHAR, S., 1961, Hydrodynamic and Hydromagnetic Stability (Oxford Univ. Press).

CHROSTOWSKI, J. and A. ZARDECKI, 1979, Opt. Commun. **29,** 230.

COLLET, P. and J.-P. ECKMANN, 1980, Iterated Maps on the Interval as Dynamical Systems (Birkhäuser, Boston).

COLLET, P., J.-P. ECKMANN and O. E. LANDFORD III, 1980, Comm. Math. Phys. **76,** 211.

COLLET, P., J.-P. ECKMANN and H. KOCH, 1981, J. Stat. Phys. **25,** 1.

DEGIORGIO, V. and L. A. LUGIATO, 1980, Phys. Lett. **77A,** 167.

DELLEDONNE, M., P. H. RICHTER and J. ROSS, 1981, Z. Phys. **B42,** 271.

DERSTINE, M. W., H. M. GIBBS, F. A. HOPF and D. L. KAPLAN, 1982, Phys. Rev. **A26,** 3720.

DRUMMOND, P. D., 1981a, IEEE J. Quantum Electron. **QE-17,** 301.

DRUMMOND, P. D., 1981b, in: Optical Bistability, eds. C. M. Bowden, M. Ciftan and H. R. Robl (Plenum, New York).

DRUMMOND, P. D., 1982, Opt. Commun. **40,** 224.

DRUMMOND, P. D. and C. W. GARDINER, 1980, J. Phys. **A13,** 2553.

DRUMMOND, P. D. and D. F. WALLS, 1980, J. Phys. **A13,** 725.

DRUMMOND, P. D. and D. F. WALLS, 1981, Phys. Rev. **A23,** 2563.

ECKMANN, J.-P., 1981, Rev. Mod. Phys. **53,** 643.

ENGLUND, J. C., 1984, Thesis, University of Texas at Austin.

ENGLUND, J. C., W. C. SCHIEVE, W. ZUREK and R. F. GRAGG, 1981, in: Optical Bistability, eds. C. M. Bowden, M. Ciftan and H. R. Robl (Plenum, New York).

ENGLUND, J. C., W. C. SCHIEVE, R. F. GRAGG and W. ZUREK, 1982, in: Instabilities, Bifurcations and Fluctuations in Chemical Systems, eds. L. E. Reichl and W. C. Schieve (Univ. of Texas Press, Austin, TX).

ERNEUX, T. and P. MANDEL, 1983, Phys. Rev. **A28,** 896.

FARINA, J. D., L. M. NARDUCCI, J. M. YUAN and L. A. LUGIATO, 1980, Opt. Eng. **19,** 469.

FARINA, J. D., L. M. NARDUCCI, J. M. YUAN and L. A. LUGIATO, 1981, in: Optical Bistability, eds. C. M. Bowden, M. Ciftan and H. R. Robl (Plenum, New York).

FEIGENBAUM, M. J., 1978, J. Stat. Phys. **19,** 25.

FEIGENBAUM, M. J., 1979, J. Stat. Phys. **21,** 669.

FEIGENBAUM, M. J., 1980, Los Alamos Sci. **1,** 4.

FELBER, F. S. and J. H. MARBURGER, 1976, Appl. Phys. Lett. **28,** 731.

FIRTH, W. J., 1981, Opt. Commun. **39,** 343.

FIRTH, W. J. and E. M. WRIGHT, 1982, Phys. Lett. **92A,** 211.

GAO, J. Y., J. M. YUAN and L. M. NARDUCCI, 1983, Opt. Commun. **44,** 201.
GERBER, P. R. and M. BÜTTIKER, 1979, Z. Phys. **B33,** 219.
GIBBON, J. D. and M. J. MCGUINESS, 1980, Phys. Lett. **77A,** 295.
GIBBON, J. D. and M. J. MCGUINESS, 1982, Physica **5D,** 108.
GIBBS, H. M., S. L. MCCALL and T. N. C. VENKATESAN, 1976, Phys. Rev. Lett. **36,** 1135.
GIBBS, H. M., S. L. MCCALL and T. N. C. VENKATESAN, 1978, in: Coherence in Spectroscopy and Modern Physics, eds. F. T. Arecchi, R. Bonifacio and M. O. Scully (Plenum, New York).
GIBBS, H. M., F. A. HOPF, D. L. KAPLAN and R. L. SHOEMAKER, 1981, Phys. Rev. Lett. **46,** 474.
GLAUBER, R. J., 1963a, Phys. Rev. **130,** 2529.
GLAUBER, R. J., 1963b, Phys. Rev. **131,** 2766.
GORDON, J. P., 1967, Phys. Rev. **161,** 367.
GRAGG, R. F., 1981, Thesis, University of Texas at Austin.
GRAHAM, R., 1976, Phys. Lett. **58A,** 440.
GRAHAM, R. and H. HAKEN, 1968, Z. Phys. **213,** 420.
GRAHAM, R. and A. SCHENZLE, 1981a, Phys. Rev. **A23,** 1302.
GRAHAM, R. and A. SCHENZLE, 1981b, in: Optical Bistability, eds. C. M. Bowden, M. Ciftan and H. R. Robl (Plenum, New York).
GRANT, D. E. and H. J. KIMBLE, 1982, Opt. Lett. **7,** 353; erratum, 1983, Opt. Lett. **8,** 66.
GRANT, D. E. and H. J. KIMBLE, 1983, Opt. Commun. **44,** 415.
GRASIUK, A. Z. and A. N. ORAEVSKII, 1964, in: Proc. Int. School of Physics "Enrico Fermi", Course XXXI: Quantum Electronics and Coherent Light, ed. P. A. Miles (Academic, New York).
GRONCHI, M. and L. A. LUGIATO, 1978, Lett. Nuovo Cim. **23,** 593; erratum, 1979, Lett. Nuovo Cim. **24,** 240.
GRONCHI, M. and L. A. LUGIATO, 1980, Opt. Lett. **5,** 108.
GRONCHI, M., V. BENZA, L. A. LUGIATO, P. MEYSTRE and M. SARGENT III, 1981, Phys. Rev. **A24,** 1419.
GROSSMANN, S. and S. THOMAE, 1977, Z. Naturf. **32a,** 1353.
HAKEN, H., 1966, Z. Phys. **190,** 327.
HAKEN, H., 1970, Handbuch der Physik, Vol. XXV/2c (Springer, New York).
HAKEN, H., 1975a, Phys. Lett. **53A,** 77.
HAKEN, H., 1975b, Z. Phys. **B20,** 413.
HAKEN, H., 1975c, Z. Phys. **B21,** 105.
HAKEN, H., 1975d, Z. Phys. **B22,** 69.
HAKEN, H., 1975e, Rev. Mod. Phys. **47,** 67.
HAKEN, H., 1983, Synergetics: An Introduction, 3rd Ed. (Springer, New York).
HAKEN, H. and H. OHNO, 1976, Opt. Commun. **16,** 205.
HANGGI, P., A. R. BULSARA and R. JANDA, 1980, Phys. Rev. **A22,** 671.
HASEGAWA, H., T. NAKAGOMI, M. MABUCHI and K. KONDO, 1980, J. Stat. Phys. **23,** 281.
HASSAN, S. S. and D. F. WALLS, 1978, J. Phys. **A11,** L87.
HASSAN, S. S., P. D. DRUMMOND and D. F. WALLS, 1978, Opt. Commun. **27,** 480.
HELLEMAN, R., 1980, in: Fundamental Problems in Statistical Mechanics, vol. 5, ed. E. G. D. Cohen (North-Holland, Amsterdam).
HELLEMAN, R., 1982, in: Long Time Predictions in Dynamics, eds. W. Horton, L. Reichl and V. Szebehely (Wiley, New York).
HERMANN, J. A., 1980, Opt. Acta **27,** 159.
HOPF, F. A. and P. MEYSTRE, 1979, Opt. Commun. **29,** 235.
HOPF, F. A. and P. MEYSTRE, 1980, Opt. Commun. **33,** 225.
HOPF, F. A., P. MEYSTRE, P. D. DRUMMOND and D. F. WALLS, 1979, Opt. Commun. **31,** 245.
HOPF, F. A., D. L. KAPLAN, H. M. GIBBS and R. L. SHOEMAKER, 1982, Phys. Rev. **A25,** 2172.

HORSTHEMKE, W. and L. BRENIG, 1977, Z. Phys. **B27,** 341.
HORSTHEMKE, W., M. MALEK-MANSOUR and L. BRENIG, 1977, Z. Phys. **B28,** 135.
IKEDA, K., 1979, Opt. Commun. **30,** 257.
IKEDA, K. and O. AKIMOTO, 1982, Phys. Rev. Lett. **48,** 617.
IKEDA, K., H. DAIDO and O. AKIMOTO, 1980, Phys. Rev. Lett. **45,** 709.
IKEDA, K., K. KONDO and O. AKIMOTO, 1982, Phys. Rev. Lett. **49,** 1467.
JULIA, G., 1918, J. de Math. (Sér. 7) **4,** 47.
KITANO, M., T. YABUZAKI and T. OGAWA, 1981, Phys. Rev. Lett. **46,** 926.
KONDO, K., M. MABUCHI and H. HASEGAWA, 1980, Opt. Commun. **32,** 136.
KRAMERS, H. A., 1940, Physica (Utrecht) **7,** 284.
KUBO, R., K. MATSUO and K. KITHARA, 1973, J. Stat. Phys. **9,** 51.
LAMB, W. E., Jr., 1964, Phys. Rev. **134,** A1429.
LANDAUER, R. and J. A. SWANSON, 1961, Phys. Rev. **121,** 1668.
LAX, M., 1968, in: Statistical Physics, Phase Transitions and Superfluidity, eds. M. Chretien, E. P. Gross and S. Deser (Gordon and Breach, New York).
LORENZ, E. N., 1963, J. Atmos. Sci. **20,** 130.
LOUISELL, W. H., 1973, Quantum Statistical Properties of Radiation (Wiley, New York).
LUGIATO, L. A., 1978, Lett. Nuovo Cim. **23,** 609.
LUGIATO, L. A., 1979, Nuovo Cim. **50B,** 89.
LUGIATO, L. A., 1980a, Lett. Nuovo Cim. **29,** 375.
LUGIATO, L. A., 1980b, Opt. Commun. **33,** 108.
LUGIATO, L. A. and R. BONIFACIO, 1978, in: Coherence in Spectroscopy and Modern Physics, eds. F. T. Arecchi, R. Bonifacio and M. O. Scully (Plenum, New York).
LUGIATO, L. A. and M. MILANI, 1983, Opt. Commun. **46,** 57.
LUGIATO, L. A., P. MANDEL, S. T. DEMBINSKI and A. KOSSAKOWSKI, 1978, Phys. Rev. **A18,** 238.
LUGIATO, L. A., J. D. FARINA and L. M. NARDUCCI, 1980, Phys. Rev. **A22,** 253.
LUGIATO, L. A., V. BENZA, L. M. NARDUCCI and J. D. FARINA, 1981, Opt. Commun. **39,** 405.
LUGIATO, L. A., M. L. ASQUINI and L. M. NARDUCCI, 1982, Opt. Commun. **41,** 450.
LUGIATO, L. A., F. CASAGRANDE and L. PIZZUTO, 1982, Phys. Rev. **A26,** 3438.
LUGIATO, L. A., L. M. NARDUCCI, D. K. BANDY and C. A. PENNISE, 1982, Opt. Commun. **43,** 281.
MAEDA, M. and N. B. ABRAHAM, 1982, Phys. Rev. **A26,** 3395.
MANDEL, P. and G. P. AGRAWAL, 1982, Opt. Commun. **42,** 269.
MANDEL, P. and T. ERNEUX, 1982a, Opt. Acta **29,** 7.
MANDEL, P. and T. ERNEUX, 1982b, Opt. Commun. **42,** 362.
MANNEVILLE, P. and Y. POMEAU, 1980, Physica **1D,** 219.
MARBURGER, J. H. and F. S. FELBER, 1978, Phys. Rev. **A17,** 335.
MARDEN, M., 1949, The Geometry of the Zeros of a Polynomial in a Complex Variable (American Mathematical Surveys III) (American Mathematical Society, New York).
MAY, R. M., 1976, Nature **261,** 459.
MAYR, M., H. RISKEN and H. D. VOLLMER, 1981, Opt. Commun. **36,** 480.
MCCALL, S. L. and H. M. GIBBS, 1980, Opt. Commun. **33,** 335.
METROPOLIS, N., M. L. STEIN and P. R. STEIN, 1973, J. Comb. Theory (A) **15,** 25.
MEYSTRE, P., 1978, Opt. Commun. **26,** 277.
MILONNI, P. W., J. R. ACKERHALT and H. W. GALBRAITH, 1983, Phys. Rev. Lett. **50,** 966.
MOLLOW, B. R., 1969, Phys. Rev. **188,** 1969.
MOORE, S. M., 1983, The Effect of a Positive Correlation Time on Optical Bistability, to appear in Physica **A**.

MURINA, T. A. and N. N. ROZANOV, 1981, Kvantovaya Elektron. **18,** 1186; Sov. J. Quantum Electron. **11,** 711.

NAKATSUKA, H., S. ASAKA, H. ITOH, K. IKEDA and M. MATSUOKA, 1983, Phys. Rev. Lett. **50,** 109.

NARDUCCI, L. M., R. GILMORE, D. H. FENG and G. S. AGARWAL, 1978, Opt. Lett. **2,** 88.

NARDUCCI, L. M., R. GILMORE, D. H. FENG and G. S. AGARWAL, 1979, Phys. Rev. **A20,** 545.

NARDUCCI, L. M., D. K. BANDY, C. A. PENNISE and L. A. LUGIATO, 1983, Opt. Commun. **44,** 207.

NICOLIS, G. and I. PRIGOGINE, 1977, Self-Organization in Nonequilibrium Systems (Wiley, New York).

OHNO, H. and H. HAKEN, 1976, Phys. Lett. **59A,** 261.

ORAEVSKII, A. N., 1981, Kvantovaya Elektron. **8,** 130; Sov. J. Quantum Electron. **11** (1), 71.

OTT, E., 1981, Rev. Mod. Phys. **53,** 655.

POMEAU, Y. and P. MANNEVILLE, 1980, Comm. Math. Phys. **74,** 189.

RABINOVICH, M. I., 1978, Usp. Fiz. Nauk **125,** 123; Sov. Phys. Usp. **21,** 443.

RISKEN, H., 1970, in: Progress in Optics VIII, ed. E. Wolf (North-Holland, Amsterdam).

RISKEN, H. and K. NUMMEDAL, 1968a, Phys. Lett. **26A,** 275.

RISKEN, H. and K. NUMMEDAL, 1968b, J. Appl. Phys. **39,** 4662.

RISKEN, H., C. SCHMID and W. WEIDLICH, 1966, Z. Phys. **194,** 337.

ROY, R. and M. S. ZUBAIRY, 1980, Phys. Rev. **A21,** 274.

RUELLE, D., 1979, Proc. New York Acad. Sci. **316,** 408.

RUELLE, D. and F. TAKENS, 1971, Comm. Math. Phys. **20,** 167.

SARGENT, M., III, M. O. SCULLY and W. E. LAMB, Jr., 1974, Laser Physics (Addison–Wesley, Reading, MA).

SATTINGER, D. H., 1973, Topics in Stability and Bifurcation Theory (Lecture Notes in Mathematics, vol. 309) (Springer, New York).

SAVAGE, C. M. and D. F. WALLS, 1983, Optical Chaos in Sub/Second Harmonic Generation, preprint (University of Waikato, Hamilton, New Zealand).

SAVAGE, C. M., H. J. CARMICHAEL and D. F. WALLS, 1982, Opt. Commun. **42,** 211.

SCHENZLE, A. and H. BRAND, 1978, Opt. Commun. **27,** 485.

SCHENZLE, A. and H. BRAND, 1979, Opt. Commun. **31,** 401.

SCHIEVE, W. C. and P. ALLEN, eds., 1981, Self-Organization and Dissipative Structures in the Social and Physical Sciences (Univ. of Texas Press, Austin, TX).

SCHIEVE, W. C., J. C. ENGLUND, R. F. GRAGG and R. R. SNAPP, 1981, Fluctuations and Instabilities in Laser-like Systems, Proc. 1st Escuela Mexicana de Fisica Estadistica, ed. Peralta Fabi (Soc. Mex. Fisica, 1983).

SCHOLZ, H. J., T. YAMADA, H. BRAND and R. GRAHAM, 1981, Phys. Lett. **82A,** 321.

SCHWENDIMANN, P., 1979, J. Phys. **A12,** L39.

SELLONI, A. and P. SCHWENDIMANN, 1979, Opt. Acta **26,** 1541.

SELLONI, A., A. QUATTROPANI, P. SCHWENDIMANN and H. P. BALTES, 1981, Opt. Acta **28,** 125.

SHAW, R., 1981, Z. Naturf. **36a,** 80.

SILBERBERG, Y. and I. BAR JOSEPH, 1982, Phys. Rev. Lett. **48,** 1541.

SNAPP, R. R., H. J. CARMICHAEL and W. C. SCHIEVE, 1981, Opt. Commun. **40,** 68.

SPENCER, M. B. and W. E. LAMB, Jr., 1972, Phys. Rev. **A5,** 884.

SUDARSHAN, E. C. G., 1963, Phys. Rev. Lett. **10,** 277.

SWINNEY, H. L. and J. B. GOLLUB, 1981, Hydrodynamic Instabilities and the Transition to Turbulence (Springer, Berlin).

SZÖKE, A., V. DANEU, J. GOLDHAR and N. A. KURNIT, 1969, Appl. Phys. Lett. **15,** 376.

TAI, K., H. M. GIBBS and J. V. MOLONEY, 1982, Opt. Commun. **43,** 297.

TEWARI, S. P., 1980, Opt. Commun. **34,** 273.

TOMITA, S., 1982, Phys. Rep. **86,** 113.
VAN KAMPEN, N. G., 1969, Adv. Chem. Phys. **15,** 65.
VAN KAMPEN, N. G., 1981, J. Stat. Phys. **25,** 431.
VELARDE, M. G. and J. C. ANTORANZ, 1979, Phys. Lett. **72A,** 123.
VELARDE, M. G. and J. C. ANTORANZ, 1980, Phys. Lett. **80A,** 220.
VELARDE, M. G. and J. C. ANTORANZ, 1981, Prog. Theor. Phys. **66,** 717.
WEIDLICH, W., H. RISKEN and H. HAKEN, 1967, Z. Phys. **201,** 396.
WEISS, C. H., 1966, Adv. Chem. Phys. **13,** 1.
WILLIS, C. R., 1977, Opt. Commun. **23,** 151.
WILLIS, C. R., 1978, Opt. Commun. **26,** 62.
WILLIS, C. R., 1983, Phys. Rev. **A27,** 375.
WILLIS, C. R. and J. DAY, 1979, Opt. Commun. **28,** 137.
WINFUL, H. G. and G. D. COOPERMAN, 1982, Appl. Phys. Lett. **40,** 298.
YAMADA, T. and R. GRAHAM, 1980, Phys. Rev. Lett. **45,** 1322.
ZARDECKI, A., 1980, Phys. Rev. **A22,** 1664.
ZARDECKI, A., 1981, Phys. Rev. **A23,** 1281.

AUTHOR INDEX

SUBJECT INDEX

CUMULATIVE INDEX - VOLUMES I–XXI